Thermal Design

Thermal Design
Heat Sinks, Thermoelectrics, Heat Pipes, Compact Heat Exchangers, and Solar Cells

HoSung Lee

WILEY

JOHN WILEY & SONS, INC.

Library of Congress Cataloging-in-Publication Data:

Lee, Ho Sung.
    Thermal design : heat sinks, thermoelectrics, heat pipes, compact heat exchangers, and solar cells / Ho Sung Lee.
        p. cm.
    Includes index.
    ISBN 978-0-470-49662-6 (hardback); 978-0-470-94997-9 (ebk); 978-0-470-95160-6 (ebk); 978-0-470-95177-4 (ebk); 978-1-118-00468-5 (ebk); 978-111-8-00470-8 9 (ebk); 978-111-8-00471-5 (ebk)
    1. Heat engineering–Materials. 2. Heat-transfer media. 3. Thermodynamics. 4. Thermoelectric apparatus and appliances. I. Title.
    TJ255.5.L44 2010
    621.402′5–dc22
                                                                                                    2010018381
ISBN: 978-0-470-49662-6

Printed in the United States of America

10 9 8 7 6 5 4 3 2 1

Contents

Preface

This book is written as a senior undergraduate or a first-year graduate textbook, covering modern thermal devices such as heat sinks, thermoelectric generators and coolers, heat pipes, compact heat exchangers, and solar cells as design components. These devices are becoming increasingly important and fundamental in thermal design in such diverse areas as microelectronic cooling, green or thermal energy conversion, thermal control and management in space, and so on. However, there is no textbook available that includes these topics, which is the rationale for the writing of this book. This book may be used as a capstone design course after students have finished the fundamental courses in areas such as thermodynamics, fluid mechanics, and heat transfer. The concept of this book is to give the student first an understanding of the physical mechanisms of the devices with detailed derivations, and second, practice in designing the devices with use of mathematical modeling, graphical optimization, and occasionally computational-fluid-dynamic (CFD) simulation. This is done through pertinent design examples developed using a commercial software, MathCAD. In other words, the design concept is embodied through the sample problems. The graphical presentation generally provides designers or students with rich and flexible solutions giving the optimal design.

This book is unique as a textbook of *thermal design* with the present topics and design methodology. It has been developed from the author's lecture notes. Since this book exhibits the fundamental framework of thermal design using modern thermal devices, the applications to the thermal systems associated with these devices are unlimited.

This book is self-contained. For example, an introduction to thermal radiation was added prior to the section of *fin design in space* for the readers who are not familiar with this subject. Many appropriate charts and tables were attached in the appendices so that readers need not look at other reference books. Detailed tutorials appropriate for use in CFD and MathCAD homework problems are also included in the appendices to help students.

Particular effort was made to create figures representing key concepts of the subject matter, keeping in mind that "one good figure is better than a thousand words." Needless to say, figures are important learning tools, focus attention and stimulate curiosity and interest.

In the past decade, a good deal of attention has been given to critically assessing traditional pedagogy and to exploring means by which students' learning may be enhanced. With respect to the development of educational tools and curricula, this assessment has stimulated serious consideration of *learning objectives* and the means of determining the extent to which prescribed objectives are being met. This textbook has three learning objectives:

1. The students should delineate physical mechanisms and transport phenomena for any process or system particularly associated with thermal devices such

as heat sinks, thermoelectric generators and coolers, heat pipes, compact heat exchangers, and solar cells.

2. The students should be able to develop mathematical models and graphical optimization for any process or system associated with the thermal devices or systems.

3. The students should be able to professionally design the thermal devices or systems using design tools.

As mentioned before, this book is self-contained. An attempt was made to include necessary background in thermodynamics, fluid mechanics, and heat transfer. Chapter 1 focuses on material needed in later chapters. Students can use this chapter as reference or review. The first and second laws of thermodynamics, internal and external convection flow, and heat transfer mechanisms were presented with essential formulas and empirical correlations.

Chapter 2 is devoted to *heat sinks*, which are the most common thermal devices for use in the electronics industry. They are used to improve the thermal control of electronic components, assemblies, and modulus by enhancing their exterior surface area through the use of fins. The governing formulas on heat dissipation and efficiency for single and multiple fins are derived and incorporated into the modeling and optimization of fin design. Particular effort was given to creating appropriate examples to reflect the design concept, which involves mathematical modeling and graphical optimization. Also in the chapter, fin design with thermal radiation (in space) was explored with two design examples.

Chapter 3 provides the fundamentals of the design of *thermoelectric generators* and *coolers*. The field of thermoelectrics has grown dramatically in recent years, but in spite of this resurgence of interest, there are very few books available. This may be the first book to deal with the design of thermoelectrics and heat pipes, providing the physical principles and fundamental formulas, which lead to mathematical modeling and graphical optimization. This chapter may prompt students to look for the waste energy recovery from the exhaust gases in automotive vehicles and power systems for spacecrafts using radioisotope thermoelectric generators (RTG). A design example at the end of the chapter was conceptualized and developed from a commercial product, which consists of two heat sinks, two fans and a thermoelectric cooler (TEC) module as a thermal system.

Chapter 4 is devoted to the design of *heat pipes*, which have been recently employed in numerous applications ranging from temperature control of the permafrost layer under the Alaska pipeline to the thermal control of optical surfaces in spacecraft. Today every laptop computer has a heat-pipe related cooling system. This book gives a clear understanding of the fundamentals of heat pipes, including the formulas, which allow modeling and optimization in design. This chapter deals with various heat pipes such as variable conductance heat pipe, loop heat pipe, micro heat pipe, and heat pipe in space. The end-of-chapter design example discusses the detailed design aspects: selecting materials and working fluid, sizing the heat pipe, selecting the wick, and performance map.

Chapter 5 discusses the design of *compact heat exchangers* including plate heat exchangers, finned-tube heat exchangers, and plate-fin heat exchangers. In order to discuss these complex exchangers, simpler heat exchangers such as a double-pipe heat exchanger and a shell-and-tube heat exchanger are also introduced. Usually, it takes a semester to cover the entire material of this chapter. However, the complex geometry and time-consuming work are incorporated into the illustrated models. This saves a lot of the students' time. Rather, students can put their efforts either into improving the model or implementing into the system design. Design tools such as MathCAD enable easy and precise minimizing of human errors in calculations.

Chapter 6 is devoted to *solar cell design*. It is often said that solar energy will be the energy in the future. Solar cells need to be developed in order to meet the formidable requirements of high efficiency and low cost. A solar cell is a technology-dependent device, wherein efficiency is a key issue in performance and design. The author believes that solar cells should be dealt with in undergraduate programs, when we consider their importance and huge demand in the future. Unfortunately, a solar cell involves many disciplines: physics, chemistry, materials, electronics, and mechanics (heat transfer). This book was written to provide the fundamentals of solar cells including both the physics and design with a ready-to-use model so that we may achieve our goal in a minimum of time.

Students or teachers should feel free to copy the ready-to-use models to suit their own purposes. This book was designed to provide profound theories and derivation of formulas so that students can easily make modifications or improvements according to their ability. For example, a student who is not strong in physics but is creative can produce a novel design using the ready-to-use models presented in this book.

Except for Chapter 1 ("Introduction"), each chapter is independent from other chapters so that it may be taught separately. Considering the volume of material in each chapter, any combination of three chapters would be appropriate for a semester's material. Chapter 1 may be skipped. For example, a thermal-oriented course would use Chapter 2, "Heat Sinks," Chapter 4, "Heat Pipes," and Chapter 5, "Compact Heat Exchangers." An electronic cooling–oriented course would use Chapter 2, "Heat Sinks," Chapter 3, "Thermoelectrics," and Chapter 4, "Heat Pipes." Intensive design can be sought with Chapter 5, "Compact Heat Exchangers," and Chapter 6, "Solar Cells." This book can be taught for one or two semesters. Note that many problems at the end of each chapter may require at least a week of work by students.

I would like to acknowledge the many suggestions, the inspiration, and the help provided by undergraduate/graduate students in the thermal design classes over the years. I also thank the College of Engineering and Applied Sciences at Western Michigan University for providing me the opportunity to teach the thermal design courses. Particular thanks to the Wiley staff for their support and their editing of the material. I also wish to thank anonymous reviewers for their suggestions and critiques that greatly improved the quality of this book. I am immensely indebted to Professor Emeritus Herman Merte, Jr. for his advice and support in the preparation of this book. My sincere appreciation is extended to Dr. Stanley L. Rajnak for his suggestions and his review in preparation of the manuscript. He reviewed the entire manuscript with his invaluable endeavors particularly finding errors in mathematics. Lastly I am truly indebted to my beloved wife, Young-Ae, for her support and forbearance during the preparation of this book, and also to my lovely daughter, Yujin, for her encouragement and assistance particularly in drawings. Without the supports mentioned above, this book would not be possible.

Kalamazoo, Michigan   HoSung Lee

1

Introduction

1.1 INTRODUCTION

Thermal design is a branch of *engineering design*, a counterpart to *machine design*, typically involving energy, fluid flow, thermodynamics, and heat transfer. Traditionally, *thermal design* has been developed as *thermal system design*, which deals with *modeling, simulation*, and *optimization*, and proper selection of the components. It typically avoids the design of the components themselves [1–7]. The components would be pumps, fans, heat exchangers, and refrigeration units, for example. This forms a definite field of thermal design with a great deal of industrial applications. However, often the traditional system design could not meet the needs of a radically changing technology. For example, traditional fan-cooling methods of electronics in computers no longer meet today's formidable heat duty. Today, every laptop computer has *heat-pipe* cooling systems. Traditional refrigeration units are too large and noisy for small systems. These are typically replaced by thermoelectric coolers, which have no moving parts and are easily controlled by electric power. A *compact heat exchanger* in a *fuel cell* is an essential component because of its high efficiency. Since it is not independent of the system, it should be custom designed to meet the requirements and the constraints of the system. *Solar cells* will require compact and efficient coolers. As a result, a new concept of *thermal design* with the component design of the novel devices briefly addressed becomes inevitable and essential. This book studies thermal design in light of the problems mentioned here.

1.2 HUMANS AND ENERGY

At the early Stone Age (about 2 millions year ago), our ancestors wandered from one place to another to find food and shelter. One of the discoveries by humankind was fire, which gave warmth and light to their shelters and protected them from dangerous animals. This fire would be the first use of energy by humankind and also a turning point in early human life, probably comparable to the discovery of electricity in the nineteenth century. About 5,000 years ago, people began using the wind to sail from one place to another. It was only 2,500 year ago that people began using windmills and waterwheels to grind grain. About 2,300 years ago, the ancient Greek mathematician Archimedes supposedly burned the ships of attacking Syracuse by using mirrors to concentrate the sun's rays on their sails. This would have been the first recorded use of focused solar energy in history. More recently, we learned to use resources such as fossil fuels (coals, oil, and natural gas), renewable energy (sunlight, wind, biofuel, tide, and geothermal heat), and nuclear energy. Solar energy is the primary source of energy for wood growing on Earth and for algae and plankton multiplying in the sea through photosynthesis. Wood turned into coal and algae and plankton turned into oil and natural gas over hundreds of millions of years, resulting in solar energy storage for use today.

Energy consumption has been drastically increased since the discovery of fossil fuels and electricity. Fossil fuels in 2007 produce 81 percent of the world energy consumption, while nuclear power and renewable energy produce the rest (from the Energy Information Administration (EIA) Annual Energy Review 2007). The emissions from fossil fuel have unfortunately resulted in global warming, global weather change, and air pollution. Nations are advised to reduce the hazardous emissions from the fossil fuel. An alternative energy (nonhydrocarbon fuel) is in demand, of which renewable energy is the most viable resource based on technology-dependent devices that require high efficiency and low cost of manufacturing. Solar cells, fuel cells, and wind turbines are examples.

It is not very difficult to predict that solar energy will be the energy of the future. Some day, every unit of community or family will have its own inexpensive power station producing electricity, heat, and nonhydrocarbon fuel (hydrogen) from either natural gases or solar-cell panels. Each automobile will run on a fuel cell powered with hydrogen-delivering water that will be recycled (zero air pollution), and will have its own small power station so that an appreciable amount of green fuel (hydrogen) can be produced during working hours. No grid and large cables will be seen on the streets. Deserts, for example, in the Middle East and North Africa, will be huge sources of green fuel (hydrogen) and electricity from solar energy. This book is intended to put a first step toward the thermal design of the novel devices for our future energy solutions.

1.3   THERMODYNAMICS

1.3.1   Energy, Heat, and Work

Energy is the capacity to perform work. Energy can exist as a numerous forms such as kinetic, potential, electric, chemical, and nuclear. Any form of energy can be transformed into another form, but the total energy always remains the same. This principle, *the conservation of energy*, was first postulated in the early nineteenth century. It applies to any isolated system. *Heat* (or *heat transferred*) is *thermal energy* that is transferred between two systems by virtue of a temperature difference. *Work* is the energy transfer associated with a force acting through a distance. Work can exist as a numerous forms such as piston work, shaft work, and electrical work. The rate of work is called *power*.

1.3.2   The First Law of Thermodynamics

The first law of thermodynamics, also known as *the conservation of energy*, provides a sound basis for studying the relationship among the various forms of energy, total energy, heat, and work. This states that energy can be neither created nor destroyed

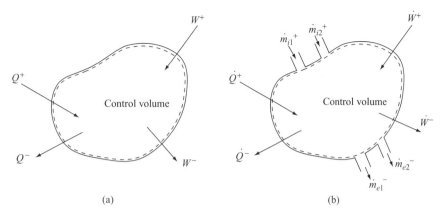

Figure 1.1 Control volumes and sign convention for (a) closed system and (b) open system

during a process; it can only change forms. The sign convention is illustrated in Figure 1.1(a).

$$Q > 0 \quad \text{Heat transferred to system} \tag{1.1a}$$

$$Q < 0 \quad \text{Heat transferred from system} \tag{1.1b}$$

$$W > 0 \quad \text{Work done on system} \tag{1.1c}$$

$$W < 0 \quad \text{Work done by system} \tag{1.1d}$$

The first law of thermodynamics for a closed system in Figure 1.1(a) is given by

$$\Delta E = Q + W \tag{1.2}$$

where ΔE is the *total energy* change in the system, Q the *heat* transferred, and W the *work* done. Caution is given to the sign of work in Equations (1.1c) and (1.1d); some books define work as positive when it is done by the system. The reason is that many engineering applications focus on the work done by a particular heat engine, and so it is helpful to define that as positive. In that case, Equation (1.2) becomes that $\Delta E = Q - W$.

Total energy, E, is the extensive property and presents all the energy such as kinetic, potential, thermal, latent, chemical, nuclear, and others. All energy except kinetic and potential energy is called *internal energy*, U. The *kinetic energy*, KE, is associated with the motion of the system as a whole. The potential energy, PE, is associated with the position of the system as a whole. In the case of a stationary system, $\Delta E = \Delta U$. The *potential energy* may be present in a variety of fields (gravity, electric, or magnetic). The total energy is expressed as

$$E = U + KE + PE \tag{1.3}$$

The internal energy is

$$U = \text{Thermal} + \text{Latent} + \text{Chemical} + \text{Nuclear} + \text{Others} \tag{1.4}$$

The change of internal energy U can be expressed thermodynamically:

$$\Delta U = \begin{cases} mc_p\Delta T & \text{for liquids and solids} \\ mc_v\Delta T & \text{for gases or air} \end{cases} \tag{1.5}$$

where m is the mass and c_p and c_v are the specific heat at constant pressure and at constant volume, respectively. ΔT is the temperature change during the process ($\Delta T = T_2 - T_1$). In a rate form, the first law of thermodynamics can be given by

$$\Delta \dot{E} = \frac{dE_{sys}}{dt} = \dot{Q} + \dot{W} \tag{1.6}$$

where $\Delta \dot{E}$ is the rate of change of total energy in the system (being the same for both the system and the control volume for a closed system). In steady-state conditions (no change with time), the rate of change of energy in the system is zero ($\Delta \dot{E} = 0$). Equation (1.6) then shows that the work $\dot{W}$ becomes equal to the heat $\dot{Q}$. The relationship between work and heat was not clear before the discovery of the first law of thermodynamics.

Now consider the *mass flow rate*, $\dot{m}$, into and out of the system, as shown in Figure 1.1(b). The mass flow internally involves pressure work, and the enthalpy (flow energy) is the sum of the internal energy (u) and the pressure work (pv), which is called enthalpy ($h = u + pv$). For a control volume, CV, in Figure 1.1(b), the first law of thermodynamics with mass flow rates is expressed by

$$\Delta \dot{E}_{CV} = \frac{dE_{CV}}{dt} = \dot{Q} + \dot{W} + \sum_{in} \dot{m}\left(h + \frac{v^2}{2} + gz\right) - \sum_{out} \dot{m}\left(h + \frac{v^2}{2} + gz\right) \tag{1.7}$$

This is equivalent to the *energy balance* in heat transfer, which is expressed as

$$\Delta \dot{E}_{st} = \dot{E}_{in} - \dot{E}_{out} + \dot{E}_g \tag{1.7a}$$

where $\Delta \dot{E}_{st}$ is the energy stored and equal to $\Delta \dot{E}_{CV}$. $\dot{E}_{in}$ is the energy entering into the control volume, $\dot{E}_{out}$ the energy exiting from the control volume and $\dot{E}_g$ the heat generation (electricity or nuclear reaction). This equation is particularly useful in heat transfer calculations. Mass must be conserved. Hence, *the conservation of mass* is expressed as

$$\frac{dm_{CV}}{dt} = \sum_{in} \dot{m} - \sum_{out} \dot{m} \tag{1.8}$$

In the case of a steady flow from Equations (1.7) and (1.8), we note that

$$\frac{dE_{CV}}{dt} = 0 \quad \text{and} \quad \frac{dm_{CV}}{dt} = 0 \tag{1.9}$$

Equation (1.7), *the first law of thermodynamics*, is the basis of fluid-thermal engineering and is widely used. However, there is an important aspect to the equation. Consider a steady-state process in a container where electrical work $\dot{W}_e$ is applied while allowing heat transfer on the surface of the container, as shown in Figure 1.2(a). You may consider a heating element for the electrical work. In accordance with the first law of thermodynamics, $\dot{W}_e$ must turn completely into $\dot{Q}$. Intuitively, this process

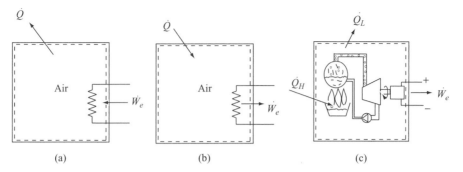

Figure 1.2 Steady-state systems with heat and electrical work, (a) possible process (100 % efficiency), (b) impossible process, and (c) steam power plant (about 60 % efficiency)

is possible. Now consider the opposite process, as shown in Figure 1.2(b). Heat flows into the system and the electrical work is produced. Do you think that this process is possible? The answer is no. Heat cannot be readily converted to the useful mechanical work or electricity. It is an impossible process. We find here a very important aspect that the first law of thermodynamics provides the quantity but not necessarily the quality. Thermodynamically, it is an *irreversible process* that typically involves *heat transfer* and *friction*. Hence, a reversible process is ideal and possesses the maximum efficiency in devices. The irreversibility is treated with a quantitative property that is referred to as *entropy* (degree of disorder). *The second law of thermodynamics* deals with the irreversibility, the quality of process, or the entropy.

Can we produce electricity from heat (or fuels)? The answer is yes. If we think of a steam power plant (typically about 60 percent thermal efficiency) as shown in Figure 1.2(c), the answer is obvious: 100 percent of $\dot{Q}_H$ (fuel) enters and produces 60 percent work of the entered energy as electricity, which leads to delivering 40 percent energy $\dot{Q}_L$ as waste heat, where $\dot{Q}_H$ and $\dot{Q}_L$ are the rate of heat transferred from the high temperature source to the low temperature sink, respectively. Diesel engines typically have about 50 percent thermal efficiency. Gasoline engines have about 30 percent thermal efficiency. Solar cells have about 15 percent thermal efficiency. Thus, the gain in electrical work from heat depends on devices and technology. We know that electricity is the form of energy most easily convertible to other forms of energy. The most efficient form of energy is called *exergy*. Exergy is a measure of the quality of energy. Thus, electricity has the highest level of exergy of any form of energy. It is interesting to note that the entities in the first law of thermodynamics can be realistic or ideal, depending on whether the thermal efficiency is included in the equation or not since the thermal efficiency reflects irreversibility and the second law of thermodynamics. The determination of a device's efficiency or effectiveness is therefore essential in thermal design. For another example, the incorporation of *friction* (or friction factor) into the equation allows engineers to use the equation in their design.

1.3.3 Heat Engines, Refrigerators, and Heat Pumps

Devices that can contain the energy of heat and make it do work are called *heat engines*. Steam and gas turbine, diesel engine, gasoline engine, and ocean thermal energy conversion (OTEC) are examples of heat engines. Most of our energy comes from the burning of fossil fuels and from nuclear reactions: both of these supply energy as heat. However, we need mechanical energy to operate machines or propel vehicles. Exergy transformation from a low exergy (heat) to a high exergy (mechanical energy or electricity) is always achieved in the expense of the low thermal efficiency or vice versa. Fuel cells are a good example that the chemical energy of a fuel is directly converted to electricity without the process to heat. This is a reason why fuel cells have a higher efficiency than conventional combustion engines. A heat engine can produce mechanical work between a high temperature source and a low temperature sink, as shown in Figure 1.3.

If we apply the first law of thermodynamics to the heat engine in Figure 1.3(a), we come up with

$$\dot{W}_{net} = \dot{Q}_H - \dot{Q}_L \qquad (1.10)$$

where $\dot{W}_{net}$ is the net work, $\dot{Q}_H$ the heat transferred to the heat engine from the high temperature source, and $\dot{Q}_L$ the heat transferred to the low temperature sink from the heat engine. Using Equation (1.10), the thermal efficiency in the heat engine is expressed as

$$\eta_{th} = \frac{Output}{Input} = \frac{\dot{W}_{net}}{\dot{Q}_H} = \frac{\dot{Q}_H - \dot{Q}_L}{\dot{Q}_H} = 1 - \frac{\dot{Q}_L}{\dot{Q}_H} \qquad (1.11)$$

Note that the thermal efficiency is less than unity. If we apply the first law of thermodynamics to the *refrigerator* in Figure 1.3(b), we have the same relationship with Equation (1.10) as

$$\dot{W}_{net} = \dot{Q}_H - \dot{Q}_L \qquad (1.12)$$

The efficiency of the refrigerator is expressed in terms of *coefficient of performance* (COP), denoted by COP_R.

$$COP_R = \frac{Output}{Input} = \frac{\dot{Q}_L}{\dot{W}_{net}} = \frac{\dot{Q}_L}{\dot{Q}_H - \dot{Q}_L} = \frac{1}{\dot{Q}_H / \dot{Q}_L - 1} \qquad (1.13)$$

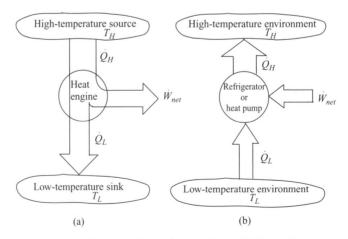

(a) (b)

Figure 1.3 Schematic flow diagrams of (a) a heat engine and (b) a refrigerator or a heat pump

Notice that the value of COP_R can be greater than unity. That is, the amount of heat removed from the low-temperature environment can be greater than the amount of work input. Most refrigerators have COP_R of $2.5 \sim 3.0$.

A device that transfers heat from a low temperature body to high temperature body is called a *heat pump*. The *heat pump cycle* is identical to a *refrigerator cycle* in principle, but differs in that the primary purpose of the heat pump is to supply heat rather than to remove it from an enclosed space. The measure of performance is also expressed in terms of coefficient of performance COP_{HP}, defined as

$$COP_{HP} = \frac{\text{Output}}{\text{Input}} = \frac{\dot{Q}_H}{\dot{W}_{net}} = \frac{\dot{Q}_H}{\dot{Q}_H - \dot{Q}_L} = \frac{1}{1 - \dot{Q}_L/\dot{Q}_H} \qquad (1.14)$$

Most heat pumps have COP_{HP} of $3.0 \sim 5.0$.

1.3.4   The Second Law of Thermodynamics

As mentioned previously, the second law of thermodynamics is associated with the irreversibility, the quality, or the entropy of the process. The classical statements for *the second law of thermodynamics* are restated here in several different ways.

Statements:

1. Heat is naturally transferred always from a high temperature body to a low temperature body, not vice versa.
2. Work is required in order to make heat transfer from a cold temperature body to a high temperature body.
3. It is impossible to construct a heat engine that operates 100 percent thermal efficiency.
4. It is impossible to construct a heat engine without a heat loss to the environment.

For example, if a person claims to invent a machine that produces 100 percent work from the energy of heat entered, this satisfies the first law of thermodynamics but violates the second law of thermodynamics. Perpetual motion machines touted in the nineteenth and early twentieth centuries violated the second law of thermodynamics. Of course, the machines were never functional.

1.3.5   Carnot Cycle

If the efficiency of all heat engines is less than 100 percent, what is the most efficient cycle we can have? The *Carnot cycle* is the most efficient cycle. A *heat engine* that operates in the Carnot cycle was proposed in 1824 by French engineer *Sadi Carnot*. Heat engines are cyclic devices and the working fluid of a heat engine returns to its initial state at the end of each cycle.

The *Carnot cycle* is ideal, so the ideal gas is used as a working fluid and the entire process is reversible. There is no friction and no heat transfer except the isothermal processes (heat transfer). The most efficient heat-transfer process is known to be an *isothermal process* where the high- or low-source temperature remains constant during the process. And the most efficient process between two-temperature

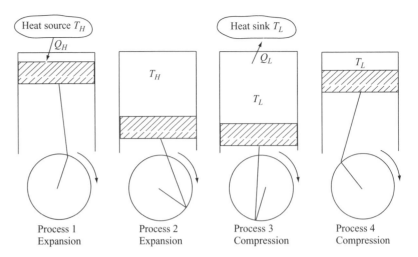

Figure 1.4 Execution of the Carnot cycle in a closed system

sources is a reversible adiabatic process. We want to construct a Carnot cycle based on these two hypotheses. A simple piston cycle giving the Carnot cycle is shown in Figure 1.4.

Process 1–2 (reversible isothermal expansion): During this process, the gas in the cylinder expands while heat Q_H is transferred from a high temperature source at T_H, to the cylinder where T_H is the temperature at the high temperature source and T_L is the temperature at the low temperature sink. Hence, the gas temperature remains constant at T_H. During this process, heat must be added in order to compensate the temperature drop due to the expansion. State 2 is determined inasmuch as the amount of work done by the piston equals the amount of heat Q_H.

Process 2–3 (reversible adiabatic expansion): During this process, the gas continues to expand until the gas temperature T_H turns into T_L while the supply of heat is stopped. The cylinder is insulated. State 3 is determined when the gas temperature reaches T_L.

Process 3–4 (reversible isothermal compression): During this process, the gas is compressed by the piston while heat Q_L is allowed to transfer from the cylinder to the low-temperature sink at T_L. Hence, the gas temperature remains constant at T_L. During this process, heat must be removed in order to compensate the temperature rise due to the compression. State 4 is determined inasmuch as the amount of work done on the piston equals the amount of heat Q_L.

Process 4–1 (reversible adiabatic compression): During this process, the gas continues to be compressed until the gas temperature T_L turns into T_H while the transfer of heat is stopped and the cylinder is insulated. State 1 is determined when the gas temperature T_L reaches T_H.

The *Carnot cycle* is an ideal cycle and also the most efficient heat cycle. The four processes mentioned here are plotted in a P-V diagram in Figure 1.5. It is interesting to note that the Carnot cycle must have a heat loss Q_L, although the cycle is reversible. Actually, this satisfies *the second law of thermodynamics*. In order words, no heat

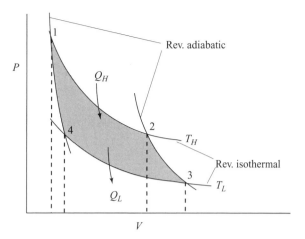

Figure 1.5   P-V diagram of the Carnot cycle

engine can be constructed without Q_L. Therefore, no 100 percent efficient heat engine is possible.

From Equation (1.11), we find that the thermal efficiency η_{th} is a function of the ratio of Q_L to Q_H since Q_L/Q_H is the same as $\dot{Q}_L/\dot{Q}_H$. The ratio is examined further here. The ideal gas law for an isothermal process at T_H gives

$$P = \frac{mRT_H}{V} \tag{1.15}$$

where P is the pressure, V the volume, and R the *gas constant*. If we apply the first law of thermodynamics to process 1–2, the isothermal process ($\Delta T = 0$), we have

$$\Delta E = \Delta U = mc_v\Delta T = Q_H + {}_1W_2 \tag{1.16}$$

Then we come up with the fact that the heat transferred must be equal to the work done by the system as

$$Q_H = -{}_1W_2 = -\int_1^2 PdV = -\int_1^2 \frac{mRT_H}{V}dV = -mRT_H \ln\frac{V_2}{V_1} \tag{1.17}$$

Similarity for process 3–4,

$$Q_L = {}_3W_4 = \int_3^4 PdV = \int_3^4 \frac{mRT_L}{V}dV = mRT_L \ln\frac{V_4}{V_3} = -mRT_L \ln\frac{V_3}{V_4} \tag{1.18}$$

The ratio of the two quantities is thus

$$\frac{Q_L}{Q_H} = \frac{-mRT_L \ln \dfrac{V_3}{V_4}}{-mRT_H \ln \dfrac{V_2}{V_1}} = \frac{T_L}{T_H} \frac{\ln \dfrac{V_3}{V_4}}{\ln \dfrac{V_2}{V_1}} \tag{1.19}$$

From process 1–2 and process 3–4, we have

$$P_1 V_1 = P_2 V_2 \tag{1.20}$$

$$P_3 V_3 = P_4 V_4 \tag{1.21}$$

Since both process 2–3 and process 4–1 are reversible and adiabatic (isentropic), we can apply the polytropic relation with $k = 1.4$ to the ideal gas in the cylinder as

$$PV^k = \text{constant} \tag{1.22}$$

Therefore, we have

$$P_2 V_2^k = P_3 V_3^k \tag{1.23}$$

$$P_4 V_4^k = P_1 V_1^k \tag{1.24}$$

Multiplying Equations (1.20), (1.21), (1.23) and (1.24) and canceling the factor $P_1 P_2 P_3 P_4$ gives

$$V_1 V_2^k V_3 V_4^k = V_2 V_3^k V_4 V_1^k \tag{1.25}$$

Rearranging,

$$(V_2 V_4)^{k-1} = (V_3 V_1)^{k-1} \tag{1.26}$$

Finally, we have

$$\frac{V_2}{V_1} = \frac{V_3}{V_4} \tag{1.27}$$

From Equation (1.19), we derive an important relationship between the ratio of heat and the ratio of temperature.

$$\frac{Q_L}{Q_H} = \frac{T_L}{T_H} \tag{1.28}$$

From Equation (1.11), the thermal efficiency η_c for a Carnot cycle engine is

$$\eta_c = 1 - \frac{\dot{Q}_L}{\dot{Q}_H} = 1 - \frac{T_L}{T_H} \tag{1.29}$$

This surprisingly simple result says that the efficiency of a Carnot cycle engine depends only on the temperatures of the two heat sources. When the difference of the two is large, the efficiency is nearly unity; when it is small, the efficiency is much less than unity. Equation (1.28) can be also applied to Equation (1.13) for refrigerators or Equation (1.14) for heat pumps for its Carnot cycle efficiency.

1.4   HEAT TRANSFER

1.4.1   Introduction

Heat transfer is *thermal energy* in transit due to a spatial temperature difference. The second law of thermodynamics tells us that heat is naturally transferred from a high-temperature body to a low-temperature body. Heat transfer has three modes or mechanisms: conduction, convection, and radiation.

Thermal energy involves motions such as the rotation, translation, and vibration of molecules. *Temperature* measures the average kinetic energy of molecules. Molecules in a warmer object have greater kinetic energy than molecules in a cooler object. When a hot object comes in contact with a cold one, the molecules in the hot object transfer some of their kinetic energy during collisions with molecules of the cold object. This transfer of kinetic energy of molecules is the mechanism of conduction and convection heat transfer. *Conduction* usually takes place in a solid.

For example, when we touch a metal spoon in a hot coffee, we feel the hotness. This is due to the conduction in the spoon. *Convection* occurs on the surface of an object with fluid in motion. As another example, think of bare skin, which usually experiences a heat loss to the ambient air by convection heat transfer.

Suppose that the surface temperature of our skin is greater than that of the ambient air. Then air adjacent to the skin gets warm, becoming lighter in density and rising to induce a flow near the skin. This is the mechanism of *free convection*. If our skin is exposed to wind, the mechanism of the heat transfer is called *forced convection*. *Radiation*, or *thermal radiation*, is somewhat different from conduction and convection. Radiation is energy emitted by matter that is at a nonzero temperature. It is *electromagnetic waves* (or *photons*) effectively propagating even in a vacuum. Radiation is generated when the movement of charged particles (molecules and atoms can be easily charged or ionized when they lose electrons) is converted to electromagnetic waves. For example, why do we feel the coldness of outside in winter when we are in a room where the air temperature remains constant through the year? Because the wall's temperature is lower in winter due to conduction. Our body interacts not only by convection with the air but also by the radiation exchange with the walls. Therefore, we have more heat loss in winter than in summer and feel colder. Note that heat transfer always has a direction and a magnitude like a vector, which is very important in engineering calculations.

Conduction occurs as energy is transported from substance to substance. The *Fourier law of heat conduction* can be expressed as

$$q_x = -kA\frac{dT}{dx} \tag{1.30}$$

where q_x is the heat transfer in x direction, k the *thermal conductivity* (W/mK), A the heat transfer area (m^2), and dT/dx the temperature gradient. The minus sign takes the heat flow positive where the temperature gradient is mathematically negative with respect to x coordinate.

Convection occurs as the energy is transported by the motion of fluid. Although the convective heat transfer appears complicated because of the complex motion of the fluid, the resultant behavior can be expressed in a simple way.

$$q = hA\,(T_s - T_\infty) \tag{1.31}$$

where h is the *convection heat transfer coefficient* (W/m²K), A the heat transfer area (m²), T_s the surface temperature, and T_∞ the fluid temperature. This equation is known as *Newton's law of cooling*. Note that the sign convention is such that the heat transfer is positive when the heat flows outward from the surface to the fluid.

Radiation is energy emitted by matter in the form of the electromagnetic waves. Hence, radiation between two objects exchanges their energies. One of the typical radiation problems is a relatively small object in large *surroundings*. In this case, the *radiation exchange* can be expressed as

$$q = A\varepsilon\sigma\left(T_s^4 - T_{sur}^4\right) \tag{1.32}$$

where A is the heat transfer area (m²), ε the *emissivity*, σ the *Stefan-Boltzmann constant* ($\sigma = 5.67 \times 10^{-8}$ W/m² · K⁴), T_s the absolute temperature (K) of the surface, and T_{sur} the absolute temperature (K) of the surroundings.

1.4.2   Conduction

Consider one-dimensional steady-state conduction in a plane wall where the inner and outer surfaces are in contact with air, as illustrated in Figure 1.6(a). Radiation is neglected. There are three components in the system: inner convection, conduction, and outer convection. Note that the heat transfer in each component must be the same.

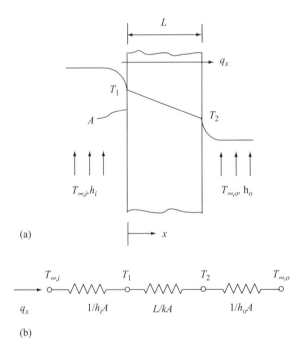

Figure 1.6   Heat transfer through a plane wall: (a) Temperature distribution, (b) The thermal circuit

The convection heat transfer to the plane wall is expressed considering the sign convention discussed previously.

$$q = h_i A \left(T_{\infty,i} - T_1 \right) = \frac{T_{\infty,i} - T}{1/h_i A} = \frac{T_{\infty,i} - T}{R_{conv,i}} \tag{1.33}$$

where $R_{conv,i}$ is the *convection thermal resistance* that is defined as $\Delta T/q$. The convection resistance in the inner convection component, $R_{conv,i}$, can be expressed as

$$R_{conv,i} = \frac{1}{h_i A} \tag{1.34}$$

The concept of thermal resistance is similar to that of electrical resistance and is widely used in thermal design. Now consider the conduction in the plane wall. The conduction heat transfer for one-dimensional steady-state conditions is expressed as

$$q = -kA \frac{dT}{dx} = -kA \frac{T_2 - T_1}{L} = \frac{T_1 - T_2}{L/kA} = \frac{T_1 - T_2}{R_{conv,i}} \tag{1.35}$$

The conduction resistance in the wall is

$$R_{cond} = \frac{L}{kA} \tag{1.36}$$

The convection heat transfer exiting from the wall is

$$q = h_o A \left(T_2 - T_{\infty,o} \right) = \frac{T_2 - T_{\infty,o}}{1/h_o A} = \frac{T_2 - T_{\infty,o}}{R_{conv,o}} \tag{1.37}$$

The convection resistance in the outer component is

$$R_{conv,o} = \frac{1}{h_o A} \tag{1.38}$$

Since the system is composed of three components in series, the thermal circuit can be constructed, which is shown in Figure 1.6(b). The *total thermal resistance* will be the sum of the three components.

$$R_{total} = \frac{1}{h_i A} + \frac{L}{kA} + \frac{1}{h_o A} = \frac{\Delta T}{q} \tag{1.39}$$

Now the heat transfer across the wall can be expressed using the total thermal resistance as,

$$q = \frac{T_{\infty,i} - T_{\infty,o}}{\dfrac{1}{h_i A} + \dfrac{L}{kA} + \dfrac{1}{h_o A}} = \frac{T_{\infty,i} - T_{\infty,o}}{R_{total}} \tag{1.40}$$

Note that the calculation of the heat transfer across the wall does not require the inner and outer surface temperatures of the wall. An overall heat transfer coefficient is defined in the following equation as

$$q = UA\Delta T = \frac{\Delta T}{1/UA} \tag{1.41}$$

The overall heat transfer coefficient U is expressed as

$$U = \frac{1}{R_{total}A} \qquad (1.42)$$

Sometimes a *UA value* that is the inverse of the total thermal resistance used in heat exchanger design.

$$UA = \frac{1}{R_{total}} \qquad (1.43)$$

Consider a thick tube, where the inner and outer surfaces experience the flows shown in Figure 1.7(a), we find an expression for the heat transfer in a similar way as the plane wall.

$$q = \frac{T_{\infty,i} - T_{\infty,o}}{\dfrac{1}{h_i\,(2\pi r_i L)} + \dfrac{\ln(r_o/r_i)}{2\pi k L} + \dfrac{1}{h_o\,(2\pi r_o L)}} = \frac{T_{\infty,i} - T_{\infty,o}}{R_{total}} \qquad (1.44)$$

Note that the temperature distribution in the tube is no longer linear as for the one in plane wall because the heat transfer area *A* increases with increasing the cylindrical coordinate *r*.

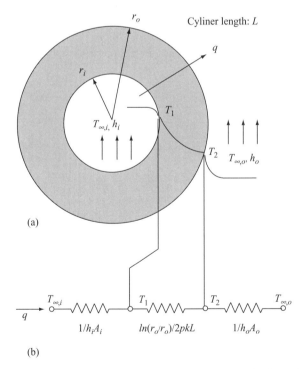

Figure 1.7 Heat transfer through a cylinder: (a) Temperature distribution, (b) Equivalent thermal circuit

1.4.3 Convection

Convection involves not only the heat also the fluid flow. Therefore, energy equations combined with the Navier-Stokes equations should be used to solve the convection problem. *The Navier-Stokes equations* that are composed of the conservation of mass and the conservation of momentum are the fundamental governing equations of fluid flow. These equations were formulated in the early nineteenth century [11, 12], one of the biggest breakthroughs in the field of applied mathematics. However, it is very difficult to solve the equations. A parallel flow over a flat plate (simple but fundamental) was solved by Prandtl [12] in 1904 and Blasius [13] in 1908. Based on those solutions, essential parameters and semiempirical correlations were able to be provided for other cases, which are very useful in thermal design and presented here.

1.4.3.1 *Parallel Flow on an Isothermal Plate*

Parallel flow on an isothermal flat plate is illustrated in Figure 1.8(a). Assume that uniform *laminar flow* enters from the left. The *boundary layer* physically develops over the plate from the leading edge due to the shear stress (or friction), starting as a laminar boundary layer and at a critical point turning into the turbulent boundary layer as shown. The velocity boundary layer thickness $\delta(x)$ is determined by the velocity, the distance, and the properties of the fluid. A nondimensional number with respect to the distance x along the plate is the *Reynolds number, Re*, defined as

$$Re_x = \frac{\rho U_\infty x}{\mu} = \frac{U_\infty x}{\nu} \tag{1.45}$$

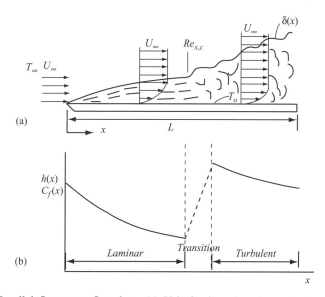

Figure 1.8 Parallel flow on a flat plate: (a) Velocity boundary layer development, (b) Heat transfer coefficient $h(x)$ and friction coefficient $C_f(x)$

In terms of the length L,

$$Re_L = \frac{\rho U_\infty L}{\mu} = \frac{U_\infty L}{\nu} \tag{1.46}$$

where U_∞ is the entering uniform velocity, ρ the density of the fluid, μ the dynamic viscosity, and L the distance from the leading edge. The meaning of the Reynolds number is the inertia force over the viscous force, which can also be interpreted as a dimensionless velocity. At a critical point, the flow pattern changes from laminar to turbulent flow. *The critical Reynolds number* for a smooth surface has been obtained by measurements, which is

$$Re_{x,c} = 5 \times 10^5 \tag{1.47}$$

As for the velocity boundary layer thickness, the *thermal boundary layer thickness* $\delta_t(x)$ forms in a similar way. The most important dimensionless number (or property) for the thermal boundary layer is the *Prandtl number*, Pr, which is defined as

$$Pr = \frac{\nu}{\alpha} = \frac{\mu c_p}{k} \tag{1.48}$$

where α is the thermal diffusivity, μ the dynamic viscosity, c_{p} the specific heat at the constant pressure, and k the thermal conductivity of the fluid.

The thermal boundary layer analysis suggests a dimensionless parameter termed *the Nusselt number* Nu_L:

$$Nu_L = \frac{hL}{k} \tag{1.49}$$

The Nusselt number is equal to the dimensionless temperature gradient at the surface or can be interpreted as a dimensionless convection heat transfer coefficient.

There are three important aspects of the convective flow over a flat plate. The first aspect is that the velocity boundary layer thickness is typically very thin (order of millimeters) and the flow beyond the boundary layer behaves as inviscid flow. The velocity adjacent to the surface within the boundary layer causes a shear stress or a friction force on the plate. This is a reason why potential flow (inviscid flow) is still a powerful tool in engineering design. The second aspect is that the transition to *turbulent flow* from *laminar flow* is a very common physical phenomenon (e.g., cigarette's smoke rising becoming turbulent, or turbulence of ship's smoke on the sea—turbulence is due to the infinite distance from the leading edge). If we measure the velocity profile in the turbulent region, we find that the bulk velocity is quite uniform even within the boundary layer as shown, although microscopic eddies prevail all around. The turbulence eventually increases the *friction coefficient* $C_{f,x}$ to some extent, as shown in Figure 1.8(b). This uniformity of the velocity in the boundary layer renders modeling of the turbulent flow feasible. The last aspect is that turbulence causes the convective heat transfer to increase significantly, as shown in Figure 1.8(b). This is an important point for thermal design. For example, turbulent flow could be an option for the enhancement in heat transfer.

Laminar Flow ($Re_L \leq 5 \times 10^5$) The analytical solutions of Prandtl and Blasius provided the following correlations for laminar flow and also the basis for the correlations for turbulent flow. The local friction coefficient $C_{f,x}$ is

$$C_{f,x} = \frac{\tau_x}{\rho U_\infty^2/2} = 0.664 Re_x^{\frac{-1}{2}} \tag{1.50}$$

The average friction coefficient is

$$\overline{C}_f = \frac{\overline{\tau}}{\rho U_\infty^2/2} = 1.328 Re_L^{\frac{-1}{2}} \tag{1.51}$$

The local Nusselt number is

$$Nu_x = \frac{h_x x}{k_f} = 0.332 Re_x^{\frac{1}{2}} Pr^{\frac{1}{3}} \quad Pr \geq 0.6 \tag{1.52}$$

The average Nusselt number is

$$\overline{Nu}_L = \frac{\overline{h}L}{k_f} = 0.664 Re_L^{\frac{1}{2}} Pr^{\frac{1}{3}} \quad Pr \geq 0.6 \tag{1.53}$$

Turbulent Flow ($Re_x > 5 \times 10^5$) A turbulent flow is very difficult to solve analytically. Hence, the following correlations mostly rely on experiment. However, the analysis of the laminar flow was very helpful in formulating the correlations for turbulent flow. The *local friction coefficient* is expressed as

$$C_{f,x} = \frac{\tau_x}{\rho U_\infty^2/2} = 0.0592 Re_x^{\frac{-1}{5}} \quad 5 \times 10^5 \leq Re_x \leq 10^8 \tag{1.54}$$

The *average friction coefficient* is

$$\overline{C}_f = \frac{\overline{\tau}}{\rho U_\infty^2/2} = 0.074 Re_L^{\frac{-1}{5}} \quad 5 \times 10^5 \leq Re_x \leq 10^8 \tag{1.55}$$

The *local Nusselt number* is

$$Nu_x = \frac{h_x L}{k_f} = 0.0296 Re_x^{\frac{4}{5}} Pr^{\frac{1}{3}} \quad 0.6 \leq Pr \leq 60 \text{ and } 5 \times 10^5 \leq Re_x \leq 10^8 \tag{1.56}$$

The *average Nusselt number* for an average heat transfer coefficient for turbulent flow

$$\overline{Nu}_L = \frac{\overline{h}L}{k_f} = 0.037 Re_L^{\frac{4}{5}} Pr^{\frac{1}{3}} \quad 0.6 \leq Pr \leq 60 \text{ and } 5 \times 10^5 \leq Re_x \leq 10^8 \tag{1.57}$$

Mixed Boundary Layer Conditions If the plate has regions where both the laminar and turbulent flows significantly contribute to the heat transfer on the plate, we may use the following correlation:

$$\overline{Nu}_L = \left(0.037 Re_L^{\frac{4}{5}} - 871\right) Pr^{\frac{1}{3}} \quad 0.6 \leq Pr \leq 60 \text{ and } 5 \times 10^5 \leq Re_x \leq 10^8 \tag{1.58}$$

1.4.3.2 A Cylinder in Cross Flow

For a circular cylinder, the characteristic length is the diameter. In this case, the *Reynolds number* is defined as

$$Re_D = \frac{\rho U_\infty D}{\mu} = \frac{U_\infty D}{\nu} \tag{1.59}$$

A *Nusselt number* is proposed by Churchill and Bernstein [14] for all Re_D and Pr as

$$\overline{Nu}_D = \frac{\overline{h}D}{k} = 0.3 + \frac{0.62 Re_D^{1/2} Pr^{1/3}}{\left[1 + (0.4/Pr)^{2/3}\right]^{1/4}} \left[1 + \left(\frac{Re_D}{282,000}\right)^{5/8}\right]^{4/5} \tag{1.60}$$

1.4.3.3 Flow in Ducts

Consider laminar flow in a duct, as shown in Figure 1.9, where fluid enters the duct with a uniform velocity. We know that when the fluid makes contact with the surface, viscous effects become important, and a boundary layer develops with increasing x. This development leads to a boundary layer merger at the centerline. After this merger, the velocity profile no longer changes with increasing x. The flow is said to be *fully developed*. The distance from the entrance to the merger is termed the *hydrodynamic entry length*, $x_{fd,h}$.

The fully developed velocity profile is parabolic for laminar flow in a duct. For turbulent flow, the profile is flatter due to turbulent mixing. When dealing with internal flows, it is important to be cognizant of the extent of the entry region, which depends on whether the flow is laminar or turbulent.

The Reynolds number for flow in a duct is defined as

$$Re_D = \frac{\rho u_m D_h}{\mu} = \frac{u_m D_h}{\nu} \tag{1.61}$$

where u_m is the *mean fluid velocity* over the duct cross section and D_h is the *hydraulic diameter*. The hydraulic diameter is used for noncircular ducts to treat as a circular duct and defined as

$$D_h = \frac{4A_c}{P_{wet}} \tag{1.62}$$

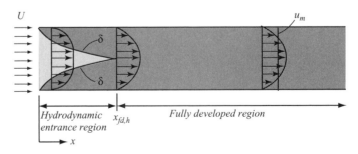

Figure 1.9 Entrance region and fully developed region of laminar flow in a duct

where A_c and P_{wet} are the flow cross-sectional area and the *wetted perimeter*, respectively. The *equivalent diameter* that is used for the heat transfer (*Nusselt number*) calculations is defined as

$$D_e = \frac{4A_c}{P_{heated}} \tag{1.62a}$$

where P_{heated} is the heated perimeter.

The critical Reynolds number corresponding to the onset of turbulence is

$$Re_{D,c} = 2300 \tag{1.63}$$

Although much larger Reynolds numbers ($Re_D = 10,000$) are needed to achieve fully turbulent conditions. The transition to turbulence is likely to begin in the developing boundary layer of the entrance region.

Friction Factor in Fully Developed Flow   The engineer is frequently interested in the pressure drop needed to sustain an internal flow because this parameter determines pump or fan power requirements. To determine the pressure drop, it is convenient to work with the *friction factor*, which is a dimensionless parameter.

The *Fanning friction factor* is defined as

$$f_{Fanning} = \frac{\tau_w}{\frac{1}{2}\rho u_m^2} \tag{1.64}$$

where τ_w is the shear stress at wall and u_m is the mean velocity in the duct. The *Darcy friction factor* is defined as

$$f_{Darcy} = \frac{4\tau_w}{\frac{1}{2}\rho u_m^2} \tag{1.65}$$

We adopt herein the Fanning friction factor as is usual in the field of thermal design. Do not confuse it with the Darcy friction factor, often used in other textbooks. For laminar flow, the friction factor is found analytically as

$$f = \frac{16}{Re_D} \tag{1.66}$$

The Fanning friction factor f is presented graphically in Figure 1.10, where the friction factor for laminar flow and turbulent flow are plotted with the various roughness of the surface of the duct. The transition at $Re_D = 2,300$ from the laminar region to the turbulent region is apparent in the figure.

For smooth circular ducts in turbulent flow, the friction factor suggested by Filonenko [15] for $10^4 < Re_D < 10^7$ is given by

$$f = (1.58 \ln (Re_D) - 3.28)^{-2} \tag{1.67}$$

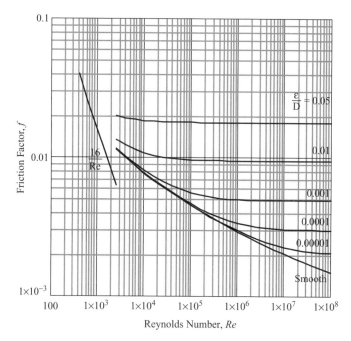

Figure 1.10   Fanning friction factor as a function of Reynolds number for pipe flow

Pressure Drop   The friction causes the pressure drop that can be easily derived from taking the force balance on a differential element in pipe flow. The pressure drop is given by

$$\Delta P = \frac{4fL}{D_h}\frac{\rho u_m^2}{2} = \frac{2fL}{D_h}\frac{G^2}{\rho} \tag{1.68}$$

where G is the mass velocity,

$$G = \rho u_m \tag{1.69}$$

Pump or Fan Power   The power of a fan or pump may be calculated by

$$\dot{W}_{actual} = \frac{dW}{dt} = \frac{\dot{m}}{\eta_p \rho}\Delta P \tag{1.70}$$

where η_p is the pump efficiency and $\dot{m}$ the mass flow rate.

Laminar Flow ($Re_D < 2300$)   For laminar flow, the hydrodynamic entry length is approximately calculated as

$$\left(\frac{x_{fd,h}}{D_h}\right)_{lam} \approx 0.05\,Re_D \tag{1.71}$$

For laminar flow, the thermal entry length may be expressed as

$$\left(\frac{x_{fd,t}}{D_h}\right)_{lam} \approx 0.05\,Re_D\,Pr \tag{1.72}$$

A general relation for the combined entry length for the hydrodynamic and thermal entry and fully developed laminar flow is given by Sieder and Tate [16]. The average Nusselt number is

$$\overline{Nu}_D = \frac{\overline{h}D}{k} = 1.86 \left(\frac{Re_D Pr}{L/D} \right)^{\frac{1}{3}} \left(\frac{\mu}{\mu_s} \right)^{0.14} \quad 0.6 < Pr < 5 \quad (1.73)$$

where all the properties are evaluated at mean temperature except μ_s, which is the absolute viscosity evaluated at the wall temperature.

The Nusselt number Nu and the friction factor f are given in Table 1.1 for fully developed laminar flow in ducts of various cross-sections.

Turbulent Flow ($Re_D > 2{,}300$) Since turbulent flow is ubiquitous in fluid flow, greater emphasis is placed on determining empirical correlations. For turbulent fully developed flow, the Dittus-Boelter equation [23] based on the Colburn equation has been widely used. The Nusselt number is given by

$$Nu_D = \frac{hD}{k} = 0.023 Re_D^{4/5} Pr^n \quad 0.6 \leq Pr \leq 160 \text{ and } Re_D \geq 10{,}000 \quad (1.74)$$

where $n = 0.4$ for $T_s > T_{mean}$ and $n = 0.3$ for $T_s < T_{mean}$. This equation has been confirmed experimentally for a range of conditions.

Although Equation (1.74) is easily applied and is mostly satisfactory, errors may be as large as 25 percent. Such errors may be reduced to less than 10 percent through the use of a more recent correlation. Gnielinski [17] recommended the following correlation valid over a large Reynolds number range, including the transition region. The *Nusselt number* for turbulent is

$$Nu_D = \frac{hD}{k} = \frac{(f/2)(Re_D - 1000)Pr}{1 + 12.7(f/2)^{1/2}(Pr^{2/3} - 1)} \quad (1.75)$$

$$3000 < Re_D < 5 \times 10^6 \text{ and } 0.5 \leq Pr < 2{,}000$$

where the friction factor f is obtain from Equation (1.67).

Table 1.1 Nusselt Numbers and Friction Factors for Fully Developed Laminar Flow in Tubes of Differing Cross-sections

Cross-section	Aspect ratio	$Nu = h \cdot D/k$ Uniform Q''	Uniform T_o	Friction factor $f \cdot Re_D$
○	—	4.36	3.66	16
△	—	2.47	3.11	13
□	1.0	3.61	2.98	14
▭	1.43	3.73	3.08	15
▭	2.0	4.12	3.39	16
▭	3.0	4.79	3.96	17
▭	4.0	5.33	4.44	18
▭	8.0	6.49	5.6	20
▭	∞	8.23	7.54	24

1.4.3.4 *Free Convection*

Free convection, also referred to as *natural convection*, is an important mechanism responsible for heat transfer in our ordinary life. Candlelight would not be possible without free convection. The hot air adjacent to the flame rises due to buoyancy (the warm air is lighter than the cold air) and the cold air replaces the hot air (this supplies oxygen) so that the candle continues burning.

There was actually an experiment for candlelight in space. The result showed that the flame was extinguished after a short period of time. For another example, you would wait a long time in order to have a boiled egg if there were no free convection. Heat transfer from the bodies of animals and human beings are free convection. Today, the cooling of house appliances such as refrigerators and TV (electronics) is based on free convection. It is interesting to note that the driving force of free convection is buoyancy, which results from gravity. We want to here consider a simple, but important, geometry, a vertical plate.

Isothermal Vertical Plate Consider a heated vertical plate at T_s in quiescent air at T_∞ where $T_s > T_\infty$, as shown in Figure 1.11. As for the flow over a flat plate, this problem involves the Navier-Stokes equation and the energy equation with a gravity effect. These equations were solved in the 1930s using the *similarity variable method* and turning them into two ordinary equations that are solvable. The schematic profiles for the temperature and velocity from the solution are illustrated in the figure. Today every heat transfer book includes these results, which are in a good agreement with measurements. More important is that the analysis provided very important parameters. These parameters are the *Grashof number* and *Prandtl number*. The product of the

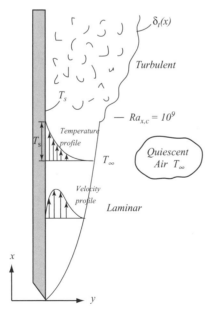

Figure 1.11 Free convection boundary layer transition on a vertical plate

Grashof and Prandtl numbers is the *Rayleigh number*, which is defined as

$$Ra_L = Gr \cdot Pr = \frac{g\beta(|T_s - T_\infty|)L^3}{\nu^2} \cdot \frac{\nu}{\alpha} = \frac{g\beta(|T_s - T_\infty|)L^3}{\nu\alpha} \tag{1.76}$$

where g is the gravity, β the *expansion coefficient*, ν the *kinematic viscosity*, and α the *thermal diffusivity*. The number should be positive always.

The results from the solution for the governing equations can be further simplified. Then, the average Nusselt number for an isothermal vertical plate is expressed by two separate equations as

$$\overline{Nu} = \frac{\overline{h}L}{k} = 0.59 Ra_L^{\frac{1}{4}} \quad \text{for } Ra_L \leq 10^9 \tag{1.77a}$$

$$\overline{Nu} = \frac{\overline{h}L}{k} = 0.10 Ra_L^{\frac{1}{3}} \quad \text{for } Ra_L > 10^9 \tag{1.77b}$$

In the special case of air, the following correlation was suggested by LeFevere [18]:

$$\overline{Nu} = \frac{\overline{h}L}{k} = 0.517 Ra_L^{\frac{1}{4}} \quad \text{for } Pr = 0.72 \text{ (air)} \tag{1.77c}$$

A correlation that may be applied over the entire range of Ra_L has been recommended by Churchill and Chu [19]. It is of the form

$$\overline{Nu} = \frac{\overline{h}L}{k} = \left\{ 0.825 + \frac{0.387 Ra_L^{\frac{1}{6}}}{\left[1 + (0.492/Pr)^{9/16}\right]^{4/9}} \right\}^2 \tag{1.78}$$

Isothermal Parallel Vertical Plates   Vertical fins are a very common geometry for heat sinks. We want to consider two isothermal vertical plates—a small spacing channel and a large spacing channel. When the spacing is close, the flow will be fully developed with a parabolic profile, as shown in Figure 1.12(a). On the one hand, if the spacing is too close, there will be no flow because of the friction on each plate. On the other hand, when the spacing is large, each velocity boundary layer will be isolated and behave as if the fluid flows over a single vertical plate, as shown in Figure 1.12(b).

In the design of a heat sink (array of fins), an increase in the heat transfer area, which is equivalent to increasing the number of fins within a given base area, is required in order to reduce the thermal resistance (or decrease the base temperature). The spacing z decreases with an increased number of fins, leading to a decrease in the convection coefficient due to the closeness. Therefore, an optimum spacing exists. Here we want to provide a correlation for the average Nusselt number.

Elenbaas [20] provided the experimental data, which are shown in Figure 1.13, where the Elenbaas number based on the Rayleigh number was used. The Rayleigh number is based on plate spacing:

$$Ra_z = \frac{g\beta(|T_s - T_\infty|)z^3}{\nu\alpha} \tag{1.79}$$

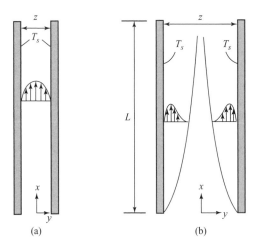

Figure 1.12 Natural convection for fully developed flow in (a) small spacing channel and (b) large spacing channel

where z is the fin spacing. Caution is given for the spacing z to be used in place of length L (see Equation (1.76)). Then, the Elenbaas number El is defined by

$$El = Gr \cdot Pr \frac{z}{L} = Ra_z \frac{z}{L} = \frac{g\beta(|T_s - T_\infty|)z^4}{\nu\alpha L} \quad (1.80)$$

Using the composite relation method, the Elenbaas data were curve fitted. The average Nusselt number is expressed as [21]

$$\overline{Nu}_z = \frac{\overline{h}_z z}{k} = \left[\frac{576}{El^2} + \frac{2.873}{El^{1/2}}\right]^{-\frac{1}{2}} \quad (1.81)$$

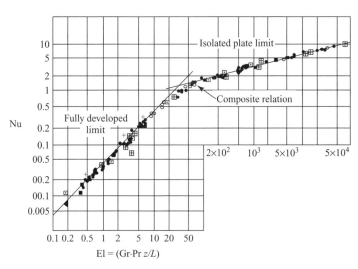

Figure 1.13 Nusselt number variation for vertical isothermal plates (data points from Elenbaas, 1942 [20])

Horizontal Cylinder   For a long horizontal isothermal cylinder, Churchill and Chu [22] suggested a correlation for the average Nusselt number. Note that the Rayleigh number Ra_D uses the cylinder diameter D.

$$Ra_D = Gr \cdot Pr = \frac{g\beta(|T_s - T_\infty|)D^3}{\nu^2} \cdot \frac{\nu}{\alpha} = \frac{g\beta(|T_s - T_\infty|)D^3}{\nu\alpha} \tag{1.82}$$

The average Nusselt number for the free convection of the horizontal cylinder is

$$\overline{Nu}_D = \frac{\overline{h}D}{k} = \left\{ 0.60 + \frac{0.387Ra_D^{1/6}}{\left[1 + (0.559/Pr)^{9/16}\right]^{8/27}} \right\}^2 \tag{1.83}$$

1.4.4   Radiation

1.4.4.1   Thermal Radiation

Radiation is electromagnetic waves (or photons), propagating through a transparent medium (air) or even effectively in a vacuum as one of the heat transfer mechanisms. Radiation is a spectrum with a wide range of the wavelengths, from microwaves to gamma rays. Thermal radiation encompasses a range from infrared to ultraviolet, which includes visible light. A rainbow assures us that the light is indeed a spectrum. The surface of any matter emits electromagnetic radiation if the temperature of the surface is greater than absolute zero. It also absorbs radiation from the surroundings, which is called *irradiation*. The *emissivity* and *absorptivity*, lying between 0 and 1, of real bodies vary, depending on the finishes of the surfaces and the nature of the irradiation. This feature provides engineers with potential for improvement in the control of heat transfer. A perfect emitter, perfect absorber, and also perfect diffuser is called a *blackbody*. Two properties of radiation that have importance here are its *spectrum* and *directionality*.

The spectral emissive powers of a blackbody and a real surface are illustrated in Figure 1.14, where the emissive power of the real surface is always less than that of the blackbody.

The second property of thermal radiation relates to its directionality, as shown in Figure 1.15. A *blackbody* has a *diffuse* property.

Blackbody   A *blackbody* is a perfect emitter and absorber. A blackbody emits radiation energy uniformly in all directions (i.e., a *diffusive emitter*). *Plank's law* provides the spectral distribution of blackbody emission:

$$E_{\lambda,b}(\lambda, T) = \frac{C_1}{\lambda^5 \left[\exp\left(C_2/\lambda T\right) - 1\right]} \tag{1.84}$$

where $C_1 = 3.472 \times 10^8 \ W \cdot \mu m^4/m^2$ and $C_2 = 1.439 \times 10^4 \ \mu m \cdot K$. This is a function of both wavelength and temperature, as shown in Figure 1.16, where the spectral emissive power distribution is dependent on the temperature and wavelength.

If we integrate the spectral distribution, Equation (1.84), with respect to wavelength λ over the entire range from $\lambda = 0$ to $\lambda = \infty$, we have the total emissive power of the blackbody, which is known as the *Stefan-Boltzmann law*. The *total emissive power* is

$$E_b = \sigma T^4 \tag{1.85}$$

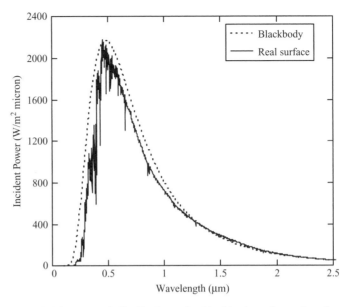

Figure 1.14 Spectral distributions of a blackbody and a real surface

where E_b is the total emissive power of a blackbody and T is the surface temperature of the body and $\sigma = 5.67 \times 10^{-8}\,\text{W/m}^2 \cdot \text{K}^4$, which is termed the *Stefan-Boltzmann constant*.

Therefore, the emissive power of a real surface is

$$E = \varepsilon \sigma T^4 \tag{1.86}$$

where ε is the *emissivity*.

Irradiation G   *Irradiation G* is an incident radiation, which may originate from other surfaces or from the surroundings, being independent of the finishes of the intercepting surface. Irradiation will have spectral and directional properties. Irradiation could be

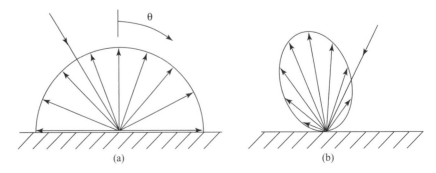

Figure 1.15 Comparison of blackbody and real surface (a) diffuse, and (b) directional

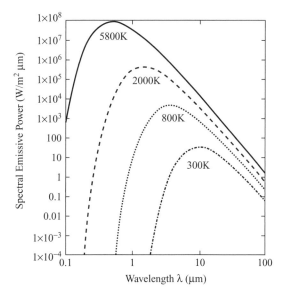

Figure 1.16 Spectral blackbody emissive power

partially absorbed, reflected, or transmitted, as shown in Figure 1.17.

$$\rho + \alpha + \gamma = 1 \tag{1.87}$$

where ρ is the *reflectivity*, α is the *absorptivity*, and γ is the *transmissivity*.
    For instance, if the surface is opaque, the transmissibility γ is zero.

$$\rho + \alpha = 1 \tag{1.88}$$

Radiosity J   *Radiosity J* is all the radiant energy leaving a surface, consisting of the direct emission εE_b from the surface and the reflected portion ρG of the irradiation, as shown in Figure 1.17. The radiosity is

$$J = \varepsilon E_b + \rho G \tag{1.89}$$

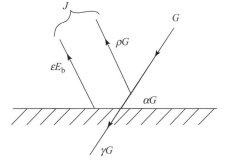

Figure 1.17 Surface radiosity J includes emission εE_b and reflection ρG; irradiation G includes absorption αG, reflection ρG, and transmission γG

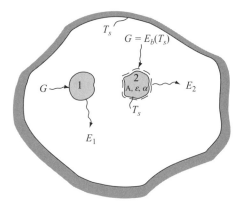

Figure 1.18 Radiation exchange in an isothermal enclosure

Kirchhoff's Law Consider a large, isothermal enclosure at surface temperature T_s, within which several small bodies are confined, as shown in Figure 1.18. Under steady-state conditions, thermal equilibrium must exist between the bodies and the enclosure. Regardless of the orientation and the surface properties of the enclosure, the irradiation experienced by any body in the enclosure must be diffuse and equal to emission from a blackbody at T_s.

The irradiation in the *enclosure* behaves like a blackbody.

$$G = E_b\left(T_s\right) \tag{1.90}$$

The absorbed irradiation is

$$G_{abs} = \alpha G = \alpha \dot{E}_b\left(T_s\right) \tag{1.91}$$

At thermal equilibrium, the energy absorbed by the body must be equal to the energy emitted. Otherwise, there would be an energy flow into or out of the body that would raise or lower its temperature. For a body denoted by 2, the energy balance gives

$$A_2\alpha_2 G - A_2 E_2 = 0 \tag{1.92}$$

Or equivalently,

$$\alpha_2 E_b(T_s) - \varepsilon_2 E_b(T_s) = 0 \tag{1.93}$$

Eventually,

$$\alpha_2 = \varepsilon_2 \tag{1.94}$$

The same thing happens for the other bodies. Therefore, in an enclosure, we have

$$\alpha = \varepsilon \tag{1.95}$$

The total emissivity of any surface is equal to its total absorptivity, which is known as the *Kirchhoff's law*. Since we know that radiation inherently has spectral and directional properties, $\alpha_{\lambda,\theta} = \varepsilon_{\lambda,\theta}$, we conclude that radiation in an enclosure is independent of the spectral and directional distributions of the emitted and incident radiation. Note that this relation is derived under the condition that the surface temperature of the body is equal to the temperature of the source of the irradiation. However, since ε and α vary weakly with temperature, this relation may be assumed and greatly simplifies the radiation calculation.

Gray Surface   A *gray surface* is defined as one for which α_λ and ε_λ are independent of wavelength λ over the spectral range of the irradiation and the surface emission.

Diffuse-Gray Surface   A common assumption in enclosure calculations is that surfaces are *diffuse-gray*. The diffuse-gray surface emits and absorbs a fraction of radiation for all directions and all wavelengths. For a diffuse-gray surface, total emissivity and absorptivity are equal. The total absorptivity is independent of the nature of the incident radiation.

Radiation Exchange for a Single Surface in an Enclosure   Consider a *diffuse-gray surface* at T_s in an enclosure (surroundings) at T_{sur}. Equation (1.95) can be applied with the assumption of the diffuse-gray surface in the enclosure. The net radiation exchange of the surface is obtained as

$$q = A\varepsilon\sigma\left(T_s^4 - T_{sur}^4\right) \tag{1.96}$$

1.4.4.2   *View Factor*

The *view factor* F_{ij} is defined as the fraction of the radiation leaving surface i that is intercepted by surface j, as shown in Figure 1.19. Some typical view factors are illustrated in Table 1.2. Consider radiative heat transfer between two or more surfaces, which is often the primary quantity of interest in fin design. Two important relations are suggested:

Reciprocity relation:

$$A_i F_{ij} = A_j F_{ji} \tag{1.97}$$

Summation rule:

$$\sum_{j=1}^{N} F_{ij} = 1 \tag{1.98}$$

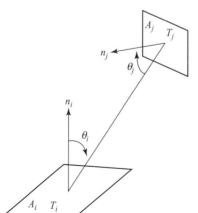

Figure 1.19   View factor associated with radiation exchange between two finite surfaces

Table 1.2 View Factors for Two- and Three-dimensional Geometries

Configuration	View factor

1. Small Object in a Large Cavity

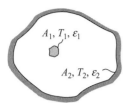

$$\frac{A_1}{A_2} = 0 \quad F_{12} = 1 \tag{1.99}$$

2. Concentric Cylinder (infinite length)

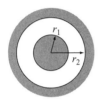

$$\frac{A_1}{A_2} = \frac{r_1}{r_2} \quad F_{12} = 1$$

$$F_{21} = \frac{A_1}{A_2} F_{12} \tag{1.100}$$

$$F_{22} = 1 - \frac{A_1}{A_2} F_{12}$$

3. Large (infinite) Parallel Plate

$$A_1 = A_2 = A \quad F_{12} = 1 \tag{1.101}$$

4. Perpendicular plates with a common edge

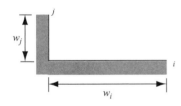

$$F_{ij} = \frac{1 + w_j/w_i - \left[1 + \left(w_j/w_i\right)^2\right]^{1/2}}{2} \tag{1.102}$$

5. Hemispherical concentric cylinder

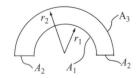

$$F_{13} = \frac{1}{2} + \frac{1}{\pi}\left(\frac{r_2}{r_1}\right)\left[1 - \left(\frac{r_1}{r_2}\right)^2\right]^{1/2} + \frac{1}{\pi}\sin^{-1}\left(\frac{r_1}{r_2}\right) - \frac{1}{\pi}\frac{r_2 - r_1}{r_1} \tag{1.103}$$

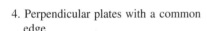

Table 1.2 (*continued*)

Configuration	View factor

6. Concentric cylinder (infinite length)
   with variable angle α

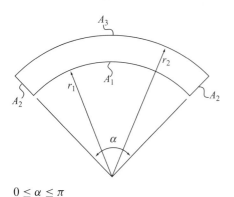

$0 \leq \alpha \leq \pi$

If $\alpha \geq \cos^{-1}\left(\dfrac{r_1}{r_2}\right)$ 　　　　　　　(1.104)

$$F_{13} = 1 - \frac{\pi}{2\alpha} + \frac{r_2}{\alpha r_1}\left[1 - \left(\frac{r_1}{r_2}\right)^2\right]^{1/2} + \frac{1}{\alpha}\cdot\sin^{-1}\left(\frac{r_1}{r_2}\right) - \frac{r_2 - r_1}{\alpha r_1}$$

$$F_{22} = \frac{2\sqrt{r_2^2 - r_1^2} + 2r_1\left[\alpha - \dfrac{\pi}{2} + \sin^{-1}\left(\dfrac{r_1}{r_2}\right)\right] - \alpha\,(r_2 + r_1)}{4\,(r_2 - r_1)}$$

Otherwise

$$F_{13} = \frac{\sqrt{r_2^2 + r_1^2 - 2r_2 r_1 \cos(\alpha)} - (r_2 - r_1)}{\alpha r_1}$$

$$F_{22} = \frac{2\sqrt{r_2^2 + r_1^2 - 2r_2 r_1 \cos(\alpha)} - \alpha\,(r_2 + r_1)}{4\,(r_2 - r_1)}$$

1.4.4.3 *Radiation Exchange between Diffuse-Gray Surfaces*

The net rate of heat transfer at a surface is obtained by applying heat balance on the control volume at the surface (see Figure 1.20).

The *diffuse-gray surface* in an enclosure has

$$\varepsilon_i = \alpha_i \qquad (1.105)$$

The heat balance at the surface gives

$$q_i = A_i(\varepsilon_i E_{bi} - \alpha_i G_i) \qquad (1.106)$$

The net rate of heat transfer is also equal to the difference between the surface radiosity and the irradiation.

$$q_i = A_i(J_i - G_i) \qquad (1.107)$$

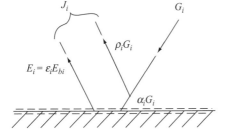

Figure 1.20 Radiation exchange on an opaque, diffuse-gray surface

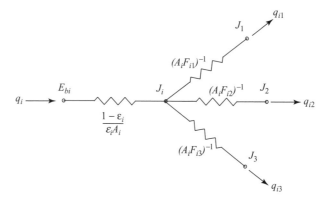

Figure 1.21 Radiation network between surface i and the remaining surfaces of the enclosure

Combining these three equations and eliminating the irradiation G_i gives

$$q_i = \frac{E_{bi} - J_i}{(1 - \varepsilon_i)/\varepsilon_i A_i} \tag{1.108}$$

The *net radiation exchange* between the radiosities in the enclosure is

$$q_{ij} = \frac{J_i - J_j}{\left(A_i F_{ij} \right)^{-1}} \tag{1.109}$$

The net rate of heat transfer at surface i is connected to the number of radiosities in the enclosure and forms a radiation network (Figure 1.21).

$$q_i = \sum_{j=1}^{N} \frac{J_i - J_j}{\left(A_i F_{ij} \right)^{-1}} \tag{1.110}$$

Combining Equations (1.108) and (1.110), we obtain

$$\frac{E_{bi} - J_i}{(1 - \varepsilon_i)/\varepsilon_i A_i} = \sum_{j=1}^{N} \frac{J_i - J_j}{\left(A_i F_{ij} \right)^{-1}} \tag{1.111}$$

REFERENCES

1. W.F. Stoecker, *Design of Thermal Systems* (3rd ed.). New York: McGraw-Hill, 1989.
2. Y. Jaluria, *Design and Optimization of Thermal Systems*. New York: McGraw-Hill, 1998.
3. A. Bejan, G. Tsatsaronis, and M. Moran, *Thermal Design and Optimization*. New York: John Wiley and Sons, 1996.
4. B.K. Hodge and R.P. Taylor, *Analysis and Design of Energy Systems* (3rd ed.). Englewood Cliffs, NJ: Prentice Hall, 1999.
5. L.C. Burmeister, *Elements of Thermal-Fluid System Design*. Englewood Cliffs, NJ: Prentice Hall, 1998.

6. W.S. Janna, *Design of Fluid Thermal Systems* (2nd ed.). Boston: PWS Publishing Company, 1998.

7. N.V. Suryanarayana and O. Arici, *Design of Simulation of Thermal Systems*. New York: McGraw-Hill, 2003.

8. A. Ertas and J.C. Jones, *The Engineering Design Process* (2nd ed.). New York: John Wiley and Sons, 1993.

9. C.L. Dym and P.L. Little, *Engineering Design: A Projected-Based Introduction.* New York: John Wiley and Sons, 2000.

10. C.L.M.H. Navier, Memoire sur les lois du movement des fluids, *Mem. Acad. R. Sci.* Paris, 6 (1823): 389–416.

11. G.G. Stokes, On the Theories of Internal Friction of Fluids in Motion, *Trans. Cambridge Phil. Soc.*, 8 (1845): pp. 145–239.

12. L. Prandtl, Über Flussigkeitsbewegung bei she kleiner Reibung, *Proc. 3rd Int. Math. Congr.* (Heidelberg 1904); NACA TM 452, 1928.

13. H. Blasius, Grenzschichten in Flüssigkeiten mit kleiner Reibung, *Z. Math. Phys.*, 56 (1908): 1–37; also in English as The Boundary Layers in Fluids with Little Friction, NACA TM 1256.

14. S.W. Churchill and M. Bernstein, *J. Heat Transfer.*, 99 (1977): 300.

15. G.K. Filonenko, Hydraulic Resistance in Pipes (in Russian), *Teplonergetika*, Vol. 1, pp. 40–44, 1954.

16. E.N. Sieder and G.E. Tite, Heat Transfer and Pressure Drop of Liquids in Tubes, *Ind. Eng. Chem.*, 28 (1936): 1429.

17. V. Gnielinski, New Equation for Heat and Mass Transfer in Turbulent Pipe and Channel Flow. *Int. Chem., Eng.*, 16 (1976): 359–368.

18. E.J. LeFevere, Laminar Free Convection from a Vertical Plane Surface, *Proc. 9th int. Congr. Applied Mech.* Brussels, 4 (1956): 168–174.

19. S.W. Churchill and H.H.S. Chu, Correlating Equations for Laminar and Turbulent Free Convection from a Vertical Plate, *Int. J. Heat Mass Transfer*, 18 (1975): 1323–1329.

20. W. Elenbaas, Heat Distribution of Parallel Plates by Free Convection, *Physica*, 9 (1) (1942): 665–671.

21. A.D. Klaus and A. Bar-Cohen, *Design and Analysis of Heat Sinks*. New York: John Wiley and Sons, 1995.

22. S.W. Churchill and H.H.S. Chu, *Int. J. Heat Mass Transfer*, 18 (1975): 1049.

23. R.H. Winterton, *Int. J. Heat Mass Transfer*, 41 (1998): 809.

2

Heat Sinks

A *heat sink* is a device to effectively absorb or dissipate heat (thermal energy) from the surroundings (air) using extended surfaces such as fins and spines. Heat sinks are used in a wide range of applications where efficient heat dissipation is required; major examples include refrigeration, heat engine, and cooling electronic devices. The most common design of a heat sink is a metal device with many cooling fins, which is referred to as a *fin array*. The heat sink performance is improved by increasing either the thermal conductivity of the fins, the surface area of the fins, or the heat transfer coefficient. The profiles of longitudinal fins include rectangular, triangular, and parabolic fins. The rectangular profile is fundamental and widely used particularly with multiple-fin arrays. The optimization of both a single fin and a multiple-fin array with forced convection and natural convection are herein presented in detail for the rectangular profile. At last, a radial fin array is discussed with a design example.

2.1 LONGITUDINAL FIN OF RECTANGULAR PROFILE

We consider a *rectangular fin*, from which the heat is transferred to the ambient at T_∞, as shown in Figure 2.1. To analyze its thermal behavior, it is necessary to be more specific about the geometry. We begin with the simplest case of a straight rectangular fin of uniform cross-section, as shown in Figure 2.1. The fin is attached to a base at a constant temperature of T_{base} and extends into a fluid of temperature T_∞. For the prescribed fin, both the cross-sectional area A_c and the profile area A_p are constant with $A_p = bt$ and $A_c = Lt$. The perimeter P of the fin is $P = 2(L + t)$, where b is the profile length, t the fin thickness, and L the fin width.

It may be assumed, because of the thinness of the fin, that the temperature of the fin is dependent only on the location x as $T = T(x)$, which is valid when the *Biot number* (ht/k) is much smaller than unity. Take a small control volume (differential element) at an arbitrary location of x with the heat flows across the control surface. An amount of conduction heat transfer q_x enters the control volume through the cross-sectional area A_c and leaves with a differential change as $(dq_x/dx)dx$. And there is also a differential convection heat transfer dq_{conv} on the circumferential surfaces of the control volume. The convection heat transfer area is a product of P and dx as shown in Figure 2.1. Considering the energy balance for the control volume gives

$$q_x - \left(q_x + \frac{dq_x}{dx}dx \right) - dq_{conv} = 0 \tag{2.1}$$

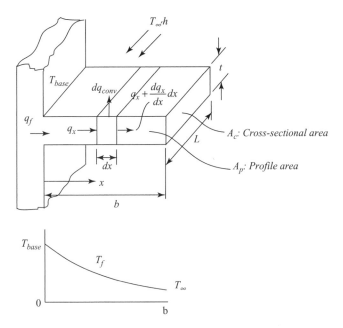

Figure 2.1   Rectangular fin of uniform cross section and the fin temperature distribution

which reduces

$$-\frac{dq_x}{dx}dx - dq_{conv} = 0 \tag{2.2}$$

The Fourier's law of conduction gives

$$q_x = -A_c k \frac{dT}{dx} \tag{2.3}$$

And the *Newton's law of cooling* (convection) gives

$$dq_{conv} = hPdx(T - T_\infty) \tag{2.4}$$

Substituting Equations (2.3) and (2.4) into Equation (2.2) yields

$$-\frac{d}{dx}\left(-A_c k \frac{dT}{dx}\right)dx - hPdx(T - T_\infty) = 0 \tag{2.5}$$

This reduces to the ordinary differential equation:

$$\frac{d^2 T}{dx^2} - \frac{hP}{kA_c}(T - T_\infty) = 0 \tag{2.6}$$

The boundary conditions at $x = 0$ and $x = b$ will be

$$T(0) = T_{base} \tag{2.6a}$$

$$A_c h_o (T(b) - T_\infty) = -kA_c \frac{dT(b)}{dx} \tag{2.6b}$$

where h_o is the local heat transfer coefficient at the fin tip, which is not confused with h over the entire fin surface. For mathematical convenience, let $\theta = T - T_\infty$ and $m^2 = hP/kA_c$. It is noted that $\frac{dT}{dx} = \frac{d\theta}{dx}$ since T_∞ is a constant. Then, rewrite Equations (2.6), (2.6a), and (2.6b).

$$\frac{d^2\theta}{dx^2} - m^2\theta = 0 \tag{2.7}$$

where $m = \sqrt{hP/kA_c} = \sqrt{\frac{h2(L+t)}{kLt}} = \sqrt{\frac{2h}{kt}\left(1 + \frac{t}{L}\right)} \cong \sqrt{\frac{2h}{kt}}$ \tag{2.7a}

The boundary conditions can be rewritten as

$$\theta(0) = \theta_b \tag{2.7b}$$

where $\theta_b = T_{base} - T_\infty$

$$h_o\theta(b) = -k\frac{d\theta(b)}{dx} \tag{2.7c}$$

The general solution of the ordinary differential equation, Equation (2.7), is

$$\theta(x) = C_1 e^{mx} + C_2 e^{-mx} \tag{2.8}$$

Solving Equation (2.8) for C_1 and C_2 with the two boundary conditions of Equations (2.7b) and (2.7c) yields

$$\theta = \theta_b \frac{\cosh m(b-x) + \dfrac{h_o}{mk}\sinh m(b-x)}{\cosh mb + \dfrac{h_o}{mk}\sinh mb} \tag{2.9}$$

2.2 HEAT TRANSFER FROM FIN

As shown in Figure 2.1, the heat transfer conducted from the base to the fin is defined as q_f, which is the same as the heat dissipation to the ambient.

$$q_f = -kA_c\frac{dT(0)}{dx} = -kA_c\frac{d\theta(0)}{dx} \tag{2.10}$$

Taking derivative of Equation (2.9) gives

$$\frac{d\theta(x)}{dx} = \theta_b \frac{-m\sinh m(b-x) + \dfrac{h_o}{mk}(-m)\cosh m(b-x)}{\cosh mb + \dfrac{h_o}{mk}\sinh mb} \tag{2.11}$$

with $x = 0$,

$$q_f = (-kA_c)(-m\theta_b)\frac{\sinh mb + \dfrac{h_o}{mk}\cosh mb}{\cosh mb + \dfrac{h_o}{mk}\sinh mb} \tag{2.12}$$

where

$$m = \sqrt{\frac{hP}{kA_c}} \tag{2.12a}$$

Then, we have

$$q_f = \sqrt{hPkA_c}\,\theta_b \frac{\sinh mb + \dfrac{h_o}{mk}\cosh mb}{\cosh mb + \dfrac{h_o}{mk}\sinh mb} \tag{2.13}$$

It is noted that Equation (2.13) is a general solution for the arbitrary shape of cross-sectional area, which includes a pin fin as long as the cross-sectional area is uniform along the fin.

If the heat loss at the fin tip is negligible, which is usually a good assumption for the most cases because of the thinness of the fin, it requires that $dT(b)/dx = 0$ or equivalently $h_o = 0$. Using Equation (2.13) with $h_o = 0$, we have

$$q_f = \sqrt{hPkA_c}\,\theta_b \tanh mb \tag{2.14}$$

Equation (2.14) is the solution for the heat transfer rate from the fin with an adiabatic tip.

2.3   FIN EFFECTIVENESS

It is sometimes questioned whether or not the use of fins is justified in terms of the heat transfer gain by comparing its performance with that of the original surface without fins. The fin effectiveness, ε_f, is defined as the ratio of the fin heat transfer rate to the heat transfer rate that would exist without the fin (Figure 2.1).

$$\varepsilon_f = \frac{q_f}{hA_c\theta_b} \tag{2.15}$$

For justification of the fin, the fin effectiveness should be

$$\varepsilon_f \geq 2 \tag{2.16}$$

Using Equation (2.14), Equation (2.15) becomes

$$\varepsilon_f = \frac{\sqrt{hPkA_c}\,\theta_b \tanh mb}{hA_c\theta_b} \tag{2.17}$$

Rearranging gives

$$\varepsilon_f = \sqrt{\left(\frac{k}{h}\right)\left(\frac{P}{A_c}\right)} \tanh mb \tag{2.18}$$

Discussion of Equation (2.18):

- $\frac{k}{h}$: if h is large such as water cooling, the fin is not effective. So small h such as with air cooling usually justifies the fin.
- $\frac{P}{A_c}$: increasing the perimeter and decreasing the cross-sectional area will increase the fin effectiveness. Thin fin design is desirable.

2.4 FIN EFFICIENCY

The fin efficiency η_f is another measure of fin performance. It is defined as the ratio of the heat transfer from a fin to the maximum possible heat transfer from the fin. The maximum rate at which a fin could dissipate energy is the rate that would exist if the entire fin surface were at the base temperature.

$$\eta_f = \frac{q_f}{q_{max}} = \frac{q_f}{hPb\theta_b} \tag{2.19}$$

$$\eta_f = \frac{\sqrt{hPkA_c}\,\theta_b \tanh ml}{hPb\theta_b} = \sqrt{\frac{kA_c}{hP}}\left(\frac{1}{b}\right) \tanh mb \tag{2.20}$$

Since

$$mb = \left(\sqrt{\frac{hP}{kA_c}}\right) b \tag{2.21}$$

$$\eta_f = \frac{\tanh mb}{mb} \tag{2.22}$$

The fin efficiency can be expected to lie between 0.5 and 0.7 for a well-designed fin.

$$0.5 \le \eta_f \le 0.7 \tag{2.23}$$

2.5 CORRECTED PROFILE LENGTH

Reasonably accurate predictions may be obtained with a *corrected fin length* b_c for a fin with a *convective tip* (Figure 2.1).

Define the corrected fin length as

$$b_c = b + \frac{t}{2} \tag{2.24}$$

By definition

$$A_p = b_c t \tag{2.25}$$

$$P \cong 2L \quad \text{since} \quad L \gg t \tag{2.26}$$

Knowing $A_c = Lt$, we have

$$mb_c = \left(\frac{hP}{kA_c}\right)^{\frac{1}{2}} b_c = \left(\frac{h2L}{kLt}\right)^{\frac{1}{2}} b_c = \left(\frac{2h}{kb_c t}\right)^{\frac{1}{2}} b_c^{\frac{1}{2}} b_c = \sqrt{2}\left(\frac{h}{kA_p}\right)^{\frac{1}{2}} b_c^{\frac{3}{2}} \tag{2.27}$$

The corrected fin efficiency is approximated as

$$\eta_f = \frac{\tanh mb_c}{mb_c} \tag{2.28}$$

In Figure 2.2, η_f is plotted with respect to a dimensionless number of $b_c^{\frac{3}{2}}\left(\frac{h}{kA_p}\right)^{\frac{1}{2}}$. Errors associated with the approximation are negligible if the *Biot number* is [10]

$$\frac{ht}{k} < 0.0625 \tag{2.30}$$

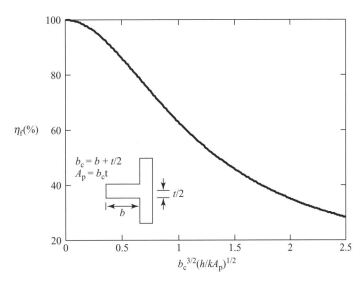

Figure 2.2 Efficiency of a rectangular fin

2.6 OPTIMIZATIONS

We want to maximize the heat transfer rate of a fin with respect to a fin thickness and a profile length. We know that the heat transfer will asymptotically increase with increasing the fin thickness. However, an engineering problem is what would be the *optimum fin thickness* if the volume of the fin ($Volume = btL$) is constant, or, equivalently, the profile area is constant ($A_p = bt$).

2.6.1 Constant Profile Area A_p

The heat transfer rate from a fin with the *adiabatic tip* is given by Equation (2.14), which is rewritten here

$$q_f = \sqrt{hPkA_c}\,\theta_b \tanh mb \tag{2.14}$$

Assuming that $L \gg t$ and using $P \cong 2L$, $A_c = Lt$, $A_p = bt$ and $mb = \left(\sqrt{\frac{hP}{kA_c}}\right)b$, we have

$$q_f = \sqrt{h2LkLt}\,\theta_b \tanh \sqrt{\frac{h2L}{kLt}}\,b \tag{2.31}$$

Rearranging,

$$q_f = (2hk)^{\frac{1}{2}} L\theta_b t^{\frac{1}{2}} \tanh \left(\frac{2h}{kt}\right)^{\frac{1}{2}} b \tag{2.32}$$

In terms of A_p

$$mb = \left(\sqrt{\frac{hP}{kA_c}}\right)b = b\left(\frac{2h}{kt}\right)^{\frac{1}{2}} = A_p \frac{b}{bt}\left(\frac{2h}{kt}\right)^{\frac{1}{2}} = A_p \left(\frac{2h}{kt^3}\right)^{\frac{1}{2}} \tag{2.33}$$

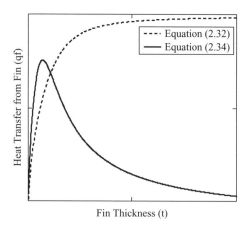

Figure 2.3   Heat transfer rates calculated from Equations (2.32) and (2.34), respectively

Rewrite Equation (2.32) using Equation (2.33):

$$q_f = (2hk)^{\frac{1}{2}} L\theta_b t^{\frac{1}{2}} \tanh A_p \left(\frac{2h}{kt^3}\right)^{\frac{1}{2}} \tag{2.34}$$

Equation (2.32), the heat transfer rate from a fin, reveals that the curve asymptotically increases, not showing the optimum value, but Equation (2.34) with a constant profile area does show the optimum value, which is illustrated in Figure 2.3.

For a constant profile area, we wish to find the optimum dimensions where it requires

$$\frac{dq_f}{dt} = 0 \tag{2.35}$$

Using the chain rule of differentiation as $\frac{d}{dx}(uv) = u\frac{dv}{dx} + v\frac{du}{dx}$, it is given by

$$\frac{dq_f}{dt} = (2hk)^{\frac{1}{2}} w\theta_b \left(t^{\frac{1}{2}} \frac{d}{dt}\left(\tanh\left(\frac{2h}{k}\right)^{\frac{1}{2}} t^{-\frac{3}{2}} A_p \right) + \tanh\left(\frac{2h}{k}\right)^{\frac{1}{2}} t^{-\frac{3}{2}} A_p \frac{d}{dt}\left(t^{\frac{1}{2}}\right) \right) \tag{2.36}$$

Using the chain rule as $\frac{d}{dx}\tanh u = \frac{1}{\cosh^2 u}\frac{du}{dx}$ and Equation (2.35)

$$t^{\frac{1}{2}} \left(\frac{2h}{k}\right)^{\frac{1}{2}} t^{-\frac{3}{2}} A_p \left(\cosh^2 \left(\frac{2h}{k}\right)^{\frac{1}{2}} t^{-\frac{3}{2}} A_p \right)^{-1} + \frac{1}{2} t^{-\frac{1}{2}} \tanh\left(\frac{2h}{k}\right)^{\frac{1}{2}} t^{-\frac{3}{2}} A_p = 0 \tag{2.37}$$

Using the trigonometry of $\sinh 2x = 2\sinh x \cosh x$

$$6 \left(\frac{2h}{k}\right)^{\frac{1}{2}} t^{-\frac{3}{2}} A_p = \sinh 2 \left(\frac{2h}{k}\right)^{\frac{1}{2}} t^{-\frac{3}{2}} A_p \tag{2.38}$$

Defining

$$\beta = mb = \left(\frac{2h}{k}\right)^{\frac{1}{2}} t^{-\frac{3}{2}} A_p \tag{2.39}$$

Equation (2.38) becomes

$$6\beta = \sinh 2\beta \tag{2.40}$$

By solving Equation (2.40), we find that $\beta = 1.4192$, which has an important meaning for the optimum heat transfer rate dissipated from the fin for a constant profile area.

$$\beta = 1.4192 \tag{2.41}$$

Note that the optimum fin efficiency is 0.627 using Equations (2.20) and (2.41). We are obviously interested in the profile fin length b_0 and fin thickness t_0 for the particular (optimum) condition.

Substituting $A_p = b_0 t_0$ into Equation (2.39)

$$\beta = mb_0 = \left(\frac{2h}{k}\right)^{\frac{1}{2}} t_0^{-\frac{3}{2}} b_0 t_0 \tag{2.42}$$

Solving for t_0 yields the *optimum fin thickness* for a constant profile area as

$$t_0 = \left(\frac{2h}{k}\right)\left(\frac{b_0}{\beta}\right)^2 \tag{2.43}$$

Solving for b_0 yields the *optimum fin profile length* for a constant profile area as

$$b_0 = \beta \left(\frac{kt_0}{2h}\right)^{\frac{1}{2}} \tag{2.44}$$

2.6.2   Constant Heat Transfer from a Fin

If the heat transfer rate q_f is constant, Equation (2.32) is written by.

$$q_f = (2hk)^{\frac{1}{2}} L\theta_b t^{\frac{1}{2}} \tanh \beta \tag{2.45}$$

$$\tanh \beta = \tanh 1.4192 = 0.8894 \tag{2.46}$$

Solving for t gives the optimum fin thickness t_0 for a constant heat transfer from a fin.

$$t_0 = \left(\frac{q_f}{(2hk)^{1/2} L\theta_b 0.8894}\right)^2$$

Rearranging,

$$t_0 = \frac{0.632}{hkL^2}\left(\frac{q_f}{\theta_b}\right)^2 \tag{2.47}$$

because $A_p = b_0 t_0$

Using Equation (2.33)

$$\beta = mb_0 = \left(\frac{2h}{kt_0}\right)^{\frac{1}{2}} b_0 \tag{2.48}$$

Rearranging,

$$b_0 = \beta \left(\frac{k}{2h} \right)^{\frac{1}{2}} t_0^{\frac{1}{2}} \tag{2.49}$$

Substituting Equation (2.47) into Equation (2.49) yields

$$b_0 = 1.4192 \left(\frac{k}{2h} \right)^{\frac{1}{2}} \left(\frac{0.632}{hkL^2} \left(\frac{q_f}{\theta_b} \right)^2 \right)^{\frac{1}{2}} \tag{2.50}$$

Finally, we have

$$b_0 = \frac{0.798}{hL} \frac{q_f}{\theta_b} \tag{2.51}$$

2.6.3   Constant Fin Volume or Mass

Mass is a product of density and volume. Assuming that the density does not change, only the volume is here considered. From Figure 2.1, the volume of the fin is given by

$$V = Lbt \tag{2.52}$$

and the profile area is

$$A_p = bt = \frac{V}{L} \tag{2.53}$$

Actually, the width L is not parameter, so this problem is basically the same as the constant profile area. Therefore, Equation (2.41) ($\beta = 1.4192$) can be applied to this problem. Using Equation (2.34) with Equation (2.53), we have

$$q_f = (2hk)^{\frac{1}{2}} L\theta_b t^{\frac{1}{2}} \tanh \left(\frac{V}{L} \right) \left(\frac{2h}{kt^3} \right)^{\frac{1}{2}} \tag{2.54}$$

$$\beta = mb_0 = b_0 \left(\frac{2h}{kt_0} \right)^{\frac{1}{2}} = V \frac{b_0}{b_0 t_0 L} \left(\frac{2h}{k} \right)^{\frac{1}{2}} t_0^{-\frac{1}{2}} \tag{2.55}$$

The optimum fin thickness with a constant fin volume can be obtained.

$$t_0 \cong \left(\frac{V}{L} \right)^{\frac{2}{3}} \left(\frac{h}{k} \right)^{\frac{1}{3}} \tag{2.56}$$

Using $t_0 = \frac{V}{b_0 L}$, the optimum fin length with a constant fin volume may be obtained as

$$b_0 \cong \left(\frac{V}{L} \frac{k}{h} \right)^{\frac{1}{3}} \tag{2.57}$$

Example 2.1 Rectangular Fin Problem   An aluminum rectangular fin of width L = 2.5 cm is cooled by forced air at $U_\infty = 1.5$ m/s and $T_\infty = 22°C$ as shown in Figure E2.1.1. The fin base temperature T_b is 85°C. Only 1.2 g of aluminum for the

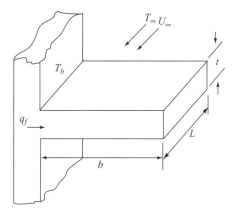

Figure E2.1.1   Rectangular fin

fin are required. Determine the fin thickness t and profile length b that maximize the heat transfer rate to the ambient air, and also determine the maximized heat transfer rate. Air properties at a film temperature of 326K: kinematic viscosity $v = 23.42 \times 10^{-6}$ m^2/s, thermal conductivity $k_{air} = 0.0283$ W/m K, and Prandtl number Pr $= 0.703$. Aluminum properties: the density $\rho_{al} = 2702$ kg/m^3 and the thermal conductivity $k_{al} = 237$ W/mK.

Solution   This is a fixed mass or volume problem. Hence, using Equation (2.56) with the present notations, we have

$$t_0 \cong \left(\frac{V}{L}\right)^{\frac{2}{3}} \left(\frac{\bar{h}}{k_{al}}\right)^{\frac{1}{3}}$$

The volume can be obtained

$$V = \frac{m}{\rho} = \frac{1.2 \times 10^{-3} \text{ kg}}{2702 \text{ kg/m}^3} = 4.441 \times 10^{-7} \text{m}^3 \text{ (fixed)}$$

For the heat transfer coefficient h, check first the Reynolds number for the flow pattern

$$\text{Re}_L = \frac{U_\infty L}{v} = (2.5 \text{ m/s})(2.5 \times 10^{-2} \text{ m})/23.42 \times 10^{-6} \text{ m}^2/\text{s} = 1601$$

Since this is less than the critical Reynolds number of 5×10^5, the flow is laminar. The average Nusselt number for the flow is given in Equation (1.53).

$$\overline{Nu}_L = \frac{\bar{h}L}{k_{air}} = 0.664 \, \text{Re}_L^{\frac{1}{2}} \, \text{Pr}^{\frac{1}{3}}$$

$$\bar{h} = \frac{k_{air}}{L} 0.664 \, \text{Re}_L^{\frac{1}{2}} \, \text{Pr}^{\frac{1}{3}} = \frac{0.0283 \text{ W/mK}}{2.5 \times 10^{-2} \text{ m}} 0.664 \, (1601)^{\frac{1}{2}} \, (0.703)^{\frac{1}{3}} = 26.65 \text{ W/m}^2\text{K}$$

The optimal fin thickness is

$$t_0 = \left(\frac{4.441 \times 10^{-7} \text{ m}^3}{2.5 \times 10^{-2} \text{ m}} \right)^{\frac{2}{3}} \left(\frac{26.65 \text{ W/m}^2\text{K}}{237 \text{ W/mK}} \right)^{\frac{1}{3}} = 3.28 \times 10^{-4} \text{ m} = \underline{0.328 \text{ mm}}$$

From Equation (2.57), the optimum fin profile length is

$$b_0 \cong \left(\frac{V}{L} \frac{k}{h} \right)^{\frac{1}{3}} = \left(\frac{4.441 \times 10^{-7} \text{ m}^3 \times 237 \text{ W/mK}}{2.5 \times 10^{-2} \text{ m} \times 26.65 \text{ W/m}^2\text{k}} \right)^{\frac{1}{3}} = 5.40 \times 10^{-2} \text{ m} = \underline{5.40 \text{ cm}}$$

For the maximized heat transfer rate, we may use Equations (2.14) and (2.54) with an assumption of the adiabatic tip.

$$q_f = \sqrt{h P k A_c}\, \theta_b \tanh mb = \left(\overline{h} 2 \left(L + t_0 \right) k_{al} L t_0 \right)^{\frac{1}{2}} \theta_b \tanh \left(\frac{V}{L} \right) \left(\frac{2\overline{h}}{k_{al} t_0^3} \right)^{\frac{1}{2}}$$

$$\left(2\overline{h} \left(L + t_0 \right) k_{al} L t_0 \right)^{\frac{1}{2}} \theta_b$$
$$= \left(2 \times 26.65 \times \left(2.5 \times 10^{-2} + 3.28 \times 10^{-4} \right) \times 237 \times 2.5 \times 10^{-2} \right.$$
$$\left. \times 3.28 \times 10^{-4} \right)^{1/2} (85 - 22) = 3.227$$

$$mb = \left(\frac{V}{L} \right) \left(\frac{2\overline{h}}{k_{al} t_0^3} \right)^{\frac{1}{2}} = \left(\frac{4.441 \times 10^{-7}}{2.5 \times 10^{-2}} \right) \left(\frac{2 \times 26.65}{237 \times \left(3.28 \times 10^{-4} \right)^3} \right)^{\frac{1}{2}} = 1.4181$$

which should be the same as $mb = \beta = 1.4192$ (Equation (2.41)) except for the rounding errors. The maximized heat transfer rate with an adiabatic tip is

$$q_f = 3.227 \times \tanh 1.4181 = \underline{2.87W} \text{ per fin}$$

Or we can also compute the maximized heat transfer rate using Equations (2.19) and (2.22). The corresponding fin efficiency will be

$$\eta_f = \frac{\tanh mb}{mb} = \frac{\tanh 1.4192}{1.4192} = 0.627$$

Thus, the maximized heat transfer rate with an adiabatic tip will be

$$q_f = \eta_f \overline{h} \left[2 \left(L + t_0 \right) b_0 \right] \left(T_b - T_\infty \right)$$
$$= 0.627 \times 26.65 \times 2 \times \left(2.5 \times 10^{-2} + 3.28 \times 10^{-4} \right) \times 5.4 \times 10^{-2} \times (85 - 22)$$
$$= \underline{2.87 \text{ W}}$$

Now we want to compute the maximized heat transfer rate with a corrected fin length as shown in Equation (2.24) to compare the results.

$$\text{Biot number } \frac{\overline{h} t_0}{k_{al}} = \frac{26.65 \times 3.28 \times 10^{-4}}{237} = 3.69 \times 10^{-5} < 0.0625$$

(See Equation (2.30)

Which indicates that the error associated with the assumptions used in this analysis is negligible. However,

$$m = \left(\frac{2\bar{h}}{k_{al}t_0}\right)^{\frac{1}{2}} = \left(\frac{2 \times 26.65}{237 \times 3.28 \times 10^{-4}}\right)^{\frac{1}{2}} = 26.185 \text{ m}^{-1}$$

The corrected profile length is

$$b_c = b_0 + \frac{t_0}{2} = 5.4 \times 10^{-2} + \frac{3.28 \times 10^{-4}}{2} = 0.054164 \text{ m}$$

$$mb_c = 26.185 \times 0.054164 = 1.4183$$

$$\eta_f = \frac{\tanh mb_c}{mb_c} = \frac{\tanh 1.4183}{1.4183} = 0.627$$

Using Equation (2.19), we have the heat transfer rate with a convective tip

$$q_f = \eta_f h P b \theta_b = \eta_f \bar{h} 2(L + t_0) b_c (T_b - T_\infty)$$

$$= 0.627 \times 26.65 \times 2 \times \left(2.5 \times 10^{-2} + 3.28 \times 10^{-4}\right) \times 0.054164 \times (85 - 22)$$

$$= \underline{2.88 \text{ W}}$$

The difference between the adiabatic tip and convective tip is about 1 %.

2.7   MULTIPLE FIN ARRAY I

2.7.1   Free (Natural) Convection Cooling

The purpose of heat sinks is to maximize the heat transfer rate from the fins, which is sometimes critical in avionic and electronic package design. The previous sections cover the single fin with the optimal fin thickness and fin profile length, which can also be applied to the multiple fins, as shown in Figure 2.4. However, it is necessary in the multiple fin arrays to determine the fin spacing z or the number of fins for a given plate area $L \times W$. Note that the thickness of fin is usually much smaller than the fin spacing z.

2.7.1.1   *Small Spacing Channel*

Consider a small hot parallel plate channel as in Case I, Figure 2.5, with a constant wall temperature T_o. Cold air at T_∞ enters through the bottom of the channel by natural convection. Assume a fully developed flow because of the closeness of the channel. Also assume that the length of the channel is long enough that the air outlet temperature is the same as T_o. Using the geometry in Figure 2.4, the number of fins n can be calculated by

$$n = \frac{W}{z + t} \qquad (2.58)$$

We assume only for the current analysis that

$$z \gg t \qquad (2.59)$$

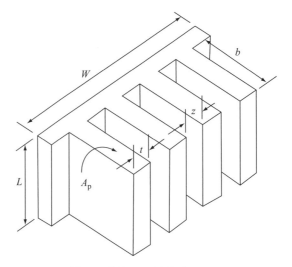

Figure 2.4 Multiple fin array

Hence, Equation (2.58) becomes approximately

$$n \cong \frac{W}{z} \tag{2.60}$$

The heat transfer rate for the channel, using an enthalpy flow, is given by

$$q_f = \dot{m}c_p(T_o - T_\infty) \tag{2.61}$$

$\dot{m}$ is the mass flow rate (kg/s) and can be obtained as

$$\dot{m} = \rho \bar{\upsilon} A_{cf} \tag{2.62}$$

ρ is the air density, $\bar{\upsilon}$ is the average velocity of the fully developed flow, and A_{cf} is the total cross-sectional flow area, which is obtained by

$$A_{cf} = (zb)\left(\frac{W}{z}\right) = Wb \tag{2.63}$$

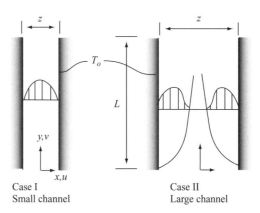

Figure 2.5 Small and large spacing channels

Now we want to derive the average velocity $\bar{v}$ in the small channel for Case I in Figure 2.5. The momentum equation in y direction is of a form

$$\rho\left(u\frac{\partial v}{\partial x} + v\frac{\partial v}{\partial y}\right) = -\frac{\partial P}{\partial y} + \mu\nabla^2 v - \rho g \qquad (2.64)$$

For a fully developed flow, we have

$$u = 0 \quad \text{and} \quad \frac{\partial v}{\partial y} = 0 \qquad (2.65)$$

Since both ends of the channel are open to the ambient of density ρ_∞, the pressure gradient will be

$$\frac{\partial P}{\partial y} = -\rho_\infty g \qquad (2.66)$$

Using Equations (2.65) and (2.66), Equation (2.64) becomes

$$0 = -(-\rho_\infty g) + \mu\frac{\partial^2 v}{\partial y^2} - \rho g \qquad (2.67)$$

Rearranging,

$$\frac{\partial^2 v}{\partial y^2} = -\frac{(\rho_\infty - \rho)g}{\mu} \qquad (2.68)$$

The thermal expansion coefficient is known to be

$$\beta = -\frac{1}{\rho}\left(\frac{\partial\rho}{\partial T}\right)_P \qquad (2.69)$$

The Boussinesq approximation of Equation (2.69) gives

$$\rho_\infty - \rho = \rho_\infty\beta(T - T_\infty) \qquad (2.70)$$

It is a reasonable approximation that $(T - T_\infty) = (T_o - T_\infty)$ [3]. Using the Boussinesq approximation of Equation (2.70), Equation (2.68) becomes

$$\frac{\partial^2 v}{\partial y^2} = -\frac{g\beta(T_o - T_\infty)}{\nu} \qquad (2.71)$$

where ν is the kinematic viscosity ($\nu = \mu/\rho$). Since the right term of Equation (2.71) is a constant, this can be integrated twice with two boundary conditions of nonslip and symmetry of velocity profile, yielding a parabolic velocity profile.

$$v = \frac{g\beta(T_o - T_\infty)}{8\nu}\left[1 - \left(\frac{x}{z/2}\right)^2\right] \qquad (2.72)$$

Upon integration of this along the spacing z, the average velocity can be obtained as

$$\bar{v} = \frac{g\beta(T_o - T_\infty)z^2}{12\nu} \qquad (2.73)$$

The mass flow rate in the small channel can be obtained using Equations (2.62), (2.63), and (2.73).

$$\dot{m} = \frac{\rho g \beta (T_o - T_\infty) z^2}{12\nu} W b \tag{2.74}$$

Now using Equation (2.61), the heat transfer rate from the fins for the small channels can be obtained by

$$q_{f,small} = \rho c_p W b \frac{g \beta (T_o - T_\infty)^2 z^2}{12\nu} \tag{2.75}$$

Notably,

$$q_{f,small} \propto z^2 \tag{2.75a}$$

2.7.1.2 *Large Spacing Channel*

Consider a large spacing channel as in Case II, Figure 2.5, large enough for each boundary layer to be isolated between two isothermal plates. Assume laminar flow. The Nusselt number for natural convection in air for a vertical plate (see Equation (1.77c)) gives [6]

$$\overline{Nu} = \frac{\overline{h}L}{k_{air}} = 0.517 Ra_L^{\frac{1}{4}} \tag{2.76}$$

where the Rayleigh number is defined in Equation (1.76) by

$$Ra_L = \frac{g \beta (|T_o - T_\infty|) L^3}{\nu \alpha} \tag{2.77}$$

In Equation (2.77), Ra_L is the Rayleigh number (for a flow over a plate of length L), β is the volumetric expansion coefficient, α is the thermal diffusivity, ν is the kinematic viscosity, and g is gravity. The expansion coefficient of air is $\beta = 1/T_f$, where T_f is the film temperature.

Using Newton's law of cooling, the heat transfer rate from the fins for the large channels can be written as

$$q_{f,large} = A_s \overline{h}(T_o - T_\infty) \tag{2.78}$$

Using Equation (2.60), the heat transfer area A_s is obtained as

$$A_s = 2bLn = 2bL \left(\frac{W}{z}\right) \tag{2.79}$$

Inserting Equations (2.76) and (2.79) into Equation (2.78) yields

$$q_{f,large} = 1.034 \frac{W}{z} b k_{air} Ra_L^{\frac{1}{4}} (T_o - T_\infty) \tag{2.80}$$

Hence,

$$q_{f,large} \propto \frac{1}{z} \tag{2.80a}$$

2.7.1.3 Optimum Fin Spacing

The optimal spacing can be obtained by equating Equations (2.75) with (2.80)[3] (see Figure 2.6):

$$q_{f,small} = q_{f,large}$$

$$\frac{z_{opt}}{L} \cong 2.3 \left[\frac{g\beta\,(T_o - T_\infty)\,L^3}{\alpha\nu} \right]^{\frac{-1}{4}} = 2.3 Ra_L^{\frac{-1}{4}} \qquad (2.81)$$

More recently, Bar-Cohen and Rohsenow suggested a correlation that has been widely used [7]:

$$\frac{z_{opt}}{L} = 2.714 Ra_L^{\frac{-1}{4}} \qquad (2.82)$$

2.7.2 Forced Convection Cooling

A multiple fin array for forced convection flow is illustrated in Figure 2.7. Consider two cases: a small and large spacing.

2.7.2.1 Small Spacing Channel

Flow between two parallel plates in the case of a small spacing channel is known as Couett-Poiseulli flow, which is a well-defined theory in fluid mechanics.

Couett-Poiseuilli flow

The momentum equation in x direction for the small spacing channel in Figure 2.8 is of the form

$$\rho \left(u \frac{\partial u}{\partial x} + v \frac{\partial u}{\partial y} \right) = -\frac{\partial P}{\partial x} + \mu \left(\frac{\partial^2 u}{\partial x^2} + \frac{\partial^2 u}{\partial y^2} \right) \qquad (2.83)$$

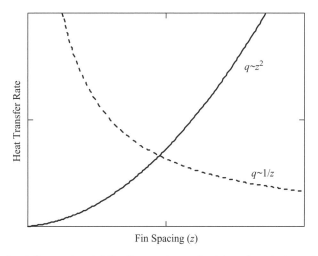

Figure 2.6 Optimal fin spacing (z) for the maximum heat transfer rate by natural convection between two vertical isothermal plates

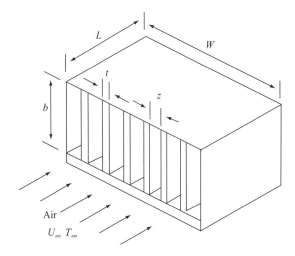

Figure 2.7   Multiple fin array for forced convection flow

For a fully developed flow, we have

$$v = 0, \quad \frac{\partial u}{\partial x} = 0, \quad \text{and} \quad u = u(y) \tag{2.84}$$

The pressure gradient along the channel is a constant because the entrance and exit are open to the ambient air.

$$\frac{\partial P}{\partial x} = \frac{\Delta P}{L} \tag{2.85}$$

Using Equations (2.84) and (2.85), Equation (2.83) becomes

$$\mu \frac{\partial^2 u}{\partial y^2} = \frac{\Delta P}{L} \tag{2.86}$$

The velocity profile can be obtained by integrating twice with two boundary conditions.

$$u\left(\frac{z}{2}\right) = 0 \quad \text{and} \quad \frac{\partial u(0)}{\partial y} = 0 \tag{2.86a}$$

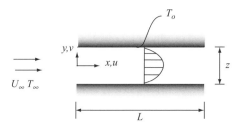

Figure 2.8   Fully developed flow for the small spacing channel

$$u = \frac{\Delta P z^2}{8\mu L}\left[1 - \left(\frac{y}{z/2}\right)^2\right]$$

(2.87)

The average velocity can be obtained using the definition as

$$\bar{u} = \frac{1}{A}\int_A u\, dA$$

(2.88)

$$\bar{u} = \frac{z^2}{12\mu}\frac{\Delta P}{L}$$

(2.89)

The heat transfer rate from the fins is of the form assuming that the fluid temperature at the exit is at T_0.

$$q_{f,small} = \dot{m}c_p(T_o - T_\infty)$$

(2.90)

The mass flow rate is obtained similarly using Equation (2.62)

$$\dot{m} = \rho\left(\frac{z^2}{12\mu}\frac{\Delta P}{L}\right)\left(zb\frac{W}{z}\right)$$

(2.91)

$$q_{f,small} = \rho W b\frac{z^2}{12\mu}\frac{\Delta P}{L}c_p(T_o - T_\infty)$$

(2.92)

This indicates that the heat transfer rate from the plates is proportional to z^2.

$$q_{f,small} \propto z^2$$

(2.93)

2.7.2.2   *Large Spacing Channel*

Consider a large spacing channel in Figure 2.9. Recall that each boundary layer is isolated. The pressure drop ΔP is assumed to be constant. For a laminar flow, the

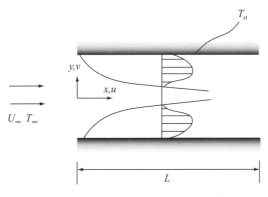

Figure 2.9   Isolated flows over the upper and lower plates for a large spacing channel

Nusselt number given in Equation (1.53) is written here.

$$\overline{Nu}_L = \frac{\overline{h}L}{k_{air}} = 0.664\,\mathrm{Re}_L^{\frac{1}{2}}\,\mathrm{Pr}^{\frac{1}{3}} \quad \text{for Pr} > 0.5 \tag{2.94}$$

where $\mathrm{Re}_L = \dfrac{U_\infty L}{\nu}$ is the Reynolds number. $\tag{2.95}$

$$q_{f,large} = A_s \overline{h}(T_o - T_\infty) \tag{2.96}$$

$$q_{f,large} = 2\frac{W}{z}bL\frac{k_{air}}{L}0.664\mathrm{Pr}^{\frac{1}{3}}\,\mathrm{Re}^{\frac{1}{2}}(T_o - T_\infty) \tag{2.97}$$

$$q_{f,large} = 1.328\frac{W}{z}bk_{air}\mathrm{Pr}^{\frac{1}{3}}\left(\frac{U_\infty L}{\nu}\right)^{\frac{1}{2}}(T_o - T_\infty) \tag{2.98}$$

Since ΔP was assumed to be constant, U_∞ may be expressed in terms of ΔP and z. Consider again the geometry in Figure 2.7 and apply the force balance across the channel box, imaging that the shear force is the root cause of the pressure drop and considering that each fin has two faces for the average shear stress $\overline{\tau}$.

$$\overline{\tau}2\left(\frac{W}{z}\right)(bL) = \Delta P\,(Wb) \tag{2.99}$$

The definition of friction coefficient for laminar flow is given in Equation (1.51).

$$\overline{C}_f = \frac{\overline{\tau}}{\frac{1}{2}\rho U_\infty^2} = 1.328\,\mathrm{Re}_L^{\frac{-1}{2}} \tag{2.100}$$

Combining Equations (2.99) and (2.100) and eliminating the average shear stress $\overline{\tau}$ gives

$$U_\infty = \left(\frac{\Delta P z}{1.328\rho L^{\frac{1}{2}}\nu^{\frac{1}{2}}}\right)^{\frac{2}{3}} \tag{2.101}$$

Inserting Equation (2.101) into (2.98) yields

$$q_{f,large} = 1.208 W b k_{air}\left(\frac{\mathrm{Pr}\,\Delta P L}{\rho \nu^2}\right)^{\frac{1}{3}} z^{\frac{-2}{3}}(T_o - T_\infty) \tag{2.102}$$

From this, we find

$$q_{f,large} \propto z^{\frac{-2}{3}} \tag{2.103}$$

Considering Equations (2.93) and (2.103), we learn that there exists an optimal fin spacing for the maximum heat transfer rate from the fins by equating these expressions. The definitions of thermal diffusivity, kinematic viscosity, and Prandtl number are used.

$$\alpha = \frac{k_f}{\rho c_p} \tag{2.104}$$

$$\nu = \frac{\mu}{\rho} \tag{2.105}$$

$$\Pr = \frac{\nu}{\alpha} \tag{2.106}$$

Combining Equations (2.92) and (2.102) with some algebra provides

$$\frac{z_{opt}}{L} = 2.725 \left(\frac{\mu\alpha}{\Delta P L^2} \right)^{\frac{1}{4}} \tag{2.107}$$

Here z_{opt} is the optimized spacing. Insert Equation (2.107) into Equation (2.92) or Equation (2.102), and $q_{f,small}$ becomes q_{max} since it is optimized at z_{opt}.

$$\frac{q_{max}}{WLb} = 0.619 \frac{k_{air}}{L^2} (T_o - T_\infty) \left(\frac{\Delta P L}{\mu \nu} \right)^{\frac{1}{2}} \tag{2.108}$$

The pressure drop may be estimated with an assumption of $\frac{fL}{D_h} \approx 1$.

$$\Delta P = \frac{fL}{D_h} \frac{\rho U_\infty^2}{2} \approx \frac{\rho U_\infty^2}{2} \tag{2.109}$$

Inserting Equation (2.109) into Equation (2.107) with Equation (2.95) yields

$$\frac{z_{opt}}{L} = 3.24 \, \mathrm{Re}_L^{\frac{-1}{2}} \, \mathrm{Pr}^{\frac{-1}{4}} \tag{2.110}$$

2.8 MULTIPLE FIN ARRAY II

Consider the heat transfer from a single fin in Equation (2.14). It is noted that n is the number of fins.

$$q_{s,f} = \sqrt{h P k A_c} \, \theta_b \tanh mb \tag{2.14}$$

with

$$A_p = bt, \, A_c = Lt, \, P \cong 2L, \text{ and } \beta = mb \tag{2.14a}$$

where

$$m = \sqrt{\frac{hP}{kA_c}} = \left(\frac{2h}{kt} \right)^{\frac{1}{2}} \tag{2.111}$$

$$q_{s,f} = \sqrt{h \, (2L) \, k \, (Lt)} \theta_b \tanh \beta$$

$$\frac{q_{s,f}}{L} = \left(\frac{2hkA_p}{b} \right)^{\frac{1}{2}} \theta_b \tanh \beta \tag{2.112}$$

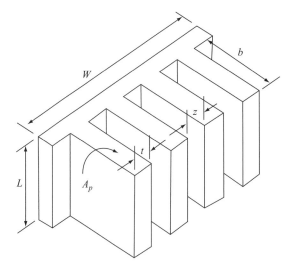

Figure 2.10   Multiple fins array

Consider the heat transfer rate q_w at the base wall between two fins,

$$q_w = h_w L z \theta_b \tag{2.113}$$

where h_w is the heat transfer coefficient at the base wall. In terms of the geometry of the base plate (See Figure 2.10),

$$z = \frac{W}{n} - t = \frac{W}{n} - \frac{A_p}{b}$$

where n is the number of fins.

$$\frac{q_w}{L} = h_w \theta_b \left(\frac{W}{n} - \frac{A_p}{b} \right) \tag{2.114}$$

The total heat transfer rate from the multiple fin array is

$$\frac{q_{total}}{L} = n \left[\left(\frac{2hkA_p}{b} \right)^{\frac{1}{2}} \theta_b \tanh \beta + h_w \theta_b \left(\frac{W}{n} - \frac{A_p}{b} \right) \right] \tag{2.115}$$

where h_w is the heat transfer coefficient along interfins surfaces. To find the maximum total heat transfer with respect to b, we set the derivative equal to zero.

$$\frac{dq_{total}}{db} = 0 \tag{2.116}$$

Performing the similar procedure to Section 2.6, we have

$$\beta \tanh \beta - 3\beta^2 \operatorname{sech}^2 \beta = \frac{2h_w b}{k} \tag{2.117}$$

If h_w equals zero, Equation (2.117) reduces to $\beta_{single,opt} = 1.4192$ as shown for a single fin in Equation (2.41) as

$$\beta \tanh \beta - 3\beta^2 \operatorname{sech}^2 \beta = 0 \tag{2.118}$$

$$3\beta = \sinh \beta \cosh \beta$$

Using $\sinh 2\beta = 2 \sinh \beta \cosh \beta$

$$6\beta = \sinh 2\beta \tag{2.119}$$

The solution for Equation (2.119) is $\beta = 1.4192$. However, if we plot the $\beta_{array,opt}$ as a function of $h_w b / k$, we find that Equation (2.117) may be approximated by the linear relation [2]

$$\beta_{array,opt} = 1.4192 + 1.125 \frac{h_w b}{k} \tag{2.120}$$

where k is the thermal conductivity of the fin material.

For instance,

$$\beta_{array,opt} = 1.420 \text{ for } b = 3 \text{ cm. } k = 175 \text{ W/mk (aluminum), } h = 6 \text{ W/m}^2\text{K.} \tag{2.121}$$

$$\beta_{array,opt} = 1.438 \text{ for } b = 3 \text{ cm. } k = 175 \text{ W/mk (aluminum), } h = 100 \text{ W/m}^2\text{K} \tag{2.122}$$

$$\beta_{array,opt} = 1.464 \text{ for } b = 20 \text{ cm. } k = 30 \text{ W/mk (cast iron), } h = 6 \text{ W/m}^2\text{K.} \tag{2.123}$$

The difference may often be sufficiently small to justify the use, in commercial practice, of the single optimum fin values.

2.8.1 Natural (Free) Convection Cooling

Elenbaas [6] suggested in 1942 that the optimum Nusselt number is 1.25 for two vertical plates, which is found in Figure 1.13.

$$\overline{Nu}_{opt} = \frac{\overline{h}_{opt} z_{opt}}{k_{fluid}} = 1.25 \tag{2.124}$$

Thus, the optimum heat transfer coefficient can be expressed

$$\overline{h}_{opt} = 1.25 \frac{k_{fluid}}{z_{opt}} \tag{2.125}$$

The optimum single-fin heat transfer using Equations (2.14) and (2.56) with $\beta = 1.4192$ becomes

$$\frac{q_{s,f}}{L} = (2hkt_0)^{\frac{1}{2}} \theta_b \tanh \beta = 1.258\theta_b \left(h^2 A_p k\right)^{\frac{1}{2}} \tag{2.126}$$

The total optimum heat transfer rate for the fin array is

$$\frac{q_{total}}{L} = n\left(\frac{q_{s,f}}{L} + \frac{q_w}{L}\right) = \frac{W}{z_{opt} + t_0}\left(1.258\theta_b \left(h_{opt}^2 A_p k\right)^{\frac{1}{2}} + h_{opt} z_{opt}\theta_b\right) \tag{2.127}$$

Using Equations (2.44), (2.125), and (2.82) with $A_p = b_0 t_0$, Equation (2.127) becomes, after some algebra:

$$\frac{q_{total}}{LW\theta_b} = \frac{0.8063 \left(\dfrac{k_{fluid} k t_0}{Ra_L^{-\frac{1}{4}} L} \right)^{\frac{1}{2}} + 1.25 k_{fluid}}{2.714 Ra_L^{-\frac{1}{4}} L + t_0} \tag{2.128}$$

where $Ra_L = \frac{g\beta(T_o - T_\infty)L^3}{\alpha\nu}$

With the approximation that the heat transfer from the interfin area between the fins is negligible, Equation (2.128) simplifies to

$$\frac{q_{total}}{LW\theta_b} \cong \frac{1.258 \left(\overline{h}_{opt} k t_0 \right)^{\frac{1}{2}}}{z_{opt} + t_0} \tag{2.129}$$

Differentiating Equation (2.129) with respect to t_0 and setting the derivative to zero provides the optimum spacing for the maximized heat transfer rate as

$$z_{opt} \cong t_0 \tag{2.130}$$

This reveals that the optimum spacing equals the maximum optimum fin thickness. This is an important relationship for the maximized optimum heat transfer rate for free convection in a vertical fin array. And it is meaningful information for design purposes. However, it is noted that the geometry of the fin array at the maximum heat transfer rate might not be best desirable in practice. For example, the optimized fins may be too large to be accommodated. Then, thinner or smaller fins may be tried and the corresponding heat transfer rates may decrease comparing to the maximum value.

2.9 THERMAL RESISTANCE AND OVERALL SURFACE EFFICIENCY

For single fin efficiency, Equation (2.19) gives

$$q_f = \eta_f h A_f \theta_b \tag{2.131}$$

The thermal resistance R_t of a single fin is defined by

$$R_t = \frac{\theta_b}{q_f} = \frac{1}{\eta_f h A_f} \tag{2.132}$$

where the unit of the thermal resistance is $^\circ C/W$ or equivalently K/W.

The overall efficiency η_o of a multiple fin array can be expressed by

$$\eta_o = \frac{q_{total}}{q_{max}} \tag{2.133}$$

Or equivalently,

$$q_{total} = \eta_o h A_t \theta_b \tag{2.133a}$$

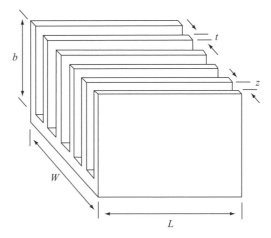

Figure 2.11   Geometry of a multiple fins array

where A_t is the total heat transfer area, including the fins and interfins area.

$$A_t = n \left(A_f + A_w \right) = n \left[2(L + t)b + Lz \right] \tag{2.134}$$

where n is the number of fins. Dimensions are shown in Figure 2.11.

Now the total heat transfer rate is the sum of the heat transfer rates from the fins area and the interfins area.

$$q_{total} = n \left(q_f + q_w \right) \tag{2.135}$$

$$q_{total} = n \left(\eta_f h A_f \theta_b + h A_w \theta_b \right) \tag{2.136}$$

where $A_w = \frac{A_t}{n} - A_f$ from Equation (2.134)

Inserting this into (2.136) provides

$$q_{total} = h\theta_b \left(n\eta_f A_f + A_t - nA_f \right) \tag{2.137}$$

$$q_{total} = A_t h\theta_b \left(n\eta_f \frac{A_f}{A_t} + 1 - n\frac{A_f}{A_t} \right)$$

$$q_{total} = A_t h\theta_b \left[1 - n\frac{A_f}{A_t} \left(1 - \eta_f \right) \right] \tag{2.138}$$

Combining Equation (2.133) and (2.138) gives an expression of the overall surface efficiency as

$$\eta_o = 1 - n\frac{A_f}{A_t} \left(1 - \eta_f \right) \tag{2.139}$$

The overall thermal resistance of a multiple fins array is defined by

$$R_{t,o} = \frac{1}{\eta_o h A_t} \tag{2.140}$$

Table 2.1 Efficiencies of Various Fin Shapes

	Shape	Efficiency	Equation
Rectangular fins $A_p = b_c t$ $b_c = b + \dfrac{t}{2}$ $A_c = Lt$		$\eta_f = \dfrac{\tanh mb_c}{mb_c}$ where $m = \left(\dfrac{2h}{kt}\right)^{1/2}$	(2.28)
Triangular fins $A_p = \dfrac{bt}{2}$		$\eta_f = \dfrac{1}{mb}\dfrac{I_1(2mb)}{I_0(2mb)}$ where $m = \left(\dfrac{2h}{kt}\right)^{1/2}$	(2.141)
Parabolic fins $A_p = \dfrac{Lt}{3}$	$y = (t/2)(1-x/b)^2$	$\eta_f = \dfrac{2}{\left[4(mb)^2 + 1\right]^{1/2} + 1}$ where $m = \left(\dfrac{2h}{kt}\right)^{1/2}$	(2.142)
Radial fins $r_{2c} = r_2 + \dfrac{t}{2}$		$\eta_f = C_1 \dfrac{K_1(mr_1) I_1(mr_{2c}) - I_1(mr_1) K_1(mr_{2c})}{I_0(mr_1) K_1(mr_{2c}) + K_0(mr_1) I_1(mr_{2c})}$ $C_1 = \dfrac{2r_1/m}{r_{2c}^2 - r_1^2}$ where $m = \left(\dfrac{2h}{kt}\right)^{1/2}$	(2.143)
Pin fins $b_c = b + \dfrac{d}{4}$		$\eta_f = \dfrac{\tanh mb_c}{mb_c}$ where $m = \left(\dfrac{4h}{kd}\right)^{1/2}$	(2.144)

Table 2.1   (*Continued*)

	Shape	Efficiency	Equation
Conical fins		$\eta_f = \dfrac{2}{mb}\dfrac{I_2(2mb)}{I_1(2mb)}$ where $m = \left(\dfrac{4h}{kd}\right)^{1/2}$	(2.145)
Concave parabolic fins	$y = (d/2)(1-x/b^2)$	$\eta_f = \dfrac{2}{\left[\frac{4}{9}(mb)^2 + 1\right]^{1/2} + 1}$ where $m = \left(\dfrac{4h}{kd}\right)^{1/2}$	(2.146)

*I_0, I_1, I_2, K_0, and K_1 are the Bessel functions.

where $R_{t,o}$ is an effective resistance that accounts for the parallel flow path by conduction/convection in the fins and by convection from the interfin surface (or called the prime surface). Table 2.1 illustrates the efficiencies of various fin shapes.

Example 2.2 Forced Convection Multiple Fins Array   A heat sink for an avionic electronic package cooling is to be designed to maintain the base temperature below 100°C. The base cold plate has a width $W = 50$ mm and Length $L = 30$ mm, as shown in Figure E2.2.1. Forced air is drawn by a fan at a velocity of $U = 1.5$ m/s and at the ambient air temperature of 20°C. Non-optimum design is allowed to accommodate these requirements. For aluminum properties, use density of 2702 kg/m³ and thermal conductivity of 177 W/mK.

(a) Determine the maximum optimum dimensions for the heat sink providing the number of fins, the fin thickness, the fin spacing, the profile length, the heat dissipation, overall efficiency and thermal resistance.

(b) The device now requires a mass of fins to be less than 15 grams and the profile length to be less than 20 mm. With these constraints, design the fin array again.

MathCAD Format Solution   Information given:

$$W_b := 50 \text{ mm} \qquad L_b := 30 \text{ mm} \qquad U_{air} := 1.5\frac{m}{s}$$
$$T_{base} := 100°C \qquad T_{inf} := 20°C \qquad \theta_b := T_{base} - T_{inf}$$

Tips for MathCAD: The use of subscript for all letter symbols is suggested to avoid the conflict with the built-in units. For instance, the letter symbol "W" is already defined as watts, so it shouldn't be used as a variable. Therefore, you can define W_b.

Design Concept   We want to first examine an optimum design without the constraints for the given geometry and conditions. And then we consider whether the optimum design is satisfactory with the requirements. If not, redesign the cold plate with the constraints in order to meet the requirements.

(a) Optimum Design without Constraints   The properties of air at the film temperature $(20 + 100)/2 = 60°C = 333K$ (Table A.2) are

$$k_{air} := 0.0292 \frac{W}{m\,K} \quad \nu_{air} := 19.2\,10^{-6} \frac{m^2}{s} \quad Pr := 0.707$$

Properties of aluminum:

$$\rho_{al} := 2702 \frac{kg}{m^3} \quad k_{al} := 177 \frac{W}{m\,K}$$

Reynolds number:

$$Re_L := \frac{U_{air}L_b}{\nu_{air}} \quad Re_L = 2.344 \times 10^3 \tag{E2.2.1}$$

which is a laminar flow because Re_L is less than 5×10^5 (see Equation (1.47)). The optimum spacing is found in Equation (2.110)

$$z_{opt} := L_b\,3.24\,Re_L^{\frac{-1}{2}}\,Pr^{\frac{-1}{4}} \quad z_{opt} = 2.19\,mm \tag{E2.2.2}$$

Using Equation (1.53), the average heat transfer coefficient for laminar flow over a flat plate is

$$h_L := \frac{k_{air}}{L_b}0.664\,Re_L^{\frac{1}{2}}\,Pr^{\frac{1}{3}} \quad h_L = 27.874 \frac{W}{m^2\,K} \tag{E2.2.3}$$

We may use Equation (2.120) for a multiple optimum fin array. However, assuming that the difference for β between the single fin and the multiple fins is sufficiently small,

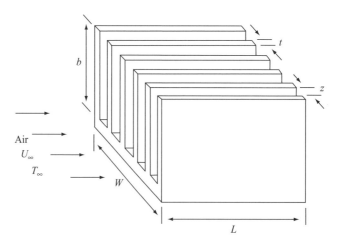

Figure E2.2.1   Forced convection multiple fin array

we develop a solution for the heat transfer rate for a fin array with $\beta = 1.4192$ shown in Equation (2.41) with the fin thickness as a parameter.

$$\beta := 1.4192 \quad \text{and the fin efficiency} \quad \eta_f := \frac{\tanh(\beta)}{\beta} \quad \eta_f = 0.627 \qquad (E2.2.4)$$

From Equation (2.44), the optimum profile length b_0 is a function of the optimum fin thickness t_0. We want to optimize the total heat transfer rate as a function of t_0.

$$b_0(t_0) := \beta \left(\frac{k_{al} \, t_0}{2 \, h_L} \right)^{\frac{1}{2}} \qquad (E2.2.5)$$

The volume of a single fin (aluminum) is also a function of t_0.

$$V_{al}(t_0) := L_b \, b_0(t_0) \, t_0 \qquad (E2.2.6)$$

The number of fins is a function of t_0.

$$n_f(t_0) := \frac{W_b}{z_{opt} + t_0} \qquad (E2.2.7)$$

The total mass of aluminum for the fin array is a function of t_0.

$$m_{al}(t_0) := n_f(t_0) \rho_{al} V_{al}(t_0) \qquad (E2.2.8)$$

The heat transfer rate of the single fin, q_{sf}, using Equation (2.14) is of the form

$$q_{sf}(t_0) := [h_L \, 2 \, (L_b + t_0) k_{al} \, L_b \, t_0]^{\frac{1}{2}} \, \theta_b \tanh(\beta) \qquad (E2.2.9)$$

The heat transfer rate at an interfin area between the two fins, q_{sw}, is

$$q_{sw} := h_L \, z_{opt} \, L_b \, \theta_b \qquad (E2.2.10)$$

Thus, the total heat transfer rate is the sum of q_{sf} and q_{sw} multiplied by the number of fins.

$$q_{total}(t_0) := n_f(t_0)(q_{sf}(t_0) + q_{sw}) \qquad (E2.2.11)$$

The heat transfer rate at the interfins area is only 2.149 % of that at the single fin surface for $t_0 = 1$ mm. This just shows how q_{sw} is small compared to q_{sf}.

$$\frac{q_{sw}}{q_{sf}(1 \text{ mm})} = 2.149 \, \% \qquad (E2.2.12)$$

The fin effectiveness is obtained using Equation (2.15):

$$\varepsilon_f(t_0) := \frac{q_{sf}(t_0)}{h_L \, L_b \, t_0 \, \theta_b} \qquad (E2.2.13)$$

The total area is obtained using Equation (2.134)

$$A_t(t_0) := n_f(t_0) \, [2(L_b + t_0)b_0(t_0) + L_b \, z_{opt}] \qquad (E2.2.14)$$

The overall surface efficiency is obtained using Equation (2.139):

$$A_f(t_0) := 2(L_b + t_0)b_0(t_0) \tag{E2.2.15}$$

$$\eta_0(t_0) := 1 - n_f(t_0)\frac{A_f(t_0)}{A_t(t_0)}(1 - \eta_f) \tag{E2.2.16}$$

The overall thermal resistance is obtained using Equation (2.140):

$$R_{to}(t_0) := \frac{1}{\eta_0(t_0)\,h_L\,A_t(t_0)} \tag{E2.2.17}$$

The results are graphed in Figures E2.2.2–E2.2.5. The thickness range for the graphs must be given as

$$t_0 := 0.01\text{ mm}, 0.02\text{ mm}.. 10\text{ mm} \tag{E2.2.17a}$$

The maximum fin thickness is obtained from Figure E2.2.2, but it can be calculated using a MathCAD function.

Initial guess $t_0 := 1$ mm

$$t_0 := \text{Maximize}(q_{total}, t_0) \quad t_0 = 2.48\text{ mm} \tag{E2.2.17b}$$

where an initial guess '$t_0 = 1$ mm' is required for the 'Maximize' function. The following results are calculated for a summary table.

$$t_0 = 2.48\text{ mm} \qquad m_{al}(t_0) = 271.077\text{ gm}$$
$$b_0(t_0) = 125.934\text{ mm} \qquad \beta = 1.419$$
$$z_{opt} = 2.19\text{ mm} \qquad \eta_f = 0.6267$$
$$h_L = 27.874\,\frac{W}{m^2\,K} \qquad \varepsilon_f(t_0) = 66.228 \tag{E2.2.17c}$$
$$q_{total}(t_0) = 119.218\text{W} \qquad \eta_0(t_0) = 0.63$$
$$n_f(t_0) = 10.708 \qquad R_{to}(t_0) = 0.645\frac{K}{W}$$

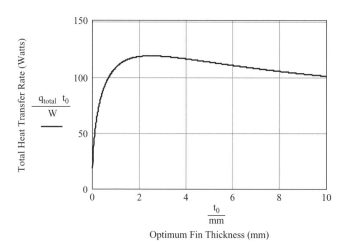

Figure E2.2.2   Total heat transfer rate versus fin thickness

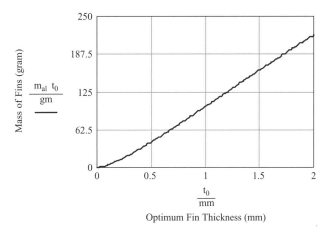

Figure E2.2.3   Mass of fins versus fin thickness

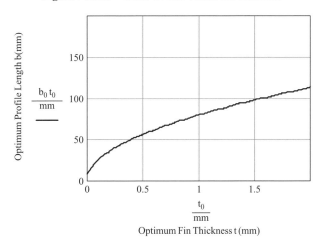

Figure E2.2.4   Optimum profile thickness versus fin thickness

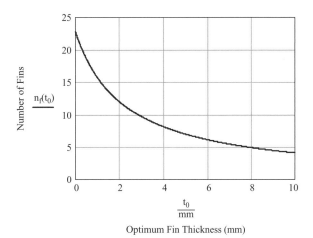

Figure E2.2.5   Number of fins versus fin thickness

These are the optimum results for the maximum heat transfer rate for the given base size (W × L = 50 mm × 30 mm). Note that both the profile length b_0 and the mass of aluminum m_{al} are considerably large. In Equation (E2.2.4), we assumed that the discrepancy of β between a single fin and multiple-fin array is small. We want to check the validation of the assumption. The calculations were repeated with Equation (2.120) in place of (2.41), finally showing that the discrepancy was less than 2 % in comparison between the two. Therefore, the assumption is justified particularly in heat sinks with forced convection.

(b) Optimum Design with Constraints   Now we want to obtain an optimum heat sink with the constraints which are the profile length $b = 20$ mm and the mass of aluminum fins = 15 grams. We seek an optimum thickness t_1 as a function with the given profile length b_1.

$$b_1 := 20 \text{ mm} \tag{E2.2.18}$$

We are to develop an expression for the total heat transfer rate as a function of t, so that we can graphically obtain the optimum thickness satisfying the design requirements.

The number of fins is expressed as

$$n_f(t_1) := \frac{W_b}{z_{opt} + t_1} \tag{E2.2.19}$$

The mass of the aluminum fins is then obtained as

$$m_{al}(t_1) := n_f(t_1)\rho_{al} L_b b_1 t_1 \tag{E2.2.20}$$

We use the optimum relationship in Equation (2.55) for the optimum β.

$$\beta(t_1) := b_1 \left(\frac{2 h_L}{k_{al} t_1} \right)^{\frac{1}{2}} \tag{E2.2.21}$$

The single fin efficiency using Equations (2.22) and (2.39) is

$$\eta_f(t_1) := \frac{\tanh(\beta(t_1))}{\beta(t_1)} \tag{E2.2.22}$$

The heat transfer rate of the single fin using Equation (2.14) is

$$q_{sf}(t_1) := [h_L 2(L_b + t_1)k_{al} L_b t_1]^{\frac{1}{2}} \theta_b \tanh(\beta(t_1)) \tag{E2.2.23}$$

The heat transfer rate at the interfins area with Equation (E2.2.2) is

$$q_{sw} := h_L z_{opt} L_b \theta_b$$

The total heat transfer rate is the sum of the above two which is multiplied by the number of fins.

$$q_{total}(t_1) := n_f(t_1)(q_{sf}(t_1) + q_{sw}) \tag{E2.2.24}$$

The fin effectiveness is obtained using Equation (2.15).

$$\varepsilon_f(t_1) := \frac{q_{sf}(t_1)}{h_L \, L_b \, t_1 \, \theta_b} \tag{E2.2.25}$$

The total area and single fin area are obtained using Equation (2.134).

$$A_t(t_1) := n_f(t_1)[2(L_b + t_1)b_1 + L_b \, z_{opt}] \tag{E2.2.26}$$

$$A_f(t_1) := 2(L_b + t_1)b_1 \tag{E2.2.27}$$

The overall surface efficiency is obtained using Equation (2.139).

$$\eta_0(t_1) := 1 - n_f(t_1)\frac{A_f(t_1)}{A_t(t_1)}(1 - \eta_f(t_1)) \tag{E2.2.28}$$

The overall thermal resistance is obtained using Equation (2.140).

$$R_{to}(t_1) := \frac{1}{\eta_0(t_1) \, h_L \, A_t \, (t_1)} \tag{E2.2.29}$$

The total heat transfer rate is plotted with respect to the fin thickness in Figure E2.2.6.

$$t_1 := 0.01 \text{ mm}, 0.02 \text{ mm}.. \ 1 \text{ mm}$$

The maximum fin thickness is found using a MathCAD function

$$\begin{aligned} &\text{Initial guess} &&t_1 := 1 \text{ mm} \\ &t_1 := \text{Maximize}(q_{total}, t_1) &&t_1 = 0.291 \text{ mm} \end{aligned} \tag{E2.2.30}$$

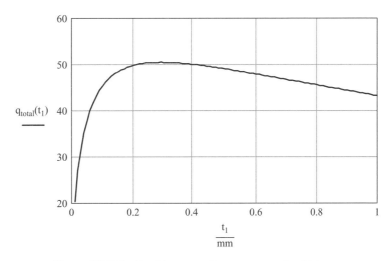

Figure E2.2.6 Total heat transfer rate versus fin thickness

The results are calculated for a summary table.

$$
\begin{array}{ll}
t_1 = 0.291 \text{ mm} & m_{al}(t_1) = 9.504 \text{ gm} \\[4pt]
b_1 = 20 \text{ mm} & \beta(t_1) = 0.658 \\[4pt]
z_{opt} = 2.19 \text{ mm} & \eta_f(t_1) = 0.8769 \\[4pt]
h_L = 27.874 \dfrac{W}{m^2 K} & \varepsilon_f(t_1) = 121.19 \\[4pt]
q_{total}(t_1) = 50.481 W & \eta_0(t_1) = 0.883 \\[4pt]
n_f(t_1) = 20.158 & R_{to}(t_1) = 1.578 \dfrac{K}{W}
\end{array}
\qquad\text{(E2.2.31)}
$$

For comparison purposes, we tabulate the parameters in Table E2.2.1 with three fin thicknesses: 0.1 mm, 0.291 mm, and 0.5 mm.

In conclusion, although the optimum array I in Table E2.2.1 has the superb performance with the maximum heat transfer rate, it does not meet the requirements of the profile length of 20 mm and the mass of fins less than 15 gram. To meet these requirements, the optimum arrays with constraints are obtained as shown in the table.

Table E2.2.1   Summary of Results of Forced Convection Fins Array

Parameter	Optimum array I without constraints	Optimum array II with constraints	Optimum array III with constraints	Optimum array IV with constraints
Fin thickness, t (mm)	2	0.5	0.291	0.1
Profile length, b (mm)	113	20	20	20
Spacing, z (mm)	2.19	2.19	2.19	2.19
Heat transfer coefficient, h (W/m²K)	27.9	27.9	27.87	27.9
Total heat transfer rate, q_{total} (W)	118.6	49.0	50.48	45.3
Number of fins	12	19	21	22
Mass of fins (g)	219	15	9.5	3.54
$\beta = mb$	1.4192	0.503	0.658	1.122
Fin efficiency, η_f	0.627	0.924	0.88	0.72
Fin effectiveness, ε_f	73.2	74.5	121	289
Overall efficiency, η_0	0.630	0.928	0.883	0.735
Overall thermal resistance, $R_{t,o}$ (°C/W)	0.653	1.618	1.578	1.761

Base plate geometry W = 50 mm and L = 30 mm, Aluminum properties: density $\rho = 2702$ kg/m³, thermal conductivity k = 177 W/mK, $\theta_b = 80$ K, U = 1.5 m/s

It is of interest to compare the mass of fins, the profile length, and the heat transfer rate between the optimum arrays without/with constraints. The thin thickness of 0.291 mm provides the mass of 9.5 gram that compares with the mass of 219 gram for the Optimum Array I. The mass of fins is particularly important in avionics.

Example 2.3 Natural Convection Multiple Fins Array The rear wall of a plate on which electronic components are mounted is passively cooled by natural convection using a multiple-fin array as shown in Figure E2.3.1. The plate geometry has $W = 50$ cm and $L = 18$ cm. The ambient air is at 25°C and at atmospheric pressure. The temperature of the rear wall must not exceed a surface temperature of 75°C. For the properties of aluminum, the density is 2700 kg/m³ and the thermal conductivity is 196 W/mK. The rear wall must dissipate the heat transfer rate up to 175 W.

(a) Show whether or not the heat dissipation can be achieved by natural convection without fins.

(b) Design the optimum heat sink without the constraint of the heat dissipation of 175 W for the given geometry and information to maximize the heat dissipation, providing the fin thickness, the profile length, the spacing, the number of fins, overall surface efficiency, and overall thermal resistance.

(c) Optimize the heat sink with the constraint of the heat dissipation of 175 W.

MathCAD Format Solution: Information given:

$$W_b := 50 \text{ cm} \qquad L_b := 18 \text{ cm}$$
$$T_{base} := 75°C \qquad T_{inf} := 25°C \qquad \theta_b := T_{base} - T_{inf}$$

Properties of aluminum:

$$\rho_{al} := 2700 \frac{\text{kg}}{\text{m}^3} \qquad k_{al} := 196 \frac{\text{W}}{\text{m K}}$$

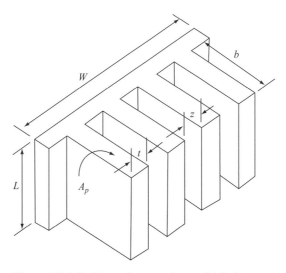

Figure E2.3.1 Natural convection multiple-fin array

Properties of air at a film temperature $(25 + 75)/2 = 50°C = 323K$ (Table A.2):

$$k_{air} := 0.0274 \frac{W}{m\,K} \quad \nu := 19.2\ 10^{-6} \frac{m^2}{s} \quad \alpha := 26.0\ 10^{-6} \frac{m^2}{s} \quad \beta := \frac{1}{323K}$$

(a) Heat Transfer without Fins First, determine whether fins should be employed. We want to compute the heat transfer rate from the rear wall without fins, and then we can compare it with the requirement.

From Equation (1.77c) developed for air, the average Nusselt number for a vertical plate is given by

$$Nu_L = \frac{h_L\, L_b}{k_{air}} = 0.517\, Ra_L^{\frac{1}{4}}$$

$$Ra_L := \frac{g\,\beta\,\theta_b\, L_b^3}{\alpha\nu} \qquad Ra_L = 1.773 \times 10^7 \quad <10^9 \text{ Laminar flow} \qquad \text{(E2.3.1)}$$

$$h_L := \frac{k_{air}}{L_b} 0.517\, Ra_L^{\frac{1}{4}} \qquad h_L = 5.107 \frac{W}{m^2\,K}$$

Using Newton's law of cooling, we can calculate the heat transfer rate without fins as

$$q_L := h_L(W_b\, L_b)\theta_b \qquad q_L = 22.982W \qquad \text{(E2.3.2)}$$

$$22.98\,W < 175\,W \qquad \text{(E2.3.3)}$$

Since the heat transfer rate of 22.98 W is much less than the required heat dissipation of 175 W, we conclude that fins must be used. Now we design the multiple fin array to maximize the heat dissipation.

(b) Optimum Design without Constraints We want to develop the optimum design of a multiple-fin array without constraints based on single fin analysis. Using Equation (2.82), the optimum fin spacing can be obtained.

$$z_{opt} := L_b\, 2.714\, Ra_L^{\frac{-1}{4}} \qquad z_{opt} = 7.528\ mm \qquad \text{(E2.3.4)}$$

Calculate the heat transfer coefficient for a vertical parallel flow between two fins. From the Elenbaas' equation (1.81), we have

$$Nu_z = \frac{h_z\, z_{opt}}{k_{air}} = \left(\frac{576}{El^2} + \frac{2.873}{El^{\frac{1}{2}}} \right)^{\frac{-1}{2}} \qquad \text{(E2.3.5)}$$

Care should be taken in using spacing z, not L, in the Rayleigh number of Equation (1.79) for the Elenbaas equation.

$$Ra_z := \frac{g\,\beta\,\theta_b\, z_{opt}^3}{\alpha\,\nu} \qquad \text{(E2.3.6)}$$

The Elenbaas number is shown in Equation (1.80).

$$El := Ra_z \frac{z_{opt}}{L_b} \tag{E2.3.7}$$

$$h_z := \frac{k_{air}}{z_{opt}} \left(\frac{576}{El^2} + \frac{2.873}{El^{\frac{1}{2}}} \right)^{\frac{-1}{2}} \qquad h_z = 4.756 \frac{W}{m^2\,K} \tag{E2.3.8}$$

Note that the heat transfer coefficient of 4.756 W/m²K is close to 5.107 W/m²K in Equation (E2.3.1). The optimum spacing appears adequate. Note that we do not assume this time that the discrepancy in β between a single fin and a multiple-fin array is small as we did in Example 2.2 because the interfins heat transfer may be significant in natural convection. We want to develop a solution for the heat transfer rate for the fin array by taking the fin thickness as a parameter. Equation (2.120) gives

$$\beta = 1.4192 + 1.125 \frac{h_z b_0}{k_{al}} \tag{E2.3.9}$$

From Equation (2.44), the profile length b is a function of the fin thickness t. It is noted that b and t are optimum values with the equation. Denote that b_0 and t_0 are optimum profile length and optimum fin thickness, respectively.

$$b_0 = \beta \left(\frac{k\, t_0}{2\, h} \right)^{\frac{1}{2}} \tag{E2.3.10}$$

Combining the above two equation and solving for b_0 gives

$$b_0(t_0) := \frac{1.4192 \left(\dfrac{k_{al}\, t_0}{2\, h_z} \right)^{\frac{1}{2}}}{\left[1 - 1.125 \left(\dfrac{k_{al}\, t_0}{2\, h_z} \right)^{\frac{1}{2}} \dfrac{h_z}{k_{al}} \right]} \tag{E2.3.11}$$

Then the β is now a function of t_0.

$$\beta(t_0) := 1.4192 + 1.125 \frac{h_z\, b_0(t_0)}{k_{al}} \tag{E2.3.12}$$

The volume of a single fin (aluminum) is a function of t_0.

$$V_{al}(t_0) := L_b\, b_0(t_0) t_0 \tag{E2.3.13}$$

The number of fins is also a function of t_0.

$$n_f(t_0) := \frac{W_b}{z_{opt} + t_0} \tag{E2.3.14}$$

The total mass of material for the fins is a function of t_0.

$$m_{al}(t_0) := n_f(t_0) \rho_{al} V_{al}(t_0) \tag{E2.3.15}$$

The single fin efficiency is obtained using Equations (2.7a), (2.21) and (2.22).

$$mb(t_0) := b_0(t_0) \sqrt{\frac{h_z 2(L_b + t_0)}{k_{al} L_b t_0}} \qquad \text{(E2.3.16)}$$

$$\eta_f(t_0) := \frac{\tanh(mb(t_0))}{mb(t_0)} \qquad \text{(E2.3.17)}$$

The single fin heat transfer is

$$q_f(t_0) := \eta_f(t_0) h_z 2(L_b + t_0) b_0(t_0) \theta_b \qquad \text{(E2.3.18)}$$

The single fin effectiveness is obtained using Equations (2.15) and (2.19).

$$\varepsilon_f(t_0) := \frac{q_f(t_0)}{h_z L_b t_0 \theta_b} \qquad \text{(E2.3.19)}$$

The total area is obtained using Equation (2.134).

$$A_t(t_0) := n_f(t_0) \left[2(L_b + t_0) b_0(t_0) + L_b z_{opt} \right] \qquad \text{(E2.3.20)}$$

The overall efficiency is obtained using Equation (2.139) with a single fin surface area.

$$A_f(t_0) := 2(L_b + t_0) b_0(t_0) \qquad \text{(E2.3.21)}$$

$$\eta_0(t_0) := 1 - n_f(t_0) \frac{A_f(t_0)}{A_t(t_0)} (1 - \eta_f(t_0)) \qquad \text{(E2.3.22)}$$

The total heat transfer rate is

$$q_{total}(t_0) := \eta_0(t_0) h_z A_t(t_0) \theta_b \qquad \text{(E2.3.23)}$$

The overall thermal resistance is obtained using Equation (1.140)

$$R_{to}(t_0) := \frac{1}{\eta_0(t_0) h_z A_t(t_0)} \qquad \text{(E2.3.24)}$$

The total heat transfer rate is plotted with respect to the fin thickness in Figure E2.3.2 to Figure E2.3.5.

$$t_0 := 0.01 \text{ mm}, 0.02 \text{ mm} .. \ 20 \text{ mm}$$

The maximum fin thickness is obtained using a MathCAD function.

$$\begin{array}{ll} \text{Initial guess} & t_0 := 1 \text{ mm} \\ t_0 := \text{Maximize}(q_{total}, t_0) & t_0 = 8.343 \text{ mm} \end{array} \qquad \text{(E2.3.25)}$$

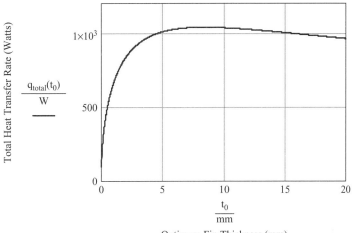

Figure E2.3.2   Total heat transfer rate versus fin thickness

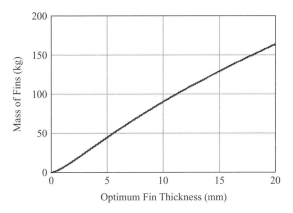

Figure E2.3.3   Mass of fins versus fin thickness

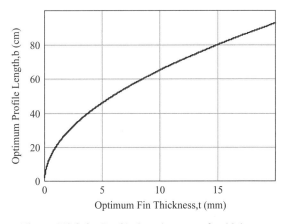

Figure E2.3.4   Profile length versus fin thickness

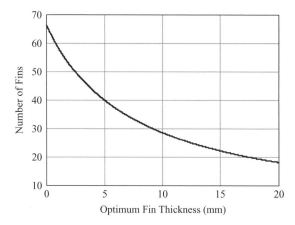

Figure E2.3.5 Number of fins versus fin thickness

The results are calculated as a summary.

$$t_0 = 8.343 \text{ mm} \qquad\qquad m_{al}(t_0) = 7.602 \times 10^4 \text{ gm}$$

$$b_0(t_0) = 595.164 \text{ mm} \qquad\qquad \beta(t_0) = 1.435$$

$$z_{opt} = 7.528 \text{ mm} \qquad\qquad \eta_f(t_0) = 0.6124$$

$$h_L = 5.107 \frac{W}{m^2 K} \qquad\qquad \varepsilon_f(t_0) = 91.432 \qquad\qquad \text{(E2.3.26)}$$

$$q_{total}(t_0) = 1.039 \times 10^3 \text{ W} \qquad\qquad \eta_0(t_0) = 0.615$$

$$n_f(t_0) = 31.505 \qquad\qquad R_{to}(t_0) = 0.048 \frac{K}{W}$$

The maximum total heat transfer rate of about 1039 W in Figure E2.3.2 occurs at $t_0 = 8.343$ mm, which compares to the optimum fin spacing of 7.528 mm in Equation (E2.3.4). Although a discrepancy between them exists, the optimum fin thickness was analytically predicted to be equal to the optimum spacing as shown in Equation (2.130). The optimum profile length of 595 mm seems too large to accommodate in the plate (500 mm × 180 mm). The result appears impractical. Note that the heat transfer rate of 1039 W is much greater than that of 175 W that is in requirement.

(c) Optimum Design with Constraints We now want to determine the optimum design with the constraint of 175 W as a function of two variables, t_1 and b_1.

The number of fins as a function of t_1:

$$n_f(t_1) := \frac{W_b}{z_{opt} + t_1} \qquad\qquad \text{(E2.3.27)}$$

The volume of the fins is

$$V_{al}(t_1, b_1) := n_f(t_1) L_b b_1 t_1 \qquad\qquad \text{(E2.3.28)}$$

The mass of the fins is

$$m_{al}(t_1, b_1) := \rho_{al} V_{al}(t_1, b_1) \qquad (E2.3.29)$$

The single fin efficiency is obtained using Equation (2.21) and (2.22).

$$mb(t_1, b_1) := b_1 \sqrt{\frac{h_z \, 2 \, (L_b + t_1)}{k_{al} \, L_b \, t_1}} \qquad (E2.3.30)$$

$$\eta_f(t_1, b_1) := \frac{\tanh(mb(t_1, b_1))}{mb(t_1, b_1)} \qquad (E2.3.31)$$

$$\underset{\sim}{\beta}(t_1, b_1) := mb(t_1, b_1) \qquad (E2.3.32)$$

The single fin effectiveness is obtained using Equations (2.15) and (2.19).

$$q_f(t_1, b_1) := \eta_f(t_1, b_1) \, h_z \, 2 \, (L_b + t_1) \, b_1 \, \theta_b \qquad (E2.3.33)$$

$$\varepsilon_f(t_1, b_1) := \frac{q_f(t_1, b_1)}{h_z \, L_b \, t_1 \, \theta_b} \qquad (E2.3.34)$$

The total area is obtained using Equation (2.134).

$$A_t(t_1, b_1) := n_f(t_1) \left[2(L_b + t_1) \, b_1 + L_b \, z_{opt} \right] \qquad (E2.3.35)$$

The overall efficiency is obtained using Equation (2.139) with a single fin surface.

$$A_f(t_1, b_1) := 2 \, (L_b + t_1) \, b_1 \qquad (E2.3.36)$$

$$\eta_0(t_1, b_1) := 1 - n_f(t_1) \frac{A_f(t_1, b_1)}{A_t(t_1, b_1)} (1 - \eta_f(t_1, b_1)) \qquad (E2.3.37)$$

The total heat transfer rate is

$$q_{total}(t_1, b_1) := \eta_0(t_1, b_1) h_z A_t(t_1, b_1) \theta_b \qquad (E2.3.38)$$

The overall thermal resistance is obtained using Equation (2.140).

$$R_{to}(t_1, b_1) := \frac{1}{\eta_0(t_1, b_1) h_z A_t(t_1, b_1)} \qquad (E2.3.39)$$

The total heat transfer rate is plotted with respect to the fin thickness as a function of the profile length.

$$t_1 := 0.01 \text{ mm}, 0.02 \text{ mm..} \, 5 \text{ mm}$$

Choose the profile length of 3 cm for the optimum design from Figure E2.3.6.

$$b_1 := 3 \text{ cm} \qquad (E2.3.40)$$

Redefine the total heat transfer rate for the 'Maximize' function as

$$\underset{\sim}{q_{total}}(t_1) := q_{total}(t_1, b_1) \qquad (E2.3.41)$$

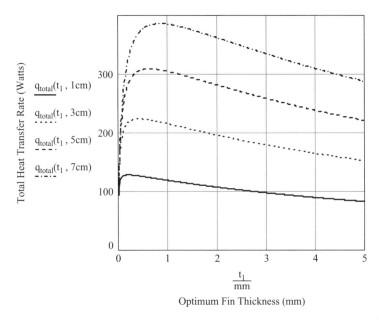

Total Heat Transfer Rate (Watts)

$q_{total}(t_1, 1cm)$

$q_{total}(t_1, 3cm)$

$q_{total}(t_1, 5cm)$

$q_{total}(t_1, 7cm)$

$\dfrac{t_1}{mm}$

Optimum Fin Thickness (mm)

Figure E2.3.6   Total heat transfer rate as functions of optimum fin thickness and profile length

The maximum fin thickness is obtained using a MathCAD function. Keep in mind that the initial guess should be entered with a value close to the solution.

$$\text{Initial guess} \qquad\qquad t_1 := 0.5 \text{ mm}$$

$$t_1 := \text{Maximize}(q_{total}, t_1) \qquad t_1 = 0.314 \text{ mm} \tag{E2.3.42}$$

The results are calculated.

$$t_1 = 0.314 \text{ mm} \qquad\qquad m_{al}(t_1, b_1) = 291.807 \text{ gm}$$

$$b_1 = 30 \text{ mm} \qquad\qquad \beta(t_1, b_1) = 0.373$$

$$z_{opt} = 7.528 \text{ mm} \qquad\qquad \eta_f(t_1, b_1) = 0.956$$

$$h_L = 5.107 \frac{W}{m^2 K} \qquad\qquad \varepsilon_f(t_1, b_1) = 183.053 \tag{E2.3.43}$$

$$q_{total}(t_1) = 177.359 W \qquad\qquad \eta_0(t_1, b_1) = 0.961$$

$$n_f(t_1) = 63.761 \qquad\qquad R_{to}(t_1, b_1) = 0.282 \frac{K}{W}$$

We learn from Figure E2.3.6 that the profile length of 3 cm approximately meets the heat transfer rate of 177 W at a fin thickness of 0.314 mm, which satisfies the requirement of 175 W.

The results of the calculations for three sets of the fin sizes are summaries in Table E2.3.1. Optimum array I without constraint can take the maximum heat dissipation up to 1039 Watts. Obviously, Optimum array I has an impractical oversize and is not necessary. Both Optimum Array II and III with constraint satisfy the requirement of 175 W. Optimum array III is chosen as the recommended design.

Table E2.3.1   Summary of Results of Multiple-fin Array Design for Natural Convection

Parameter	Optimum array I without constraints	Optimum array II with constraints	Optimum array III with constraints
Fin thickness, t (mm)	8.343	0.424	0.314
Profile length, b (cm)	59.5	4	3
Spacing, z (mm)	7.5	7.5	7.5
Total heat transfer rate, q_{total} (W)	1039	223.7	177.3
Number of fins	32	63	64
Mass of fins (kg)	76.0	0.518	0.292
β	1.435	0.429	0.373
Fin effectiveness, ε_f	91.4	178	183
Overall efficiency, η_o	0.615	0.948	0.961
Overall thermal resistance, $R_{t,o}$(°C/W)	0.048	0.223	0.282

Base plate geometry: $W = 50$ cm and $L = 18$ cm, Aluminum properties: density $\rho = 2700$ kg/m^3, thermal conductivity $k = 196$ W/mK, $\theta_b = 50$ K

Note that the mass of the fins is substantially reduced as the fin thickness is reduced. Fin efficiency seems not important in the present analysis although the recommended criteria lies between 0.5 ~ 0.7 in Equation (2.23).

Example 2.4 Forced Convection Multiple Square Fin Array   A 20 × 20-mm square aluminum fin array is to be designed to maintain a base temperature below 80°C on a base pipe with a radius r_1 of 4 mm, as shown in Figure E2.4.1. The base pipe has a width $W = 40$ mm as shown in the figure. Forced air is induced by a fan at a velocity of $U = 2.3$ m/s and at an ambient temperature of 30°C. For the aluminum properties, use density of 2702 kg/m^3 and thermal conductivity of 177 W/mK. Estimate the number of fins, the fin thickness, the fin spacing, the heat dissipation, the overall efficiency, and the thermal resistance for the optimum design.

MathCAD Format Solution   Information given:

$$W_b := 40 \text{ mm} \qquad U_{air} := 2.3 \frac{m}{s}$$

$$T_{base} := 80°C \qquad T_{inf} := 30°C \qquad \theta_b := T_{base} - T_{inf}$$

Properties of air at a film temperature (30 + 80)/2 = 55°C = 328 K

$$k_{air} := 0.0292 \frac{W}{m\,K} \qquad \nu_{air} := 19.2 \ 10^{-6} \frac{m^2}{s} \qquad Pr := 0.707$$

Properties of aluminum:

$$\rho_{al} := 2702 \frac{kg}{m^3} \qquad k_{al} := 177 \frac{W}{m\,K}$$

Assumption   Treat the square fin as an equivalent radial fin as shown in Figure E2.4.2 and Figure E2.4.3.

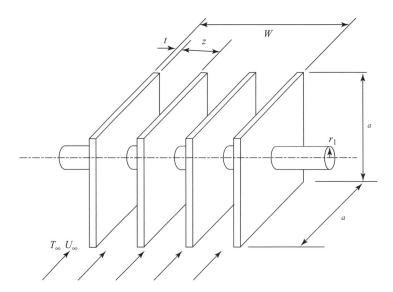

Figure E2.4.1 Square fin array

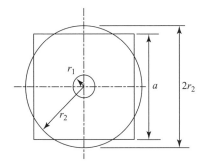

Figure E2.4.2 Square fin and radial fin

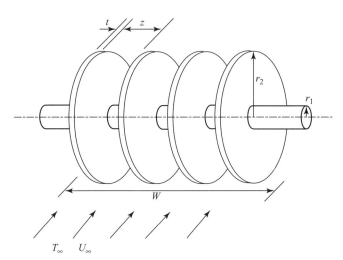

Figure E2.4.3 Radial fin array

Dimensions:

$$r_1 := 4 \text{ mm} \quad a := 20 \text{ mm} \quad \text{since} \quad a^2 = \pi\, r_2^2$$

$$r_2 := \frac{a}{\sqrt{\pi}} \quad r_2 = 0.011 \text{ m} \quad \text{Equivalent radius} \tag{E2.4.1}$$

For the optimum spacing,

$$L_b := a \quad L_b = 0.02 \text{ m} \tag{E2.4.2}$$

Reynolds number:

$$Re_L := \frac{U_{air}\, L_b}{\nu_{air}} \quad Re_L = 2.396 \times 10^3 \tag{E2.4.3}$$

which is less than 5×10^5, so it is laminar flow. The optimum spacing is given in Equation (2.110).

$$z_{opt} := L_b\, 3.24\, Re_L^{\frac{-1}{2}}\, Pr^{\frac{-1}{4}} \quad z_{opt} = 1.444 \text{ mm} \tag{E2.4.4}$$

Using Equation (1.53), the average heat transfer coefficient for laminar flow is

$$h_L := \frac{k_{air}}{L_b} 0.664\, Re_L^{\frac{1}{2}} Pr^{\frac{1}{3}} \quad h_L = 42.272 \frac{W}{m^2 K} \tag{E2.4.5}$$

The heat transfer coefficient is independent of the fin thickness. Using Equation (2.7a),

$$m1(t) := \sqrt{\frac{2\, h_L}{k_{al}\, t}} \tag{E2.4.6}$$

Define

$$mr1(t) := r_1 \sqrt{\frac{2\, h_L}{k_{al}\, t}} \quad mr2(t) := r_2 \sqrt{\frac{2\, h_L}{k_{al}\, t}} \tag{E2.4.7}$$

The radial fin efficiency is obtained in terms of Bessel functions using Equation (2.143).

$$\eta_f(t) := \frac{2\, r_1}{m1(t)\, \left(r_2^2 - r_1^2\right)} \frac{K1(mr1(t))I1(mr2(t)) - I1(mr1(t))K1(mr2(t))}{I0(mr1(t))K1(mr2(t)) + K0(mr1(t))I1(mr2(t))} \tag{E2.4.8}$$

The number of fins is also a function of t:

$$n_f(t) := \frac{W_b}{z_{opt} + t} \tag{E2.4.9}$$

The volume of a single fin is obtained as a function of t.

$$V_f(t) := \pi \left(r_2^2 - r_1^2\right) t \tag{E2.4.10}$$

The total mass of the material for the fin array is a function of t.

$$m_{al}(t) := n_f(t) \rho_{al} V_f(t) \tag{E2.4.11}$$

The single fin area is

$$A_f(t) := 2\pi \left(r_2^2 - r_1^2\right) + \pi\, 2r_2 t \tag{E2.4.12}$$

The total area including the interfin base area is

$$A_t(t) := n_f(t)(A_f(t) + \pi\, 2r_1\, z_{opt}) \tag{E2.4.13}$$

The overall efficiency is obtained using Equation (2.139).

$$\eta_o(t) := 1 - n_f(t)\frac{A_f(t)}{A_t(t)}(1 - \eta_f(t)) \tag{E2.4.14}$$

The total heat transfer rate is

$$q_{total}(t) := \eta_o(t)\, h_L\, A_t(t)\, \theta_b \tag{E2.4.15}$$

The thermal resistance is obtained using Equation (2.140).

$$R_{to}(t) := \frac{1}{\eta_o(t)\, h_L\, A_t(t)} \tag{E2.4.16}$$

A range for the graph of the fin thickness for Figures (E2.4.4) to (E2.4.7) is

$$t := 0.001\,\text{mm}, 0.002\,\text{mm}..\ 1\,\text{mm}$$

We find the maximum heat transfer rate at fin thickness for a range of 0.1-0.2 mm. We tabulate the parameters for three fin thickness: 0.1 mm, 0.15 mm, and 0.2 mm. And we summarize the results in Table E2.4.1.

$$q_{total}(0.1\,\text{mm}) = 35.928\,\text{W} \quad m_{al}(0.1\,\text{mm}) = 2.449\,\text{gm} \quad \eta_o(0.1\,\text{mm}) = 0.883$$

$$R_{to}(0.1\,\text{mm}) = 1.392\frac{\text{K}}{\text{W}}$$

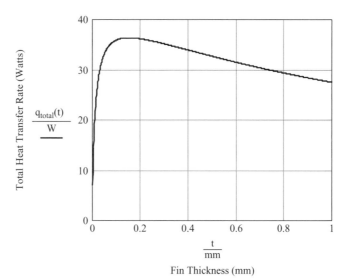

Figure E2.4.4   Total heat transfer rate versus fin thickness

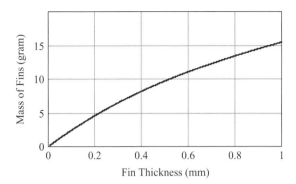

Figure E2.4.5   Mass of fins versus Fin thickness

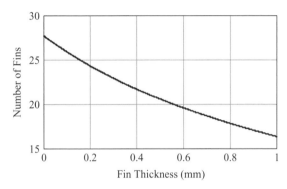

Figure E2.4.6   Number of fins versus fin thickness

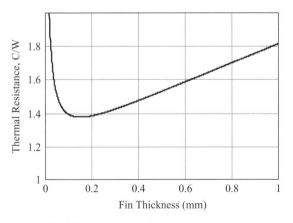

Figure E2.4.7   Thermal resistance versus fin thickness

Table E2.4.1   Summary of Results of a Square Fin Array Design

Parameter	Optimum array I	Optimum array II	Optimum array III
Fin thickness, t (mm)	0.1	0.15	0.2
Square Fin (mm)	20	20	20
Spacing, z (mm)	1.444	1.444	1.444
Heat transfer coefficient, h (W/m^{2}K)	42.27	42.27	42.27
Total heat transfer rate, q_{total} (W)	35.928	36.362	36.153
Number of fins	26	25	25
Mass of fins (g)	2.45	3.56	4.60
Overall efficiency, η_o	0.883	0.918	0.937
Overall thermal resistance, $R_{t,o}$(°C/W)	1.392	1.375	1.383

Array geometry: $W = 40$ mm, $a = 20$ mm, aluminum properties: density $= 2707$ kg/m^3, thermal conductivity $= 177$ W/mK, $T_{base} = 80°$C, $T_{inf} = 30°$C, and $U = 2.3$ m/s.

$$q_{total}(0.15\,\text{mm}) = 36.362\text{W} \quad m_{al}(0.15\,\text{mm}) = 3.558\,\text{gm} \quad \eta_o(0.15\,\text{mm}) = 0.918$$

$$R_{to}(0.15\,\text{mm}) = 1.375\frac{\text{K}}{\text{W}}$$

$$q_{total}(0.2\,\text{mm}) = 36.153\text{W} \quad m_{al}(0.2\,\text{mm}) = 4.599\,\text{gm} \quad \eta_o(0.2\,\text{mm}) = 0.937$$

$$R_{to}(0.2\,\text{mm}) = 1.383\frac{\text{K}}{\text{W}}$$

$$n_f(0.1\,\text{mm}) = 25.911 \quad n_f(0.15\,\text{mm}) = 25.098 \quad n_f(0.2\,\text{mm}) = 24.335$$

We select optimum array II as a best result for the given geometry and conditions. We find then that the heat transfer rate of about 36 Watts can be dissipated through the fin array, the fin thickness is 0.15 mm (surprisingly thin), and the number of fins is 25 within width $W = 40$ mm. Note that these results are based on the assumption that treats square fins as radial fins. However, the heat transfer area of the radial fins was designated to match the area of the square fin. If we perform the numerical computations for the exact square fin array previously obtained, the uncertainty between them may be verified. Keep in mind that the numerical computations do not readily provide the optimum design, such as the number of fins, fin thickness, and optimum spacing.

2.10   FIN DESIGN WITH THERMAL RADIATION

In this section, we study two specific problems in space, where radiation is only the heat transfer mechanism. One problem is a longitudinal fin with radiation from space, the Sun, and Earth. The other problem is an axial fin array on a cylindrical pipe in space at a distance away from the Sun and Earth. It is suggested that readers or students go over the introduction in section 1.4.4 (Radiation) before reading this section, particularly view factor.

2.10.1   Single Longitudinal Fin with Radiation

Consider that a longitudinal fin is exposed to an environment at a temperature of zero Kelvin in space (actually 3 K of dark matter), where no air for the convection heat transfer exists, as shown in Figure 2.12. Furthermore, radiation from the surroundings, such as the Sun, Earth, or a planet, is incident on the plate surfaces. The fluxes absorbed on the top and bottom sides are q''_{top} and q''_{bot}, respectively. This problem can be solved numerically. However, we wish to develop some analytical equations to optimize the fin design with optimum fin thickness and profile length that should maximize the heat dissipation. Take a small differential element in the plate and apply energy balance to it, as shown in Figure 2.12. In this analysis, the fin plate length L is 1 meter for simple calculations.

We assume that the base temperature is constant and that the material properties do not change with temperature. The environment in space is at zero Kelvin, so the irradiation from space is zero. The radiation fluxes from the Sun and Earth are treated as the boundary conditions.

Applying heat balance to the control volume yields

$$q_x - \left(q_x + \frac{dq_x}{dx} dx \right) - dq_{rad} + \left(q''_{top} + q''_{bot} \right) dx = 0 \qquad (2.147)$$

where q_x is a conduction heat transfer,

$$q_x = -kt \frac{dT}{dx} \qquad (2.148)$$

and q_{rad} is the emissive power (radiation) from both sides of the plate surface

$$dq_{rad} = 2dx\varepsilon\sigma T^4 \qquad (2.149)$$

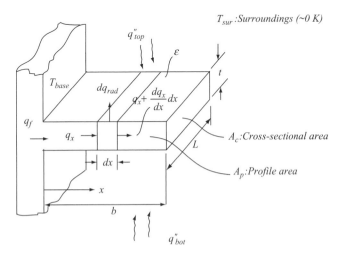

Figure 2.12   Single longitudinal fin with the irradiation by surroundings and the fluxes at the top and bottom surfaces

Equation (2.147) reduces to

$$-kt\frac{d^2T}{dx^2} + 2\sigma\varepsilon T^4 = q''_{top} + q''_{bot} \tag{2.150}$$

Equation (2.150) is a nonlinear differential equation, which are usually difficult to solve and for which the solution would vary depending on the boundary conditions. Numerical methods using computational software may be applied, but optimization usually requires a great amount of work and time. We here try to obtain an optimum design with minimal fin weight or volume for a given heat load using an analytical approach.

Boundary conditions for the thin plate are $T = T_{base}$ specified at $x = 0$ and, from an assumption of adiabatic tip, $dT/dx = 0$ at $x = b$. To find $T(x)$, multiply Equation (2.150) by dT/dx and integrate from x to b.

$$\int_x^b \left(-kt\frac{d^2T}{dx^2} + 2\sigma\varepsilon T^4\right)\frac{dT}{dx}dx = \int_x^b \left(q''_{top} + q''_{bot}\right)\frac{dT}{dx}dx \tag{2.151}$$

Let $dv = \frac{d^2T}{dx^2}dx$, then $v = \frac{dT}{dx}$ after integration. Note that $v = 0$ at $x = b$ and $v = v$ at $x = x$. From the first term, we conduct a special integration by substitution.

$$\int_x^b \left(\frac{dT}{dx}\right)\left(\frac{d^2T}{dx^2}\right)dx = \int_v^0 v\,dv = -\frac{1}{2}\left(\frac{dT}{dx}\right)^2 \tag{2.152}$$

Finally, Equation (2.151) leads to

$$-\frac{kt}{2}\left(\frac{dT}{dx}\right)^2 + \frac{2}{5}\varepsilon\sigma\left[T^5 - T(b)^5\right] = (q''_t + q''_t)[T - T(b)] \tag{2.153}$$

where $T = T(b)$ at $x = b$. Solving for dT/dx yields

$$\frac{dT}{dx} = -\left(\frac{4\varepsilon\alpha}{5kt}\right)^{1/2}\left\{T^5 - T(b)^5 - \frac{5}{2\varepsilon\sigma}\left(q''_{top} + q''_{bot}\right)[T - T(b)]\right\}^{1/2} \tag{2.154}$$

The minus sign was chosen for the square root because $T(x)$ must be decreasing with x.

The heat transfer rate at the entrance of the radiating fin using the Fourier's law is,

$$q_f = -kt\left(\frac{dT}{dx}\right)_{x=0} \tag{2.155}$$

Using the temperature gradient dT/dx at the entrance of the plate $x = 0$ with Equation (2.154), where $T = T_{base}$, we obtain

$$q_f = kt\left(\frac{4\varepsilon\alpha}{5kt}\right)^{1/2}\left\{T_{base}^5 - T(b)^5 - \frac{5}{2\varepsilon\sigma}\left(q''_{top} + q''_{bot}\right)[T_{base} - T(b)]\right\}^{1/2} \tag{2.156}$$

Equation (2.156) contains four unknowns, t, b, q_f and $T(b)$.

From Equation (2.154), separate variables and integrate again to obtain

$$x = \left(\frac{5kt}{4\varepsilon\sigma}\right)^{1/2} \int_T^{T_{base}} \frac{dT}{\left\{T^5 - T(b)^5 - (5/2\varepsilon\sigma)\left(q''_{top} + q''_{bot}\right)(T - T(b))\right\}^{1/2}} \qquad (2.157)$$

which satisfies $T = T_{base}$ at $x = 0$ and $T = T(b)$ at $x = b$. For $x = b$, we have

$$b = \left(\frac{5kt}{4\varepsilon\sigma}\right)^{1/2} \int_{T(b)}^{T_{base}} \frac{dT}{\left\{T^5 - T(b)^5 - (5/2\varepsilon\sigma)\left(q''_{top} + q''_{bot}\right)(T - T(b))\right\}^{1/2}} \qquad (2.158)$$

Equation (2.158) also contains four unknowns, t, b, q_f and $T(b)$. Therefore, we have Equations (2.156) and (2.158) with four unknowns. There are obviously an infinite number of the solutions. An idea is to find a minimal profile area (or volume of fin) for the optimum design. See Example 2.5 for the solution. The temperature distribution with the obtained optimum thickness t and profile length b is then found by evaluating the integral in Equation (2.157) numerically to find x for various T values (in the lower limit of the integral) between T_{base} and $T(b)$. Equations (2.157) and (2.158) were originally addressed in [9].

Example 2.5 Single Rectangular Fin with Radiation A single rectangular fin in a radiator is used to dissipate energy in orbit as shown in Figure E2.5.1. Both sides of the fin are exposed to an environment at $T_e \approx 0\,K$ and it also receives radiation from the sun at top and the earth at bottom, $q''_{top} = 80\,\text{W}/\text{m}^2$ and $q''_{bottom} = 30\,\text{W}/\text{m}^2$, respectively. The fin is diffuse-gray with emissivity $\varepsilon = 0.8$ on both sides and has a constant thermal conductivity of 28.5 W/mK. The analysis can be done for a width of $L = 1$ m. Design the fin for the optimum (minimum) volume by providing the fin thicknesses, profile lengths, volumes, temperatures at the tip, and efficiencies for

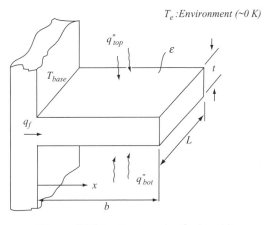

Figure E2.5.1 Rectangular fin in orbit

three heat dissipations per length: 150 W/m, 200 W/m, and 250 W/m. Also provide a temperature distribution along the fin for the median dissipation of 200 W/m.

Design Concept: The formulation of the governing equation ends up with a non-linear differential equation, Equation (2.150), which is difficult to solve analytically. Numerical solutions using computational software are always available these days, but optimization requires a great amount of work and time.

Recall Equations (2.156) and (2.158) with four unknowns, t, b, q_f, and $T(b)$. We denote $T(b)$ in Equation (2.156) as $T_b(t, q_f)$, because the unknown $T(b)$ is a function of both t and q_f. In the same way, $b(t, q_f)$ in Equation (2.158) is eventually a function of both t and q_f. MathCAD allow us to functionally solve $T_b(t, q_f)$ and $b(t, q_f)$ with the two equations, where t and q_f are independent variables.

MathCAD Format Solution: We assume that the base temperature is constant and that the material properties do not change with temperature. The environment in orbit is at zero Kelvin, so the irradiation from the space is zero. The radiation fluxes from the Sun and Earth are treated as the boundary conditions.

Properties and information given:

$$T_{base} := 400\,K \quad k := 28.5\frac{W}{m\,K} \quad \varepsilon := 0.8 \quad \sigma := 5.67\ 10^{-8}\frac{W}{m^2 K^4}$$

$$q_t := 80\frac{W}{m^2} \quad q_b := 30\frac{W}{m^2}$$

Initial guesses are required for a "Find" function, which is a built-in function, where "Given" is also a required command function in connection with the Find function.

Values for the initial guess:

$$T_b := 300\,K \quad b := 14\,cm$$

A Given function is written as a command. Using Equations (2.156) and (2.158), we obtain

Given

$$q_f = (k\,t)\left(\frac{4\,\varepsilon\,\sigma}{5\,k\,t}\right)^{\frac{1}{2}}\left[T_{base}^5 - T_b^5 - \frac{5}{2\,\varepsilon\,\sigma}(q_t + q_b)(T_{base} - T_b)\right]^{\frac{1}{2}} \tag{E2.5.1}$$

$$b = \left(\frac{5\,k\,t}{4\,\varepsilon\,\sigma}\right)^{\frac{1}{2}}\int_{T_b}^{T_{base}}\left[T^5 - T_b^5 - \frac{5}{2\,\varepsilon\,\sigma}(q_t + q_b)(T - T_b)\right]^{\frac{-1}{2}}dT \tag{E2.5.2}$$

Obtain the functional solution as:

$$\binom{T_b(t, q_f)}{b(t, q_f)} := \text{Find}(T_b, b) \tag{E2.5.3}$$

We express the profile area, which is a product of the fin thickness and profile length.

$$A_p(t, q_f) := t\,b(t, q_f) \tag{E2.5.4}$$

We define three required heat loads per unit length as:

$$q_{f1} := 150\frac{W}{m} \quad q_{f2} := 200\frac{W}{m} \quad q_{f3} := 250\frac{W}{m} \tag{E2.5.5}$$

We try to plot the profile areas along the fin thicknesses to find the minimal area.

$$t := 1\,mm, 1.1\,mm.. \ 12\,mm$$

The profile area curves in Figure E2.5.2 show very steep slopes near the minimum values, which indicate the existence of singularities (values explode). The region beyond the singularity is of no practical interest. However, after several graphical explorations due to the nonlinearity of the governing equation, we can eventually figure out the optimum fin thicknesses with the minimal profile areas for the three heat loads: 150 W/m, 200 W/m, and 250 W/m, which are approximately 3 mm, 6 mm, and 9 mm, respectively. Recall that each curve represents a designated heat load. Finding the optimum thicknesses leads to finding the corresponding optimum profile lengths and fin temperatures at the tips. The profile length is plotted in Figure E2.5.3 for curiosity. The profile length appears to be not changing much in the regions after the optimum fin thicknesses.

Now we try to plot the temperature distribution specific for a heat load of 200 W/m. The specific optimum thickness for the load is approximately:

$$t := 6\,mm \tag{E2.5.6}$$

In Equation (2.157), x is obviously a function of the fin temperature. However, it can be thought of inversely. We try to find the fin temperature as a function of x and q_f using the 'root' function, which is also a built-in function in MathCAD.

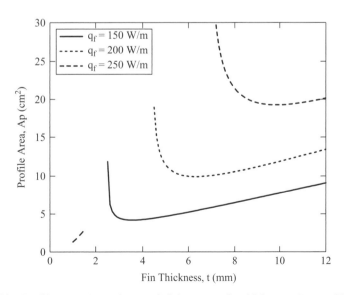

Figure E2.5.2 Profile area (or volume of fin) versus fin thickness along with three heat dissipations

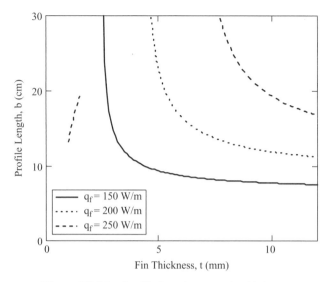

Figure E2.5.3 Profile length versus fin thickness

Initial guess for the following "root" function appears very sensitive to the convergence of the solution. (Usually any numerical value would be accepted as a guess.) An attempt was made to determine the closest value for the solution, which was found to be the fin temperature at the tip T_b.

Initial guess:

$$\underset{\sim}{T} := T_b(t, q_{f2}) \tag{E2.5.7}$$

$$T(x, q_f) := \text{root}\left[\left(\frac{5kt}{4\varepsilon\sigma}\right)^{\frac{1}{2}} \int_T^{T_{base}} \left[T^5 - T_b(t, q_f)^5\right.\right.$$

$$\left.\left. - \frac{5}{2\varepsilon\sigma}(q_t + q_b)(T - T_b(t, q_f))\right]^{\frac{-1}{2}} dT - x, T\right] \tag{E2.5.8}$$

We plotted the solution (the temperature distribution along the profile length) in Figure E2.5.4, which took several minutes on a typical PC. The fin temperature monotonically decreases as expected. Note that the fin temperature at the tip is approximately 320K for the given load of 200 W/m.

$$x := 0.1\,\text{cm}, 1.4\,\text{cm}.. 25\,\text{cm}$$

The corresponding values with the optimum fin thicknesses are sought for the summary results. Define the environment temperature, T_e, for readers and the three optimum fin thicknesses:

$$T_e := 0\text{K} \tag{E2.5.9}$$

$$t_1 := 3\,\text{mm} \quad t_2 := 6\,\text{mm} \quad t_3 := 9\,\text{mm} \tag{E2.5.10}$$

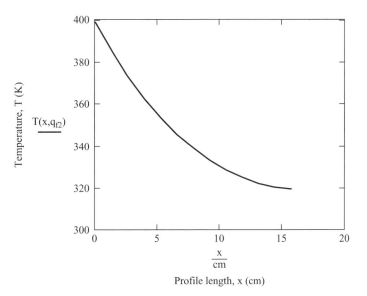

Figure E2.5.4   Temperature distribution for $q_f = 200\,\mathrm{W/m}$

Fin effectiveness is the ratio of the active fin heat transfer to the heat transfer without the fin:

$$\varepsilon_f(t, q_f) := \frac{q_f}{t\,\varepsilon\,\sigma\,\left(T_{base}^4 - T_e^4\right)} \tag{E2.5.11}$$

$$\varepsilon_f(t_1, q_{f1}) = 43.058 \quad \varepsilon_f(t_2, q_{f2}) = 28.706 \quad \varepsilon_f(t_3, q_{f3}) = 23.921 \tag{E2.5.12}$$

These values of effectiveness indicate that adding the fins increases the heat loads. Fin efficiency is the ratio of the active fin heat transfer to the maximum possible heat transfer, which occurs when the fin temperature is constant at the base temperature:

$$\eta_f(t, q_f) := \frac{q_f}{2(b(t, q_f) + t)\,\varepsilon\,\sigma\,(T_{base}^4 - T_e^4)} \tag{E2.5.13}$$

$$\eta_f(t_1, q_{f1}) = 0.43 \quad \eta_f(t_2, q_{f2}) = 0.504 \quad \eta_f(t_3, q_{f3}) = 0.476 \tag{E2.5.14}$$

Now we calculate the optimum values with the optimum fin thicknesses.

$$A_p(t_1, q_{f1}) = 4.413\,\mathrm{cm}^2 \quad A_p(t_2, q_{f2}) = 9.893\,\mathrm{cm}^2 \quad A_p(t_3, q_{f3}) = 19.523\,\mathrm{cm}^2$$

$$b(t_1, q_{f1}) = 14.711\,\mathrm{cm} \quad b(t_2, q_{f2}) = 16.488\,\mathrm{cm} \quad b(t_3, q_{f3}) = 21.692\,\mathrm{cm}$$

$$T_b(t_1, q_{f1}) = 298.585\,\mathrm{K} \quad T_b(t_2, q_{f2}) = 318.988\,\mathrm{K} \quad T_b(t_3, q_{f3}) = 312.849\,\mathrm{K}$$
$$\tag{E2.5.15}$$

Properties and information given:

$$k = 28.5\frac{\mathrm{W}}{\mathrm{m\,K}} \quad \varepsilon = 0.8 \quad T_{base} = 400\,\mathrm{K} \quad q_t = 80\frac{\mathrm{W}}{\mathrm{m}^2} \quad q_b = 30\frac{\mathrm{W}}{\mathrm{m}^2}$$

Table E2.5.1   Summary of Optimum Design of the Rectangular
Fin for the Minimal volume of the Fin

	Optimum I	Optimum II	Optimum III
q_f (W/m)	150	200	250
t (mm)	3	6	9
b (cm)	14.7	16.5	21.7
A_p (cm^2)	4.4	9.9	19.5
T_b(K)	298.6	319.0	312.8
ε_f	43.0	28.7	23.9
η_f	0.43	0.504	0.476

A summary of the optimum design of the rectangular fin is tabulated in
Table E2.5.1. This would suggest that the mathematical methods in this problem save
enormous work and time compared to the numerical methods. Note that this problem
was optimized for the minimum weight of fins, not for the maximal fin efficiency.

Example 2.6 Axial Fin Array on a Cylindrical Pipe   A long axial fin array on a
cylindrical pipe is designed to dissipate energy from the pipe, in which a number of
radioisotope thermoelectric generators (RTG) produce electricity for the power of a
spacecraft (Figure E2.6.1). It is widely known that adding fins on a plane does not
increase the radiative dissipation, which means that adding fins on a plane provides
additional weight and complexity with no gain in the heat load.

A question arises as to whether adding fins on a *cylindrical pipe* will provide
additional heat load. If it provides additional heat load, what is the optimal number
of fins? We wish to study this problem mathematically. The radiative dissipation as

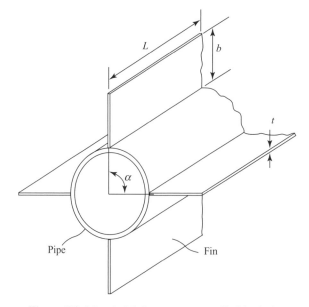

Figure E2.6.1   Axial fin array on a cylindrical pipe

a heat load by the 30-cm diameter pipe with fins, if justified, should be 1.2 kW per meter, at least. The base temperature in the cylindrical pipe is maintained at 400 K, while the fin temperatures are assumed to be constant at the three different temperatures of 280 K, 290 K, and 300 K. The axial fin array is diffuse-gray with $\varepsilon = 0.8$ and exposed to an environment at $T_e = 3$ K. Design the fin array for its minimal weight.

Design Concept   The exact solution for this problem seems complex, requiring significant numerical computations since the fin temperature will decrease along the fin. However, we attempt here to approach the problem mathematically with an *assumption* that the surface temperatures of the fins are constant and diffuse-gray, which greatly simplifies the problem. The concept of view factor accounts for the radiation exchange between the surfaces of the fins and the pipe. The radiation network allows us to easily link the surfaces. One of the difficulties in this problem is to find the optimal angle between the fins, which determines the optimal number of fins. The author developed the view factors in the circular sector with a variable angle for the present purpose. They are listed in item No. 6 in Table 1.2. A circular sector with a variable angle is shown in Figure E2.6.2, where the system may be viewed as a three-surface enclosure, with the third surface being the cold environment denoted as a dotted line of surface area A_3. Surfaces A_1 and A_2 of the cylinder pipe are real surfaces, while surface A_3 of the environment is considered as a blackbody according to the Kirchhoff's law.

The radiation network is constructed by first identifying nodes associated with the radiosities of each surface, as shown in Figure E2.6.3. The relation between a real surface and radiosity is given in Equation (1.108) and the relation between radiosities is given in Equation (1.109). Since the environment is regarded as a blackbody in an enclosure, we have $J_3 = E_{b3}$ with $\varepsilon = 1$. Two more relations are obtained using Equation (1.111). Three equations with three unknowns provide the solutions for J_1, J_2, and J_3, leading to q_1 and q_2 and $q_3 = q_1 + q_2$. From the radiation network in Figure E2.6.3, we need to obtain the three view factors: F_{12}, F_{13}, and F_{23}.

Viewing from surface A_1 in Figure E2.6.2 and using the summation rule of Equation (1.98), we have

$$F_{12} + F_{13} = 1 \quad \text{or} \quad F_{12} = 1 - F_{13} \qquad (E2.6.1)$$

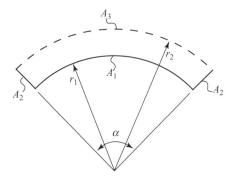

Figure E2.6.2   Circular sector

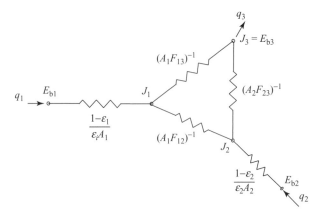

Figure E2.6.3   Radiation network for a circular sector

Note that surface 1 in this case cannot see itself. Using the reciprocity relation of Equation (1.97) between surfaces A_1 and A_2, we have

$$A_1 F_{12} = A_2 F_{21} \quad \text{or} \quad F_{21} = \frac{A_1}{A_2} F_{12} = \frac{A_1}{A_2} (1 - F_{13}) \qquad \text{(E2.6.2)}$$

Viewing from surface 2 in Figure E2.6.2 and using the summation rule, we have

$$F_{21} + F_{22} + F_{23} = 1 \quad \text{or} \quad F_{23} = 1 - F_{21} - F_{22} \qquad \text{(E2.6.3)}$$

where F_{22} means that surface A_2 sees its own surface A_2. But this happens only when the angle is small enough so that the curvature of the pipe between the fins no longer hides the two pieces of surface A_2 from one another.

Substituting Equation (E2.6.2) into (E2.6.3) yields

$$F_{23} = 1 - \frac{A_1}{A_2} (1 - F_{13}) - F_{22} \qquad \text{(E2.6.4)}$$

From this equation, we need to find two view factors: F_{13} and F_{22}, which are found in Table 1.2.

MathCAD Format Solution   Properties and information given:

$$\sigma := 5.67 \ 10^{-8} \frac{\text{W}}{\text{m}^2\text{K}^4} \qquad \varepsilon_1 := 0.8 \qquad \varepsilon_2 := 0.8 \qquad \text{(E2.6.5)}$$

$$T_1 := 400 \text{ K} \qquad T_2 := 270 \text{ K} \qquad T_e := 3 \text{ K} \qquad \text{(E2.6.6)}$$

For a single fin design, we borrow the optimum design data for the minimum volume from Example 2.5. We assume that the heat load of each fin is 200 W/m, which can produce 1.2 kW of the present requirement with six fins. The design data showed that the minimal volume of the fin was obtained with the fin thickness of

6 mm and the profile length of 16.5 mm. Considering the geometry in Figure E2.6.2, we have

$$b := 16.5\,\text{cm} \quad t := 6\,\text{mm} \quad \underset{\sim}{L} := 1\,\text{m} \tag{E2.6.7}$$

$$r_1 := 15\,\text{cm} \quad r_2 := r_1 + b \tag{E2.6.8}$$

Define the circular angle in terms of the number of fins (n).

$$\alpha(n) := \frac{2\,\pi}{n} \tag{E2.6.9}$$

Define the blackbody radiation for surfaces A_1 and A_2 (see Equation (1.85)).

$$E_{b1} := \sigma\,T_1^4 \quad E_{b2}(T_2) := \sigma\,T_2^4 \tag{E2.6.10}$$

The areas of each surface are defined.

$$A_1(n) := \frac{2\,\pi\,r_1\,L}{n} \quad A_2 := 2\,b\,L \quad A_3(n) := \frac{2\,\pi\,r_2\,L}{n} \tag{E2.6.11}$$

The view factors according to Table 1.2 are

$$F_{13}(n) := 1 - \frac{\pi}{2\,\alpha(n)} + \frac{r_2}{\alpha(n)\,r_1}\left[1 - \left(\frac{r_1}{r_2}\right)^2\right]^{0.5} + \frac{1}{\alpha(n)}\,\text{asin}\left(\frac{r_1}{r_2}\right) - \frac{(r_2 - r_1)}{\alpha(n)\,r_1}$$

$$\tag{E2.6.12}$$

$$F_{13}(n) := \left| \begin{array}{l} F_{13}(n) \quad \text{if} \quad \alpha(n) > \text{acos}\left(\dfrac{r_1}{r_2}\right) \\[2mm] \dfrac{\sqrt{r_2^2 + r_1^2 - 2\,r_2\,r_1\,\cos(\alpha(n))} - (r_2 - r_1)}{\alpha(n)r_1} \quad \text{otherwise} \end{array} \right. \tag{E2.6.13}$$

$$F_{22}(n) := \frac{2\sqrt{r_2^2 - r_1^2} + 2\,r_1\left(\alpha(n) - \dfrac{\pi}{2} + \text{asin}\left(\dfrac{r_1}{r_2}\right)\right) - \alpha(n)\,(r_2 + r_1)}{4\,(r_2 - r_1)} \tag{E2.6.14}$$

$$F_{22}(n) := \left| \begin{array}{l} F_{22}(n) \quad \text{if} \quad \alpha(n) > \text{acos}\left(\dfrac{r_1}{r_2}\right) \\[2mm] \dfrac{2\sqrt{r_2^2 + r_1^2 - 2\,r_2\,r_1\,\cos(\alpha(n))} - \alpha(n)(r_2 + r_1)}{4(r_2 - r_1)} \quad \text{otherwise} \end{array} \right. \tag{E2.6.15}$$

$$F_{22}(n) := \left| \begin{array}{l} F_{22}(n) \quad \text{if} \quad F_{22}(n) > 0 \\ 0 \ \text{otherwise} \end{array} \right. \tag{E2.6.16}$$

$$F_{12}(n) := 1 - F_{13}(n) \tag{E2.6.17}$$

$$F_{23}(n) := 1 - \frac{A_1(n)}{A_2}(1 - F_{13}(n)) - F_{22}(n) \tag{E2.6.18}$$

Using Equation (1.111), we construct three equations from the radiation network of Figure E2.6.3 for J_1, J_2, and J_3.

Initial guesses are as follows:

$$J_1 := 100\,\frac{W}{m^2} \quad J_2 := 100\,\frac{W}{m^2} \quad J_3 := 100\,\frac{W}{m^2} \tag{E2.6.19}$$

Given

$$\frac{E_{b1} - J_1}{\left(\dfrac{1 - \varepsilon_1}{\varepsilon_1\,A_1(n)}\right)} = \frac{J_1 - J_3}{(A_1(n)\,F_{13}(n))^{-1}} + \frac{J_1 - J_2}{(A_1(n)\,F_{12}(n))^{-1}} \tag{E2.6.20}$$

$$\frac{E_{b2}(T_2) - J_2}{\left(\dfrac{1 - \varepsilon_2}{\varepsilon_2\,A_2}\right)} = \frac{J_2 - J_1}{(A_1(n)\,F_{12}(n))^{-1}} + \frac{J_2 - J_3}{(A_2\,F_{23}(n))^{-1}} \tag{E2.6.21}$$

$$J_3 = \sigma\,T_e^4 \tag{E2.6.22}$$

Solve for J_1, J_2, and J_3 as functions of n and T_2.

$$\begin{pmatrix} J_1(n, T_2) \\ J_2(n, T_2) \\ J_3(n, T_2) \end{pmatrix} := \mathrm{Find}(J_1, J_2, J_3) \tag{E2.6.23}$$

Define the total heat dissipation from both the cylinder pipe and two fins in the circular sector.

$$q_{total}(n, T_2) := n\left[\frac{E_{b1} - J_1(n, T_2)}{\dfrac{1 - \varepsilon_1}{\varepsilon_1\,A_1(n)}} + \frac{E_{b2}(T_2) - J_2(n, T_2)}{\left(\dfrac{1 - \varepsilon_2}{\varepsilon_2\,A_2}\right)}\right] \tag{E2.6.24}$$

First of all, we want to see how the view factors vary with the number of fins (or the circular angle).

$$n := 1, 1.1.. \ 10$$

Recall that surface $A_1 =$ cylinder pipe, surface $A_2 =$ two fins each side, and surface $A_3 =$ blackbody environment. It is interesting to note that view factor F_{22} starts increasing at $n = 4.5$, because surface A_2 does not see its other side until the number of fins reaches 4.5. At this value, F_{23} starts decreasing, the total heat dissipation drops, as seen in Figure E2.6.4. Practically the number of fins must be integer, not fraction.

Now we inspect the total heat dissipation for three different fin surface temperatures of surface A_2 as:

$$T_{2a} := 280\,K \quad T_{2b} := 290\,K \quad T_{2c} := 300\,K \tag{E2.6.25}$$

Obviously, we see the effect of the increase of F_{22} on the total heat dissipation of Figure E2.6.5, which becomes significant as lowering the fin temperatures of surface A_2. When the base temperature of the cylinder pipe is at 400 K, if the fin temperature is lower than 300 K, there is not much gain in the heat load with increasing the number of fins.

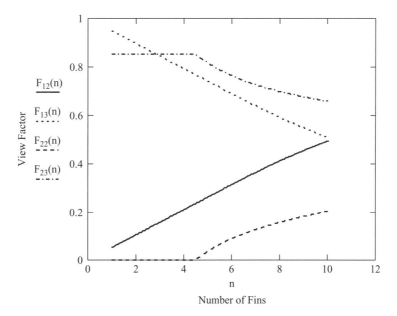

Figure E2.6.4 View factors versus number of fins

The optimum number of fins seems dependent on the average fin temperature. If the fin temperature is not high enough, the number of fins is restricted by five (5). If the fin temperature is designed high enough, the number of fins may be increased by more than five (5). The heat dissipations for all the three fin temperatures actually satisfy the least design requirement of 1.2 kW.

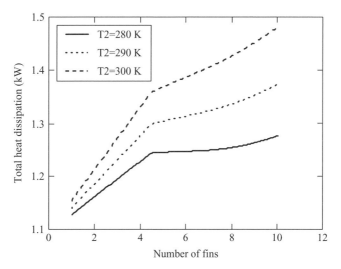

Figure E2.6.5 Total heat dissipation versus number of fins along with three average fin temperatures

REFERENCES

1. F.P. Incropera, D.P. Dewitt, T.L. Bergman, and A.S. Lavine, *Fundamentals of Heat and Mass Transfer* (6th ed.). Hoboken, NJ: John Wiley & Sons, 2007.

2. A.D. Klaus and A. Bar-Cohen, *Design and Analysis of Heat Sinks*. New York: John Wiley and Sons, 1995.

3. A. Bejan, G. Tsatsaronis, and M. Moran, *Thermal Design and Optimization*. New York: John Wiley and Sons, 1996.

4. F.P. Incropera, *Liquid Cooling of Electronic Devices by Single-Phase Convection*. New York: John Wiley and Sons, 1999.

5. W. Elenbaas, Heat Distribution of Parallel Plates by Free Convection, *Physica*, 9 (1), 665-671.

6. E.J. LeFevere, Laminar Free Convection from a Vertical Plane Surface, *Proc. 9th int. Congr. Applied Mech.*, vol. 4. (Brussels, 1956), pp. 168–174.

7. A. Bar-Cohen and W.M. Rohsenow, Thermally Optimum Spacing of Vertical, Natural Convection Cooled, Parallel Plates, *J. Heat Trans.*, vol. 106 (1984), 116–122.

8. E.N. Sieder, and G.E. Tite, Heat Transfer and Pressure Drop of Liquids in Tubes, *Ind. Eng. Chem.*, 28, 1429, 1936.

9. S. Siegrl and J. Howell, *Thermal Radiation Heat Transfer* (4th ed.). New York: Taylor & Francis, 2002.

10. L.T. Yeh and R.C. Chu, *Thermal Management of Microelectronic Equipment*. New York: ASME Press, 2002.

11. U. Grigull and W. Hauf, Natural Convection in Horizontal Cylindrical Annuli, *Proc. 3rd Int. Heat Transfer Conf.* (1966), vol. 2, pp. 182–195.

PROBLEMS

2.1 An aluminum rectangular fin with the base at $100°C$ is cooled by an air stream at $30°C$. The estimated heat transfer coefficient is 8.5 W/m^{2}K. The thermal conductivity of the fin is 166 W/m K. The dimensions of the fin shown in Figure P2.1 are L = 12 cm, b = 2.5 cm, and t = 1 mm. Determine the fin efficiency and the total heat transfer rate. Repeat the problem with a stainless steel fin where the thermal conductivity is 14.2 W/mK.

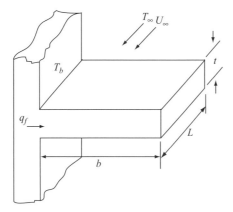

Figure P2.1 Rectangular fin

2.2 An aluminum rectangular fin of width 30 cm is cooled by forced air at 20°C with a velocity of 1.75 m/s, as shown in Figure P2.1. The fin base temperature is at 100°C. The heat dissipation to the ambient air is required to be 50 W. For aluminum properties, use the density of 2702 kg/m³ and the thermal conductivity of 175 W/mK.

  (a) Determine the fin thickness and profile length to maximize the heat transfer rate.

  (b) Determine the fin efficiency and fin effectiveness.

  (c) Compute the heat transfer rate using Equation (2.14). Does the result in part (c) satisfy the 50 W requirement?

  (d) Compute the thermal resistance (°C/W) of the fin.

2.3 An electronic package cooling is designed to maintain the base temperature below 80°C. The base plate has a width $W = 30$ mm and Length $L = 20$ mm, as shown in Figure P2.3. Forced air is induced by a fan at a velocity of $U = 1.7$ m/s and ambient air temperature of 30°C. The device requires both a mass of fins less than 15 grams and a profile length less than 20 mm. Non-optimum design is allowed to accommodate the requirements. For aluminum properties, use a density of 2,702 kg/m³ and a thermal conductivity of 177 W/mK. Estimate the number of fins, the fin thickness, and the profile length maximize the heat transfer rate.

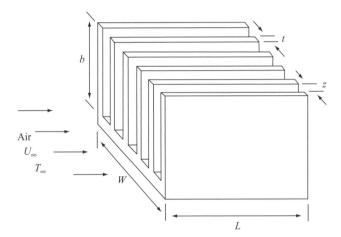

Figure P2.3   Forced convection multiple fin array

2.4 A heat sink in an application of thermoelectric cooling is designed to maintain the base cooling plate below 75°C. The base plate has a width $W = 9$ cm and Length $L = 17$ cm, as shown in Figure P2.4. Forced air is induced by a fan at a velocity of $U = 1.5$ m/s and ambient air temperature at 25°C. The device requires that the profile length be limited to 35 mm at most and that the heat dissipation be at least 90 W. A non-optimum design is allowed to accommodate the requirements. For aluminum properties, use a density of 2,700 kg/m³ and a thermal conductivity of 177 W/mK.

  (a) First, develop a MathCAD program for the optimum characteristics with the graphs of the maximum optimum heat transfer rate, the optimum mass of fins, the optimum profile length, and the number of fins.

  (b) Design the heat sink showing the optimum spacing, the number of fins, the fin thickness, the profile length, the mass of fins, the fin efficiency and fin effectiveness.

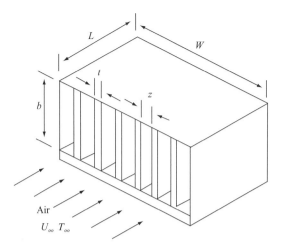

Figure P2.4 Fin array for forced convection flow 90w

2.5 The rear wall of a box containing electronic components is cooled by natural convection using a multiple fins array, as shown in Figure P2.5. The box geometry gives that $W = 15$ cm and $L = 12$ cm. The rear wall must passively dissipate 180 W to an environment at 25°C at atmospheric pressure without exceeding a surface temperature of 85°C. ~~The fin height (profile length) b is limited to 19 mm.~~ First, show whether the heat dissipation can be achieved by natural convection without fins. If necessary, design the fin array to accommodate the requirements. Estimate the fin thickness, the profile length, the spacing, the number of fins, the overall efficiency, and the thermal resistance to maximize the heat transfer rate from the box. For the properties of aluminum, the density is 2,700 kg/m^3 and the thermal conductivity is 196 W/mK.

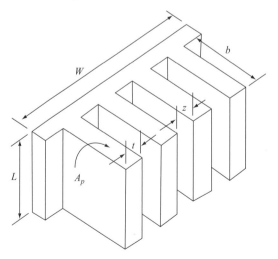

Figure P2.5 Fin array for natural convection

2.6 A radial aluminum fin array, with overall radius $r_2 = 50$ mm, is designed to maintain a temperature below 65°C on a base pipe with a radius r_1 of 10 mm, as shown in Figure P2.6. The base pipe has width $W = 180$ mm. Forced air is induced by a fan at a velocity of

$U = 1.9$ m/s and an ambient temperature of $25°C$. For the aluminum properties, use the density of $2,694$ kg/m^3 and the thermal conductivity of 182 W/mK. Estimate the number of fins, the fin thickness, the fin spacing, the heat dissipation, the overall efficiency, and the thermal resistance for the optimum design.

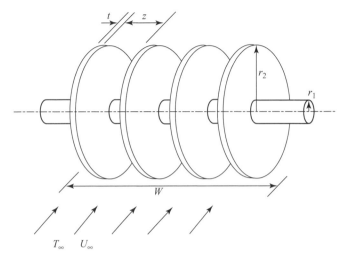

Figure P2.6   Radial fins array

2.7   A thin plate of thickness t and length $2b$ is between two tubes in a radiator used to dissipate energy in orbit as shown in Figure P2.7. The dimension is long in the direction normal to the cross section shown. Both sides of the plate have the same emissivity and radiate to the environment at $T \approx 0$ K. Radiation from the surroundings, such as from the sun, earth, or a planet, is incident on the plate surfaces and the fluxes absorbed on the top and bottom sides are $q''_{top} = 190$ W/m^2 and $q''_{bot} = 50$ W/m^2. The plate is diffuse-gray with emissivity of $\varepsilon = 0.8$ on both sides, and has constant thermal conductivity $k = 45.6$ W/mK. The base tube temperature T_{tube} is maintained at 400 K. Design the plate for a heat load of 200 W/m.

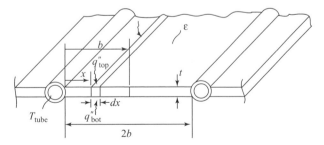

Figure P2.7   Space radiator

2.8   A long axial fin array on a cylinder pipe is designed to dissipate energy from the pipe, in which a number of radioisotope thermoelectric generators (RTG) produce electricity for the power of a spacecraft (Figure P2.8). The radiative dissipation as a heat load by the 60-cm diameter pipe with fins, if justified, should be 2.2 kW per meter, at least. The axial fins array is diffuse-gray with $\varepsilon = 0.8$ and exposed to an environment at $T_e = 4$ K. The base temperature in the cylindrical pipe is maintained at 400 K, while the fin temperatures are expected to be at 320 K. Design the fin array for its minimal weight.

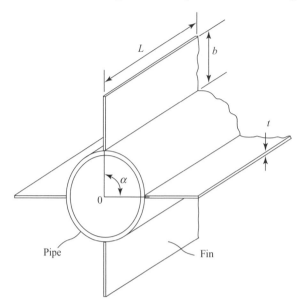

Figure P2.8   Axial fin array on a cylindrical pipe

Computer Assignments

2.9   Consider a concentric cylinder where air is at atmospheric pressure. Compute and provide the isotherms and streamlines in concentric cylinders, when the outer cylinder of 21-cm diameter is at 300 K and the inner cylinder of 7-cm diameter is at 315 K. The streamlines and isotherms are photographed as shown in Figures P2.9 (a) and (b). You may go through the tutorial in Appendix E for the CFD software FLUENT to compute the isotherms and streamlines.

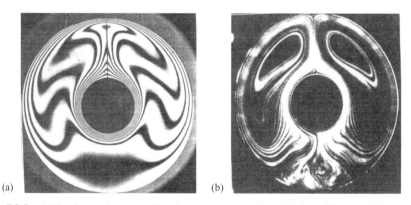

(a)   (b)

Figure P2.9   (a) Isotherms in convection between concentric cylinders, (b) streamlines in convection between concentric cylinders. The interferometer photograph on the left illustrates the temperature field. The photograph on the right shows the flow pattern visualized with cigarette smoke. (Grigull and Hauf [11]: Permission from ASME)

Project

2.10 A thermal engineer is to design a $10'' \times 7'' \times 7''$ box in which two printed circuit boards (PCB) made of aluminum are passively cooled by natural convection. An internal heat sink with a cold plate attached to one of the walls of the box is considered. All the walls are insulated except the cold plate, which is maintained at a certain temperature using a thermoelectric cooler that is not included in this project. Each board has a thermal dissipation up to 12 W. Design the box, including both the PCBs and the heat sink, by determining the cold plate temperature in order to maintain the temperature of the PCBs less than $180°$F. You may look at Tutorial II in Appendix F. The dimension of the PCB is open to students, so that every student may have a different size. Develop a MathCAD model for the internal heat sink for the given heat dissipation. Note that the fin size of the heat sink in the box should be adequate to allow sufficient air flow at the top and bottom. Design the heat sink using MathCAD. You should figure out analytically the average internal air temperature and the cold plate temperature to achieve the design requirement. Construct the three-dimensional (3-D) geometry for the box, including two PCBs and a heat sink with a cold plate, using SolidWorks as a design tool. Import the file into Gambit for meshing and then import into FLUENT for the simulations. Finally, find by trial and error a proper (optimum) temperature of the cold plate to maintain the PCB temperature below $180°$F. Note that this problem does not include the thermoelectric cooling, but the cold plate temperature must remain as high as possible to minimize the power consumption of the thermoelectric cooling device.

3

Thermoelectrics

3.1 INTRODUCTION

Thermoelectrics is literally associated with thermal and electrical phenomena. Thermoelectrics can directly convert thermal energy into electrical energy or vice versa. A *thermocouple* uses the electrical potential (electromotive force) generated between two dissimilar wires to measure temperature. Basically, thermoelectrics consists of two devices: a thermoelectric generator and a thermoelectric cooler. These devices have no moving parts and require no maintenance. Thermoelectric generators have a great potential for waste heat recovery from power plants and automotive vehicles. This device also provides reliable power in remote areas such as in space and at mountain top telecommunication sites. Thermoelectric coolers provide refrigeration and temperature control in electronic packages and medical instruments. Thermoelectrics has become increasingly important with numerous applications. Since thermoelectricity was discovered in the early nineteen century, there has not been much improvement in efficiency and material until the recent development of *nanotechnology*, which has led to remarkable improvement in performance. It is thus very important to understand the fundamentals of thermoelectrics for the development and the thermal design. We start with a brief history of thermoelectricity.

In 1821, Thomas J. Seebeck discovered that an electromotive force or a potential difference could be produced by a circuit made from two dissimilar wires when one of the junctions was heated. This is called the *Seebeck effect*.

In 1834, after 13 years, Jean Peltier discovered the reverse process, that the passage of an electric current through a thermocouple produces heating or cooling depended on its direction. This is called the *Peltier effect*. Although the above two effects were demonstrated to exist, it was very difficult to measure each effect as a property of the material because the Seebeck effect is always associated with two dissimilar wires and the Peltier effect is always followed by the additional Joule heating that is heat generation due to the electrical resistance of the passage of a current. (Joule heating was discovered in 1841 by James P. Joule.)

In 1854, William Thomson (later Lord Kelvin) discovered that if a temperature difference exists between any two points of a current-carrying conductor, heat is either liberated or absorbed depending on the direction of current and material, which is in addition to the Peltier heating. This is called the *Thomson effect*. He also studied the relationships between the above three effects thermodynamically, showing that the

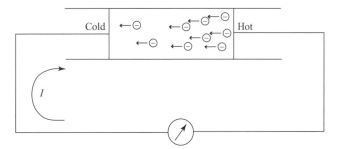

Figure 3.1 Electron concentration in a thermoelectric material

electrical Seebeck effect results from a combination of the thermal Peltier and Thomson effects. Although the Thomson effect itself is small compared to the other two, it leads to a very important and useful relationship, which is called the *Kelvin relationship*.

The mechanisms of thermoelectricity were not well understood until the discovery of electrons at the end of the nineteenth century. Now it is known that solar energy, an electric field, or thermal energy can liberate some electrons from their atomic binding even at room temperature (from the valence band to the conduction band of a conductor) where the electrons become free to move randomly. However, when a temperature difference across a conductor is applied as shown in Figure 3.1, the hot region of the conductor produces more free electrons and diffusion of the electrons (charge carriers including holes) naturally occurs from the hot region to the cold region. An electromotive force (EMF) is generated in a way that an electric current flows against the temperature gradient.

A large number of thermocouples, each of which consists of p-type and n-type semiconductor elements, are connected electrically in series and thermally in parallel by sandwiching them between two high thermal conductivity but low electrical conductivity ceramic plates to form a module, which is shown in Figure 3.2.

Consider two wires made from different metals joined at both ends, as shown in Figure 3.3, forming a close circuit. Ordinarily, nothing will happen. However, when one of the ends is heated, something interesting happens. A current flows continuously in the circuit. This is called the *Seebeck effect*, in honor of Thomas Seebeck, who made this discovery in 1821. The circuit that incorporates both thermal and electrical effects is called a *thermoelectric circuit*. The thermocouple uses the Seebeck effect to measure temperature. And the effect forms the basis of a thermoelectric generator.

In 1834, Jean Peltier discovered that the reverse of the Seebeck effect is possible by demonstrating that cooling can take place by applying a current across the junction rather than by applying heat, as in the Seebeck effect. The thermal energy converts to the electrical energy without turbine or engines. The heat pumping is possible without a refrigerator or compressor.

There are some advantages of thermoelectric devices in spite of the low thermal efficiency. There are no moving parts in the device, therefore there is less potential for failure in operation. Controllability of heating and cooling is very attractive in many applications such as avionics, medical instruments, and microelectronics.

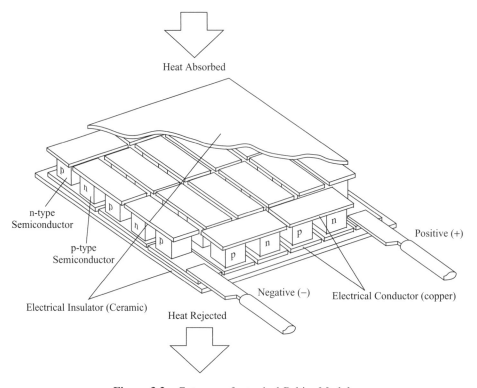

Figure 3.2   Cutaway of a typical Peltier Module

3.2   THERMOELECTRIC EFFECT

The thermoelectric effect consists of three effects: the Seebeck effect, the Peltier effect, and the Thomson effect.

3.2.1   Seebeck Effect

The *Seebeck effect* is the conversion of a temperature difference into an electric current. As shown in Figure 3.3, wire A is joined at both ends to wire B and a voltmeter is inserted in wire B. Suppose that a temperature difference is imposed between two junctions, it will generally found that a potential difference or voltage V will appear on the voltmeter. The potential difference is proportional to the temperature difference. The potential difference V is

$$V = \alpha \Delta T \qquad (3.1)$$

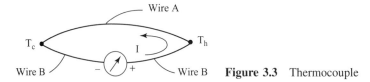

Figure 3.3   Thermocouple

where $\Delta T = T_h - T_c$ and α is called the *Seebeck coefficient* (also called *thermopower*) which is usually measured in $\mu V/K$. The sign of α is positive if the electromotive force, *emf*, tends to drive an electric current through wire A from the hot junction to the cold junction as shown in Figure 3.3. In practice one rarely measures the absolute Seebeck coefficient because the voltage meter always reads the relative Seebeck coefficient between wires A and B. The absolute Seebeck coefficient can be only calculated from the Thomson coefficient. Equation (3.1) can also be written in the form

$$V_{oc} = \alpha \Delta T \qquad (3.2)$$

where V_{oc} is the voltage of open circuit and $\Delta T = T_h - T_c$.

3.2.2 Peltier Effect

When current flows across a junction between two different wires, it is found that heat must be continuously added or subtracted at the junction in order to keep its temperature constant, which is illustrated in Figure 3.4. The heat is proportional to the current flow and changes sign when the current is reversed. More details about the Peltier effect are found in Section B.1 in Appendix B.

3.2.3 Thomson Effect

When current flows in a wire with a temperature gradient, heat is absorbed or librated across the wire depending on the material and the direction of the current. The Thomson heat is proportional to both the electric current and the temperature gradient, which is schematically shown in Figure 3.4. The Thomson coefficient is unique among the three thermoelectric coefficients because it is the only thermoelectric coefficient directly measurable for individual materials. There is other form of heat, called *Joule heating*, which is irreversible and is always generated as current flows in a wire. The Thomson heat is *reversible* between heat and electricity. This heat is not the same as Joule heating, or $I^2 R$. More details about the Thomson effect is found in Section B.1 in Appendix B.

3.2.4 Thomson (or Kelvin) Relationships

The interrelationships between the three thermoelectric effects are important in order to understand the basic phenomena. In 1854, Thomson [3] studied the relationships

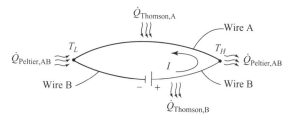

Figure 3.4   Schematic for the Peltier effect and the Thomson effect

thermodynamically and provided two relationships by applying the *first and second laws of thermodynamics* with an assumption that the reversible and irreversible processes in thermoelectricity are separable. The necessity for the assumption remained an objection to the theory until the advent of the new thermodynamics. The Thomson effect is relatively small compared to the Peltier effect, but it plays an important role to deduce the Thomson relationship. The relationships were later completely confirmed by experiments. The two relationships lead to the very useful Peltier cooling as

$$\dot{Q}_{Peltier} = \alpha T I \tag{3.3}$$

where T is the temperature at a junction between two different materials and the dot above the heat Q indicates the amount of heat transported per unit time. Detailed derivation of Equation (3.3) is found in Section B.2 in Appendix B.

3.3   THERMOELEMENT COUPLE (THERMOCOUPLE)

A thermocouple typically consists of p-type and n-type semiconductor elements. Many thermocouples are usually connected thermally in parallel and electrically in series to form a thermoelectric module. A thermocouple is illustrated in Figure 3.5. We assume herein that thermal and electrical contact resistances are negligible. We also assume the properties of the material do not vary with temperature. Hence, we consider actually ideal thermoelectric devices for simplicity. We will discuss real thermoelectric devices with those contact resistances at the last section (Section 3.10) of this chapter.

3.4   THE FIGURE OF MERIT

The performance of thermoelectric devices is measured by a *figure of merit, Z*, where the unit is $1/°C$ for the material:

$$Z = \frac{\alpha^2}{\rho k} \tag{3.4}$$

where α = the Seebeck coefficient, V/°C
ρ = the electrical resistivity, Ohm·cm
k = the thermal conductivity, W/cm°C

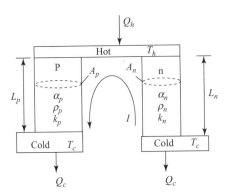

Figure 3.5   Thermoelectric couple with a pair of dissimilar elements

Table 3.1   Figures of Merit for Some Typical Semiconductors

Material	Type	Temperature (°C)	Figure of merit Z (K^{-1})	Source
Bi_2Te_3	p	25	2.5×10^{-3}	[4]
Bi_2Te_3	n	25	2.5×10^{-3}	[4]
SbBiTeSe	p	70	3.0×10^{-3}	[5]
Bi SbTe	p	150	2.5×10^{-3}	[5]
Bi_2Te_3-$74Sb_2Te_3$	n	150	3.0×10^{-3}	[6]
Bi_2Te_3-$25Bi_2Se_3$	p	150	2.7×10^{-3}	[6]
PbTe	n, p	450 (325–625)	1.3×10^{-3}	[9]
ZnSb	p	175	1.4×10^{-3}	[4]
SiGe	p	1,000	0.4×10^{-3}	[1]
SiGe	n	1,000	0.8×10^{-3}	[1]
GeTe	p	450	1.7×10^{-3}	[4]
MnTe	p	900	0.4×10^{-3}	[4]
$CeS_{1.4}$	n	1,100	1.8×10^{-3}	[4]
$AgSbTe_2$	p	400	1.3×10^{-3}	[4]
InAs	n	700	0.7×10^{-3}	[4]

A high Seebeck coefficient, a low electrical resistivity, and a low thermal conductivity are essential to achieve a satisfactory value of Z. Increasing the Seebeck coefficient or decreasing the resistivity is preferable to reducing the thermal conductivity in order increase the figure of performance [7].

Table 3.1 lists the values of the figure of merit Z for several semiconductor materials. Bithmuth telluride (Bi_2Te_3) and its alloy have been used widely for thermoelectric coolers (TEC), while lead telluride (PbTe) has been widely used for thermoelectric generators (TEG). The intrinsic properties are hidden by doping effects or an excess of the components. The most widely used thermoelectric material is a bismuth telluride alloy (Bi_2Te_3-Sb_2Te_3 and Bi_2Te_3-Bi_2Se_3), for which the figure of merit can be found up to 3.0×10^{-3}, as shown in Table 3.1.

The dimensionless figure of merit ZT_{ave} is introduced and often used as the characteristics of the materials. T_{ave} is the average temperature of the hot and cold junctions. For good materials, ZT_{ave} is equal to unity or higher than unity. A recent report reveals ZT_{ave} value between 2 and 3 at room temperature [10]. The typical properties and ZT_{ave} for Bi_2Te_3 are found in Figure B.5 and B.6 in Appendix B.

There is plenty of scope for thermoelectric refrigeration, even if new materials are not found. However, if improved figure of merit can be obtained, the possibilities seem boundless.

3.5   SIMILAR AND DISSIMILAR MATERIALS

3.5.1   Similar Materials

For semiconductor thermoelectric couples, the same basic material is customarily used to fabricate both the p-type and n-type elements. A thermoelectric couple is usually composed of a p-type and an n-type semiconductor as shown in Figure 3.5. It is customary to use the following notation.

$$\alpha = \bar{\alpha}_p - \bar{\alpha}_n \qquad (3.5a)$$

$$\rho = \bar{\rho}_p + \bar{\rho}_n \qquad (3.5b)$$

$$k = \bar{k}_p + \bar{k}_n \qquad (3.5c)$$

The bar denotes an average value over the range between the hot-junction and cold-junction temperature. The p and n properties are acquired by impurity doping that is arranged *for similar materials* to give

$$\alpha_p \cong -\alpha_n \qquad (3.6a)$$

$$\rho_p \cong \rho_n \qquad (3.6b)$$

$$k_n \cong k_p \qquad (3.6c)$$

3.5.2 Dissimilar Materials

Since most thermoelectric devices are composed of a pair of *dissimilar materials* as shown in Figure 3.5, the *material figure of merit* for an optimum geometry on the efficiency can be obtained analytically. The *figure of merit* is then expressed as

$$Z = \frac{(\alpha_p - \alpha_n)^2}{\left[(k_p \rho_p)^{\frac{1}{2}} + (k_n \rho_n)^{\frac{1}{2}} \right]^2} \qquad (3.7)$$

or

$$Z = \frac{(\alpha_p - \alpha_n)^2}{KR} \qquad (3.7a)$$

where (see Section B.3 in Appendix B)

$$KR = \left(\frac{k_p A_p}{L_p} + \frac{k_n A_n}{L_n} \right) \left(\frac{\rho_p L_p}{A_p} + \frac{\rho_n L_n}{A_n} \right) \qquad (3.7b)$$

Eventually,

$$KR = \left[(k_p \rho_p)^{\frac{1}{2}} + (k_n \rho_n)^{\frac{1}{2}} \right]^2 \qquad (3.7c)$$

Maximizing Z is equivalent to minimizing KR since α is not a function of geometry (see Section B.4 in Appendix B for the detailed optimization). The minimum KR leads to

$$\frac{L_n/A_n}{L_p/A_p} = \left(\frac{\rho_p k_n}{\rho_n k_p} \right)^{\frac{1}{2}} \qquad (3.8)$$

Equations (3.7) and (3.8) are useful in design of thermoelectric elements in terms of both geometry and materials. Interestingly, these two relations were proved analytically for both TEG and TEC [1].

3.6   THERMOELECTRIC GENERATOR (TEG)

A thermoelectric generator is a power generator, where thermal energy converts to electrical energy without moving parts. A simple electrical circuit for a thermoelectric generator is shown in Figure 3.6. The thermoelectric couple is composed of two elements: a p-type and an n-type semiconductor. The heat flow at the hot junction involves three terms: the heat associated with the Seebeck effect, the half of joule heating, and the thermal conduction.

$$\dot{Q}_h = \alpha T_h I - \frac{1}{2} I^2 R + K\,(T_h - T_c) \tag{3.9}$$

where α is the Seebeck coefficient, R the internal (electrical) resistance of the elements, K the thermal conductance, I the current, T_h the hot junction temperature, and T_c the cold junction temperature. The detailed derivation of Equation (3.9) is found in Section B.3 in Appendix B.

3.6.1   Similar and Dissimilar Materials

3.6.1.1   *Similar Materials*

The similar materials here are interpreted as having the same base materials and the same geometry for thermoelectric elements. In this case, the analysis becomes simpler as $L_p = L_n = L$ and $A_p = A_n = A$, where L is the length of the thermoelement and A is the cross-sectional area of the thermoelement. Many studies have been developed based on the similar materials because having dissimilar materials is somewhat cumbersome in handling. The properties of α, ρ, and k can be handled as shown both in Equations (3.5a, 3.5b, and 3.5c) and Equations (3.6a, 3.6b, and 3.6c).

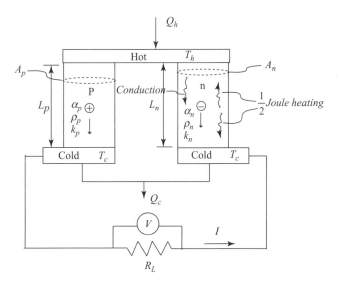

Figure 3.6   A thermoelectric power generator

We rewrite them here for readers

$$\alpha = \overline{\alpha}_p - \overline{\alpha}_n \tag{3.5a}$$

$$\rho = \overline{\rho}_p + \overline{\rho}_n \tag{3.5b}$$

$$k = \overline{k}_p + \overline{k}_n \tag{3.5c}$$

$$\alpha_p \cong -\alpha_n \tag{3.6a}$$

$$\rho_p \cong \rho_n \tag{3.6b}$$

$$k_n \cong k_p \tag{3.6c}$$

In Equation (3.9), the thermal conductance by the Fourier's law of conduction is

$$K = \frac{kA}{L} \tag{3.10}$$

The internal electrical resistance, R, of the thermoelectric couple is

$$R = \frac{\rho L}{A} \tag{3.11}$$

where ρ is the electrical resistivity.

3.6.1.2 *Dissimilar Materials*

Dissimilar materials here are interpreted as both dissimilar materials and dissimilar geometry for the thermoelectric elements. Many thermoelectric materials in commercial products are indeed dissimilar. Hence, we want to develop the analysis for the dissimilar materials. Consider an electrical circuit shown in Figure 3.6.

In Equation (3.9), α can be used as defined in Equations (3.5a) for the dissimilar materials, rewriting here

$$\alpha = \overline{\alpha}_p - \overline{\alpha}_n \tag{3.5a}$$

The thermal conductance for the dissimilar materials is

$$K = \frac{k_p A_p}{L_p} + \frac{k_n A_n}{L_n} \tag{3.12}$$

And the electrical resistance is

$$R = \frac{\rho_p L_p}{A_p} + \frac{\rho_n L_n}{A_n} \tag{3.13}$$

The analysis herein is practically the same for both the similar and dissimilar materials. The figure of merit is

$$Z = \frac{\left(\alpha_p - \alpha_n\right)^2}{\left[\left(k_p \rho_p\right)^{\frac{1}{2}} + \left(k_n \rho_n\right)^{\frac{1}{2}}\right]^2} \tag{3.7}$$

The geometric ratio is given by

$$\frac{L_n/A_n}{L_p/A_p} = \left(\frac{\rho_p k_n}{\rho_n k_p}\right)^{\frac{1}{2}} \tag{3.8}$$

3.6.2 Conversion Efficiency and Current

The power generated at the load can be calculated as

$$\dot{W} = I^2 R_L \tag{3.14}$$

where R_L = load resistance.

The conversion (or thermal) efficiency of the thermoelectric generator (TEG) is of the form

$$\eta_t = \frac{\dot{W}}{\dot{Q}_h} \tag{3.15}$$

which can be rewritten using Equations (3.9) and (3.14) as

$$\eta_t = \frac{I^2 R_L}{\alpha T_h I - \frac{1}{2} I^2 R + K(T_h - T_c)} \tag{3.15a}$$

The heat rejected at the cold junction is obtained from the first law of thermodynamics applied to the system.

$$\dot{Q}_c = \dot{Q}_h - \dot{W} \tag{3.16}$$

The voltage in the open circuit in Figure 3.7 for the thermoelectric generator will be of the form

$$V_{oc} = I(R_L + R) \tag{3.17}$$

Knowing that the voltage in Equation (3.2) is indeed the voltage of the open circuit and combining Equations (3.2) and (3.17) with $\Delta T = T_h - T_c$ gives the current in the circuit:

$$I = \frac{\alpha(T_h - T_c)}{(R_L + R)} \tag{3.18}$$

3.6.3 Maximum Conversion Efficiency

The maximum conversion efficiency is the highest thermal efficiency. Inserting Equation (3.18) into Equation (3.15a) with Equation (3.7c) gives the conversion

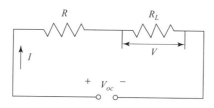

Figure 3.7 An open circuit for the thermoelectric generator

efficiency as

$$\eta_t = \eta_c \frac{\dfrac{R_L}{R}}{\left(1 + \dfrac{R_L}{R}\right) - \dfrac{\eta_c}{2} + \dfrac{\left(1 + \dfrac{R_L}{R}\right)^2}{ZT_h}} \qquad (3.19)$$

where the Carnot cycle thermal efficiency is used as defined in Equation (1.29).

$$\eta_c = 1 - \frac{T_c}{T_h} \qquad (3.20)$$

The ZT_h in Equation (3.19) can be expressed in terms of $Z\overline{T}$ and T_c/T_h as

$$ZT_h = ZT_h \frac{T_h + T_c}{2} \frac{2}{T_h + T_c} = 2Z\overline{T} \left(1 + \frac{T_c}{T_h}\right)^{-1} \qquad (3.21)$$

where the average temperature is defined as

$$\overline{T} = \frac{T_h + T_c}{2} \qquad (3.22)$$

Inserting Equation (3.21) into Equation (3.19) gives a slightly different form of the conversion efficiency as

$$\eta_t = \frac{\dfrac{R_L}{R} \left(1 - \dfrac{T_c}{T_h}\right)}{\left(1 + \dfrac{R_L}{R}\right) - \dfrac{1}{2}\left(1 - \dfrac{T_c}{T_h}\right) + \dfrac{1}{2Z\overline{T}}\left(1 + \dfrac{R_L}{R}\right)^2 \left(1 + \dfrac{T_c}{T_h}\right)} \qquad (3.23)$$

The conversion efficiency η_t is a function of three dimensionless values as $\dfrac{R_L}{R}$, $Z\overline{T}$ and T_c/T_h.

To find the maximum conversion efficiency, differentiate Equation (3.23) with respect to $\dfrac{R_L}{R}$ and set the derivative to zero as

$$\frac{d\eta_t}{d(R_L/R)} = 0 \qquad (3.24)$$

After some algebra, this leads to

$$\frac{R_L}{R} = (1 + Z\frac{T_h + T_c}{2})^{\frac{1}{2}}$$

We let the value of R_L/R be the special value M that is now the optimum ratio for the maximum conversion efficiency as

$$M = \frac{R_L}{R} = (1 + Z\overline{T})^{\frac{1}{2}} \qquad (3.25)$$

A good thermoelectric material will have a high M value. Inserting Equation (3.25) into Equation (3.23) leads to a simple form, where the efficiency becomes the maximum conversion efficiency as

$$\eta_{mc} = \eta_c \frac{\left(1 + Z\overline{T}\right)^{\frac{1}{2}} - 1}{\left(1 + Z\overline{T}\right)^{\frac{1}{2}} + \frac{T_c}{T_h}} \tag{3.26}$$

Note that

$$\eta_{mc} \to \eta_c \text{ as } Z \to \infty \tag{3.27}$$

3.6.4   Maximum Power Efficiency

The maximum power efficiency is a specific conversion efficiency where the power reaches the maximum power. Inserting Equation (3.18) into Equation (3.14) gives the power in the general form.

$$\dot{W} = \left(\frac{\alpha \Delta T}{R_L + R}\right)^2 R_L = \left(\frac{\alpha \Delta T}{\frac{R_L}{R} + 1}\right)^2 \frac{R_L}{R} \tag{3.28}$$

We want to maximize power with respect to the resistance ratio R_L/R.

$$\frac{d\dot{W}}{d\left(R_L/R\right)} = 0 \tag{3.29}$$

Using the chain rule, we have

$$\left(\frac{\alpha \Delta T}{\frac{R_L}{R} + 1}\right)^2 + \frac{R_L}{R} (\alpha \Delta T)^2 (-2) \frac{1}{\left(\frac{R_L}{R} + 1\right)^3} = 0 \tag{3.30}$$

We have an important relationship for the maximum power

$$R_L = R \tag{3.31}$$

Under this condition, Equation (3.18) becomes the maximum powered current.

$$I_{mp} = \frac{\alpha \Delta T}{2R} \tag{3.32}$$

Inserting Equation (3.31) into Equation (3.28) gives the maximum power

$$\dot{W}_{\max} = \frac{\alpha^2 \Delta T^2}{4R} \tag{3.33}$$

Note that the maximum power is proportional to the square of the temperature difference. Equation (3.19) with Equation (3.31) gives the maximum power conversion efficiency as

$$\eta_{mp} = \eta_c \frac{1}{\dfrac{4}{ZT_h} + 2 - \dfrac{\eta_c}{2}} \tag{3.34}$$

Or using Equation (3.21), we have an expression in terms of $Z\overline{T}$ and T_c/T_h.

$$\eta_{mp} = \frac{\left(1 - \dfrac{T_c}{T_h}\right)}{\dfrac{2}{Z\overline{T}}\left(1 + \dfrac{T_c}{T_h}\right) + 2 - \dfrac{1}{2}\left(1 - \dfrac{T_c}{T_h}\right)} \tag{3.34a}$$

3.6.5   Maximum Performance Parameters

Now we want to define the maximum performance parameters. The maximum power was already given in Equation (3.33). Hence, using Equation (3.28), the maximum power ratio is defined

$$\frac{\dot{W}}{\dot{W}_{\max}} = \frac{\left(\dfrac{\alpha \Delta T}{R_L + R}\right)^2 R_L}{\dfrac{\alpha^2 \Delta T^2}{4R}} = \frac{4\dfrac{R_L}{R}}{\left(\dfrac{R_L}{R} + 1\right)^2} \tag{3.35}$$

The maximum current hypothetically occurs if the load resistance is zero in Equation (3.18). Hence, the maximum current will be

$$I_{\max} = \frac{\alpha \Delta T}{R} \tag{3.36}$$

The maximum current ratio will be

$$\frac{I}{I_{\max}} = \frac{\dfrac{\alpha \Delta T}{(R_L + R)}}{\dfrac{\alpha \Delta T}{R}} = \frac{1}{\dfrac{R_L}{R} + 1} \tag{3.37}$$

The voltage of the open circuit is hypothetically the highest voltage in the circuit, so the maximum voltage is defined as the voltage of open circuit, including Equation (3.2).

$$V_{\max} = V_{oc} = I(R_L + R) = \alpha \Delta T \tag{3.38}$$

And the voltage across the load as shown in Figure 3.6 is

$$V = I R_L \tag{3.39}$$

Then, the maximum voltage ratio becomes

$$\frac{V}{V_{\max}} = \frac{I R_L}{I(R_L + R)} = \frac{\dfrac{R_L}{R}}{\dfrac{R_L}{R} + 1} \tag{3.40}$$

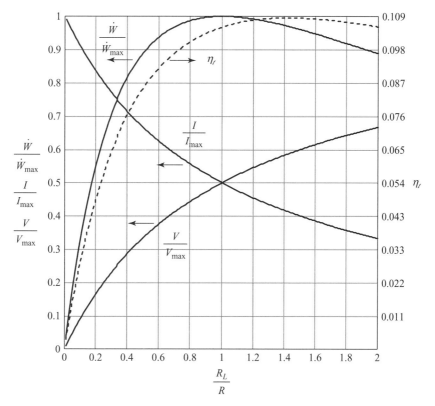

Figure 3.8 Generalized chart of TEG characteristics. It is assumed in the curve of the conversion efficiency η_t that $Z\overline{T} = 1$ and $T_c/T_h = 0.5$

The maximum ratios given in Equations (3.35), (3.37), and (3.40) are solely a function of the resistance ratio R_L/R. Hence, the dimensionless ratios of maximum power, current, and voltage can be graphed with respect to the resistance ratio R_L/R (Figure 3.8). The conversion efficiency η_t in Equation (3.23) is superimposed on the graph in order to compare the characteristics, where the dimensionless values $Z\overline{T} = 1$ and $T_c/T_h = 0.5$ were used. Notice that the maximum power reaches the top exactly at $R_L/R = 1$ as predicted in Equation (3.31), while the maximum conversion efficiency reaches the top at $R_L/R = 1.414$ either graphically or exactly using the M value in Equation (3.25). In the thermoelectric generators, the performance varies as the load resistance changes as shown in Figure 3.8, so that it is essential to arrange the load resistance for the best performance, which would lie between the maximum power and the maximum conversion efficiency.

The discrepancy between the maximum conversion efficiency and the maximum power efficiency is plotted in Figure 3.9 as a function of $Z\overline{T}$ and T_c/T_h. Also the discrepancy of the maximum power and the maximum conversion power is plotted in Figure 3.10 as a function of $Z\overline{T}$. Especially when $Z\overline{T} \sim 1$, the discrepancy between the maximum power efficiency and the maximum conversion efficiency is negligible within 3 % error, as shown in Figures 3.9 and 3.10. Hence, the maximum power may be considered as the best performance. However, when $Z\overline{T} \geq 3$, the discrepancy

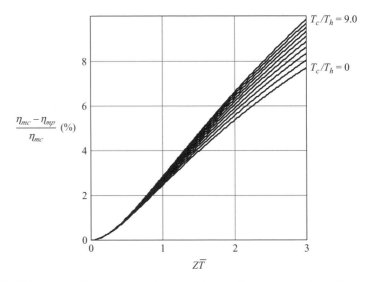

Figure 3.9   Discrepancy in percentage between the maximum conversion efficiency and the maximum power efficiency

increases to about 10 % and more. In this case, the maximum power may not be the best choice, demanding a much higher heat transfer rate at the hot junction due to the low efficiency.

It should be noted in Figure 3.8 that the maximum conversion efficiency with $Z\overline{T} = 1$ and $T_c/T_h = 0.5$ is approximately 0.109. That is relatively low compared with other

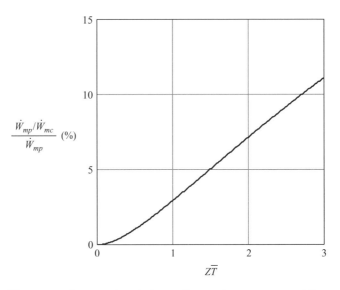

Figure 3.10   Discrepancy in percentage between the maximum power and the maximum conversion power

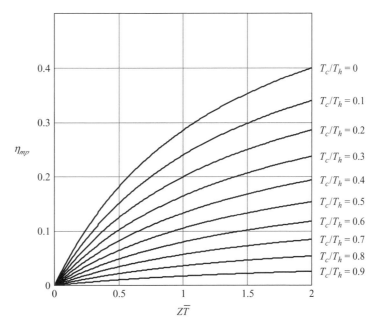

Figure 3.11   Maximum power efficiency as a function of $Z\overline{T}$ and T_c/T_h

systems such as engines or turbines. This low efficiency is one of the characteristics of the *TEG*.

Equations (3.34a) and (3.26) are plotted in Figures 3.11 and 3.12, respectively. When the maximum power is a primary concern, Figure 3.11 can be used.

If we graphically obtain $\dot{W}$ from the generalized chart of Figure 3.8 and η_{mp} from Figure 3.11 or η_{mc} from Figure 3.12, we can estimate the corresponding heat transfer rate $\dot{Q}_h$ necessary at the hot junction from the definition of the conversion efficiency given in Equation (3.15).

It should be noted that the efficiencies can be increased by either increasing the dimensionless figure of merit, $Z\overline{T}$, or reducing the cold-to-hot temperature ratio T_c/T_h. Some research reports reveal a potential of $Z\overline{T} \sim 3$ using the multilayer thin film technology with silicon germanium alloys [11].

3.6.6   Multicouple Modules

The number of multicouple is denoted by subscript n. The parameters of a thermoelectric module are given as a single couple, multiplied by n :

$$\left(\dot{W}\right)_n = n\dot{W} \tag{3.41}$$

$$\left(\dot{Q}_h\right)_n = n\dot{Q}_h \tag{3.42}$$

$$(R)_n = nR \text{ and } (R_L)_n = nR_L \tag{3.43}$$

$$(V)_n = nV \tag{3.44}$$

$$(K)_n = nK \tag{3.45}$$

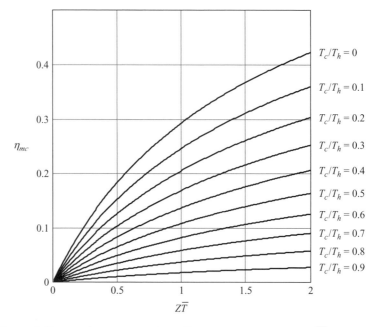

Figure 3.12 Maximum conversion efficiency as a function of $Z\overline{T}$ and T_c/T_h

The parameters that are not multiplied by n are given

$$(I)_n = I \tag{3.46}$$

$$(\eta_t)_n = \frac{(\dot{W})_n}{(\dot{Q}_h)_n} = \eta_t \tag{3.47}$$

$$\left(\frac{\dot{W}}{\dot{W}_{\max}}\right)_n = \frac{\dot{W}}{\dot{W}_{\max}} \tag{3.48a}$$

$$\left(\frac{I}{I_{\max}}\right)_n = \frac{I}{I_{\max}} \tag{3.48b}$$

$$\left(\frac{V}{V_{\max}}\right)_n = \frac{V}{V_{\max}} \tag{3.48c}$$

Some thermoelectric modules (TEM) based on bismuth telluride are illustrated in Table 3.2 and Table 3.3. The maximum performance parameters are particularly useful for design purposes. The multicouple modules in TEG are often custom fabricated to fit demands.

Example 3.1 Thermoelectric Generator A thermoelectric generator (TEG) is fabricated from p-type and n-type semiconductors (PbTe). The hot and cold junctions are 725 K and 295 K, respectively. At the mean temperature of 510 K, the properties of the material are $\alpha_n = -187 \times 10^{-6}$ V/K, $\alpha_p = 187 \times 10^{-6}$ V/K, $\rho_n = \rho_p =$

Table 3.2 Multicouple Modules of TEG

| | $T_h = 230°C$ and $T_c = 30°C$, $ZT_{ave} = 0.4$ | | | | | Dimension (mm) | | |
Module	R (Ω) (internal resistance)	$\dot{W}_{max}$* (Watts)	I_{max}** (Amps)	V_{max}*** (Volts)	Number of couples	Width	Length	Height
TEG-1a	7	2.6	1.22	8.8	126	30	30	3.0
TEG-2a	3	5.9	2.8	8.6	126	40	40	3.7
TEG-3a	1.7	10.5	4.97	8.7	126	56	56	4.0
TEG-4a	1.2	14.7	7.0	8.6	126	56	56	4.0

Table 3.3 Multiple Modules of TEG for Waste Heat Recovery

| | $T_h = 230°C$ and $T_c = 30°C$, $ZT_{ave} = 0.4$ | | | | | Dimension (mm) | | |
Module	R (Ω) (internal resistance)	$\dot{W}_{max}$* (Watts)	I_{max}** (Amps)	V_{max}** (Volts)	Number of couples	Width	Length	Height
TEG-1b	4.0	2.5	1.6	6.53	97	29	29	5.08
TEG-2b	1.15	9	5.6	6.5	97	62.7	62.7	6.5
TEG-3b	0.3	19	16	5.0	71	75	75	5.08

*The maximum power is obtained from Equation (3.33) when the load resistance matches the internal resistance.
**The maximum current occurs as shown in Equation (3.36) if the load resistance is zero.
***The maximum voltage is the voltage of the open circuit, which is shown in Equation (3.38).

$1.64 \times 10^{-3} \Omega$ cm, $k_n = k_p = 0.0146$ W/cm K. The length of the thermoelectric element is 1.0 cm, and the cross-sectional area of the element is 1.2 cm^2 (See Figure E3.1.1).

(a) Determine the maximum power, the maximum power efficiency, the heat absorbed at the hot junction, the corresponding output voltage across the load, and the current for the maximum power.

(b) Calculate the power, the maximum conversion efficiency, the heat absorbed at the hot junction, the output voltage across the load, and the current for the maximum conversion efficiency.

MathCAD Format Solution   We use commercial software MathCAD to solve the problem.

Properties of lead telluride (PbTe):

$$\alpha_n := 187 \cdot 10^{-6} \frac{V}{K} \qquad \alpha_p := -187 \cdot 10^{-6} \frac{V}{K}$$
$$\rho_n := 1.64 \cdot 10^{-3} \, \Omega \, cm \quad \rho_p := 1.64 \cdot 10^{-3} \, \Omega \, cm \qquad (E3.1.1)$$
$$k_n := 0.0146 \frac{W}{cm \, K} \qquad k_p := 0.0146 \frac{W}{cm \, K}$$

The properties for both elements can be treated as the similar materials, as shown in Equations (3.5a), (3.5b), and (3.5c).

$$\alpha := \alpha_p - \alpha_n \quad \rho := \rho_p + \rho_n \quad k := k_p + k_n \qquad (E3.1.2)$$

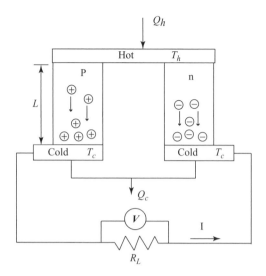

Figure E3.1.1   Thermoelectric generator

The dimensions of the element are

$$L_e := 1.0\,cm \quad A_e := 1.2\,cm^2 \tag{E3.1.3}$$

The cold and hot junction temperatures are

$$T_h := 725\,K \quad T_c := 295\,K \quad \Delta T := T_h - T_c \tag{E3.1.4}$$

The internal resistance of the couple in Equation (3.11) is

$$R_e := \frac{\rho\,L_e}{A_e} \quad R_e = 2.733 \times 10^{-3}\Omega \tag{E3.1.5}$$

(a) Maximum Power with $R_L = R_e$ as shown in Equation (3.31)   The current for maximum power is calculated using Equation (3.32).

$$I_{mp} := \frac{\alpha(T_h - T_c)}{2\,R_e} \quad I_{mp} = 29.418\,A \tag{E3.1.6}$$

From Figure 3.6, the output voltage across the load at the maximum power is obtained by

$$V_L := I_{mp}\,R_L \quad V_L = 0.08\,V \tag{E3.1.7}$$

The maximum power of the load in Equation (3.14) or Equation (3.33) is

$$w_{mp} := I_{mp}^2\,R_L \quad W_{mp} = 2.366\,W \tag{E3.1.8}$$

The heat absorbed at the hot junction is calculated using Equation (3.9) with Equation (3.10).

$$Q_{mp} := \alpha T_h\,I_{mp} - \frac{1}{2}I_{mp}^2\,R_e + k\,A_e\frac{(T_h - T_c)}{L_e} \quad Q_{mp} = 21.861\,W \tag{E3.1.9}$$

Using Equation (3.15), the conversion efficiency for the maximum power is

$$\eta_{mp} := \frac{W_{mp}}{Q_{mp}} \quad \eta_{mp} = 0.108 \tag{E3.1.10}$$

This can also be calculated using Equation (3.34).

(b) Maximum Conversion Efficiency   The figure of merit in Equation (3.4) is

$$Z := \frac{\alpha^2}{\rho k} \quad Z = 1.46 \times 10^{-3} \frac{1}{K} \tag{E3.1.11}$$

The average temperature is

$$T_{ave} := \frac{725\,K + 295\,K}{2} \quad Z\,T_{ave} = 0.745 \tag{E3.1.12}$$

Using Equations (3.25) and (3.26),

$$M := (1 + Z\,T_{ave})^{\frac{1}{2}} \quad M = 1.321 \tag{E3.1.13}$$

$$\eta_{mc} := \left(1 - \frac{T_c}{T_h}\right) \frac{(M - 1)}{M + \frac{T_c}{T_h}} \quad \eta_{mc} = 0.11 \tag{E3.1.14}$$

The load resistance for the maximum conversion efficiency is calculated using Equation (3.25).

$$R_L := R_e\,(1 + Z\,T_{ave})^{\frac{1}{2}} \quad R_L = 3.611 \times 10^{-3}\,\Omega \tag{E3.1.15}$$

Using Equation (3.18), the current is

$$I_{mc} := \frac{\alpha(T_h - T_c)}{(R_L + R_e)} \quad I_{mc} = 25.351\,A \tag{E3.1.16}$$

The voltage, power generated, heat absorbed are calculated as

$$V_L := I_{mc}\,R_L \qquad\qquad V_L = 0.092\,V \tag{E3.1.17}$$

$$W_{mc} := I_{mc}^2\,R_L \qquad\qquad W_{mc} = 2.32\,W \tag{E3.1.18}$$

$$Q_{mc} := \frac{W_{mc}}{\eta_{mc}} \qquad\qquad Q_{mc} = 21.063\,W \tag{E3.1.19}$$

or the heat absorbed can also be calculated using Equation (3.9)

$$Q_{mc} := \alpha\,T_h\,I_{mc} - \frac{1}{2}\,I_{mc}^2\,R_e + k\,A_e\,\frac{(T_h - T_c)}{L_e} \quad Q_{mc} = 21.063\,W \tag{E3.1.20}$$

The results for the thermoelectric generator are summarized in Table E3.1.1.

Example 3.2 Thermoelectric Generator—Dissimilar Materials   A   thermoelectric generator is to be designed to supply 590 W of power by a burner that operates on a gaseous fuel in a remote area as shown in Figure E3.2.1. The output voltage is preferred

Table E3.1.1 Summary of the Results for the Maximum Power and the Maximum Conversion Efficiency

	Maximum power	Maxi. conversion efficiency
Current, I (A)	29.418	25.315
Voltage across load, V	0.08	0.092
Power generated, W	2.366	2.32
Heat absorbed, $\dot{Q}_h(W)$	21.861	21.063
Thermal efficiency, $\eta_t(\%)$	10.8	11.0

to be approximately 28 V, otherwise, a DC/DC converter may be required. The cold-junction plate is cooled with heat sinks using heat pipes as shown. The hot-junction temperature is maintained at 538°C while the cold junction temperature is maintained at 163°C. The thermoelement material of lead telluride at the average temperature of 350°C has $\alpha_p = 190 \times 10^{-6}$ V/K, $\alpha_n = -170 \times 10^{-6}$ V/K, $\rho_p = 0.95 \times 10^{-3}\,\Omega$ cm, $\rho_n = 1.01 \times 10^{-3}\,\Omega$ cm, $k_p = 0.012$ W/cm K, $k_n = 0.014$ W/cm K. The cross-sectional area of the n-type element A_n is 0.6 cm^2, and the length of the elements are same for both p-type and n-type elements, $L_p = L_n = 0.8$ cm. For the design of a thermoelectric generator, answer the following questions.

1. Determine the cross-sectional area, A_p, of the p-type thermoelement for the optimum geometry.
2. Determine the performance parameters for (a) the maximum conversion efficiency and (b) the maximum power, respectively. The performance parameters includes: the number of thermocouples, the thermal efficiency, the voltage across the load, the current, the heat absorbed, and the heat dissipated. (c) Determine also the maximum performance parameters.

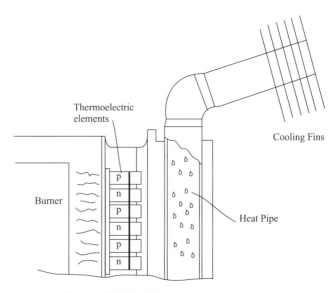

Figure E3.2.1 Thermoelectric generator

MathCAD Format Solution The properties of lead telluride used are

$$\alpha_p := 190 \cdot 10^{-6} \frac{V}{K} \qquad \alpha_n := -170 \cdot 10^{-6} \frac{V}{K}$$

$$\rho_p := 0.95 \cdot 10^{-3} \, \Omega \, cm \quad \rho_n := 1.01 \cdot 10^{-3} \, \Omega \, cm \qquad (E3.2.1)$$

$$k_p := 1.2 \cdot 10^{-2} \frac{W}{cm \, K} \qquad k_n := 1.4 \cdot 10^{-2} \frac{W}{cm \, K}$$

1. Define the relative Seebeck coefficient from Equation (2.5a) as

$$\alpha := \alpha_p - \alpha_n \qquad (E3.2.2)$$

The dimensions given are

$$L_p := 0.8 \, cm \quad L_n := L_p \quad A_n := 0.6 \, cm^2 \qquad (E3.2.3)$$

2. Determine the cross-sectional area, A_p, of the p-type thermoelement. Using Equation (3.8), the optimum geometry is obtained as

$$\frac{L_n \, A_p}{L_p \, A_n} = \left(\frac{\rho_p \, k_n}{\rho_n \, k_p}\right)^{\frac{1}{2}} \quad A_p := \frac{L_p \, A_n}{L_n} \left(\frac{\rho_p \, k_n}{\rho_n \, k_p}\right)^{\frac{1}{2}} \quad A_p = 0.629 \, cm^2 \qquad (E3.2.4)$$

Determine the performance parameters for (a) the maximum conversion efficiency and (b) the maximum power, respectively. The performance parameters includes: the number of thermocouples, the thermal efficiency, the voltage across the load, the current, the heat absorbed, and the heat dissipated. (c) Determine also the maximum performance parameters.

Using Equation (3.13), the thermocouple's electrical resistance for a dissimilar material is

$$R_e := \frac{\rho_p \, L_p}{A_p} + \frac{\rho_n \, L_n}{A_n} \quad R_e = 2.556 \times 10^{-3} \Omega \qquad (E3.2.5)$$

The power requirement is

$$W_n := 590 \, W \qquad (E3.2.6)$$

The hot- and cold-junction temperatures given are

$$T_h := 538°C \quad T_c := 163°C \qquad (E3.2.7)$$

The average temperature is defined and the temperature ratio is calculated as

$$T_{avg} := \frac{T_h + T_c}{2} \quad \frac{T_c}{T_h} = 0.538 \qquad (E3.2.8)$$

Using Equation (3.7), the figure of merit, Z, is obtained as

$$Z := \frac{(\alpha_p - \alpha_n)^2}{\left[(k_p \, \rho_p)^{\frac{1}{2}} + (k_n \, \rho_n)^{\frac{1}{2}}\right]^2} \quad Z = 2.545 \times 10^{-3} \frac{1}{K} \qquad (E3.2.9)$$

And the dimensionless figure of merit is

$$Z T_{avg} = 1.587 \tag{E3.2.9}$$

(a) Maximum Conversion Efficiency (use subscript 1)

Using Equation (3.25) at the maximum conversion efficiency, the load resistance of a thermoelement couple is

$$R_{L1} := R_e \left(1 + Z T_{avg}\right)^{\frac{1}{2}} \quad R_{L1} = 4.111 \times 10^{-3} \Omega \quad \frac{R_{L1}}{R_e} = 1.608 \tag{E3.2.10}$$

Using Equation (3.26), the maximum conversion efficiency is

$$\eta_1 := \left(1 - \frac{T_c}{T_h}\right) \frac{\left[(1 + Z T_{avg})^{\frac{1}{2}} - 1\right]}{(1 + Z T_{avg})^{\frac{1}{2}} + \frac{T_c}{T_h}} \quad \eta = 0.131 \tag{E3.2.11}$$

Using Equation (3.18), the current at the maximum conversion efficiency is

$$I_1 := \frac{\alpha (T_h - T_c)}{(R_{L1} + R_e)} \quad I_1 = 20.25 \, A \tag{E3.2.12}$$

Using Equation (3.14), the power for a thermocouple at the maximum conversion efficiency is

$$W_1 := I_1^2 R_{L1} \quad W_1 = 1.686 \, W \tag{E3.2.13}$$

The number of thermocouples is obtained by dividing the power requirement by the power of a thermocouple.

$$n_1 := \frac{W_n}{W_1} \quad n_1 = 350.003 \tag{E3.2.14}$$

From Equation (3.15), the heat absorbed on the hot junction is

$$Q_{h1n} := \frac{W_n}{\eta_1} \quad Q_{h1n} = 4.502 \times 10^3 \, W \tag{E3.2.15}$$

From Equation (3.16), the heat dissipated on the cold junction is

$$Q_{c1n} := Q_{h1n} - W_n \quad Q_{c1n} = 3.912 \times 10^3 \, W \tag{E3.2.16}$$

From Figure 3.6, the voltage across the load is

$$V_{1n} := n_1 I_1 R_{L1} \quad V_{1n} = 29.136 \, V \tag{E3.2.17}$$

The internal resistance is

$$R_{en1} := n_1 R_e \quad R_{en1} = 0.895 \Omega \tag{E3.2.18}$$

(b) Maximum Power (use subscript 2) Using Equation (3.31), we have

$$R_{L2} := R_e \tag{E3.2.19}$$

From Equation (3.34a), the maximum power efficiency is

$$\eta_2 := \frac{\left(1 - \dfrac{T_c}{T_h}\right)}{\dfrac{2}{ZT_{avg}}\left(1 + \dfrac{T_c}{T_h}\right) + 2 - \dfrac{1}{2}\left(1 - \dfrac{T_c}{T_h}\right)} \qquad \eta_2 = 0.125 \tag{E3.2.20}$$

Using Equation (3.18), the current for the maximum power is

$$I_2 := \frac{\alpha(T_h - T_c)}{2\,R_e} \qquad I_2 = 26.41\,\text{A} \tag{E3.2.21}$$

From Equation (3.14), the maximum power is

$$W_2 := I_2^2\,R_{L2} \qquad W_2 = 1.783\,\text{W} \tag{E3.2.22}$$

The number of thermocouples

$$n_2 := \frac{W_n}{W_2} \qquad n_2 = 330.962 \tag{E3.2.23}$$

The heat absorbed at the hot junction is

$$Q_{h2n} := \frac{W_n}{\eta_2} \qquad Q_{h2n} = 4.731 \times 10^3\,\text{W} \tag{E3.2.24}$$

The heat dissipated on the cold junction is

$$Q_{c2n} := Q_{h2n} - W_n \qquad Q_{c2n} = 4.141 \times 10^3\,\text{W} \tag{E3.2.25}$$

The total voltage across the load is

$$V_{2n} := n_2\,I_2\,R_{L2} \qquad V_{2n} = 22.34\,\text{V} \tag{E3.2.26}$$

The internal resistance is

$$R_{en2} := n_2\,R_e \qquad R_{en2} = 0.846\,\Omega \tag{E3.2.27}$$

(c) Maximum Performance Parameters   The maximum power is the power requirement as

$$W_{nmax} := W_n \qquad W_{nmax} = 590\,\text{W} \tag{E3.2.28}$$

The maximum current is obtained using Equation (3.36).

$$I_{max} := \frac{\alpha(T_h - T_c)}{R_e} \qquad I_{max} = 52.82\,\text{A} \tag{E3.2.29}$$

From Equation (3.38), the maximum voltage per module for the maximum conversion efficiency is

$$V_{nmax} := \eta_1\,\alpha\,(T_h - T_c) \qquad V_{nmax} = 47.25\,\text{V} \tag{E3.2.30}$$

The results for the design of the thermoelectric generator are summarized in Table E3.2.1.

Table E3.2.1   Summary of the Design of the Thermoelectric Generator

Parameters	Maximum conversion efficiency	Maximum power efficiency	Maximum performance parameters
Number of couples, n	350	331	-
R_L/R	1.608	1	-
Power, W_n(W)	590	590	590
Voltage, V_n(V)	29.13	22.34	47.25
Current, I (A)	20.25	26.41	52.82
Internal resistance, R (Ω)	0.895	0.846	-
Heat absorbed, Q_{hn} (W)	4501	4734	-
Heat dissipated, Q_{cn} (W)	3911	4140	-
Efficiency, η_t	0.131	0.125	-

Typically a thermoelectric generator is designed based on the maximum conversion efficiency (which has the minimum power) although the operating conditions range between the maximum conversion efficiency and the maximum power. Therefore, the number of thermocouples is determined to be 350. The load resistance, R_L, is determined from the resistance ratio, R_L/R, obtained. Remind that the above summary was computed based on the ideal thermoelectric generator. The real thermoelectric generator includes the thermal and electrical contact resistances and will be discussed in Section 3.10. However, a real thermoelectric generator was produced by Global Thermoelectric (Alberta, Canada), of which the model 8550 shows the power of 590 W, the number of thermocouples of 325, the voltage of 28 V, and the current of 21 A. It seems that the specification of the product looks close to the values computed in the present work. However, the comparison is not accurate unless the sizes and materials of thermoelements between them are matched. The size and material of the present thermoelement used were not confirmed with those of the product. The size of the present work (thermoelement length and cross-sectional area) has a somewhat larger intentionally than a practical size to minimize the thermal and contact resistances for the purpose of the present analysis. Detailed discussions are found in Section 3.10.

Example 3.3 Thermoelectric Generator—TEM   A thermoelectric generator is to be designed to recover the waste thermal energy from the exhaust pipeline in a vehicle as shown in Figure E3.3.1. The cold plate is cooled by circulating the vehicle's cooling water. The size for the thermoelectric modules is limited by 320 mm × 160 mm. The hot and cold junction temperatures are estimated to be 250°C and 50°C, respectively. The power generated should be no less than 150 W. The output voltage is required to be 14 V to charge the battery (12 V) in the vehicle. Design the thermoelectric generator using the TEG modules listed in Table 3.2 or Table 3.3, providing the power, the current, the voltage, the heat absorbed at the hot junction, the heat dissipated at the cold junction, the thermal efficiency, and the number of modules.

Solution   Consider eight modules of TEG-3b to meet the power requirement of 150 W. Since the ZT of the materials for the modules is 0.4, the discrepancy between the maximum power efficiency and the maximum conversion efficiency will be negligible within 1 % as shown in Figure 3.9. Hence, choose $R_L/R = 1$ for the maximum power.

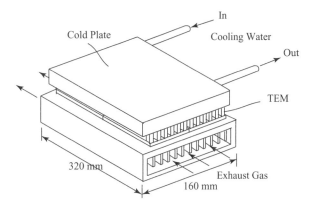

Figure E3.3.1 Thermoelectric generator device

The specifications of the selected module TEG-3b:

$$\dot{W}_{max} = 19\,\text{W}, I_{max} = 16\,\text{A}, V_{max} = 5.0\,\text{V}, ZT_{ave} = 0.4, \text{size} = 75\,\text{mm}$$

$$\times\, 75\,\text{mm} \times 5\,\text{mm}, \text{N} = 8 \qquad\qquad\qquad\qquad \text{(E3.3.1)}$$

N = number of modules

$$T_c/T_h = (50 + 273)\,\text{K}/(250 + 273)\,\text{K} = 0.62 \qquad\qquad \text{(E3.3.2)}$$

Single Module

Since $\dot{W}/\dot{W}_{max} = 1$ from Figure 3.8 with $R_L/R = 1$, $\dot{W} = \dot{W}_{max} = 19\,\text{W}$ (E3.3.3)

Since $I/I_{max} = 0.5$ from Figure 3.8, $I = 0.5 \times I_{max} = 8\,\text{A}$ (E3.3.4)

Since $V/V_{max} = 0.5$ from Figure 3.8, $V = 0.5 \times V_{max} = 2.5\,\text{V}$ (E3.3.5)

Using $T_c/T_h = 0.6$ and $Z\overline{T} = 0.4$, it is found from Figure 3.11 that

$$\eta_{mp} \approx 0.05 \qquad\qquad\qquad\qquad\qquad \text{(E3.3.6)}$$

$$\dot{Q}_h = \dot{W}/\eta_{mp} = 19\,\text{W}/0.05 = 380\,\text{W} \qquad\qquad \text{(E3.3.7)}$$

From Equation (3.16),

$$\dot{Q}_c = \dot{Q}_h - \dot{W} = 380\,\text{W} - 19\,\text{W} = 361\,\text{W} \qquad\qquad \text{(E3.3.8)}$$

Eight Modules   Eight modules are arranged electrically in series and thermally in parallel as shown in Figure E3.3.2.

The panel power is calculated as

$$\left(\dot{W}\right)_N = N\dot{W} = 8 \times 19\,\text{W} = 152\,\text{W} \qquad\qquad \text{(E3.3.9)}$$

The calculated panel power of 152 W is satisfied to meet the power requirement of 150 W.

$$(I)_N = I = \underline{8\,\text{A}} \qquad\qquad\qquad\qquad \text{(E3.3.10)}$$

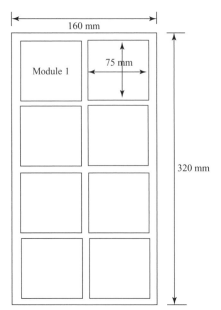

Figure E3.3.2 TEG Panel

$$(V)_N = N V = 8 \times 2.5 \, \text{V} = 20 \, \text{V} \tag{E3.3.11}$$

A DC/DC converter is used to bring down the voltage of 20 V to 14 V. The total heat absorbed at the hot junction temperature should be

$$(\dot{Q}_h)_N = N \dot{Q}_h = 8 \times 380 \, \text{W} = 3{,}040 \, \text{W} \tag{E3.3.12}$$

The heat sink in the exhaust gas passage should be properly designed to meet the heat transfer rate of 3040 W absorbed at the hot junction. The heat dissipated at the cold junction temperature is

$$(\dot{Q}_c)_N = N \dot{Q}_c = 8 \times 361 \, \text{W} = 2{,}888 \, \text{W} \tag{E3.3.13}$$

The cooling water flow rate is arranged to meet the heat dissipation of 2,888 W.

3.7 THERMOELECTRIC COOLERS (TEC)

The generation of an electric current by establishing a temperature difference across a thermoelectric material suggests that a counter effect can be created by passing an electric current through the material. This important discovery was made in 1834 by Jean Charles A. Peltier.

A simple electrical circuit for thermoelectric cooling (TEC) is shown in Figure 3.13. The amount of heat absorbed at the cold junction is associated with the Peltier cooling, the half of Joule heating, and the thermal conduction. The basic equation used in the design of the Peltier refrigerators is the determination of the net heat removed from the cold junction, namely,

$$\dot{Q}_c = \alpha T_c I - \frac{1}{2} I^2 R - K \Delta T \tag{3.49}$$

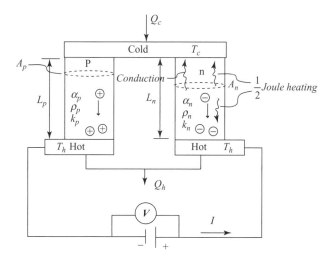

Figure 3.13   An electrical circuit for a thermoelectric cooling

where $\Delta T = T_h - T_c$ and K is the thermal conductance. The detailed derivation of Equation (3.49) can be found in Section B.3 in Appendix B.

3.7.1   Similar and Dissimilar Materials

3.7.1.1   Similar Materials

The similar materials here are interpreted as having the same base materials and the same geometry for both p-type and n-type thermoelectric elements. In this case, $L_p = L_n = L$ and $A_p = A_n = A$. The properties of α, ρ, and k can be handled as shown in Equations (3.5a), (3.5b), and (3.5c) and Equations (3.6a), (3.6b), and (3.6c).

The equations are repeated here as a convenience.

$$\alpha = \overline{\alpha}_p - \overline{\alpha}_n \tag{3.5a}$$

$$\rho = \overline{\rho}_p + \overline{\rho}_n \tag{3.5b}$$

$$k = \overline{k}_p + \overline{k}_n \tag{3.5c}$$

$$\alpha_p \cong -\alpha_n \tag{3.6a}$$

$$\rho_p \cong \rho_n \tag{3.6b}$$

$$k_n \cong k_p \tag{3.6c}$$

In Equation (3.49), the thermal conductance K by Fourier's law is shown in Equation (3.10) and rewritten here:

$$K = \frac{kA}{L} \tag{3.10}$$

where
$\quad$ k is the thermal conductivity
$\quad$ L is the length of an element, and
$\quad$ A is the cross-sectional area of a thermoelement

The internal electrical resistance of the thermoelectric couple is shown in Equation (3.11) and rewritten here:

$$R = \frac{\rho L}{A} \tag{3.11}$$

where ρ is the electrical resistivity.

3.7.1.2 Dissimilar Materials

Dissimilar materials here are interpreted as both dissimilar base materials and dissimilar geometry for the thermoelectric elements. Many thermoelectric materials in commercial products are indeed dissimilar. Hence, we want to develop the analysis for the dissimilar materials.

In Equation (3.49), α can be used as defined in Equations (3.5a) for the dissimilar materials, rewritten here

$$\alpha = \bar{\alpha}_p - \bar{\alpha}_n \tag{3.5a}$$

The thermal conductance is given in equation (3.12) for dissimilar materials and rewritten here:

$$K = \frac{k_p A_p}{L_p} + \frac{k_n A_n}{L_n} \tag{3.12}$$

The electrical resistance is shown in Equation (3.13) for the dissimilar materials and rewritten here:

$$R = \frac{\rho_p L_p}{A_p} + \frac{\rho_n L_n}{A_n} \tag{3.13}$$

For the figure of merit for the optimum geometry, Equations (3.7) can be rewritten as.

$$Z = \frac{\left(\alpha_p - \alpha_n\right)^2}{\left[\left(k_p \rho_p\right)^{\frac{1}{2}} + \left(k_n \rho_n\right)^{\frac{1}{2}}\right]^2} \tag{3.7}$$

For the optimum geometric ratio, Equation (3.8) can be rewritten here.

$$\frac{L_n/A_n}{L_p/A_p} = \left(\frac{\rho_p k_n}{\rho_n k_p}\right)^{\frac{1}{2}} \tag{3.8}$$

The following analysis after this point is practically the same for both the similar and dissimilar materials.

3.7.2 The Coefficient of Performance

The amount of heat rejected from the hot plate is also obtained as shown in Figure 3.13.

$$\dot{Q}_h = \alpha T_h I + \frac{1}{2} I^2 R - K \Delta T \tag{3.50}$$

Considering the 1st law of thermodynamics across the thermoelectric couple, we have

$$\dot{W} = \dot{Q}_h - \dot{Q}_c \tag{3.51}$$

The input electrical power (the amount of work per unit time), $\dot{W}$, across the thermocouple is obtained subtracting Equation (3.50) by (3.49).

$$\dot{W} = \alpha I \, (T_h - T_c) + I^2 R \tag{3.52}$$

where the first term is the rate of working to overcome the thermoelectric voltage, whereas the second term is the resistive loss. Since the power is $\dot{W} = IV$, the voltage across the couple in Figure 3.13 will be

$$V = \alpha \, (T_h - T_c) + IR \tag{3.53}$$

The coefficient of performance, COP, is similar to the thermal efficiency in meaning except that the value may exceed unity. The COP is defined by the ratio of the cooling rate to the input power.

$$COP = \frac{\dot{Q}_c}{\dot{W}} \tag{3.54}$$

Inserting $\dot{Q}_c$ and $\dot{W}$ given in Equations (3.49) and (3.52) into Equation (3.54) gives

$$COP = \frac{\alpha T_c I - \dfrac{1}{2} I^2 R - K \Delta T}{\alpha I \, (T_h - T_c) + I^2 R} \tag{3.55}$$

3.7.3 Optimum Current for the Maximum Cooling Rate

The design is commonly optimized for a prescribed geometry by determining the current. Differentiate Equation (3.49) with respect to I and set it to zero:

$$\frac{d\dot{Q}_c}{dI} = \alpha T_c - IR = 0 \tag{3.56}$$

Solving for I, the optimized current for maximizing the cooling rate can be obtained.

$$I_o = \frac{\alpha T_c}{R} \tag{3.57}$$

3.7.4 Maximum Performance Parameters

The maximum performance parameters below are useful in design. Equation (3.57) is virtually the maximum current for the given material and geometry. Hence, the maximum current is

$$I_{max} = \frac{\alpha T_c}{R} \tag{3.58}$$

The maximum possible temperature difference would occur when the current is equal to I_{max} and the heat removed at the cold junction is equal to zero.

$$\dot{Q}_c = 0 \text{ and } I = I_{max} \tag{3.59}$$

From Equation (3.49), the maximum possible temperature difference is

$$\Delta T_{\text{max}} = \frac{\alpha^2 T_c^2}{2KR} \tag{3.60}$$

where $KR = k\rho$ for similar materials and $KR = \left[(k_p \rho_p)^{\frac{1}{2}} + (k_n \rho_n)^{\frac{1}{2}} \right]^2$ for dissimilar materials.

Equivalently, using the figure of merit,

$$\Delta T_{\text{max}} = 0.5 \, Z T_c^2 \tag{3.61}$$

It is of interest that ΔT_{max} is related to Z and T_c, where Z is given in Equation (3.4) or (3.7). The maximum cooling power is then obtained by substituting Equation (3.58) into Equation (3.49).

$$\left(\dot{Q}_c \right)_{I,o} = \alpha T_c \left(\frac{\alpha T_c}{R} \right) - \frac{1}{2} \left(\frac{\alpha T_c}{R} \right)^2 R - K\Delta T \tag{3.62}$$

$$\left(\dot{Q}_c \right)_{I,o} = \frac{\alpha^2 T_c^2}{2R} - K\Delta T \tag{3.63}$$

This is the current-optimized cooling rate as

$$\left(\dot{Q}_c \right)_{I,o} = K \, (\Delta T_{\text{max}} - \Delta T) \tag{3.64}$$

If we let $\Delta T = 0$, this becomes the maximum possible cooling rate for the given material and geometry.

$$\left(\dot{Q}_c \right)_{\text{max}} = \frac{\alpha^2 T_c^2}{2R} = K\Delta T_{\text{max}} \tag{3.65}$$

The maximum voltage may be calculated with I_{max} and no heat load. Inserting Equations (3.58) and (3.60) into Equation (3.53) gives

$$V_{\text{max}} = \frac{\alpha^3 T_c^2}{2KR} + \alpha T_c \tag{3.66}$$

Or, equivalently,

$$V_{\text{max}} = \alpha \, (\Delta T_{\text{max}} + T_c) \tag{3.67}$$

3.7.5 Optimum Current for the Maximum *COP*

The maximum *COP* can be obtained by differentiating Equation (3.54) and setting it to zero.

$$\frac{d \, (COP)}{dI} = 0 \tag{3.68}$$

After some algebra, we have

$$I_{COP} = \frac{\alpha \, (T_h - T_c)}{R \left[(1 + Z\overline{T})^{\frac{1}{2}} - 1 \right]} \tag{3.69}$$

Inserting this into Equation (3.55) yields

$$COP_{\max} = \frac{T_c}{T_h - T_c} \frac{(1 + Z\overline{T})^{\frac{1}{2}} - \frac{T_h}{T_c}}{(1 + Z\overline{T})^{\frac{1}{2}} + 1} \tag{3.70}$$

$COP_{\max}$ approaches the Carnot limits of $\frac{T_c}{T_h - T_c}$ as $Z \to \infty$, where Z is defined in Equation (3.4) or (3.7).

3.7.6   Generalized Charts

We want to express the parameters in dimensionless forms, to obtain generalized graphs. With Equations (3.58) and (3.60), dividing Equation (3.49) by Equation (3.65) provides the maximum cooling ratio as

$$\frac{\dot{Q}_c}{\dot{Q}_{c\,\max}} = 2\left(\frac{I}{I_{\max}}\right) - \left(\frac{I}{I_{\max}}\right)^2 - \frac{\Delta T}{\Delta T_{\max}} \tag{3.71}$$

Dividing Equation (3.49) by Equation (3.52) provides the coefficient of performance of refrigerator:

$$COP_R = \frac{2\left(\dfrac{I}{I_{\max}}\right) - \left(\dfrac{I}{I_{\max}}\right)^2 - \dfrac{\Delta T}{\Delta T_{\max}}}{\dfrac{I}{I_{\max}}\left(\dfrac{\Delta T}{\Delta T_{\max}}\right) ZT_c + 2\left(\dfrac{I}{I_{\max}}\right)^2} \tag{3.72}$$

Equations (3.71) and (3.72) are graphically presented in Figure 3.14 using $ZT_c = 1$. ZT_c is a dimensionless figure of merit and an important parameter for the thermoelectric materials. Figure 3.14 shows the generalized relationship between the operating current ratio I/I_{max}, the cooling power ratio $\dot{Q}_c/\dot{Q}_{c\,max}$ and the coefficient of performance COP, as functions of the junction temperature difference ratio $\Delta T/\Delta T_{max}$. The higher the operating current ratio I/I_{max}, the greater the cooling power ratio $\dot{Q}_c/\dot{Q}_{c\,max}$. As the current ratio I/I_{max} is decreased, the COP increases and then decreases abruptly. The higher the COP for a given cooling power the lower the ratio of electrical input power/cooling power. The heat generated at the hot junctions must be dissipated by a heat sink or a heat exchanger. A suitable operating current is between the values corresponding to COP_{max} and I_{max}. It is noted that the lower the junction temperature difference $\Delta T/\Delta T_{max}$, the greater the cooling power ratio $\dot{Q}_c/\dot{Q}_{c\,max}$ and the COP.

The effect of ZT_c on the cooling rate $\dot{Q}_c$ and on the COP cannot be solely judged on Equations (3.71) and (3.72) since ΔT_{max} is also a function of ZT_c. In general, the higher ZT_c the greater the COP and the greater the cooling rate. Actually, many efforts are made to increase the value to $ZT_c > 1$ in commercial materials. Dividing Equation (3.53) by Equation (3.67) provides the terminal voltage ratio,

$$\frac{V}{V_{\max}} = \frac{\left(\dfrac{\Delta T}{\Delta T_{\max}}\right)\dfrac{ZT_c}{2} + \dfrac{I}{I_{\max}}}{\dfrac{ZT_c}{2} + 1} \tag{3.73}$$

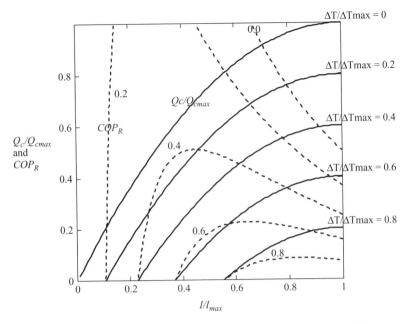

Figure 3.14 Generalized Charts for a single stage Peltier module. $ZT_c = 1$

Figure 3.15 shows the generalized relationship between the three parameters operating junction temperature difference ratio, voltage ratio V/V_{max}, and cooling power ratio Q_c/Q_{cmax}, as a function of the operating current ratio I/I_{max}.

3.7.7 Optimum Geometry for the Maximum Cooling in Similar Materials

Define the geometric ratio G,

$$G = \frac{A}{L} \tag{3.74}$$

Hence, the resistance is

$$R = \frac{\rho L}{A} = \frac{\rho}{G} \tag{3.75}$$

The cooling rate in Equation (3.49) for similar materials becomes

$$\dot{Q}_c = \alpha T_c I - \frac{1}{2} I^2 \frac{\rho}{G} - kG\Delta T \tag{3.76}$$

From Equation (3.57) and (3.75), the optimized current is

$$I_o = \frac{\alpha T_c G}{\rho} \tag{3.77}$$

Solving for G gives the current optimized geometric ratio $G_{I,o}$ as

$$G_{I,o} = \frac{\rho I_o}{\alpha T_c} \tag{3.78}$$

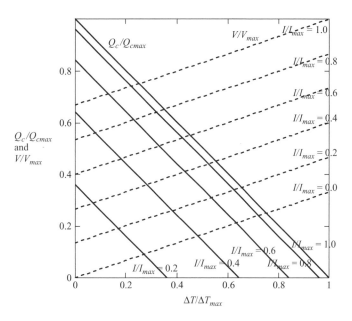

Figure 3.15 Generalized charts for a single stage Peltier module, $ZT_c = 1$

To maximize the cooling rate, differentiate it with respect to the geometric ratio G.

$$\frac{d\dot{Q}_c}{dG} = -\frac{1}{2}I^2\frac{\rho}{G^2} + k\Delta T = 0 \tag{3.79}$$

$$G_o = I\left(\frac{\rho}{2k\Delta T}\right)^{\frac{1}{2}} \tag{3.80}$$

When I is set equal to I_o, the following relationship is established:

$$\frac{G_{I,o}}{G_o} = \left(\frac{\Delta T}{\Delta T_{\max}}\right)^{\frac{1}{2}} \tag{3.81}$$

3.7.8 Thermoelectric Modules

The number of thermocouples is denoted as a subscript n. The parameters of a thermoelectric module are given as a single thermocouple multiplied by n (See Figure 3.2)

$$\left(\dot{W}\right)_n = n\dot{W} \tag{3.82}$$

$$\left(\dot{Q}_c\right)_n = n\dot{Q}_c \tag{3.83}$$

$$\left(\dot{Q}_h\right)_n = n\dot{Q}_h \tag{3.84}$$

$$(R)_n = nR \tag{3.85}$$

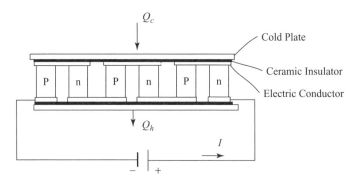

Figure 3.16 Multiple couples of a thermoelectric cooling device

$$(V)_n = nV \tag{3.86}$$

$$(K)_n = nK \tag{3.87}$$

The parameters that are not to be multiplied by n are

$$(I)_n = I \tag{3.88}$$

$$(COP)_n = \frac{(\dot{Q}_c)_n}{(\dot{W})_n} = COP \tag{3.89}$$

$$\left(\frac{\dot{Q}_c}{\dot{Q}_{\max}}\right)_n = \frac{\dot{Q}_c}{\dot{Q}_{\max}} \tag{3.90}$$

$$\left(\frac{I}{I_{\max}}\right)_n = \frac{I}{I_{\max}} \tag{3.91}$$

$$\left(\frac{V}{V_{\max}}\right)_n = \frac{V}{V_{\max}} \tag{3.92}$$

Table 3.4 Performance and Geometry of Commercial Cooling Modules for Standard TEC

| | $T_h = 25°C$ | | | | | Dimensions (mm) | | |
Module	$\Delta T_{\max}$ (°C)	$Q_{\max}$ (Watts)	$I_{\max}$ (Amps)	$V_{\max}$ (Volts)	Number of Couples	Width	Length	Height
1a	68	1.8	3.9	0.85	7	9.4	9.4	4.7
2a	67	5.3	2.5	3.8	31	15	15	3.2
3a	67	14.4	3.0	8.6	71	23	23	3.6
4a	66	20.2	2.7	15.3	127	30	30	3.6
5a	67	51.4	6.0	15.4	127	40	40	3.8
6a	67	120	14	15.4	127	62	62	4.6

Table 3.5 Performance and Geometry of Commercial Cooling Modules for High Watts Density TEC

| Module | $T_h = 25°C$ | | | | | Dimensions (mm) | | |
	ΔT_{max} (°C)	Q_{max} (Watts)	I_{max} (Amps)	V_{max} (Volts)	Number of Couples	Width	Length	Height
1b	69	69	7.9	14.4	127	25	25	1.9
2b	69	69	7.9	14.4	127	30	30	2.6
3b	69	95	11	14.4	127	30	30	2.4
4b	69	126	14.6	14.4	127	40	40	2.8

3.7.9   Commercial TEC

Multiple thermocouples can be used as shown to increase the cooling capacity as schematically shown in Figure 3.16. An electrical insulator is usually placed between the electric conductor and the cold plate. The performance and geometry of typical Peltier modules for high power density TEC are shown in Table 3.4 and Table 3.5.

3.7.10   Multistage Modules

The maximum temperature difference can be reached by a single-stage module. However, it is possible to obtain greater temperature differences by operating cooling units in a cascade shown in Figure 3.17. The first stage of the cascade provides a low-temperature heat sink for the second stage which, in turn, provides a sink at an even lower temperature for the third stage, and so on. The single-stage maximum COP is shown in Equation (3.70). Rewritten here:

$$COP_{max} = \frac{T_c}{T_h - T_c} \frac{(1 + Z\overline{T})^{\frac{1}{2}} - \frac{T_h}{T_c}}{(1 + Z\overline{T})^{\frac{1}{2}} + 1} \tag{3.70}$$

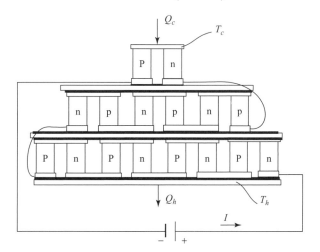

Figure 3.17 Three-stage module

Assume that each stage is to be operated with the same maximum *COP* as the single stage. Since it is optional, the *N*-stage maximum *COP* is obtained as [14]

$$(COP_{max})_N = \left[\left(1 + \frac{1}{COP_{max}}\right)^N - 1\right]^{-1} \tag{3.93}$$

where *N* is the number of the stages.

If we further assume that the temperature difference between the hot and cold temperatures is the same for each stage while maintaining the hot junction temperature at 300 K, the *N*-stage cold temperature will be

$$(T_c)_N = T_c - (N - 1)(T_h - T_c) \tag{3.94}$$

Solving for the single-stage cold temperature,

$$T_c = \frac{(T_c)_N + (N - 1)T_h}{N} \tag{3.95}$$

Substituting Equation (3.95) into Equation (3.70), Equation (3.93) is plotted in Figure 3.18. The results slightly differ from those found in Nolas et. al [14]. A conservative value of $Z = 2.5 \times 10^{-3}$ K^{-1} and a hot junction temperature of 300 K were used in the calculations.

It is clear that there is a substantial gain in $(COP)_N$ for the cold junction temperature below about 250 K if a multistage unit is used. It is, of course, essential to use more than one stage when the cold temperature is below 233 K, even when the required cooling power is negligible. Also the minimum values of the cold temperature for the multistage modules may be estimated from the Figure 3.18.

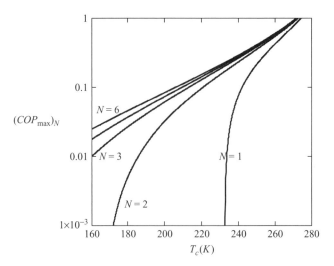

Figure 3.18   Coefficient of Performance vs. T_c for N-stage thermoelectric cascade. It is assumed that $Z = 2.5 \times 10^{-3}$ K^{-1} and $T_h = 300$ K

Table 3.6 Maximum Performance of Commercial Multistage Peltier Modules

Number of stages	ΔT_{max} (°C)	$T_h = 25°C$ Q_{cmax} (W)	I_{max} (A)	V_{max} (V)	Dimensions (mm) Top width	Top length	Bottom width	Bottom length	height
2-Stage-31-17	81	6.0	4.0	3.8	15	15	20	20	7.2
3-Stage-127-71-31	96	12.6	3.5	15.4	20	20	40	40	10.4
4-Stage-127-71-31-17	107	6.8	3.1	14.6	15	15	40	40	13.8
5-Stage- 127-71-31-17-7	118	3.4	3.0	14.5	10	10	40	40	16.9
6-Stage-127-71-31-17-7-2	131	1.2	3.0	14.5	5	5	40	40	20.1

3.7.10.1 *Commercial Multistage Peltier Modules*

Commercial multistage Peltier modules cover a range of ceramic face sizes from $3.2 \times 3.2\,mm^2$ to $62 \times 62\,mm^2$ at the top (cooling side), from $3.8 \times 3.8\,mm^2$ to $62 \times 62\,mm^2$ at the base (heated side), heights from 3.8 mm to 21.4 mm, and I_{max} from 0.7 to 9.5 A, Q_{cmax} from 0.39 W to 59 W, V_{max} from 0.8 to 14 V, and the number of stages from two to six. Table 3.6 shows the maximum performance parameters of the leading commercial multistage Peltier modules.

3.7.11 Design Options

There are usually three objectives in the application of thermoelectric modules: (1) maximum heat pumping, (2) maximum *COP*, and (3) maximum speed of response. Maximum heat pumping requires the module to operate at the current and voltage that pumps the maximum amount of heat over the specific temperature difference. The current at the maximum *COP* is generally less than the current for maximum heat pumping. The design of a module for maximum speed of response is more complex. Consideration must not only be given to the thermal and electrical characteristics of the module, but also to the thermal mass of the module and the device being cooled. In addition, the characteristics of the heat sink must be considered in the overall design [1].

Example 3.4 Thermoelectric Cooler   A thermoelectric cooler consists of n-type and p-type semiconductors as illustrated in Figure E3.4.1. The cold junction and hot junction temperatures are maintained at 270 K and 300 K, respectively. The semiconductor has dimensions of A = 0.25 cm^2 and L = 1.05 cm. The thermoelectric materials are $\alpha_n = -187 \times 10^{-6}$ V/K, $\alpha_p = 187 \times 10^{-6}$ V/K, $\rho_n = \rho_p = 1.64 \times 10^{-3}\,\Omega$ cm, $k_n = k_p = 0.0146$ W/cm K.

(a) Determine the current, the voltage, the cooling power, the heat delivered, and the *COP* for the maximum cooling power.

(b) Determine the current, the voltage, the cooling power, the heat delivered, and the power for the maximum *COP*.

(c) Determine the maximum performance parameters: the maximum current, the maximum temperature, the maximum cooling power, and the maximum voltage.

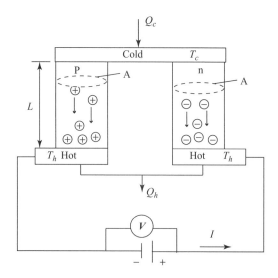

Figure E3.4.1   Schematic of a thermoelement couple

MathCAD Format Solution: The properties of the material are

$$\alpha_n := -190 \cdot 10^{-6} \frac{V}{K} \quad \alpha_p := 190 \cdot 10^{-6} \frac{V}{K}$$

$$\rho_n := 1.05 \cdot 10^{-3} \, \Omega \, cm \quad \rho_p := 1.05 \cdot 10^{-3} \, \Omega \, cm \qquad \text{(E3.4.1)}$$

$$k_n := 0.0125 \frac{W}{cm \, K} \quad k_p := 0.0125 \frac{W}{cm \, K}$$

The relative properties for the thermocouple are

$$\alpha := \alpha_p - \alpha_n \quad \rho := \rho_p + \rho_n \quad k := k_p + k_n \qquad \text{(E3.4.2)}$$

The dimension of the thermoelement is

$$L_e := 1.05 \, cm \quad A_e := 0.25 \, cm^2 \qquad \text{(E3.4.3)}$$

The cold and hot junction temperatures and the temperature difference are

$$T_h := 300 \, K \quad T_c := 270 \, K \quad \Delta T := T_h - T_c \qquad \text{(E3.4.4)}$$

The figure of merit is

$$Z := \frac{\alpha^2}{\rho \, k} \quad Z = 2.75 \times 10^{-3} \frac{1}{K} \qquad \text{(E3.4.5)}$$

The average temperature is defined and the dimensionless figure of merit is computed.

$$T_{avg} := \frac{T_h + T_c}{2} \quad Z \, T_{avg} = 0.784 \qquad \text{(E3.4.6)}$$

The internal electric resistance of the thermocouple is

$$R_e := \frac{\rho\, L_e}{A_e} \quad R_e = 8.82 \times 10^{-3}\,\Omega \qquad (E3.4.7)$$

The thermal conductance is

$$K_e := \frac{k\, A_e}{L_e} \quad K_e = 5.952 \times 10^{-3}\,\frac{W}{K} \qquad (E3.4.8)$$

(a) Maximum Cooling Power   From Equation (3.57), the maximum current for the optimum power is

$$I_o := \frac{\alpha\, T_c}{R_e} \quad I_o = 11.633\,A \qquad (E3.4.9)$$

From Equation (3.49), the cooling power for the maximum cooling power is

$$Q_{co} := \alpha\, I_o\, T_c - \frac{1}{2} I_o^2 R_e - K_e(T_h - T_c) \quad Q_{co} = 0.418\,W \qquad (E3.4.10)$$

From Equation (3.52), the electrical power for the maximum cooling power is

$$W_o := [\alpha\, I_o\,(T_h - T_c)] + I_o^2 R_e \quad W_o = 1.326\,W \qquad (E3.4.11)$$

From Equation (3.54), the COP for the maximum cooling power is

$$COP_o := \frac{Q_{co}}{W_o} \quad COP_o = 0.315 \qquad (E3.4.12)$$

From Equation (3.51), the heat dissipated on the hot junction is

$$Q_{ho} := Q_{co} + W_o \quad Q_{ho} = 1.744\,W \qquad (E3.4.13)$$

From Equation (3.53), the voltage across the thermocouple is

$$V_o := \alpha\,(T_h - T_c) + I_o R_e \quad V_o = 0.114\,V \qquad (E3.4.14)$$

(b) Maximum COP   The parameters for the maximum COP are obtained as follows
Using Equation (3.69), the optimum current for the maximum COP is

$$I_{cop} := \frac{\alpha\,(T_h - T_c)}{R_e\left[(1 + Z\, T_{avg})^{\frac{1}{2}} - 1\right]} \quad I_{cop} = 3.851\,A \qquad (E3.4.15)$$

Using Equation (3.49), the cooling power for the maximum COP is

$$Q_{cop} := \alpha\, T_c\, I_{cop} - \frac{1}{2} I_{cop}^2 R_e - K_e(T_h - T_c) \quad Q_{cop} = 0.151\,W \qquad (E3.4.16)$$

The electrical power for the maximum COP is

$$W_{cop} := [\alpha\, I_{cop}\,(T_h - T_c)] + I_{cop}^2 R_e \quad W_{cop} = 0.175\,W \qquad (E3.4.17)$$

The maximum COP is

$$\text{COP}_{\max} := \frac{Q_{\text{cop}}}{W_{\text{cop}}} \quad COP_{\max} = 0.865 \tag{E3.4.18}$$

The heat dissipated on the hot junction for the maximum COP is

$$Q_{\text{hcop}} := Q_{\text{cop}} + W_{\text{cop}} \quad Q_{\text{hcop}} = 0.326\,\text{W} \tag{E3.4.19}$$

The voltage across the thermocouple for the maximum COP is

$$V_{\text{cop}} := \alpha\,\Delta T + I_{\text{cop}}\,R_e \quad V_{\text{cop}} = 0.045\,\text{V} \tag{E3.4.20}$$

c) Maximum Performance Parameters
Using Equation (3.58), the maximum current is

$$I_{\max} := \frac{\alpha T_c}{R_e} \quad I_{\max} = 11.633\,\text{A} \tag{E3.4.21}$$

Using Equation (3.60), the maximum temperature is

$$\Delta T_{\max} := \frac{\alpha^2\,T_c^2}{2\,K_e\,R_e} \quad \Delta T_{\max} = 100.255\,\text{K} \tag{E3.4.22}$$

Using Equation (3.65), the maximum cooling power is

$$Q_{\text{cmax}} := K_e\,\Delta T_{\max} \quad Q_{\text{cmax}} = 0.597\,\text{W} \tag{E3.4.23}$$

Using Equation (3.67), the maximum voltage is

$$V_{\max} := \alpha\,(\Delta T_{\max} + T_c) \quad V_{\max} = 0.141\,\text{V} \tag{E3.4.24}$$

Two sets of the performances for the maximum cooling power and the maximum *COP* are presented in Table E3.4.1. The cooling power reveals that the maximum cooling power gains about three times of the maximum *COP*, whereas it has to consume about seven times of electrical power of the maximum *COP*. Remind that the values

Table E3.4.1 The Results for the Performances of a Thermocouple Given

Parameter	Maximum cooling power	Maximum COP	Maximum performance parameters
Current, I (A)	11.633	3.851	—
Voltage, V (V)	0.114	0.045	—
Cooling power, Q_c (W)	0.418	0.151	—
Heat rejected, Q_h (W)	1.744	0.326	—
Electrical Power, W (W)	1.326	0.175	—
COP	0.315	0.865	—
I_{max}, (A)	—	—	11.633
ΔT_{max}, (K)	—	—	100.25
Q_{cmax}, (W)	—	—	0.597
V_{max}, (V)	—	—	0.141

Operating conditions: $T_h = 300\text{K}$, $T_c = 270\,\text{K}$, $ZT_{avg} = 0.784$, $L = 1.05\,\text{cm}$, $A = 0.25\,\text{cm}^2$.

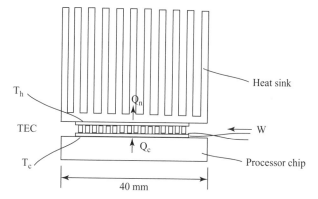

Figure E3.5.1 TEC aided cooling device with a forced air cooled heat sink

in the table were made based on the ideal thermocouple. The real thermocouple should include the thermal and electrical contact resistances, which can be found in Section 3.10. This example was developed for the purpose of analysis, not for the realistic information.

Example 3.5 Design of a Thermoelectric Cooler A 40×40 mm processor chip is cooled using a TEC with a heat sink, as shown in Figure E3.5.1. The cold and hot temperatures are required to be 56°C and 70°C, respectively. The maximal processor heat dissipation (heat load) is 25 Watts. A commercial module is to be selected from the list of modules in Table 3.4 or Table 3.5 in Section 3.7.9. Usually manufacturers provide their generalized charts for the selection of their products. However, we want to practice the design procedure using the generalized charts developed here for the ideal thermoelectric coolers as illustrated in Figures B.7 through B.9 in Appendix B.

The heat load (or cooling rate or heat pumping) is given as $Q_c = 25$ W. The cold and hot temperatures are given as $T_c = 56$°C and $T_h = 70$°C so that $\Delta T = 14$°C. Assume that $ZT_c = 1$

Design concept   We choose the optimum *COP* in order to minimize the electrical power that must be dissipated through the heat sink of the hot junction side as shown in Figure E3.5.1.

Solution   Step 1: Determine the maximum cooling rate, Q_{max}.

ΔT_{max} of the most commercial TEC is found to be approximately 68°C, as shown in Tables 3.4 and 3.5. Hence, we calculate the maximum temperature ratio $\Delta T / \Delta T_{max}$ as

$$\Delta \text{T} / \Delta T_{max} = 14°\text{C} / 68°\text{C} = 0.205$$

From Figure B.7, we find that the maximum *COP* of approximately is 1.4 and the maximum current ratio I / I_{max} is about 0.25 at the maximum temperature ratio $\Delta T / \Delta T_{max} = 0.2$.

From Figure B.8, using both $I / I_{max} = 0.25$ and $\Delta T / \Delta T_{max} = 0.2$, we find an approximate maximum cooling rate ratio $Q_c / Q_{max} \approx 0.26$. Hence,

$$Q_{max} = Q_c / 0.26 = 25/0.26 = 96.1 \text{ W}$$

Step 2: Choose a commercial TEC module with the Q_{max} just obtained.

Considering both the maximum heat load and the size of the processor chip, we choose TEC Module 3b from Table 3.5 with these maximum values $\Delta T_{max} = 69°C$, $Q_{max} = 95$ W, $I_{max} = 11$ A, and $V_{max} = 14.4$, dimension: $30 \times 30 \times 2.4$ mm, number of couples: 127.

If the commercial Q_{max} is significantly different from the obtained Q_{max}, we should recalculate the maximum heat load ratio and the maximum current ratio based on the commercial Q_{max} as

$$\frac{\dot{Q}_c}{\dot{Q}_{max}} = \frac{25 \text{ Watts}}{95 \text{ Watts}} = 0.263. \text{ From Figure B.8, } I/I_{max} \approx 0.25$$

Step 3: Determine the input power, current, voltage, and heat dissipation at the hot junction.

$$\text{Since } COP = \frac{\dot{Q}_c}{\dot{W}} \text{ and } \dot{W} = \dot{Q}_h - \dot{Q}_c$$

$$\dot{W} = \frac{\dot{Q}_c}{COP} = \frac{25}{1.4} = 17.86 \text{ Watts}$$

$$Q_h = 25 \text{ Watts} + 17.86 \text{ Watts} = 42.86 \text{ Watts}$$

This must be dissipated to the ambient air through the heat sink.

$$I = 0.25 \times I_{max} = 0.2 \times 11 \text{ A} = 2.2 \text{ A}$$

From Figure B.9, we find the maximum voltage ratio using $\Delta T/\Delta T_{max} = 0.2$ and $I/I_{max} = 0.25$

$$\frac{V}{V_{max}} \approx 0.22$$

The terminal voltage is

$$V = 0.22 \times 14.4 = 3.2 \text{ Volts}$$

When you use the general charts provided by the manufacturers, more accurate results may be obtained.

3.8 APPLICATIONS

3.8.1 Thermoelectric Generators

A thermoelectric generator is a power generator without moving parts, is silent in operation, and is reliable. Its relatively low efficiency (typically around 5 %) has restricted its use to specialized medical, military, and space applications, such as nuclear power for pacemakers, radioisotope power for deep space probes, and remote power, such as oil pipe lines and sea buoys, where cost is not a main consideration. In recent years, an increasing public awareness of environmental issues, especially, pertaining to global warming, has resulted in broad-based research into alternative commercial methods of generating electrical power. Thermoelectrics has emerged as a serious contender. It has attracted increasing attention as a green source of electricity able to meet a wide range of power requirements [8].

3.8.2 Thermoelectric Coolers

Perhaps the widest application of thermoelectric cooling has been in the control of temperature, particularly for scientific instruments and for electronic and optoelectronic systems. Sometimes accurate control of temperature is not needed. Instead, it is important to reduce the temperature of some device so that its efficiency is improved or, perhaps, can actually operate. Examples include the cooling of lasers and infrared detectors. In such cases, it is the convenience of thermoelectric refrigeration for small-scale applications that is a key factor. One of the most thorough assessments of thermoelectric air conditioning has been carried out in a train carriage on the French railway [12]. Another successful large-scale application is a 27-kW industrial water cooler [13].

3.9 DESIGN EXAMPLE

A thermoelectric air cooler is designed to pump out heat at a rate of 21 Watts from electronic components in a closed box located at a remote area, as shown in Figure 3.19. The air cooler consists of an internal fan, an internal heat sink, a thermoelectric cooler (TEC), an external heat sink, and an external fan. For the heat transfer rate given, the air temperature in the box must be maintained at 25°C when the ambient air is at 40°C.

The voltage across the thermoelectric cooler is preferred to be 12 V. The dimensions of both the internal and external heat sinks is preferred to be $W_1 = L_1 = 76$ mm, $b_1 = 30$ mm, $W_2 = L_2 = 100$ mm, and $b_2 = 30$ mm. The volume flow rates of the internal and external fans are also given to be 14 cfm (cubic feet per minute) and 25 cfm, respectively. Design the internal and external heat sinks by providing the fin thicknesses, the fin spacings, and the number of fins to meet the requirements.

Design the thermoelectric cooler (TEC) with the properties given and provide the high and low junction temperatures, T_h and T_c. The properties for the n-type and p-type semiconductors are given as $\alpha_n = -2.2 \times 10^{-4}$ V/K , $\rho_n = 1.2 \times 10^{-3}\,\Omega$ cm,

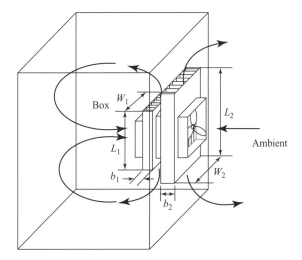

Figure 3.19 Air flow of a thermoelectric air cooler

$k_n = 1.62 \times 10^{-2}$ W/cm K, $\alpha_p = 2.2 \times 10^{-4}$ V/K, $\rho_p = 1.2 \times 10^{-3}\Omega$ cm, and $k_p = 1.62 \times 10^{-2}$ W/cm K.

3.9.1 Design Concept

In order to visualize the energy flow in the entire system, a thermal circuit is constructed, which is schematically shown in Figure 3.20 (b). R_i and R_o are the overall thermal resistances for the internal heat sink and external heat sink, respectively, which were defined in Equation (2.140). The components of the air cooler are an internal heat sink, a thermoelectric module, and an external heat sink as shown in Figure 3.20 (a). $\dot{Q}_c$ is the amount of heat transported at the internal heat sink, which is actually the design requirement (21 Watts). T_c and T_h are unknown at this moment since they depend on the design of all the components. An electric power is to be applied to the TEC to make the energy flow from a low temperature to a high temperature. Applying an energy balance or 1st law of thermodynamics to the system gives a relationship as

$$\dot{Q}_c = \dot{Q}_h - \dot{W} \tag{3.96}$$

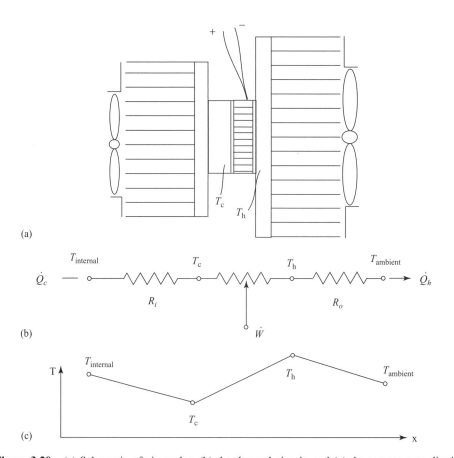

Figure 3.20 (a) Schematic of air cooler, (b) the thermal circuit, and (c) the temperature distribution

This indicates that $\dot{Q}_h$ is $\dot{Q}_c$ plus $\dot{W}$. This is the reason why the external heat sink appears larger than the internal heat sink. T_h is usually assumed to be approximately 15°C higher than the ambient temperature so that it allows the heat dissipation to be possible. We now assume that T_c is approximately 10°C lower than the internal air temperature for the same reason. However, this assumption may be improved by considering the relationship between the temperature difference $(T_h - T_c)$, $\dot{Q}_c$, and *COP* given in Figure 3.14. The smaller the temperature difference, the greater the heat flow or the *COP* is to be expected. However, this characteristic may be constrained by the heat sinks attached to both sides.

With these initial assumptions on T_c and T_h, we can move on to designing the internal and external heat sinks, eventually formulating the heat transfer equations as a function of T_c and T_h. It turned out that the assumed values for T_c and T_h negligibly affect the design of the heat sinks. Also we can design the TEC by using a commercial TEC module. It is somewhat important to obtain the properties of the TEC module, so that we approach theoretically the design problem. The calculated maximum performance parameters allow us to choose a TEC module for our purpose. Or we can design the TEC module that could be custom-fabricated. We develop the equations to represent the TEC module only as a function of T_c and T_h. Although we have completed the design of all the components of the thermoelectric air cooler, interestingly we do not know the exact performance until we have the solution by coupling the equations for the heat sinks and the TEC, which gives only two equations with the two parameters T_c and T_h. Two unknowns with two equations can be solved for T_c and T_h.

3.9.2 Design of Internal and External Heat Sinks

MathCAD Format Solution

Subscript 1 denotes the internal heat sink and subscript 2 denotes the external heat sink. Geometry of the internal and external heat sinks:

$$W_1 := 76\,\text{mm} \quad W_2 := 100\,\text{mm}$$

$$L_1 := 76\,\text{mm} \quad L_2 := 100\,\text{mm} \tag{3.97}$$

$$b_1 := 30\,\text{mm} \quad b_2 := 30\,\text{mm}$$

The volume flow rates are

$$V_{\text{fr1}} := 14\,\text{cfm} \quad V_{\text{fr2}} := 25\,\text{cfm} \tag{3.98}$$

The cross flow areas are

$$A_{c1} := b_1\,W_1 \quad A_{c2} := b_2\,W_2 \tag{3.99}$$

The velocities are (perpendicular flow divided by two)

$$U_1 := \frac{1}{2}\frac{V_{\text{fr1}}}{A_{c1}} \quad U_1 = 1.449\,\frac{\text{m}}{\text{s}} \tag{3.100}$$

$$U_2 := \frac{1}{2}\frac{V_{fr2}}{A_{c2}} \quad U_2 = 1.996\,\frac{m}{s} \qquad (3.101)$$

The internal and external air temperatures and the cold and hot junction temperatures will be estimated now and updated later.

$$T_{int} := 25°C \quad T_c := 15°C \quad \theta_1 := T_{int} - T_c \qquad (3.102)$$

$$T_{ext} := 40°C \quad T_h := 55°C \quad \theta_2 := T_h - T_{ext} \qquad (3.103)$$

The properties of air at the film temperatures $(15 + 25)/2 = 20°C$ and $(40 + 45)/2 = 42.5°C$ (Table A.2)

$$K_{air1} := 0.0263\,\frac{W}{m\,K} \quad \nu_{air1} := 15.89\,10^{-6}\,\frac{m^2}{s} \quad Pr_{air1} := 0.707 \qquad (3.104)$$

$$K_{air2} := 0.0274\,\frac{W}{m\,K} \quad \nu_{air2} := 17.40\,10^{-6}\,\frac{m^2}{s} \quad Pr_{air2} := 0.705 \qquad (3.105)$$

The properties of aluminum are obtained from Table A.3 as

$$\rho_{al} := 2702\,\frac{kg}{m^3} \quad k_{al} := 177\,\frac{W}{m\,K} \qquad (3.106)$$

The characteristic length for flow over a fin plate is divided by 2 assuming that the perpendicular flow by a fan takes half of the total length.

$$L_{c1} := \frac{L_1}{2} \quad L_{c2} := \frac{L_2}{2} \qquad (3.107)$$

The Reynolds numbers show laminar flows.

$$Re_1 := \frac{U_1\,L_{c1}}{\nu_{air1}} \quad Re_1 = 3.465 \times 10^3 \quad \text{less than } 10 \times 10^5 \text{ (laminar flow)} \qquad (3.108)$$

$$Re_2 := \frac{U_2\,L_{c2}}{\nu_{air2}} \quad Re_2 = 5.651 \times 10^3 \quad \text{less than } 10 \times 10^5 \text{ (laminar flow)} \qquad (3.109)$$

The optimum fin spacing is found using Equation (2.110)

$$z_{opt1} := L_{c1}\,3.24\,Re_1^{-\frac{1}{2}}\,Pr_{air1}^{-\frac{1}{4}} \quad z_{opt1} = 2.281\,mm \qquad (3.110)$$

$$z_{opt2} := L_{c2}\,3.24\,Re_2^{-\frac{1}{2}}\,Pr_{air2}^{-\frac{1}{4}} \quad z_{opt2} = 2.352\,mm \qquad (3.111)$$

Using Equation (1.53), the average heat transfer coefficients for laminar flow are

$$h_1 := \frac{k_{air1}}{L_{c1}}\,0.664\,Re_1^{\frac{1}{2}}\,Pr_{air1}^{\frac{1}{3}} \quad h_1 = 24.099\,\frac{W}{m^2\,K}$$

$$(3.112)$$

$$h_2 := \frac{k_{air2}}{L_{c2}}\,0.664\,Re_2^{\frac{1}{2}}\,Pr_{air2}^{\frac{1}{3}} \quad h_2 = 24.344\,\frac{W}{m^2\,K}$$

The number of fins is obtained as a function of fin thickness in order to find the optimum fin thickness.

$$n_{f1}(t_1) := \frac{W_1}{z_{opt1} + t_1} \quad n_{f2}(t_2) := \frac{W_2}{z_{opt2} + t_2} \tag{3.113}$$

The mass of fins is

$$m_{f1}(t_1) := n_{f1}(t_1)\rho_{al} L_1 b_1 t_1 \quad m_{f2}(t_2) := n_{f2}(t_2)\rho_{al} L_2 b_2 t_2 \tag{3.114}$$

Using Equation (2.55), we have

$$\beta_1(t_1) := b_1 \left(\frac{2 h_1}{k_{al} t_1}\right)^{\frac{1}{2}} \quad \beta_2(t_2) := b_2 \left(\frac{2 h_2}{k_{al} t_2}\right)^{\frac{1}{2}} \tag{3.115}$$

The single fin efficiency is found using Equation (2.22).

$$\eta_{f1}(t_1) := \frac{\tan h(\beta_1(t_1))}{\beta_1(t_1)} \quad \eta_{f2}(t_2) := \frac{\tan h(\beta_2(t_2))}{\beta_2(t_2)} \tag{3.116}$$

The single fin areas are

$$A_{f1}(t_1) := 2 (L_1 + t_1) b_1 \quad A_{f2}(t_2) := 2(L_2 + t_2) b_2 \tag{3.117}$$

The total surface areas are obtained considering the single fin area and the interfins area.

$$A_{t1}(t_1) := n_{f1}(t_1) (A_{f1}(t_1) + L_1 z_{opt1}) \tag{3.118}$$

$$A_{t2}(t_2) := n_{f2}(t_2) (A_{f2}(t_2) + L_2 z_{opt2}) \tag{3.119}$$

The overall efficiency is obtained using Equation (2.139).

$$\eta_{o1}(t_1) := 1 - n_{f1}(t_1) \frac{A_{f1}(t_1)}{A_{t1}(t_1)}(1 - \eta_{f1}(t_1)) \tag{3.120}$$

$$\eta_{o2}(t_2) := 1 - n_{f2}(t_2) \frac{A_{f2}(t_2)}{A_{t2}(t_2)}(1 - \eta_{f2}(t_2)) \tag{3.121}$$

Now the total heat transfers are obtained using Equation (2.133).

$$q_{tot1}(t_1) := \eta_{o1}(t_1) A_{t1}(t_1) h_1 \theta_1 \tag{3.122}$$

$$q_{tot2}(t_2) := \eta_{o2}(t_2) A_{t2}(t_2) h_2 \theta_2 \tag{3.123}$$

The overall thermal resistances are obtained using Equation (2.140).

$$R_{to1}(t_1) := \frac{1}{\eta_{o1}(t_1) h_1 A_{t1}(t_1)} \quad R_{to2}(t_2) := \frac{1}{\eta_{o2}(t_2) h_2 A_{t2}(t_2)} \tag{3.124}$$

A graphical range of fin thickness is

$$t_1 := 0.01 \, mm, 0.02 \, mm..2 \, mm$$

$$t_2 := 0.01 \, mm, 0.02 \, mm..2 \, mm$$

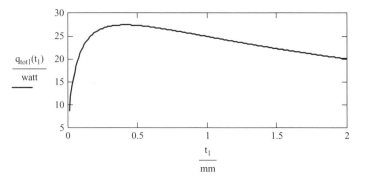

Figure 3.21 Total heat transfer rate vs. fin thickness

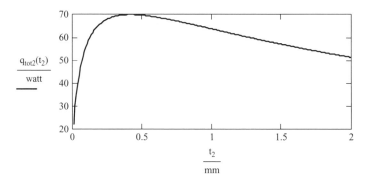

Figure 3.22 Total heat transfer rate vs. fin thickness

The optimum fin thickness is found using a MathCAD's buit-in function of 'Solve box'.

$$\text{Guess} \quad t_1 := 1 \, \text{mm} \quad t_2 := 1 \, \text{mm}$$
$$t_1 := \text{Maximize}(q_{tot1}, t_1) \quad t_1 = 0.412 \, \text{mm} \tag{3.125}$$
$$t_2 := \text{Maximize}(q_{tot2}, t_2) \quad t_2 = 0.418 \, \text{mm}$$

The heat transfer rates in the internal and external heat sinks for the optimum fin thickness based on the assumptions of T_c and T_h can be updated with the exact solution of T_c and T_h to be obtained later.

$$q_{tot1}(t_1) = 27.364 \, \text{W} \quad q_{tot2}(t_2) = 69.847 \, \text{W} \tag{3.126}$$

The number of fins is

$$n_{f1}(t_1) = 28.224 \quad n_{f2}(t_2) = 36.101 \tag{3.127}$$

The mass of fins is

$$m_{f1}(t_1) = 71.595 \, \text{gm} \quad m_{f2}(t_2) = 122.354 \, \text{gm} \tag{3.128}$$

The overall efficiency is

$$\eta_{o1}(t_1) = 0.846 \quad \eta_{o2}(t_2) = 0.846 \tag{3.129}$$

The overall thermal resistance is

$$R_{to1}(t_1) = 0.365 \frac{K}{W} \quad R_{to2}(t_2) = 0.215 \frac{K}{W} \tag{3.130}$$

3.9.3   Design of Thermoelectric Cooler (TEC)

Properties of TEC given are

$$\alpha_n := 2.2 \cdot 10^{-4} \frac{V}{K} \quad \alpha_p := -2.2 \cdot 10^{-4} \frac{V}{K} \tag{3.131}$$

$$\rho_n := 1.2 \cdot 10^{-2} \, \Omega \, cm \quad \rho_p := 1.2 \cdot 10^{-3} \, \Omega \, cm$$

$$k_n := 1.62 \cdot 10^{-2} \frac{W}{cm \, K} \quad k_p := 1.62 \cdot 10^{-2} \frac{W}{cm \, K}$$

$$\alpha := \alpha_p - \alpha_n \quad \rho := \rho_p + \rho_n \quad k := k_p + k_n \tag{3.132}$$

Geometric factor G is defined and estimated according to the manufacturer's specification

$$G_e = \frac{A_e}{L_e} \quad G_e := 0.12 \, cm \tag{3.133}$$

The same number of thermocouples is adopted as a typical module.

$$n := 127 \tag{3.134}$$

The length of the element leg is estimated, which allows us to compute its area:

$$L_e := 0.2 \, cm \quad A_e := G_e \, L_e \quad A_e = 0.024 \, cm^2 \tag{3.135}$$

The temperature difference between the cold and hot junctions is defined.

$$\Delta T := T_h - T_c \quad \Delta T = 40 \, K \tag{3.136}$$

The internal electric resistance of the couple is

$$R_e := \frac{\rho \, L_e}{A_e} \quad R_e = 0.02 \, \Omega \tag{3.137}$$

The thermal conductance is

$$K_e := \frac{k \, A_e}{L_e} \tag{3.138}$$

The figure of merit is

$$Z := \frac{\alpha^2}{\rho \, k} \quad Z = 2.49 \times 10^{-3} \frac{1}{K} \tag{3.139}$$

The maximum performance parameters for the module are obtained.

$$I_{max} := \frac{\alpha \, T_c}{R_e} \qquad I_{max} = 6.339 \, A \tag{3.140}$$

$$\Delta T_{max} := 0.5 \, Z \, T_c^2 \qquad \Delta T_{max} = 103.361 \, K \tag{3.141}$$

Table 3.7   TEC module

Module: model CP1.4-127-06L			
Qmax	51.4 Watts	Dimensions	
Imax	6 Amperes	Width	40 mm
Vmax	15.4 V	Length	40 mm
ΔTmax	67°C	Thickness	3.8 mm
Number of couple	127		

$$Q_{cmax} := n\,K_e\,\Delta T_{max} \qquad\qquad Q_{cmax} = 51.037\text{ W} \qquad (3.142)$$

$$V_{max} := n\,\alpha\,(\Delta T_{max} + T_c) \qquad\qquad V_{max} = 21.878\text{ V} \qquad (3.143)$$

Now we try to find a commercial module with the calculated maximum performance values. Table 3.7 provides one example.

The current can be expressed in terms of a desired voltage $V_o = 12$ V using Equations (3.53) and (3.86).

$$V_o := 12\text{ V} \qquad (3.144)$$

$$I_o := \frac{\dfrac{V_o}{n} - \alpha\,\Delta T}{R_e} \qquad I_o = 3.844\text{ A} \qquad (3.145)$$

The amount of heat absorbed at the cold junction is obtained using Equation (3.49).

$$Q_c := n\left[\alpha\,T_c\,I_o - \frac{1}{2}\,I_o^2\,R_e - K_e(T_h - T_c)\right] \qquad (3.146)$$

The input power is obtained using Equations (3.52) and (3.82).

$$W_o := n\left[\alpha\,I_o(T_h - T_c) + I_o^2\,R_e\right] \qquad (3.147)$$

The heat balance gives

$$Q_h := Q_c + W_o \qquad (3.148)$$

The relationship between the heat sinks and the TEC is

$$Q_c = q_{tot1} \quad \text{and} \quad Q_h = q_{tot2} \qquad (3.149)$$

3.9.4   Finding the Exact Solution for T_c and T_h

Combining Equations (3.84) to (3.149) yields two equations for two unknowns, T_c and T_h. Rewrite the given and estimated temperatures.

$$T_{int} := 25°C \quad T_{ext} := 40°C \quad T_c = 15°C \quad T_h = 55°C \qquad (3.150)$$

We may explore different values for the internal and external temperatures for Table 3.8.

$$T_{int} := 25°C \quad T_{ext} := 40°C \qquad (3.151)$$

Table 3.8   Summary of Design

	Heat Sink		TEC	
	Internal	External		Optimum
Heat load (Watts)	24.3	70.3	Cooling rate, Q_c (Watts)	24.1
Fin thickness (mm)	0.412	0.418	Input power (Watts)	45.9
Fin spacing (mm)	2.281	2.352	Heat rejected, Q_h (Watts)	70.0
Number of fins	29	36	COP	0.525
Mass of fins (gram)	71.6	122.3	Voltage (V)	12
Overall efficiency	0.846	0.846	Current (A)	3.84
Overall thermal resistance (K/V)	0.365	0.215	Q_{cmax} (Watts)	51
L (mm)	76	100	ΔT_{max} (°C)	103
W (mm)	76	100	V_{max} (V)	21.9
b (mm)	30	30	I_{max} (A)	6.34
			Dimension	$40 \times 40 \times 3.8$ mm

Guesses for T_c and T_h are not necessary here because the values were already given earlier. A MathCAD block is constructed to find the solutions as

Given

$$\frac{T_{int} - T_c}{R_{to1}(t_1)} = n \left[\alpha T_c \frac{\frac{V_o}{n} - \alpha(T_h - T_c)}{R_e} - \frac{1}{2} \left[\frac{\frac{V_o}{n} - \alpha(T_h - T_c)}{R_e} \right]^2 R_e - K_e(T_h - T_c) \right]$$

(3.152)

$$\frac{T_h - T_{ext}}{R_{to2}(t_2)} = \frac{T_{int} - T_c}{R_{to1}(t_1)} + n \left[\alpha \frac{\frac{V_o}{n} - \alpha(T_h - T_c)}{R_e}(T_h - T_c) + \left[\frac{\frac{V_o}{n} - \alpha(T_h - T_c)}{R_e} \right]^2 R_e \right]$$

(3.153)

$$\begin{pmatrix} T_C \\ T_h \end{pmatrix} := \text{Find}(T_c, T_h)$$

(3.154)

The solution provides the exact temperatures, which are close enough to the initial estimation.

$$T_c = 16.141°C \quad T_h = 55.169°C$$

(3.155)

We need to recalculate the heat transfer data with the exact temperatures

$$Q_c := n \left[\alpha T_c I_0 - \frac{1}{2} I_0^2 R_e - K_e(T_h - T_c) \right] \quad Q_c = 24.106 \, W$$

(3.156)

$$W_o := n \left[\alpha I_0(T_h - T_c) + I_0^2 R_e \right] \quad W_o = 45.924 \, W$$

(3.157)

$$Q_h := Q_c + W_o \quad Q_h = 70.03 \, W$$

(3.158)

Table 3.9 Performance Data for the Thermoelectric Air Cooler

Ambient=20°C		Ambient=40°C		Ambient=60°C	
Q_c	$T_{int} - T_{amb}$	Q_c	$T_{int} - T_{amb}$	Q_c	$T_{int} - T_{amb}$
8.7	−40	12.0	−40	15.3	−40
18.4	−20	21.7	−20	25.0	−20
28.1	0	31.3	0	34.5	0
37.8	20	40.9	20	44.0	20

$$\text{COP} := \frac{Q_c}{W_o} \quad \text{COP} = 0.525 \tag{3.159}$$

The design of the internal and external heat sinks and the thermoelectric cooler are summarized in Table 3.8. The design requirement of 21 Watts is satisfied, as a cooling rate of 24.1 Watts is obtained in the table. Note that, in this problem, the cooling rate is optimized for the optimum *COP* in order to minimize the input power and the heat rejected, so that the heat sinks have minimum sizes.

3.9.5 Performance Curves for Thermoelectric Air Cooler

The calculated cooling capacities for three ambient temperatures are tabulated as performance in Table 3.9, which is plotted in Figure 3.23.

$$\text{data\_20c} := \begin{pmatrix} 8.7 & -40 \\ 18.4 & -20 \\ 28.1 & 0 \\ 37.8 & 20 \end{pmatrix} \quad \text{data\_40c} := \begin{pmatrix} 12.0 & -40 \\ 21.7 & -20 \\ 31.3 & 0 \\ 40.9 & 20 \end{pmatrix}$$

$$\text{data\_60c} := \begin{pmatrix} 15.3 & -40 \\ 25.0 & -20 \\ 34.5 & 0 \\ 44.0 & 20 \end{pmatrix} \tag{3.160}$$

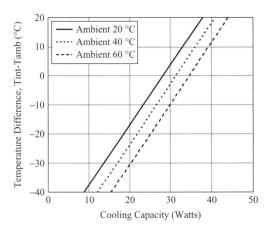

Figure 3.23 Performance curves for the thermoelectric air cooler

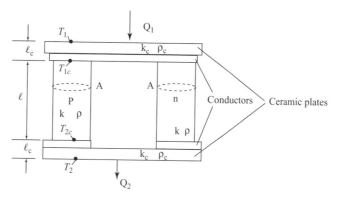

Figure 3.24   Basic configuration of a real thermoelectric couple ($T_1 > T_2$)

The cooling capacity is dependent only on the temperature difference. The cooling load (capacity) decreases as the temperature difference decreases. Under a constant external (ambient) temperature, the cooling load decreases with decreasing internal temperature in the enclosure. In general, the cooling load increases as the ambient temperature increases.

3.10   THERMOELECTRIC MODULE DESIGN

So far we have studied ideal thermoelectric devices without considering the thermal and electrical contact resistances. For real thermoelectric devices, optimum module design must include thermal and electrical resistances. Importantly, the thermoelement length determines its performance and fabrication cost. Herein we want to develop realistic formulas whereby we can optimize thermoelectric generators (TEG) and thermoelectric coolers (TEC).

3.10.1   Thermal and Electrical Contact Resistances for TEG

We now consider a section of a thermocouple in a multiple thermoelectric module to study the effect of the thermal and electrical contact resistances as shown in Figure 3.24. n-type and p-type semiconductor thermoelements are connected in series by highly conducting metal strips to form a thermocouple. Then, a number of thermocouples are connected electrically in series and sandwiched between thermally conducting but electrically insulating ceramic plates.

A heat balance at each section is given by

$$Q_1 = \frac{Ak_c}{l_c}(T_1 - T_{1c}) \tag{3.161}$$

$$Q_1 = \alpha I T_{1c} - \frac{1}{2}I^2 R + \frac{Ak}{l}(T_{1c} - T_{2c}) \tag{3.162}$$

$$Q_2 = \alpha I T_{2c} + \frac{1}{2}I^2 R + \frac{Ak}{l}(T_{1c} - T_{2c}) \tag{3.163}$$

$$Q_2 = \frac{Ak_c}{l_c}(T_{2c} - T_2) \tag{3.164}$$

where $\alpha = \alpha_p - \alpha_n$ and $k = k_p + k_n$.

   k is the thermal conductivity of the thermoelements

   k_c is the thermal contact conductivity which includes the thermal conductivity
of ceramic plates and the thermal contacts

   l is the thermoelement length, and

   l_c is the thickness of the contact layer

The electric resistance is composed of two resistances: thermocouple and electrical contact.

$$R = R_o + R_c = \frac{\rho l}{A} + \frac{\rho_c}{A} = \frac{\rho l}{A}\left(1 + \frac{\rho_c}{\rho}\frac{1}{l}\right) = \frac{\rho l}{A}\left(1 + \frac{s}{l}\right) \tag{3.165}$$

Therefore,

$$R = R_o\left(1 + \frac{s}{l}\right) \tag{3.166}$$

where ρ is the electrical resistivity, ρ_c the electrical contact resistivity, $\rho = \rho_p - \rho_n$, $s = \rho_c/\rho$ and $R_o = \rho l/A$. We develop a relationship between the real and ideal temperature differences by arranging the heat balance equations.

$$\text{Equation (3.161)} + \text{Equation (3.164)} = \text{Equation (3.162)} + \text{Equation (3.163)} \tag{3.167}$$

$$\frac{Ak_c}{l_c}(T_1 - T_{1c} + T_{2c} - T_2) = \alpha I (T_{1c} + T_{2c}) + 2\frac{Ak}{l}(T_{1c} - T_{2c}) \tag{3.168}$$

$$T_1 - T_2 = (T_{1c} - T_{2c})\left(1 + 2\frac{k}{k_c}\frac{l_c}{l}\right) + \underbrace{\frac{\alpha I l_c}{Ak_c}(T_{1c} + T_{2c})}_{(3.169a)} \tag{3.169}$$

We use $R_L = R$ at the maximum power for the most interesting operating condition. From Equation (3.18), the electrical current can be expressed as

$$I = \frac{\alpha (T_{1c} - T_{2c})}{R_L + R} = \frac{\alpha (T_{1c} - T_{2c})}{2R} \tag{3.170}$$

Then, the expression (3.169a) in Equation (3.169) becomes

$$\frac{\alpha l_c}{Ak_c}\left(\frac{\alpha (T_{1c} - T_{2c})}{2R}\right)(T_{1c} + T_{2c}) = \frac{\alpha^2}{RK}\frac{l_c}{Ak_c}\frac{Ak (T_{1c} + T_{2c})}{l}\frac{(T_{1c} - T_{2c})}{2}$$
$$= Z\overline{T}_c\frac{k}{k_c}\frac{l_c}{l}(T_{1c} - T_{2c}) \tag{3.171}$$

We assume that $Z\overline{T}_c \cong 1$. Then,

$$\frac{\alpha I l_c}{Ak_c}(T_{1c} + T_{2c}) = \frac{k}{k_c}\frac{l_c}{l}(T_{1c} - T_{2c}) \tag{3.172}$$

Inserting this into Equation (3.169) gives

$$T_{1c} - T_{2c} = (T_1 - T_2)(1 + 3rl_c/l)^{-1} \qquad (3.173)$$

where $r = k/k_c$. This is an important relationship between the real and ideal temperature differences. The temperature differences become identical at an ideal case that is either $r = 0$ or $l_c = 0$. Let $R_L = R$ as mentioned earlier. From Equation (3.18), we have

$$I = \frac{\alpha(T_{1c} - T_{2c})}{R_L + R} = \frac{\alpha(T_{1c} - T_{2c})}{2R} = \frac{\alpha(T_1 - T_2)}{2R_o\left(1 + \dfrac{s}{l}\right)\left(1 + 3r\dfrac{l_c}{l}\right)} \qquad (3.174)$$

Using $R_o = \rho l / A$, the electrical current is expressed in terms of the real temperatures and thermal and electrical contact resistances by

$$I = \frac{A\alpha(T_1 - T_2)}{2\rho l\left(1 + \dfrac{s}{l}\right)\left(1 + 3r\dfrac{l_c}{l}\right)} \qquad (3.175)$$

This may allow designers to estimate the cross-sectional area A needed for a thermoelement with a given electrical current. Using $W_n = nI^2 R$, the power output W_n for a real thermoelectric module that has a number of thermocouples, n, is expressed by

$$W_n = \frac{nA\alpha^2(T_1 - T_2)^2}{4\rho l\left(1 + \dfrac{s}{l}\right)\left(1 + 3r\dfrac{l_c}{l}\right)^2} \qquad (3.176)$$

Or equivalently the power output per unit area is

$$W_n/nA = \left(\frac{kT_1}{l}\right)\frac{Z\overline{T}_1(1 + T_2/T_1)^{-1}(1 - T_2/T_1)^2}{4\left(1 + \dfrac{s}{l}\right)\left(1 + 3r\dfrac{l_c}{l}\right)^2} \qquad (3.177)$$

Since $W_n = IV_n$, V_n can be obtained by dividing Equation (3.176) by (3.175) as

$$V_n = \frac{n\alpha(T_1 - T_2)}{2\left(1 + 3r\dfrac{l_c}{l}\right)} \qquad (3.178)$$

This may allow designers to estimate the number of thermocouples, n, with a given voltage. The conversion thermal efficiency is defined as

$$\eta_t = \frac{W_n}{nQ_1} = \frac{nI^2 R}{n\left[\alpha I T_{1c} - \dfrac{1}{2}I^2 R + \dfrac{Ak}{l}(T_{1c} - T_{2c})\right]} \qquad (3.179)$$

Substituting Equations (3.166) and (3.175) into (3.179) and rearranging yields

$$\eta_{cr} = \frac{\dfrac{T_1 - T_2}{T_1}}{\dfrac{4}{ZT_1}\left(1 + \dfrac{s}{l}\right)\left(1 + 3r\dfrac{l_c}{l}\right) + 2\left(1 + 3r\dfrac{l_c}{l}\right)\dfrac{T_{1c}}{T_1} - \dfrac{1}{2}\left(\dfrac{T_1 - T_2}{T_1}\right)} \qquad (3.180)$$

which is the contact resistance conversion efficiency, in which T_{1c}/T_1 is unknown and is to be derived herein. From Equation (3.161), we have

$$T_1 = T_{1c} + Q_1 \frac{l_c}{Ak_c} \qquad (3.181)$$

From Equation (3.162), we have

$$Q_1 = \alpha I T_{1c} - \frac{1}{2} I^2 R + \frac{Ak}{l} (T_{1c} - T_{2c}) \qquad (3.182)$$

Substituting Equation (3.175) into (3.182) gives

$$Q_1 = \frac{A\alpha^2 (T_1 - T_2) T_{1c}}{2\rho l \left(1 + \frac{s}{l}\right)\left(1 + 3r\frac{l_c}{l}\right)} - \frac{A\alpha^2 (T_1 - T_2)^2}{8\rho l \left(1 + \frac{s}{l}\right)\left(1 + 3r\frac{l_c}{l}\right)^2} + \frac{Ak}{l} \frac{(T_{1c} - T_{2c})}{\left(1 + 3r\frac{l_c}{l}\right)} \qquad (3.183)$$

Knowing that

$$ZT_1 = ZT_1 \left(\frac{T_1 + T_2}{2}\right)\left(\frac{2}{T_1 + T_2}\right) = 2Z\overline{T}\left(1 + \frac{T_2}{T_1}\right)^{-1} \qquad (3.184)$$

And inserting this into Equation (3.181) and solving for T_{1c}/T_1 gives

$$\xi_{TEG} = \frac{T_{1c}}{T_1} = \frac{1 + \frac{1}{4}r\frac{l_c}{l} \dfrac{Z\overline{T}\left(1 + \frac{T_2}{T_1}\right)^{-1}\left(1 - \frac{T_2}{T_1}\right)^2}{\left(1 + \frac{s}{l}\right)\left(1 + 3r\frac{l_c}{l}\right)^2} - \dfrac{r\frac{l_c}{l}\left(1 - \frac{T_2}{T_1}\right)}{\left(1 + 3r\frac{l_c}{l}\right)}}{1 + r\frac{l_c}{l} \dfrac{Z\overline{T}\left(1 + \frac{T_2}{T_1}\right)^{-1}\left(1 - \frac{T_2}{T_1}\right)}{\left(1 + \frac{s}{l}\right)\left(1 + 3r\frac{l_c}{l}\right)}} \qquad (3.185)$$

This becomes unity if the thermoelectric generator (TEG) is ideal ($r = 0$ or $l_c = 0$). From Equation (3.180), the contact resistance conversion efficiency is rewritten as

$$\eta_{cr} = \frac{\left(1 - \frac{T_2}{T_1}\right)}{\frac{2}{Z\overline{T}}\left(1 + \frac{T_2}{T_1}\right)\left(1 + \frac{s}{l}\right)\left(1 + 3r\frac{l_c}{l}\right) + 2\left(1 + 3r\frac{l_c}{l}\right)\xi_{TEG} - \frac{1}{2}\left(1 - \frac{T_2}{T_1}\right)} \qquad (3.186)$$

This real conversion efficiency converges to the ideal conversion efficiency for the maximum power as $r \to 0$ and $s \to 0$.

$$\eta_{mp} = \frac{\left(1 - \frac{T_2}{T_1}\right)}{\frac{2}{Z\overline{T}}\left(1 + \frac{T_2}{T_1}\right) + 2 - \frac{1}{2}\left(1 - \frac{T_2}{T_1}\right)} \qquad (3.187)$$

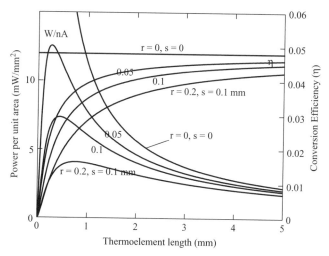

Figure 3.25 Power per unit area and conversion efficiency vs. thermoelectric length for different r. $k_p = k_n = 1.5$ W/mK, $l_c = 1$ mm, $T_1 = 400$ K, $T_2 = 300$ K, $Z = 3 \times 10^{-3}$ K^{-1}, and $Z\overline{T} = 1.05$

which was previously given in Equation (3.34a). It is found that the results of Equation (3.180) are in good agreement with Rowe and Min [22] although the formulas used appear quite different.

The effects of the thermal and electrical contact resistances are illustrated in Figure 3.25 for which Equations (3.177) and (3.186) were used. The power output and conversion efficiency are dependent upon the thermoelement length. It can be seen that in order to obtain high conversion efficiency, the module should be designed with long thermoelements. However, if a large power-per-unit-area is required, the thermoelement length should be optimized at a relatively shorter length. It is apparent that the optimum design of a thermoelectric module is likely to be a compromise between obtaining high conversion efficiency or large power output. The main objective of thermoelectric module design is to determine a set of design parameters which meet the required specifications at minimum cost. Note that although a long thermoelement reveals high conversion efficiency and the contact resistances become negligible. However, this is not an option due to the high material cost. Also note that the decrease in conversion efficiency becomes significant when the thermoelement length is below 1 mm. In general, a high conversion efficiency is required if the heat source is expensive, while a large power-per-unit-area is required if the waste heat source is available or fabrication cost is to be reduced. For commercially available thermoelectric modules, appropriate values of contact resistances are s $\sim$ 0.1 mm and r $\sim$ 0.2 [15]. From Figure 3.25, we find a useful design feature that an ideal thermoelectric model with sufficiently long thermoelement length reasonably predicts the performance of the real thermoelectric module that has a short thermoelement length.

There is found an important experimental work conducted by Min and Rowe [21] which supports the aforementioned theory of a real thermoelectric module with thermal and electrical contact resistances. They measured the electrical contact resistances for three modules (that have different thermoelement length) and found it to be constant

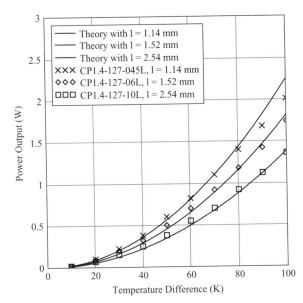

Figure 3.26   Experimental and theoretical power output of thermoelectric modules as a function of temperature difference and thermoelectric length. Equation (3.177) was used for the theory and the experimental data were adopted from Min and Rowe [21]

($s = 0.1$mm). They also measured the power output with varying temperature differences for each module, which is shown in Figure 3.26. Indeed the power-per unit-area increases with decreasing the thermoelement length as the theory predicts. The experiments and the theory show good agreement. Fine adjustment of the thermal contact conductivity for each module was necessary to provide the precise values which are shown in Table 3.10. It is also found from the comparison between the measurements and the theory that the commercial thermoelectric modules (MELCOR) reveal thermal contact conductivity ranging from $0.5k_c = 12 \sim 16$ W/mK which corresponds to $r = 0.09 \sim 0.13$ with $s = 0.1$ mm ($0.5k_c$ is the thermal contact conductivity of each element of a thermocouple).

3.10.2   Thermal and Electrical Contact Resistances for TEC

As for TEG in Section 3.10.1, we consider a section of a thermocouple in a multiple thermoelectric module for TEC to study the effect of the thermal and electrical contact

Table 3.10   Parameters for Commercial Thermoelectric Modules (MELCOR)

Modules*	l (mm)*	A/l (mm)*	$r = k/k_c$	$s = \rho_c/\rho^*$	$0.5k_c$(W/mK)**
CP1.4-127-10	2.54	0.771	0.096	0.1	15.6
CP1.4-127-06	1.52	1.289	0.122	0.1	12.3
CP1.4-127-045	1.14	1.719	0.104	0.1	14.4

*Experimental data adopted from [21].
**$0.5\ k_c$ indicates the thermal contact conductivity of each thermoelement of a thermocouple.

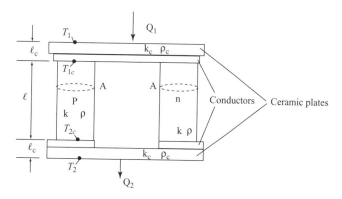

Figure 3.27 Basic configuration of a real thermoelectric couple $(T_2 > T_1)$

resistances as shown in Figure 3.27. We want to formulate the real conversion efficiency and power output which include the thermal and electrical contact resistances.

A steady state heat balance at each section is given by

$$Q_1 = \frac{Ak_c}{l_c}(T_1 - T_{1c}) \tag{3.188}$$

$$Q_1 = \alpha I T_{1c} - \frac{1}{2}I^2 R - \frac{Ak}{l}(T_{2c} - T_{1c}) \tag{3.189}$$

$$Q_2 = \alpha I T_{2c} + \frac{1}{2}I^2 R - \frac{Ak}{l}(T_{2c} - T_{1c}) \tag{3.190}$$

$$Q_2 = \frac{Ak_c}{l_c}(T_{2c} - T_2) \tag{3.191}$$

where

$\alpha = \alpha_p - \alpha_n$ and

$k = k_p + k_n$. k is the thermal conductivity of the thermoelements

k_c the thermal contact conductivity which includes the thermal conductivity of ceramic plates and the thermal contacts

l the thermoelement length, and

l_c the thickness of the contact layer.

ρ is the electrical resistivity.

The electric resistance is composed of two resistances: thermocouple and electrical contact.

$$R = R_o + R_c = \frac{\rho l}{A} + \frac{\rho_c}{A} = \frac{\rho l}{A}\left(1 + \frac{\rho_c}{\rho}\frac{1}{l}\right) = \frac{\rho l}{A}\left(1 + \frac{s}{l}\right) \tag{3.192}$$

Therefore,

$$R = \frac{\rho l}{A}\left(1 + \frac{s}{l}\right) \tag{3.193}$$

where

ρ is the electrical resistivity

ρ_c the electrical contact resistivity

$$\rho = \rho_p - \rho_n \text{ and}$$
$$s = \rho_c/\rho.$$

We develop a relationship between the real and ideal temperature differences by arranging the heat balance equations.

Equation (3.188) + Equation (3.191) = Equation (3.189) + Equation (3.190) $\qquad$ (3.194)

$$\frac{Ak_c}{l_c}(T_1 - T_{1c} + T_{2c} - T_2) = \alpha I (T_{2c} + T_{1c}) - 2\frac{Ak}{l}(T_{2c} - T_{1c}) \qquad (3.195)$$

Rearranging

$$T_2 - T_1 = (T_{2c} - T_{1c})\left(1 + 2\frac{k}{k_c}\frac{l_c}{l}\right) - \underbrace{\frac{\alpha I l_c}{Ak_c}(T_{1c} + T_{2c})}_{(3.196a)} \qquad (3.196)$$

The operating current lies between the optimum COP and the maximum cooling power depending on its application. The optimum current for the maximum COP is the lower bound while the maximum current for the maximum cooling power is the upper bound for the operating range of a thermoelectric cooler (See Equations (3.58) and (3.69)).

$$I_{COP} = \frac{\alpha (T_{2c} - T_{1c})}{R\left(\sqrt{1 + Z\overline{\overline{T}}} - 1\right)} \leq I = \frac{\alpha (T_{2c} - T_{1c})}{R\psi} \leq I_{max} = \frac{\alpha T_{1c}}{R} \qquad (3.197)$$

which leads to

$$\frac{1}{\sqrt{1 + Z\overline{\overline{T}}} - 1} \leq \frac{1}{\psi} \leq \frac{1}{\dfrac{T_{2c} - T_{1c}}{T_{1c}}} \qquad (3.198)$$

where ψ is called *the operating factor* ranging typically between 0.1 and 0.4. We define the operating factor ψ by

$$I = \frac{\alpha(T_{2c} - T_{1c})}{R\psi} \qquad (3.199)$$

Knowing that $K = Ak/l$ and $Z = \alpha^2/\rho k$ and assuming that $\overline{T}_c = \overline{T}$. Equation (3.196a) becomes

$$\frac{\alpha I l_c (T_{2c} + T_{1c})}{Ak_c}$$

$$= \frac{\alpha l_c}{Ak_c}\left(\frac{\alpha(T_{2c} - T_{1c})}{R\psi}\right)(T_{2c} + T_{1c}) = \frac{\alpha^2}{\rho k}\frac{Ak}{l}\frac{l_c}{Ak_c}\frac{(T_{2c} - T_{1c})}{\psi}\frac{(T_{2c} + T_{1c})}{(1 + s/l)}2 \qquad (3.200)$$

$$= 2\frac{Z\overline{T}_c}{\psi(1 + s/l)}\frac{k}{k_c}\frac{l_c}{l}(T_{2c} - T_{1c}) = 2\frac{Z\overline{T}}{\psi(1 + s/l)}\frac{k}{k_c}\frac{l_c}{l}(T_{2c} - T_{1c})$$

Inserting this into Equation (3.196) gives

$$T_2 - T_1 = (T_{2c} - T_{1c}) \left[1 - 2 \left(\frac{Z\overline{T}}{\psi\,(1+s/l)} - 1 \right) \frac{k}{k_c} \frac{l_c}{l} \right] \tag{3.201}$$

Or equivalently

$$T_{2c} - T_{1c} = (T_2 - T_1) \left(1 - mr\frac{l_c}{l} \right)^{-1} \tag{3.202}$$

where $r = k/k_c$ and

$$m = 2 \left(\frac{Z\overline{T}}{\psi\,(1+s/l)} - 1 \right) \tag{3.203}$$

Equation (3.202) is an important relationship between the ideal and real temperature differences. The coefficient of performance (COP) is defined as

$$COP = \frac{Q_1}{W} = \frac{\alpha I T_{1c} - \dfrac{1}{2} I^2 R - \dfrac{Ak}{l}(T_{2c} - T_{1c})}{\alpha I\,(T_{2c} - T_{1c}) + I^2 R} \tag{3.204}$$

Plugging Equation (3.199) into this gives

$$COP = \frac{\alpha \left(\dfrac{\alpha(T_{2c} - T_{1c})}{R\psi} \right) T_{1c} - \dfrac{1}{2} \left(\dfrac{\alpha(T_{2c} - T_{1c})}{R\psi} \right)^2 R - \dfrac{Ak}{l}(T_{2c} - T_{1c})}{\alpha \left(\dfrac{\alpha(T_{2c} - T_{1c})}{R\psi} \right)(T_{2c} - T_{1c}) + \left(\dfrac{\alpha(T_{2c} - T_{1c})}{R\psi} \right)^2 R} \tag{3.205}$$

which reduces to

$$COP = \frac{T_{1c}}{T_{2c} - T_{1c}} \frac{\psi - \dfrac{1}{2}\dfrac{T_{2c} - T_{1c}}{T_{1c}} - \dfrac{\psi^2}{ZT_{1c}}\left(1+\dfrac{s}{l}\right)}{\psi + 1} \tag{3.206}$$

The COP can be expressed as

$$COP = \frac{\left(\dfrac{T_{1c}}{T_1}\right)\left(1 - mr\dfrac{l_c}{l}\right)}{\dfrac{T_2}{T_1} - 1} \frac{\left[\psi - \dfrac{\dfrac{T_2}{T_1} - 1}{2\,(T_{1c}/T_1)\,(1 - mrl_c/l)} - \dfrac{\psi^2\left(\dfrac{T_2}{T_1}+1\right)\left(1+\dfrac{s}{l}\right)}{2Z\overline{T}\,(T_{1c}/T_1)} \right]}{\psi + 1} \tag{3.207}$$

where T_{1c}/T_1 is unknown and derived below.

From Equation (3.199) with Equations (3.193) and (3.202), the current becomes

$$I = \frac{A\alpha\,(T_2 - T_1)}{\psi\rho l\left(1 + \dfrac{s}{l}\right)\left(1 - mr\dfrac{l_c}{l}\right)} \tag{3.208}$$

The voltage across a thermocouple can be derived from Equation (3.53).

$$
\begin{aligned}
V &= \alpha\,(T_{2c} - T_{1c}) + IR \\
&= \alpha\,(T_{2c} - T_{1c}) + \left(\frac{\alpha(T_{2c} - T_{1c})}{R\psi}\right) R
\end{aligned} \tag{3.209}
$$

which leads to the voltage across a thermoelectric module as

$$V_n = \frac{n\alpha\,(T_2 - T_1)}{\left(1 - mr\dfrac{l_c}{l}\right)}\left(\frac{\psi + 1}{\psi}\right) \tag{3.210}$$

The power of the module is obtained since $W_n = IV_n$ as

$$W_n = \frac{nA\alpha^2\,(T_2 - T_1)^2}{\rho l\left(1 + \dfrac{s}{l}\right)\left(1 - mr\dfrac{l_c}{l}\right)^2}\left(\frac{\psi + 1}{\psi^2}\right) \tag{3.211}$$

From Equation (3.188), we have

$$T_{1c} = T_1 + Q_1\frac{l_c}{Ak_c} \tag{3.212}$$

Inserting Equation (3.208) into Equation (3.189) gives

$$
\begin{aligned}
Q_1 &= \alpha I T_{1c} - \frac{1}{2}I^2 R - \frac{Ak}{l}\,(T_{2c} - T_{1c}) \\[2mm]
&= \alpha T_{1c}\left(\frac{A\alpha\,(T_2 - T_1)}{\psi\rho l\left(1 + \dfrac{s}{l}\right)\left(1 - mr\dfrac{l_c}{l}\right)}\right) - \frac{1}{2}\left(\frac{A\alpha\,(T_2 - T_1)}{\psi\rho l\left(1 + \dfrac{s}{l}\right)\left(1 - mr\dfrac{l_c}{l}\right)}\right)^2 \\[2mm]
&\quad \times R - \frac{Ak}{l}\,(T_{2c} - T_{1c}) \\[2mm]
&= \frac{A\alpha^2 T_{1c}}{\rho l}\frac{(T_2 - T_1)}{\psi\left(1 + \dfrac{s}{l}\right)\left(1 - mr\dfrac{l_c}{l}\right)} - \frac{A\alpha^2}{2\rho l}\frac{(T_2 - T_1)^2}{\psi^2\left(1 + \dfrac{s}{l}\right)\left(1 - mr\dfrac{l_c}{l}\right)^2} \\[2mm]
&\quad - \frac{Ak}{l}\frac{(T_2 - T_1)}{\left(1 - mr\dfrac{l_c}{l}\right)}
\end{aligned} \tag{3.213}
$$

Inserting this into Equation (3.212) and solving for T_{1c}/T_1 gives

$$\xi_{TEC} = \frac{T_{1c}}{T_1} = \frac{1 + r\frac{l_c}{l}\frac{Z\overline{T}\left(1 + \frac{T_2}{T_1}\right)^{-1}\left(\frac{T_2}{T_1} - 1\right)^2}{\psi^2\left(1 + \frac{s}{l}\right)\left(1 - mr\frac{l_c}{l}\right)^2} + r\frac{l_c}{l}\frac{\left(\frac{T_2}{T_1} - 1\right)}{\left(1 - mr\frac{l_c}{l}\right)}}{1 + 2r\frac{l_c}{l}\frac{Z\overline{T}\left(1 + \frac{T_2}{T_1}\right)^{-1}\left(\frac{T_2}{T_1} - 1\right)}{\psi\left(1 + \frac{s}{l}\right)\left(1 - mr\frac{l_c}{l}\right)}} \tag{3.214}$$

From Equation (3.207), the COP is expressed in terms of the real temperatures.

$$COP = \frac{\xi_{TEC}\left(1 - mrl_c/l\right)}{\frac{T_2}{T_1} - 1}\left[\psi - \frac{\frac{T_2}{T_1} - 1}{2\xi_{TEC}\left(1 - mrl_c/l\right)} - \frac{\psi^2\left(\frac{T_2}{T_1} + 1\right)\left(1 + \frac{s}{l}\right)}{2Z\overline{T}\xi_{TEC}}\right]}{\psi + 1} \tag{3.215}$$

The cooling power per unit area is obtained from Equation (3.213) by

$$\frac{Q_{1n}}{nA} = \frac{kT_1}{l}\left[\frac{2Z\overline{T}\xi_{TEC}\left(\frac{T_2}{T_1} + 1\right)^{-1}\left(\frac{T_2}{T_1} - 1\right)}{\psi\left(1 + \frac{s}{l}\right)\left(1 - mr\frac{l_c}{l}\right)} - \frac{Z\overline{T}\left(\frac{T_2}{T_1} + 1\right)^{-1}\left(\frac{T_2}{T_1} - 1\right)^2}{\psi^2\left(1 + \frac{s}{l}\right)\left(1 - mr\frac{l_c}{l}\right)^2}\right.$$

$$\left. - \frac{\left(\frac{T_2}{T_1} - 1\right)}{\left(1 - mr\frac{l_c}{l}\right)}\right] \tag{3.216}$$

One interesting point is at the optimum COP, which is shown in Equation (3.198).

$$\psi = \sqrt{1 + Z\overline{T}} - 1 \tag{3.217}$$

It is recalled that this was derived for the ideal optimum COP (see Equation (3.69)). However, we want to see the effect of thermoelement length at the optimum COP. Equations (3.215) and (3.216) with Equation (3.217) for a given condition are plotted in Figure 3.28. Although the ideal cooling power ($r = 0$ and $s = 0$) increases without limit with decreasing thermoelement length, the real cooling power rapidly decreases beyond a certain length because of the thermal and electrical contact resistances. And also the *COP* is desired not to be very low. It is thus found in the literature that the thermoelement length is practically limited to about 1 mm due to the above two reasons, which is also associated with the manufacturability. The thermal contact conductivity

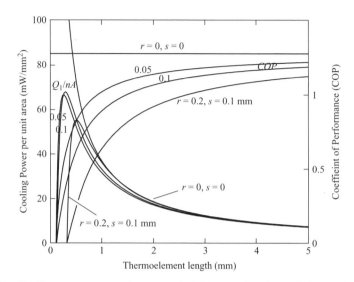

Figure 3.28 Cooling power per unit area and COP as a function of thermoelement length and r for $\psi = \sqrt{1 + Z\overline{T}} - 1 \approx 0.365$ (Optimum COP). $k_p = k_n = 1.5\,\text{W/mK}$, $l_c = 0.7\,\text{mm}$, $s = 0.1\,\text{mm}$, $T_1 = 275\,\text{K}$, $T_2 = 300\,\text{K}$, $Z = 3 \times 10^{-3}\,\text{K}^{-1}$, and $Z\overline{T} = 1.05$

ratio r is dependent on present technology and $r = 0.1$ is a typical value. Therefore, we herein consider that $r = 0.1$ in the design of a thermoelectric module, which is not impractical (see also Table 3.10 for the experimental values).

As mentioned earlier, thermoelectric coolers typically run at operating currents between the optimum *COP* and the maximum cooling power. As shown in Equation (3.198), the operating factor, ψ, cannot be readily obtained because it is expressed in terms of the ideal high and low temperatures (T_{2c} and T_{1c}) that are not known at this stage. Therefore, Equations (3.215) and (3.216) are plotted as a function of ψ and l for $r = 0.1$ to figure out the operating range of ψ, which is shown in Figure 3.29. We find from the figure that $\psi = 0.2 \sim 0.4$ for $l = 1\,\text{mm}$ between the real maximum power and the real optimum *COP*. Now we take a look at the effect of thermoelement length for the maximum power with $\psi = 0.2$, as shown in Figure 3.30. It is extremely important to secure $r = 0.1$ in order to operate such a small thermoelement length. It is interesting to note that the real value of $\psi = 0.4$ for the optimum *COP* is compared with the ideal value of $\psi = 0.365$.

In Figures 3.28 and 3.30, it is important to note that the *COPs* are significantly dependent on the thermal contact conductivity ratio, r, while the cooling powers are not. Reducing r is very important in the effective operation of thermoelectric modules, which is apparently dependent on the ability of fabrication.

Equations (3.215) and (3.216) are also plotted as a function of temperature differences for $\psi = 0.2$ and $\psi = 0.4$, which are shown in Figure 3.31. The *COP* exponentially increases with decreasing the temperature differences. The cooling power at the optimum COP with $\psi = 0.4$ forms a parabolic curve with the temperature differences. The maximum temperature difference is seen at 63.5 K. It is interesting to note that the maximum cooling power per unit area of 28 mW/mm^2 for $\psi = 0.4$ lies at the temperature difference of 31 K with its *COP* of 0.5. In a similar way, we find that the

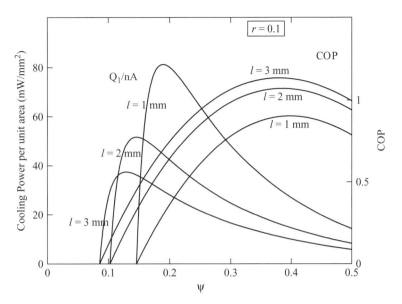

Figure 3.29   Cooling power per unit area and COP as a function of ψ and l for $r = 0.1$. $k_p = k_n = 1.5$ W/mK, $l_c = 0.7$ mm, $s = 0.1$ mm, $T_1 = 275$ K, $T_2 = 300$ K, $Z = 3 \times 10^{-3}$ K^{-1}, and $Z\overline{T} = 1.05$

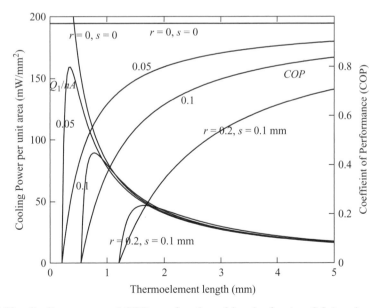

Figure 3.30   Cooling power and COP as a function of l and r for $\psi = 0.2$ (maximum cooling power). $k_p = k_n = 1.5$ W/mK, $l_c = 0.7$ mm, $s = 0.1$ mm, $T_1 = 275$ K, $T_2 = 300$ K, $Z = 3 \times 10^{-3}$ K^{-1}, and $Z\overline{T} = 1.05$

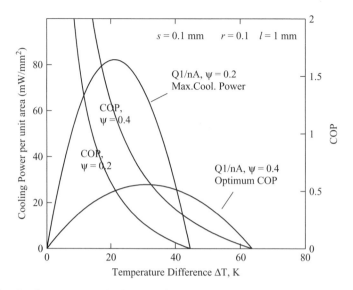

Figure 3.31 Cooling power and COP as a function of the temperature differences across the module for the optimum COP at $\psi = 0.4$ and the maximum cooling power at $\psi = 0.2$. $k_p = k_n = 1.5$ W/mK, $l_c = 0.7$ mm, $s = 0.1$ mm, $T_2 = 300$ K, $Z = 3 \times 10^{-3}$ K^{-1}, and $Z\overline{T} = 1.05$

maximum cooling power of 82 mW/mm^2 for $\psi = 0.2$ is at the temperature difference of 21 K with its COP of 0.5. We find an interesting optimum operating range of high and low temperature difference for $21 \sim 31$ K for the given conditions.

3.11   DESIGN EXAMPLE OF TEC MODULE

A TEC module for a small refrigerator is designed to provide 4 W of cooling at the maximum current of 4.4 A. The module is illustrated in Figure 3.32. The hot side of the module is maintained at 300 K and a temperature difference of 40 K must be maintained. The electrical and thermal contact resistances were predetermined by experiments and their values are listed below in Table 3.11. A particular thermoelement length of 2.8 mm is required for a specific application. The material properties of Bi$_2$Te$_3$ and the information for the thermal and electrical contact resistances are listed in the table. Determine the cross-sectional area of thermoelement, the number of thermoelement couples, and the maximum performance parameters for the module: ΔT_{max}, V_{max}, I_{max}, and Q_{max}. Also provide the recommended optimum operating conditions including currents, temperature difference, COP, and cooling powers.

3.11.1   Design Concept

We wish to design a TEC module with the aid of equations and graphs previously developed as a function of ψ and temperature difference. Since the optimum COP has the minimum cooling power, we design the number of thermoelement couple based on the minimum cooling power. As a result, the TEC module can actually provide more power than the minimum cooling power for all other conditions, but it apparently pays for the lower COP than the optimum COP.

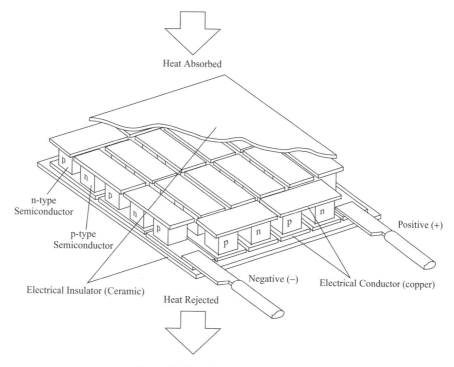

Figure 3.32   Thermoelectric module

MathCAD format solution

We use the mathematical software 'MathCAD' to solve the problem.

Given information   The properties of the material (Bi_2Te_3) are

$$\alpha_p := 200\,\frac{\mu V}{K} \qquad \alpha_n := -200\,\frac{\mu V}{K}$$

$$\rho_p := 1.0 \cdot 10^{-3}\,\Omega\,cm \quad \rho_n := 1.0 \cdot 10^{-3}\,\Omega\,cm \qquad (3.218)$$

$$k_p := 1.4\,\frac{W}{m\,K} \qquad k_n := 1.4\,\frac{W}{m\,K}$$

Table 3.11   Properties and Contact Resistances

Description	Value
Seebeck coefficient, α_p	200 $\mu V/K$
Seebeck coefficient, α_n	$-200\ \mu V/K$
Thermal conductivity, $k_p = k_n$	1.4 W/m·K
Electrical resistivity, $\rho_p = \rho_n$	1.0 mΩ·cm
Thermal contact conductivity ratio, $r = k/k_c$	0.1
Electrical contact resistivity ratio, $s = \rho_c/\rho$	0.1mm
Ceramic plate thickness, l_c	0.7 mm
Thermoelement length, l	2.8 mm

The properties of thermoelement couple is the sum of p-type and n-type thermoelements as

$$\alpha := \alpha_p - \alpha_n \quad \rho := \rho_p + \rho_n \quad k := k_p + k_n \tag{3.219}$$

The figure of merit is calculated by

$$Z := \frac{\alpha^2}{\rho k} \quad Z = 2.857 \times 10^{-3} \frac{1}{K} \tag{3.220}$$

The hot and cold junction temperatures are given and expressed as temperature ratio T_r.

$$T_1 := 260\,K \quad T_2 := 300\,K \tag{3.221}$$

$$T_r := \frac{T_2}{T_1} \quad T_r = 1.154 \tag{3.222}$$

Redefine the cold junction temperature in terms of T_r and define the dimensionless figure of merit as

$$T_1(T_r) := \frac{T_2}{T_r} \quad ZT_{avg}(T_r) := Z \frac{T_2}{2}\left(1 + \frac{1}{T_r}\right) \quad ZT_{avg}(T_r) = 0.8 \tag{3.223}$$

The electrical and thermal contact resistances are provided (see Equations (3.193) and (3.202) for s and r).

$$s := 0.1\,mm \quad r := 0.1 \tag{3.224}$$

The thickness of the ceramic plates and the thermoelement length are given as

$$l_c := 0.7\,mm \quad l_o := 2.8\,mm \tag{3.225}$$

Define the maximum current required as

$$I_{max} := 4.4A \tag{3.226}$$

Operating factor Define the operating factor for the optimum COP using Equation (3.217).

$$\psi := \sqrt{1 + ZT_{avg}(T_r)} - 1 \quad \psi = 0.342 \tag{3.227}$$

And from Equation (3.203),

$$m_z(l_o, T_r, \psi) := 2\frac{ZT_{avg}(T_r)}{\psi\left(1 + \frac{s}{l_o}\right)} - 2 \tag{3.228}$$

Cross-Sectional Area The cross-sectional area A_e is obtained using Equation (3.208) with $I_{max} = 4.4$ A.

$$A_e := \frac{I_{max}\,\psi\,\rho\,l_o\left(1 + \frac{s}{l_o}\right)\left(1 - m_z(l_o, T_r, \psi)\,r\frac{l_c}{l_o}\right)}{\alpha\,(T_2 - T_1(T_r))} \quad A_e = 5.866\,mm^2 \tag{3.229}$$

From Equation (3.214), we have

$$
\xi_{TEC}(s, r, l_o, T_r, \psi) := \frac{\left(1 + r\dfrac{l_c}{l_o} \dfrac{ZT_{avg}(T_r)\,(1 + T_r)^{-1}\,(T_r - 1)^2}{\psi^2\left(1 + \dfrac{s}{l_o}\right)\left(1 - m_z(l_o, T_r, \psi)\,r\dfrac{l_c}{l_o}\right)^2} + r\dfrac{l_c}{l_o}\dfrac{T_r - 1}{1 - m_z(l_o, T_r, \psi)\,r\dfrac{l_c}{l_o}}\right)}{1 + 2\,r\dfrac{l_c}{l_o}\dfrac{ZT_{avg}(T_r)(1 + T_r)^{-1}(T_r - 1)}{\psi\left(1 + \dfrac{s}{l_o}\right)\left(1 - m_z(l_o, T_r, \psi)\,r\dfrac{l_c}{l_o}\right)}}
$$

(3.230)

In order to reduce the size of equation, G_ψ is introduced as

$$
G_\psi(s, r, l_o, T_r, \psi) := \psi - \frac{(T_r - 1)}{2\xi_{TEC}(s, r, l_o, T_r, \psi)\left(1 - m_z(l_o, T_r, \psi)\,r\dfrac{l_c}{l_o}\right)}
$$
$$
- \frac{\psi^2(T_r + 1)\left(1 + \dfrac{s}{l_o}\right)}{2\,ZT_{avg}(T_r)\,\xi_{TEC}(s, r, l_o, T_r, \psi)}
$$

(3.231)

The real COP is obtained from Equation (3.215)

$$
COP(s, r, l_o, T_r, \psi) := \frac{\xi_{TEC}(s, r, l_o, T_r, \psi)\left(1 - m_z(l_o, T_r, \psi)\,r\dfrac{l_c}{l_o}\right)}{(T_r - 1)}\frac{G_\psi(s, r, l_o, T_r, \psi)}{\psi + 1}
$$

(3.232)

In order to eliminate some extreme data unnecessary in this analysis, the data beyond a certain value are discarded as

$$
\underset{\sim}{COP}(s, r, l_o, T_r, \psi) := \begin{vmatrix} COP(s, r, l_o, T_r, \psi) & \text{if } 1 - m_z(l_o, T_r, \psi)\,r\dfrac{l_c}{l_o} > 0.2 \\ 0 & \text{otherwise} \end{vmatrix}
$$

(3.233)

The real cooling power per thermoelement couple is obtained from Equation (3.216). Some functions are introduced to reduce the size of equation.

$$
X_1(s, r, l_o, T_r, \psi) := \frac{2\,ZT_{avg}(T_r)\,\xi_{TEC}(s, r, l_o, T_r, \psi)(1 + T_r)^{-1}(T_r - 1)}{\psi\left(1 + \dfrac{s}{l_o}\right)\left(1 - m_z(l_o, T_r, \psi)\,r\dfrac{l_c}{l_o}\right)}
$$

(3.234)

$$
X_2(s, r, l_o, T_r, \psi) := \frac{ZT_{avg}(T_r)(1 + T_r)^{-1}(T_r - 1)^2}{\psi^2\left(1 + \dfrac{s}{l_o}\right)\left(1 - m_z(l_o, T_r, \psi)\,r\dfrac{l_c}{l_o}\right)^2}
$$

(3.235)

$$
X_3(s, r, l_o, T_r, \psi) := \frac{(T_r - 1)}{\left(1 - m_z(l_o, T_r, \psi)\,r\dfrac{l_c}{l_o}\right)}
$$

(3.236)

The real cooling power per thermocouple is expressed from Equation (3.216) by

$$Q_1(s, r, l_o, T_r, \psi) := \frac{A_e \, kT_1 \, (T_r)}{l_o} \, (X_1(s, r, l_o, T_r, \psi) - X_2(s, r, l_o, T_r, \psi)$$

$$- X_3(s, r, l_o, T_r, \psi)) \qquad (3.237)$$

In order to eliminate some extreme data unnecessary in this analysis, the data beyond a certain value are discarded as

$$\underset{\sim}{Q_1}(s, r, l_o, T_r, \psi) := \left| \begin{array}{l} Q_1(s, r, l_o, T_r, \psi) \text{ if } 1 - m_z(l_o, T_r, \psi) \, r \, \dfrac{l_c}{l_o} > 0.2 \\ 0W \quad \text{otherwise} \end{array} \right. \qquad (3.238)$$

The cooling power per thermocouple is computed by

$$Q_1(s, r, l_o, T_r, \psi) = 0.129 \, W \qquad (3.239)$$

The number of thermocouple, n, is obtained by dividing the required cooling power of 4 W by the cooling power per thermocouple.

$$n := \frac{4W}{Q_1(s, r, l_o, T_r, \psi)} \qquad n = 31.114 \qquad (3.240)$$

Now we define the cooling power per module by

$$Q_{1n}(s, r, l_o, T_r, \psi) := nQ_1(s, r, l_o, T_r, \psi) \qquad (3.241)$$

Define three different thermoelement lengths for further analysis and confirm the temperature ratio.

$$l_{o1} := 1 \, mm \quad l_{o2} := 1 \, mm \quad l_{o3} := 2.8 \, mm \quad T_r = 1.154 \qquad (3.242)$$

In order to see the performance at the real optimum COP, Equations (3.232) and (3.241) are plotted for three different thermoelement lengths as a function of ψ and l in Figure 3.33. First, define a graphing range for ψ as

$$\underset{\sim}{\psi} := 0.07, 0.071..0.5 \qquad (3.243)$$

In order to determine the maximum values in Figure 3.33 for the optimum COP and the maximum cooling power, we need the following transformation as

$$COP_1(\psi) := COP(s, r, l_o, T_r, \psi) \quad Q_{cn}(\psi) := Q_{1n}(s, r, l_o, T_r, \psi) \qquad (3.244)$$

an initial guess is needed for a MathCAD function of 'Maximize' as

$$\text{initial guess} \quad \psi := 0.2 \qquad (3.245)$$

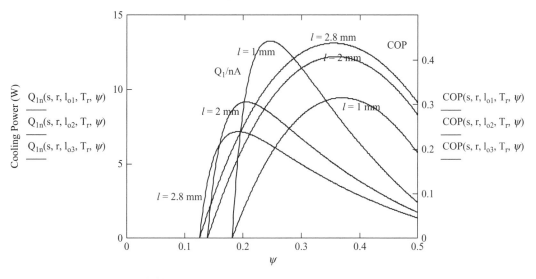

Figure 3.33 Cooling power and COP as a function of ψ and l

We find the maximum values as

$$\psi_1 := \text{Maximize}(\text{COP}_1, \psi) \quad \psi_1 = 0.355 \quad Q_{cn}(\psi_1) = 3.725\,\text{W} \quad \text{COP}_1(\psi_1) = 0.437 \tag{3.246}$$

$$\psi_2 := \text{Maximize}(Q_{cn}, \psi) \quad \psi_2 = 0.192 \quad Q_{cn}(\psi_2) = 7.159\,\text{W} \quad \text{COP}_1(\psi_2) = 0.227 \tag{3.247}$$

Note that the cooling power of 3.725 W at $\psi_1 = 0.355$ is slightly less than the required cooling power of 4 W. However, this shouldn't be a problem since it is a minimum power. Now we remind and confirm some parameters (it is necessary sometimes to avoid errors) as

$$s = 0.1\,\text{mm} \quad r = 0.1 \quad l_o = 2.8\,\text{mm} \quad \psi_1 = 0.355 \quad \psi_2 = 0.192 \tag{3.248}$$

Define the temperature difference, ΔT, in terms of the temperature ratio, T_r, in order to see the effect of performance with respect to the temperature difference rather than the temperature ratio.

$$\Delta T(T_r) := T_2 \left(1 - \frac{1}{T_r}\right) \tag{3.249}$$

Define a graphing range for T_r as

$$T_{r1} := 1, 1.002.. 1.3 \tag{3.250}$$

It is interesting to note that the curve of Q_{1n} for $\psi = 0.355$ (optimum COP) forms a parabola with a maximum value with respect to ΔT, indicating an optimum operating condition, and the COPs exponentially increase with decreasing ΔT. Decreasing ΔT would be very challenging depending on both the design of heat sinks in hot and cold junctions and the design of a thermoelectric module. However, it is important to

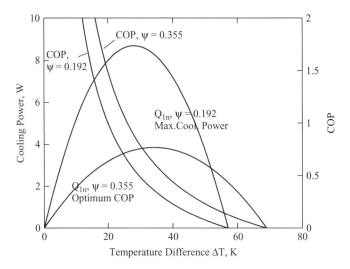

Figure 3.34   Cooling power and COP as a function of temperature difference

recognize the phenomenon that the coefficient of performance, COP, can be increased with decreasing the temperature difference. These can be seen in Figure 3.34. In order to obtain the maximum values of the performance, we need to define the followings as

$$Q_{1c}(T_r) := Q_{1n}(s, r, l_o, T_r, \psi_1) \quad Q_{2c}(T_r) := Q_{1n}(s, r, l_o, T_r, \psi_2) \tag{3.251}$$

$$COP_{\psi1}(T_r) := COP(s, r, l_o, T_r, \psi_1) \quad COP_{\psi2}(T_r) := COP(s, r, l_o, T_r, \psi_2) \tag{3.252}$$

$$\text{initial guess} \quad T_r = 1.154 \tag{3.253}$$

$$T_{r\max 1} := \text{Maximize}(Q_{1c}, T_r) \quad T_{r\max 1} = 1.128 \tag{3.254}$$

$$T_{r\max 2} := \text{Maximize}(Q_{2c}, T_r) \quad T_{r\max 2} = 1.103 \tag{3.255}$$

The maximum values for the COP and the cooling power with respect to the temperature differences are obtained as

$$\Delta T(T_{r\max 1}) = 34.12\,\text{K} \quad Q_{1c}(T_{r\max 1}) = 3.836\,\text{W} \quad COP_{\psi1}(T_{r\max 1}) = 0.617 \tag{3.256}$$

$$\Delta T(T_{r\max 2}) = 27.996\,\text{K} \quad Q_{2c}(T_{r\max 2}) = 8.682\,\text{W} \quad COP_{\psi2}(T_{r\max 2}) = 0.557 \tag{3.257}$$

It is reminded that these maximum values are very useful in optimum operation. From Equation (3.208), we compute the currents corresponding to the maximum values.

$$I_{TEC}(T_r) := \frac{A_e\,\alpha\,T_1(T_r)(T_r - 1)}{\psi_1\,\rho\,l_o\left(1 + \dfrac{s}{l_o}\right)\left(1 - m_z(l_o, T_r, \psi_1)\,r\,\dfrac{l_c}{l_o}\right)} \tag{3.258}$$

$$I_{TEC}(T_{r\max 1}) = 3.604\,\text{A} \quad I_{TEC}(T_{r\max 2}) = 2.961\,\text{A} \tag{3.259}$$

Now the maximum performance parameters are obtained. From Figure 3.34, we seek the maximum temperature difference for which the cooling power must be zero, so Equation (3.251) of the cooling power is solved using MathCAD function of 'root'. First, define an initial guess as

$$\text{initial guess} \quad T_r = 1.154 \tag{3.260}$$

$$T_{r0} := \text{root}(Q_{1c}(T_r), T_r) \quad T_{r0} = 1.298 \tag{3.261}$$

Using the temperature ratio at which the cooling power is zero, we compute the temperature difference using Equation (3.249).

$$\Delta T_{max} := \Delta T(T_{r0}) \quad \Delta T_{max} = 68.873 \, \text{K} \tag{3.262}$$

Finally, we find the temperature difference of 68.873 K which can be also seen in Figure 3.34. The same maximum temperature difference of TEC modules is often suggested in various commercial products as a specification. The maximum cooling power is then determined with this maximum temperature difference as

$$Q_{max} := \frac{n \, A_e \, k}{l_o} \, \Delta T_{max} \quad Q_{max} = 10.941 \, \text{W} \tag{3.263}$$

The maximum voltage is computed with the maximum temperature difference using Equation (3.210).

$$V_{COP}(T_r) := \frac{n \, \alpha \, T_1(T_r)(T_r - 1)}{\left(1 - m_z(l_o, T_r, \psi_1) \, r \, \dfrac{l_c}{l_o}\right)} \, \frac{\psi_1 + 1}{\psi_1} \tag{3.264}$$

$$V_{max} := V_{COP}(T_{r0}) \quad V_{max} = 3.458 \, \text{V} \tag{3.265}$$

3.11.2 Summary of Design of a TEC Module

We summarize the results for the design of a TEC module with given conditions.
Cross-sectional area of thermoelement is

$$A_e = 5.106 \, \text{mm}^2 \tag{3.266}$$

The number of thermoelement couple is

$$n = 31.114 \tag{3.267}$$

The maximum performance parameters are

$$\Delta T_{max} = 68.873 \, \text{K} \tag{3.268}$$

$$V_{max} = 3.458 \, \text{V} \tag{3.269}$$

$$I_{max} = 4.4 \, \text{A} \tag{3.270}$$

$$Q_{max} = 10.941 \, \text{W} \tag{3.271}$$

Table 3.12 Recommended Operating Range

Symbol	Maxi. Cooling Power	Optimum COP
ψ	0.192	0.355
ΔT (K)	28.0	34.2
I (A)	2.96	3.6
Q_{cn} (W)	8.6	3.8
COP	0.56	0.62

The recommended optimum operating range is illustrated in Table 3.12. Any conditions between these two points should be the optimum conditions for the operation of the TEC module. It is interesting to note that the condition for the maximum cooling power provides the best performance without a significant loss in the COP.

REFERENCES

1. D.M. Rowe, *CRC Handbook of Thermoelectrics.* Boca Raton, FL: CRC Press, 1995.
2. H.A. Sorensen, Energy Conversion Systems, John Wiley & Sons, 1983.
3. L.-T Yeh and R. Chu, Thermal Management of Microelectronic Equipment. New York: ASME Press, 2002.
4. P.E Snyder, Chemistry for Thermoelectric Materials, *Chemical and Engineering News*, March 13, 1961.
5. D.A. Wright, Materials for Direct Conversion Thermoelectric Generators, *Metall. Rev.*, vol. 15, no. 147.
6. R.R. Heikes and R.W Ure, *Thermoelectricity*, *Science and Engineering*, New York: Interscience Publishers, 1961.
7. R.E. Simon, M.J. Ellsworth, and R.C. Chu, An Assessment of Module Cooling Enhancement with Thermoelectric Coolers. *ASME Journal of Heat Transfer*, vol. 127 (January 2005), pp. 76–84.
8. S.B. Riffat and X. Ma, Thermoelectrics: A Review and Potential Applications, *Applied Thermal Engineering,* vol. 23 (2003), 913–935.
9. D.M. Rowe and C.M. Bhandari, *Modern Thermoelectrics.* London: Holt, Rinehart, and Winston, 1983.
10. R. Ventatasubramanian, E. Silvola, T. Colpitts, and B. O'Quinn, Thin-Film Thermoelectric Devices with High Room-Temperature Figures of Merit. *Nature* 413, October 2001, pp. 597–602.
11. P.M. Martin and L.C. Olsen, Scale-Up of Si/Si0.8Ge0.2 and B4C/B9 Superlattices for Harvesting of Waste Heat in Diesel Engines, Pacific Northwest National Laboratory, U.S. Department of Energy Office of FreedomCAR & Vehicle Technologies, in *Proceedings of the 2003 Diesel Engine Emissions Reduction Conference* (Newport, RI, 2003).
12. J.G. Stockholm, L. Pujol-Soulet, and P. Sternat, *Proceedings,* Third International Conference on Thermoelectric Energy Conversion (Arlington, Texas). New York: IEEE, 1982, p. 136.
13. J.E. Buffet and J.G. Stockholm, *Proceedings*, Eighteenth Intersociety Energy Conversion Engineering Conference, Orlando, Florida, American Institute of Chemical Engineers (New York, 1983), p. 253.

14. G.S. Nolas, J. Sharp, and H.J. Goldsmid, *Thermoelectrics*, Berlin-Heidelberg: Springer, 2001.

15. D.M. Rowe, *Thermoelectrics Handbook: Macro to Nano*. Boca Raton, FL: CRC Press, 2006.

16. T.J. Seebeck, Ueber den magnetismus der galvenische kette, *Abh. K. Akad. Wiss Berlin* (1864), p. 289.

17. J.C.A. Peltier, Nouvelles experiences sur la caloricite des courants electrique, *Ann. Chem. Phys.*, vol. 56 (1834), p. 371.

18. W., Thomson, Account of Researches in Thermo-Electricity, *Philos. Mag.* vol. 5, no. 8, p. 62.

19. L. Onsager, Reciprocal Relations In Irreversible Processes, I. *Phys. Rev.*, vol. 37 (Februrary 15, 1931), 405–526.

20. R.P. Benedict, Thermoelectric Thermometry, in *Fundamentals of Temperature, Pressure, and Flow Measurements* (3rd ed.). New York: John Wiley and Sons, 1984.

21. G. Min and D.M. Rowe, Optimization of Thermoelectric Module Geometry for "Waste Heat" Electric Power Generation, *Journal of Power Sources*, vol. 38 (1992), pp. 253–259.

22. D.M. Rowe and G. Min, Design Theory of Thermoelectric Modules for Electric Power Generation, *IEE Proc.-Sci. Meas. Technol.* vol. 143, no. 6 (1996), pp. 351–356.

PROBLEMS

3.1 A thermoelectric generator (TEG) is fabricated from p-type and n-type semiconductors (PbTe). The hot and cold junctions are $600\,K$ and $300\,K$, respectively. At the mean temperature of $510\,K$, the properties of the material are $\alpha_n = -190 \times 10^{-6}\,V/K$, $\alpha_p = 190 \times 10^{-6}\,V/K$, $\rho_n = \rho_p = 1.35 \times 10^{-3}\,\Omega\,cm$, $k_n = k_p = 0.014\,W/cm\,K$. The length of the thermoelectric element is $1.0\,cm$, and the cross-sectional area of the element is $1.0\,cm^2$.

(a) Determine the maximum power, the maximum power efficiency, the heat absorbed at the hot junction, the corresponding output voltage across the load, and the current for the maximum power.

(b) Calculate the power, the maximum conversion efficiency, the heat absorbed at the hot junction, the output voltage across the load, and the current for the maximum conversion efficiency.

3.2 A thermoelectric generator is to be designed to supply $240\,W$ of power by a burner which operates on a gaseous fuel in a remote area as shown in Figure P3.2. The output voltage is preferred to be approximately $24\,V$, otherwise, a DC/DC converter may be required. The cold junction plate is cooled with heat sinks using heat pipes as shown. The hot junction temperature is maintained at $450°C$ while the cold junction temperature is maintained at $100°C$. The thermolement material of lead telluride at the average temperature of $275°C$ is used having $\alpha_p = 180 \times 10^{-6}\,V/K$, $\alpha_n = -175 \times 10^{-6}\,V/K$, $\rho_p = 1.05 \times 10^{-3}\,\Omega\,cm$, $\rho_n = 0.96 \times 10^{-3}\,\Omega\,cm$, $k_p = 0.013\,W/cm\,K$, $k_n = 0.015\,W/cm\,K$. The cross-sectional area of the n-type element A_n is $0.3\,cm^2$, and the length of the elements are same for both p-type and n-type elements, $L_p = L_n = 0.6\,cm$. For the design of a thermoelectric generator, answer the following questions.

1. Determine the cross-sectional area, A_p, of the p-type thermoelement for the optimum geometry.

2. Determine the performance parameters for (a) the maximum conversion efficiency and (b) the maximum power, respectively. The performance parameters includes: the number of thermocouples, the thermal efficiency, the voltage across the load, the current, the

heat absorbed, and the heat dissipated. (c) Determine also the maximum performance
parameters.

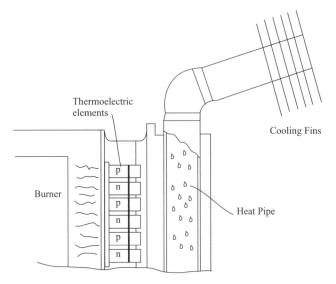

Figure P3.2   Thermoelectric generator

3.3   A thermoelectric generator is to be designed to recover the waste thermal energy from
the exhaust pipeline in a vehicle as shown in Figure P3.3. The cold plate is cooled by
circulating the vehicle's cooling water. The size for the thermoelectric modules is limited
by 120 mm × 80 mm. The hot and cold junction temperatures are estimated to be 270°C and
60°C, respectively. The power generated should be no less than 35 W. The output voltage
is required to be 24 V to charge the battery (12 V) in the vehicle. Design the thermoelectric
generator using the TEG modules listed in Table 3.2 or Table 3.3, providing the power, the
current, the voltage, the heat absorbed at the hot junction, the heat dissipated at the cold
junction, the thermal efficiency, and the number of modules.

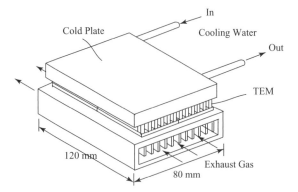

Figure P3.3   Thermoelectric generator device

3.4 A thermoelectric cooler consists of n-type and p-type semiconductors. The cold junction and hot junction temperatures are maintained at 270 K and 300 K, respectively. The semiconductor has dimensions of A = 0.08 cm² and L = 0.5 cm. The thermoelectric materials are $\alpha_n = -227 \times 10^{-6}$ V/K, $\alpha_p = 227 \times 10^{-6}$ V/K, $\rho_n = \rho_p = 1.95 \times 10^{-3}\,\Omega$ cm, $k_n = k_p = 0.0173$ W/cm K.

(a) Determine the current, the voltage, the cooling power, the heat delivered, and the *COP* for the maximum cooling power.

(b) Determine the current, the voltage, the cooling power, the heat delivered, and the power for the maximum *COP*.

(c) Determine the maximum performance parameters: the maximum current, the maximum temperature, the maximum cooling power, and the maximum voltage.

3.5 A 60 × 60 mm processor chip is cooled using a TEC with a heat sink at the hot temperature as shown in Figure P3.5. The cold and hot temperatures are required to be 65°C and 85°C, respectively. The processor heat dissipation is required to be less than 40 Watts. A commercial module is to be selected from the list of modules in Tables 3.4 or 3.5. Usually manufacturers provide their generalized charts for the selection of their products. However, we want to practice the design procedure using the generalized charts shown here in Figures B.7 through B.9 for estimation assuming $ZT_c = 1$.

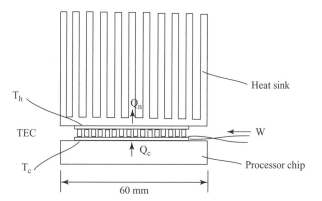

Figure P3.5 TEC cooling device with a heat sink

3.6 A thermoelectric air cooler is designed to pump out heat at a rate of 52 Watts from electronic components in a closed box located at a remote area, as shown in Figure P3.6. The air cooler consists of an internal fan, an internal heat sink, a thermoelectric cooler (TEC), an external heat sink, and an external fan. For the heat transfer rate given, the air temperature in the box must be maintained at 20°C, when the ambient air is at 40°C. The voltage across the thermoelectric cooler is preferred to be 12 V. The dimensions of both the internal and external heat sinks is preferred to $W_1 = 117$ mm, $L_1 = 162$ mm, $b_1 = 50$ mm, $W_2 = 140$ mm, $L_2 = 254$ mm, and $b_2 = 50$ mm. The volume flow rates of the internal and external fans are also given to be 35 cfm (cubic feet per minute) and 50 cfm, respectively. Design the internal and external heat sinks by providing the fin thicknesses, the fin spacings, and the number of fins to meet the requirements. Design the thermoelectric cooler (TEC) with the properties given and provide the high and low junction temperatures, T_h and T_c. The properties for the n-type and p-type semiconductors are given as $\alpha_n = -2.2 \times 10^{-4}$ V/K, $\rho_n = 1.2 \times 10^{-3} \Omega$ cm, $k_n = 1.62 \times 10^{-2}$ W/cm K, $\alpha_p = 2.2 \times 10^{-4}$ V/K, $\rho_p = 1.2 \times 10^{-3} \Omega$ cm, and $k_p = 1.62 \times 10^{-2}$ W/cm K.

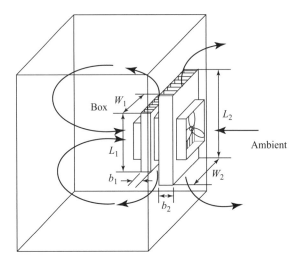

Figure P3.6 Air flow of a thermoelectric air cooler

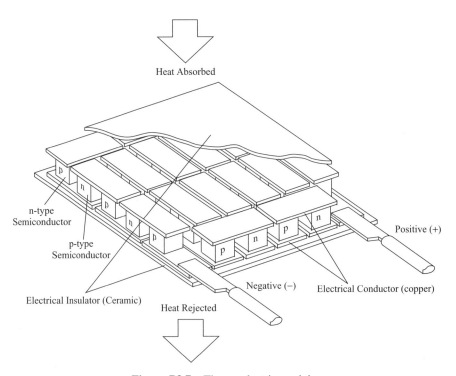

Figure P3.7 Thermoelectric module

3.7 A TEC module for a small refrigerator is designed to provide 10 W of cooling at the maximum current of 5 A. The module is illustrated in Figure P3.7. The hot side of the module is maintained at 300 K and a temperature difference of 40 K must be maintained. The electrical and thermal contact resistances were predetermined by experiments and their values are listed below in the following table. It is determined that a particular thermoelement length of 1 mm is required. The material properties of Bi_2Te_3 and the information for the thermal and electrical contact resistances are listed in the table. Determine the cross-sectional area of thermoelement, the number of thermoelement couples, and the maximum performance parameters for the module: ΔT_{max}, V_{max}, I_{max}, and Q_{max}. Also provide the recommended optimum operating conditions including currents, temperature difference, COP, and cooling powers.

Properties and contact resistances

Description	Value
Seebeck coefficient, α_p	200 μV/K
Seebeck coefficient, α_n	-200 μV/K
Thermal conductivity, $k_p = k_n$	1.4 W/m·K
Electrical resistivity, $\rho_p = \rho n$	1.0 mΩ·cm
Thermal contact conductivity ratio, $r = k/k_c$	0.1
Electrical contact resistivity ratio, $s = \rho_c/\rho$	0.1mm
Ceramic plate thickness, l_c	0.7 mm
Thermoelement length, l	1 mm

4

Heat Pipes

4.1 OPERATION OF HEAT PIPE

A heat pipe is a simple device with no moving parts that can transfer large quantities of heat over fairly large distances essentially at a constant temperature without requiring any power input. A heat pipe is basically a sealed slender tube containing a wick structure lined on the inner surface and a small amount of fluid such as water at the saturated state, as shown in Figure 4.1. It is composed of three sections: evaporator section at one end, where heat is absorbed and the fluid is vaporized; a condenser section at the other end, where the vapor is condensed and heat is rejected; and the adiabatic section in between, where the vapor and the liquid phases of the fluid flow in opposite directions through the core and the wick, respectively, to complete the cycle with no significant heat transfer between the fluid and the surrounding medium.

The operation of a heat pipe is based on the thermodynamic properties of a fluid vaporizing at one end and condensing at the other end. Initially, a wick of the heat pipe is saturated with liquid and the core section is filled with vapor, as shown in Figure 4.1. When the evaporator end of the heat pipe is brought into contact with a hot surface or is placed into a hot environment, heat will flow into the heat pipe.

Being at a saturated state, the liquid in the evaporator end of the heat pipe will vaporize as a result of this heat transfer, causing the vapor pressure there to rise. This resulting pressure difference drives the vapor through the core of the heat pipe from the evaporator toward the condenser section. The condenser end of the heat pipe is in a cooler environment, and thus, its surface is slightly cooler. The vapor that comes into contact with this cooler surface condenses, releasing the heat a vaporization, which is rejected to the surrounding medium. The liquid then returns to the evaporator end of the heat pipe through the wick as a result of capillary action in the wick, completing the cycle. As a result, heat is absorbed at one end of the heat pipe and is rejected at the other end, with the fluid inside serving as a transport medium for heat.

The boiling and condensation processes are associated with extremely high heat transfer coefficients, and thus it is natural to expect the heat pipe to be an extremely effective heat transfer device, since its operation is based on alternate boiling and condensation of working fluid. A simple heat pipe with water as the working fluid has an effective thermal conductivity of the order of 100,000 W/m·K compared with about 400 W/m·K for copper. For a heat pipe, it is not unusual to have an effective conductivity of 400,000 W/m·K, which is a thousand times that of copper. A 15-cm long, 0.6-cm-diameter horizontal cylindrical heat pipe with water, for example, can transfer heat at a rate of 300 W. Therefore, heat pipes are preferred in some critical applications, despite their high initial cost [2].

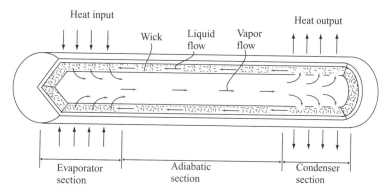

Figure 4.1   Cutaway view of a heat pipe

Heat pipes and thermosyphons both operate on a closed two-phase cycle and utilize the latent heat of vaporization to transfer heat with very small temperature gradients. Thermosyphons, however, rely solely on gravitational force to return the liquid phase of the working fluid from the condenser to the evaporator, while heat pipes utilize some sort of capillary wicking structure to promote the flow of liquid from the condenser to the evaporator. As a result of the capillary pumping occurring in the wick, heat pipes can be used in horizontal orientation, microgravity environments, or even applications where the capillary structure must pump the liquid against gravity from the condenser to the evaporator.

4.2   SURFACE TENSION

Surface tension is an important phenomenon in heat pipes. Hence, we want to start considering a vapor bubble at rest in liquid, where the pressure in the bubble is denoted as Pressure I and the outside pressure adjacent to the interface is denoted as Pressure II, as shown in Figure 4.2(a). We can imagine a half bubble with the surface tension on the cut interface and the pressures on the inner interface, which is something like a free body diagram as shown in Figure 4.2(b).

The force exerted on the circumference in Figure 4.2(b) is nothing more than the surface tension along the circumference as

$$F = \sigma (2\pi r) \tag{4.1}$$

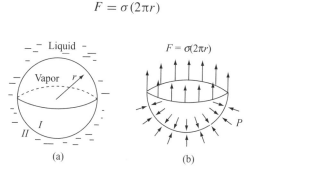

Figure 4.2   (a) A vapor bubble in liquid, (b) Cutaway of a half bubble

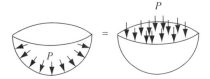

Figure 4.3 Equivalence of the pressure on the internal surface and on the circular plane of the hemisphere

Geometry tells us that the vertical components of the pressures on the inner interface are equal to the pressures on the circular middle plane of the half bubble, which is shown in Figure 4.3. Now we can also speculate that since the bubble is at rest, all the forces on the half bubble, including the pressure on the both sides of the interface, must be balanced. Hence,

$$F = (P_I - P_{II})\pi r^2 \qquad (4.2)$$

Combining Equations (4.1) and (4.2) gives us the well-known *Young-Laplace equation* for a sphere:

$$P_I - P_{II} = \frac{2\sigma}{r} \qquad (4.3)$$

This indicates that the inside pressure is always greater than the outside pressure because σ and r are positive values and the pressure difference become very large when the radius is very small.

Consider three different capillary tubes at atmospheric pressure as shown in Figure 4.4(a). We learn at a glance that the tube radius is related to the height of liquid in the tube. The contact angle is defined as the angle between the liquid interface and the solid surface, as denoted θ in Figure 4.4(b). The contact angle is the property of the liquid on the solid surface.

Pressures I, II, and III are seen with the points in Figure 4.4(b) to indicate the locations. Pressures I and II are at the meniscus across the interface, while Pressure III is at the bulk liquid level. Notice that Pressure I is equal to Pressure III, neglecting the static pressure in air. Hence,

$$P_I = P_{III} \qquad (4.4)$$

We know that the y-component of the force due to the surface tension gives

$$F_y = \sigma \cos(\theta) 2\pi r \qquad (4.5)$$

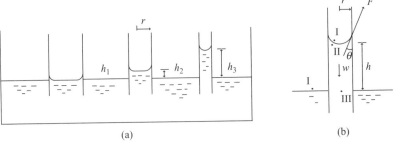

Figure 4.4 Capillary rise (a) three different rises of tubes, (b) Contact angle

The weight of the portion of liquid in the tube above the bulk liquid level is obtained as

$$W = mg = \rho V g = \rho(\pi r^2 h)g \tag{4.6}$$

where g is the gravity, V is the volume, m is the mass, and ρ is the density of the liquid.

The force due to the surface tension is balanced by the weight W of the portion of the liquid above the water-free level, which is shown in Figure 4.4(b). Hence, combining Equations (4.5) and (4.6) gives

$$\rho g h = \frac{2\sigma \cos(\theta)}{r} \tag{4.7}$$

The pressure difference between Pressure III and Pressure II is static. Hence, assuming the gas density is much smaller than the liquid density, we have

$$P_{III} - P_{II} = \rho g h \tag{4.8}$$

Using Equation (4.4), (4.7), and (4.8), we have an equivalent expression of the Young-Laplace equation:

$$P_I - P_{II} = \frac{2\sigma \cos(\theta)}{r} \tag{4.9}$$

In the case of a sphere, as θ is zero and consequently $\cos(0) = 1$, Equation (4.9) becomes Equation (4.3).

From this, we learn that the surface tension naturally contributes a contact angle that forms a meniscus and causes Pressure II to drop. This, in turn, pulls the liquid upward until the equilibrium is achieved, which is called the *capillary rise*.

Contact Angle (θ)   As mentioned before, the contact angle is defined as the angle in the liquid side between the vapor-liquid interface and the solid surface, as illustrated in Figure 4.5(a) and (b). The pressure difference between locations I and II can be found using Equation (4.9).

When $0 < \theta < 90°$, a liquid is termed *wetting*. When $90° < \theta < 180°$, the liquid is termed *nonwetting*. Thus, the contact angle is a direct index of wettability of the liquid. The contact angle varies with the solid materials. For example, the contact angle of water on a gold surface is zero, while the contact angle of water on Teflon is 112°.

4.3 HEAT TRANSFER LIMITATIONS

Heat pipes undergo various heat transfer limitations, depending on the working fluid, the wick structure, the dimensions of heat pipes, and the operational temperatures. Heat pipes can be limited in one of four ways:

1. *Capillary limitation*. There is insufficient capillary pressure to pump the working fluid back to the condenser.

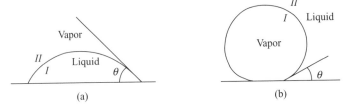

Figure 4.5   (a) A liquid droplet on the solid surface, (b) A vapor bubble on the solid surface

2. *Sonic limitation*. The vapor velocity inside of the heat pipe reaches the sonic velocity limitation, causing choked flow.
3. *Entrainment limitation*. Liquid droplets are entrained in the vapor flow and carried back to the condenser.
4. *Boiling limitation*. The evaporator heat flux is so high that boiling occurs in the evaporator wick.

4.3.1   Capillary Limitation

Although heat pipe performance and operation are strongly dependent on shape, working fluid, and wick structure, the fundamental phenomenon that governs the operation arises from the difference in the capillary pressure across the liquid-vapor interfaces in the evaporator and condenser sections [1].

Vaporization from the evaporator section of a heat pipe causes the meniscus to recede into the wick, and *condensation* in the condenser causes flooding. The combined effect of this vaporization and condensation process results in a meniscus radius of

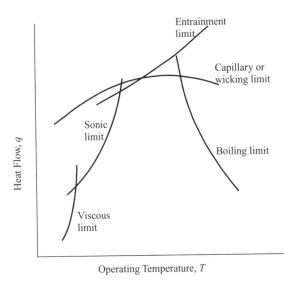

Figure 4.6   Schematic of limits on heat pipe performance

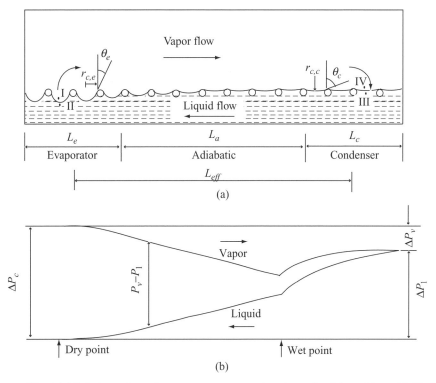

Figure 4.7   (a) Variation of meniscus curvature as a function of axial position, (b) Typical liquid and vapor pressure distributions in a heat pipe

curvature that varies along the axial length of the heat pipe, as shown in Figure 4.7(a). The point at which the meniscus has a minimum radius of curvature is typically referred to as the *dry point* and usually occurs in the evaporator at the point farthest from the condenser region. The *wet point* occurs at that point where the vapor pressure and liquid pressure are approximately equal, or where the radius of curvature is at a maximum. Typically, the wet point is found near the end of the condenser farthest from the evaporator.

The capillary pressure difference is virtually the pressure difference between the pressures at point III and point II in the liquid, which is seen in Figure 4.7(a). The relationship between the liquid and vapor pressures in an operating heat pipe is illustrated in Figure 4.7(b). Notice that the vapor–liquid pressure difference appears greatest in the evaporator region, while the pressure difference becomes smallest in the condenser region.

4.3.1.1   *Maximum Capillary Pressure Difference*

For a heat pipe to function properly, the net capillary pressure difference (or drop) between the evaporator and condenser must be greater than the summation of all the pressure losses occurring throughout the liquid and vapor flow paths. The maximum capillary pressure difference will be

$$\Delta P_{c,m} = \int_{Leff} \frac{\partial P_v}{\partial x} dx + \int_{Leff} \frac{\partial P_l}{\partial x} dx + \Delta P_{ph,e} + \Delta P_{ph,c} + \Delta P_{norm} + \Delta P_{axial} \quad (4.10)$$

where

$$L_{eff} = \frac{L_e}{2} + L_a + \frac{L_c}{2} \quad (4.11)$$

$\Delta P_{c,m}$ = maximum capillary pressure difference generated within capillary wicking structure

$\frac{\partial P_v}{\partial x}$ = sum of inertial and viscous pressure drops in vapor phase

$\frac{\partial P_l}{\partial x}$ = sum of inertial and viscous pressure drops in liquid phase

$\Delta P_{ph,e}$ = pressure gradient across phase transition in evaporator

$\Delta P_{ph,l}$ = pressure gradient across phase transition in condenser

ΔP_{norm} = normal hydrostatic pressure drop

ΔP_{axial} = axial hydrostatic pressure drop

Although calculating the liquid pressure drop occurring in the wicking structure of a heat pipe is relatively straight forward, calculating the vapor pressure gradient is significantly more difficult.

Because of the large difference in the density of these two phases, the vapor velocity must be significantly higher than the velocity of the liquid phase.

Initial pressure gradient in the liquid phase is typically negligible and therefore has been omitted. However, the inertial pressure gradient in the vapor phase may make a significant contribution to the flow characteristics.

The phase transition pressure drop may be important in microheat pipes, but typically, it is neglected. Exceptions are cases in porous wicks at very low temperatures. Equation (4.10) may be simplified. The maximum capillary pressure limit will become

$$\Delta P_{c,m} = \Delta P_v + \Delta P_l + \Delta P_{norm} + \Delta P_{axial} \quad (4.12)$$

The pressure difference between point I and point II at the evaporator in Figure 4.7(a) using Equation (4.9) is

$$P_I - P_{II} = \frac{2\sigma \cos (\theta_e)}{r_{c,e}} \quad (4.13)$$

where $r_{c,e}$ is the capillary radius in evaporator.

The pressure difference at the condenser is

$$P_{IV} - P_{III} = \frac{2\sigma \cos (\theta_c)}{r_{c,c}} \quad (4.14)$$

where $r_{c,c}$ is the capillary radius in condenser.

The maximum capillary pressure would occur when

$$\theta_e = 0 \quad \text{and} \quad \theta_c = \frac{\pi}{2} \Rightarrow r_{c,c} = \infty \tag{4.15}$$

$$P_I \cong P_{IV} \tag{4.16}$$

where θ_e and θ_c are the contact angles at the evaporator and condenser, respectively, as shown in Figure 4.7.

Subtracting Equation (4.14) from Equation (4.13) using Equations (4.15) and (4.16) yields

$$\Delta P_{c,m} = P_{III} - P_{II} = \frac{2\sigma}{r_{c,e}} - \frac{2\sigma}{r_{c,c}} \cong \frac{2\sigma}{r_{c,e}} \tag{4.17}$$

Hence, the maximum capillary pressure difference becomes a simple form of

$$\Delta P_{c,m} = \frac{2\sigma}{r_{c,e}} \tag{4.18}$$

Values for the effective capillary radius r_c are given in Table 4.1 for some of the more common wicking structures.

4.3.1.2   *Vapor Pressure Drop*

As discussed earlier, the vapor velocity is significantly higher than the velocity of the liquid. For this reason, in addition to the pressure gradient resulting from frictional drag, we should consider the pressure gradient due to the dynamic pressure. Usually, the vapor-phase pressure difference (or drop) is generally much smaller than that for the

Table 4.1   Effective Capillary Radius for Several Wick Structures

Structure	r_c	Data
Circular cylinder (artery or tunnel wick)	r	r = radius of liquid flow passage
Rectangular groove	ω	ω = groove width
Triangular groove	$\omega/\cos\beta$	ω = groove width β = half-included angle
Parallel wires	ω	ω = wire spacing
Wire screens	$(\omega + d_\omega)/2 = 1/2N$	d = wire diameter N = screen mesh number per inch ω = wire spacing
Packed spheres	$0.41 r_s$	r_s = sphere radius
Sintered metal fibers*	$d/2\,(1-\varepsilon)$	d = fiber diameter ε = porosity (ratio of pore volume to total volume; use manufacturer's data)
Trapezoidal microheat pipe*	$\omega/(\cos\alpha\cos\theta)$	ω = grooove width α = half included angle θ = liquid wetting angle

*Reference [26]

liquid phase, as shown in Figure 4.7(b). Applying the one-dimensional approximation to the momentum equation, a widely used simple form of the vapor pressure difference was obtained by Chi [5]:

$$\Delta P_v = \frac{C(f_v \, \mathrm{Re}_v)\,\mu_v}{2r_v^2 A_v \rho_v h_{fg}} L_{eff}\, q \tag{4.19}$$

where f_v = friction factor of vapor $\qquad$ (4.19a)
Re_v = Reynolds number of vapor
μ_v = absolute viscosity of vapor
r_v = hydraulic radius of the vapor path
A_v = cross-sectional area of vapor path
ρ_v = vapor density
h_{fg} = latent heat of vaporization
q = heat transfer rate (see Equation (4.25))

When $\mathrm{Re}_v < 2300$ and $Ma_v < 0.2$ $\qquad$ (4.20)

$f_v \, \mathrm{Re}_v = 16$
$\quad C = 1.0$
$\quad Ma_v$ = Mach number of vapor (see Equation (4.24))

When $\mathrm{Re}_v < 2300$ and $Ma_v > 0.2$ $\qquad$ (4.21)

$f_v \, \mathrm{Re}_v = 16$
$$C = \left[1 + \left(\frac{\gamma_v - 1}{2}\right) Ma_v^2\right]^{-1/2}$$

When $\mathrm{Re}_v > 2300$ and $Ma_v < 0.2$ $\qquad$ (4.22)

$$f_v \mathrm{Re}_v = 0.038 \left[\frac{2r_v q}{A_v \mu_v h_{fg}}\right]^{3/4}$$
$\quad C = 1.0$

γ_v = specific heat ratio $\left(\dfrac{c_{po}}{c_{vo}}\right)$ (or called k) $\qquad$ (4.23)

1.67 for monatomic (single atom) vapors
1.40 for diatomic (two atoms) vapors or air
1.33 for polyatomic (more than two) vapors
$\qquad R_v$ = gas constant
Ethanol = 180.5 J/kgK
Methanol = 259.5 J/kgK
Water = 461.5 J/kgK
Nitrogen = 296.8 J/kgK

When $\mathrm{Re}_v > 2300$ and $Ma_v > 0.2$ [26]   (4.23a)

$f_v \mathrm{Re}_v = 0.038$

$$C = \left[1 + \left(\frac{\gamma_v - 1}{2}\right) Ma_v^2\right]^{-1/2} \left(\frac{2r_v q}{A_v \mu_v h_{fg}}\right)^{3/4}$$

$$Ma_v = \frac{V}{c} = \frac{\dot{m}}{\rho_v A_v \sqrt{\gamma_v R_v T_v}} = \frac{q}{\rho_v A_v h_{fg} \sqrt{\gamma_v R_v T_v}} \tag{4.24}$$

where Ma_v is the Mach number of vapor, $\dot{m}$ is the mass flow rate, R_v is the gas constant, and q is the rate of heat transfer.

Notice that the rate of heat transfer with phase change in either the evaporator or the condenser is a product of mass flow rate $\dot{m}$ and the latent heat of vaporization or condensation h_{fg} as

$$q = \dot{m} h_{fg} \tag{4.25}$$

The Reynolds number of vapor is expressed in terms of the rate of heat transfer.

$$\mathrm{Re}_v = \frac{\rho_v V d_v}{\mu_v} = \frac{4\dot{m}}{\pi d_v \mu_v} = \frac{4q}{\pi d_v \mu_v h_{fg}} \tag{4.26}$$

4.3.1.3   *Liquid Pressure Drop*

While the capillary pumping pressure promotes the flow of liquid through the wicking structure, the viscous forces in the liquid result in a pressure drop. Hence, the liquid pressure drop is obtained as a form of

$$\Delta P_l = \left(\frac{\mu_l}{K A_w h_{fg} \rho_l}\right) L_{eff}\, q \tag{4.27}$$

where μ_l = absolute viscosity of liquid
$\quad\;\; K$ = wick (or artery) permeability
$\quad\; A_w$ = wick (or artery) cross-sectional area
$\quad\; h_{fg}$ = latent heat of vaporization
$\quad\;\; \rho_l$ = liquid density

4.3.1.4   *Normal Hydrostatic Pressure Drop*

The hydrostatic pressure gradient occurring in heat pipes is due to gravitational or body forces within the vapor and liquid phases. The normal hydrostatic pressure drop occurs only in heat pipes in which circumferential communication of the liquid in the wick is possible. It is the result of the body force acting perpendicular to the longitudinal axis of the heat pipe, which is illustrated in Figure 4.8.

$$\Delta P_{norm} = \rho_l g d_v \cos\psi \tag{4.28}$$

where d_v is the diameter of the vapor portion of the pipe and Ψ is the angle the heat pipe makes with respect to the horizontal.

Table 4.2 Wick Permeability for Several Wick Structures [5]

Structure	K	Data
Circular cylinder (artery or tunnel wick)	$r^2/8$	r = radius of liquid flow passage
Open rectangular grooves	$2\varepsilon\,(r_{h,l})^2/(f_l\,\mathrm{Re}_l)$	ε = wick porosity ω = groove width s = groove pitch δ = groove depth $r_{h,l} = 2\omega\delta/(\omega+2\delta)$
Circular annular wick	$2\,(r_{h,l})^2/(f_l\,\mathrm{Re}_l)$	$r_{h,l} = r_1 - r_2$
Wrapped screen wick	$\dfrac{d_w^2\varepsilon^3}{122\,(1-\varepsilon)^2}$	d_w = wire diameter in inch $\varepsilon = 1 - (\pi N d_w/4)$ N = mesh number per inch ε = porosity (ratio of pore volume to total volume)
Packed sphere	$\dfrac{r_s^2\varepsilon^3}{37.5\,(1-\varepsilon)^2}$	r_s = sphere radius ε = porosity (dependent on packing mode)
Sintered metal fibers	$C_1\dfrac{y^2-1}{y^2+1}$ $y = 1 + \dfrac{C_2 d^2\varepsilon^3}{(1-\varepsilon)^2}$ $C_1 = 6.0\times10^{10}\,\mathrm{m}^2$ $C_2 = 3.3\times10^3\,1/\mathrm{m}^2$	d = fiber diameter ε = porosity (ratio of pore volume to total volume; use manufacturers data)
Trapezoidal microheat pipe*	$\dfrac{2\varepsilon\,(r_{h,l})^2}{f_l\,\mathrm{Re}_l}$	$r_{h,l} = r_1 - r_2$
Rectangular artery***	$\dfrac{2r_{h,l}^2}{f_l\,\mathrm{Re}_l}$	ω = arterial width δ = arterial depth $r_{h,l} = \dfrac{\delta\omega}{\delta+\omega}$ $A_w = \delta\omega N$ N = number of arteries $f_l\mathrm{Re}_l = 16$ for laminar $f_l\mathrm{Re}_l = \dfrac{\mathrm{Re}_l}{4\,(0.79\ln(\mathrm{Re}_l)-1.64)^2}$ for turbulent [32]

*Reference [26]
**Reference [32]
***Author revised.

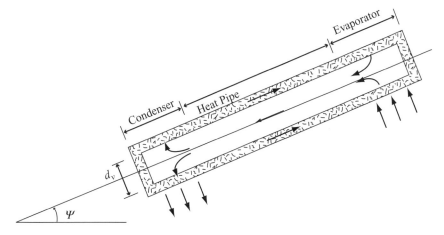

Figure 4.8   Orientation of a heat pipe

4.3.1.5   *Axial Hydrostatic Pressure Drop*

The axial hydrostatic pressure drop results from the body force acting along the longitudinal axis, which is illustrated in Figure 4.8.

$$\Delta P_{axial} = \rho_l g L \sin \psi \tag{4.29}$$

where L is the overall length of the pipe.

4.3.2   Approximation for Capillary Pressure Difference

For most heat pipes, the maximum heat transfer rate due to capillary limitation, neglecting the vapor pressure drop and the normal hydrostatic pressure drop, can be expressed [5]. In that case, Equation (4.12) reduces to

$$\Delta P_{c,m} = \Delta P_l + \Delta P_{axial} \tag{4.31}$$

Hence, using Equations (4.18), (4.27), and (4.28), we have

$$\frac{2\sigma}{r_{c,e}} = \left(\frac{\mu}{K A_w h_{fg} \rho_l} \right) L_{eff}\, q + \rho_l g L \sin \psi \tag{4.32}$$

Solving for the heat transfer rate q finally gives an expression that has been widely used as

$$q_{c,m} = \left(\frac{\rho_l \sigma h_{fg}}{\mu_l} \right) \left(\frac{A_w K}{L_{eff}} \right) \left(\frac{2}{r_{c,e}} - \frac{\rho_l}{\sigma} g L \sin \psi \right) \tag{4.33}$$

4.3.3 Sonic Limitation

The sonic limit in heat pipes is analogous to that in converging-diverging nozzles, where the mass flow rate is constant and the velocity varies. In heat pipes, the area is constant and the velocity varies because of the evaporation and condensation along the heat pipe.

The sonic limitation also can occur in heat pipes during start-up at low temperatures. The low temperature reduces the speed of sound in vapor core. A sufficiently high mass flow rate can cause sonic flow conditions and generate a shock wave that chokes the flow and restricts a pipe's ability to transfer heat to the condenser. Levy [8] developed this expression:

$$q_{s,\max} = A_v \rho_v h_{fg} \left[\frac{\gamma_v R_v T_v}{2(\gamma_v + 1)} \right]^{\frac{1}{2}} \tag{4.34}$$

where γ_v is the specific ratio (c_{po}/c_{vo}).

Alternatively, Busse [9] presented an expression using one-dimensional laminar flow for the sonic limitation, which agrees well with experimental data.

$$q_{s,\max} = 0.474 h_{fg} A_v \left(\rho_v P_v \right)^{\frac{1}{2}} \tag{4.35}$$

4.3.4 Entrainment Limitation

As a result of the high vapor velocities, liquid droplets may be picked up or entrained in the vapor flow and cause excess liquid accumulation in the condenser and hence dryout of the evaporator wick. Cotter [10] presented a method to determine the entrainment limitations using the Weber number.

The *Weber number* is defined as the ratio of the viscous shear force to the force resulting from the surface tension.

$$We = \frac{2r_w \rho_v V_v^2}{\sigma} \tag{4.36}$$

First relate the vapor velocity and the heat transport capacity to the axial heat flow as

$$V_v = \frac{q}{A_v \rho_v h_{fg}} \tag{4.37}$$

and assume that to prevent entrainment of liquid droplets in the vapor flow, the Weber number must be less than unity. Then the maximum heat transfer rate due to entrainment is obtained as

$$q_{e,\max} = A_v h_{fg} \left(\frac{\sigma \rho_v}{2r_{h,w}} \right)^{\frac{1}{2}} \tag{4.38}$$

where $r_{h,w}$ is the hydraulic radius of the wick $(r_{h,w} = D_h/2)$.

The hydraulic diameter is defined as

$$D_h = \frac{4A}{P} \tag{4.39}$$

where A is the cross-flow area and P is the wetted perimeter.

Table 4.3 Effective Thermal Conductivity for Liquid-saturated Wick [5, 22]

Wick structure	k_{eff}
Wick and liquid in series	$\dfrac{k_l k_w}{\varepsilon k_w + k_l (1 - \varepsilon)}$
Wick and liquid in parallel	$\varepsilon k_l + k_w (1 - \varepsilon)$
Wrapped screen	$\dfrac{k_l \left[(k_l + k_w) - (1 - \varepsilon)(k_l - k_w)\right]}{(k_l + k_w) + (1 - \varepsilon)(k_l - k_w)}$
Packed sphere	$\dfrac{k_l \left[(2k_l + k_w) - 2(1 - \varepsilon)(k_l - k_w)\right]}{(2k_l + k_w) + (1 - \varepsilon)(k_l - k_w)}$
Rectangular grooves	$\dfrac{\left(w_f k_l k_w \delta\right) + w k_l \left(0.185 w_f k_w + \delta k_l\right)}{(w + w_f)\left(0.185 w_f k_w + \delta k_l\right)}$
Sintered metal fibers*	$\varepsilon^2 k_l + (1 - \varepsilon)^2 k_w + \dfrac{4\varepsilon (1 - \varepsilon) k_l k_w}{k_l + k_s}$

*Revised from the original table [5].

4.3.5 Boiling Limitation

At very high heat fluxes, nucleate boiling may occur in the wicking structure and bubbles may become trapped in the wick, blocking the liquid return and resulting in evaporator dryout. Chi [5] developed the limit as follows:

$$q_{b,\max} = \frac{4\pi L_e k_{eff} T_v \sigma}{h_{fg} \rho_v \ln\left(\dfrac{r_i}{r_v}\right)} \left(\frac{1}{r_n} - \frac{1}{r_{c,e}}\right) \tag{4.40}$$

where r_n is the nucleation cavity radius, which is approximately 0.254 μm.

4.3.6 Viscous Limitation

At very low operating temperatures, the vapor pressure difference between the evaporator and the condenser regions of a heat pipe or thermosyphon may be extremely small. In some cases, the viscous forces within the vapor region may actually be larger than the pressure gradients caused by the imposed temperature field. When this occurs, the pressure gradients within the vapor region may not be sufficient to generate flow, and the vapor may stagnate. Busse [9] carried out a two-dimensional analysis, finding that the radial velocity component had a significant effect. He derived the following equation with good experimental agreement [9, 21]:

$$q_{vis,\max} = \frac{A_v r_v^2 h_{fg} \rho_v P_v}{16 \mu_v L_e} \tag{4.41}$$

Alternatively, a relationship between the overall pressure drop in the vapor phase and the absolute pressure in the vapor phase was suggested as

$$\frac{\Delta P_v}{P_v} < 0.1 \tag{4.42}$$

Example 4.1 Heat Pipe A heat pipe for an electronics application uses ethanol as the working fluid. The evaporator is 2 cm long, the condenser is 3 cm long, and the heat pipe has no adiabatic section, as shown in the figure. The diameter of the vapor space is 3 mm, and the wicking structure consists of three layers of #500 mesh 304 stainless steel screen (wire diameter = 0.00085 in.). If the heat pipe operates at 30°C in a horizontal position, determine the capillary, sonic, entrainment, and boiling limits.

Solution Thermodynamic properties of ethanol at 30°C from Table A.4 are as follows:

$$P_v = 10{,}000 \text{ Pa}, \; h_{fg} = 888{,}600 \text{ J/kg}, \; \rho_l = 781 \text{ kg/m}^3,$$

$$\rho_v = 0.38 \text{ kg/m}^3, \; k_l = 0.168 \text{ W/m} \cdot \text{K}$$

$$\mu_l = 1.02 \times 10^{-3} \text{ kg/m·s}, \; \mu_v = 0.91 \times 10^{-5} \text{ kg/m·s},$$

$$\sigma = 2.44 \times 10^{-2} \text{ N/m}, \; k_w = 14.9 \text{ W/m·K}$$

Given dimensions

$$L_e = 0.02 \text{ m}, \; L_c = 0.03 \text{ m, and}$$

$$L_{eff} = \frac{L_e}{2} + L_a + \frac{L_c}{2} = 0.01 + 0.015 = 0.025 \text{ m}$$

$$d_v = 0.003 \text{ m}, \; r_v = d_v/2 = 0.0015 \text{ m}$$

$$A_v = \pi r_v^2 = 7.069 \times 10^{-6} \text{ m}^2$$

$$d_w = 0.00085 \text{ in.} = 2.159 \times 10^{-5} \text{ m}$$

$$t_w = 3\,(2d_w) = 12.954 \times 10^{-5} \text{ m (to be assumed)}$$

$$d_i = d_v + 2t_w = 0.003259 \text{ m}$$

$$N = 500 \text{ in.}^{-1} = 500/\text{in.} \times 1 \text{ in.}/2.54 \times 10^{-2} \text{ m} = 19{,}685 \text{ m}^{-1}$$

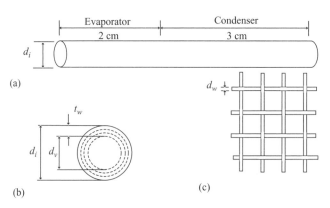

Figure E4.1.1 (a) A heat pipe, (b) The cross-section of the heat pipe, and (c) #500 mesh screen

Capillary Limit   Since the heat pipe operates in a horizontal position, the axial hydro-static pressure drop will be zero. Neglecting the phase transition pressure gradient and the initial pressure gradient in the evaporator, Equation (4.12) gives

$$\Delta P_{c,m} = \Delta P_v + \Delta P_l + \Delta P_{norm} \qquad \text{(E4.1.1)}$$

The problem concept is to determine the heat transfer rate for the maximum capillary pressure difference. Hence, from Equation (4.18),

$$\Delta P_{c,m} = \frac{2\sigma}{r_{c,e}} = \frac{2\left(2.44 \times 10^{-2}\right)}{2.54 \times 10^{-5}} = 1{,}921 \, \text{Pa} \qquad \text{(E4.1.2)}$$

where the capillary radius in the evaporator $r_{c,e}$ is found in Table 4.1 as

$$r_{c,e} = \frac{1}{2N} = \frac{1}{2\left(500/\text{in.}\right)} = 0.001 \, \text{in.} = 2.54 \times 10^{-5} \, \text{m} \qquad \text{(E4.1.3)}$$

The vapor pressure drop using Equation (4.19) is

$$\Delta P_v = \frac{C\left(f_v \text{Re}_v\right)\mu_v}{2r_v^2 A_v \rho_v h_{fg}} L_{eff} q \qquad \text{(E4.1.4)}$$

The Reynolds number and the Mach number are not known at this time because the vapor velocity is not known. Hence, we first assume that $Re_v < 2{,}300$ for laminar flow and $Ma_v < 0.2$ and later verify the assumption.

From Equation (4.20),

$$f_v \text{Re}_v = 16 \qquad \text{(E4.1.5)}$$

$$\Delta P_v = \frac{(1)\,(16)\left(0.91 \times 10^{-5}\right) \, L_{eff} \, q}{2\,(0.0015)^2 \left(7.069 \times 10^{-6}\right) (0.38)\,(888{,}600)} = \overset{.3387 q}{\cancel{13.55 \, \text{Pa}}} \qquad \text{(E4.1.6)}$$

The liquid pressure drop can be obtained using Equation (4.27)

$$\Delta P_l = \left(\frac{\mu_l}{K A_w h_{fg} \rho_l}\right) L_{eff} q \qquad \text{(E4.1.7)}$$

where A_w is the cross-sectional area of the wick, so

$$A_w = \frac{\pi\left(d_i^2 - d_v^2\right)}{4} = \frac{\pi\left(0.003259^2 - 0.003^2\right)}{4} = 1.273 \times 10^{-6} \, \text{m}^2 \qquad \text{(E4.1.8)}$$

where d_i and d_v are the inner diameter of the heat pipe and the vapor-path diameter of the heat pipe, respectively, which are illustrated in Figure E4.1.1. K is the wick permeability from Table 4.2 for wrapped screen wick.

$$K = \frac{d_w^2 \varepsilon^3}{122\,(1-\varepsilon)^2} = \frac{\left(2.159 \times 10^{-5}\right)^2 (0.6495)^3}{122\,(1 - 0.6495)^2} = 8.52 \times 10^{-12} \qquad \text{(E4.1.9)}$$

where

$$\varepsilon = 1 - \frac{\pi N d_w}{4} = 1 - \frac{1.05\pi\,(19{,}685)\,(2.159 \times 10^{-5})}{4} = 0.6495 \qquad (E4.1.10)$$

$$\Delta P_l = \frac{(1.02 \times 10^{-3})\,(0.025)\,q}{(8.52 \times 10^{-12})\,(1.273 \times 10^{-6})\,(888{,}600)\,(781)} = 3{,}387q \qquad (E4.1.11)$$

The normal hydrostatic pressure drop is

$$\Delta P_{norm} = \rho_l g d_v \cos\psi = (781)(9.81)(0.003)\cos(0) = 22.98\,\text{Pa} \qquad (E4.1.12)$$

Now inserting Equations (E4.1.2), (E4.1.6), (E4.1.11), and (E4.1.12) into Equation (E4.1.1) yields

$$1{,}921 = \cancel{13.55} .3387q + 3{,}387q + 22.98 \qquad (E4.1.13)$$

Solving for q gives the heat transfer rate for the maximum capillary pressure drop as

$$q = q_{c,m} = 0.56\,\text{W} \qquad (E4.1.14)$$

which is actually the capillary limit. $.3387q$

Note that the vapor pressure drop of 13.55 Pa is negligible compared with the liquid pressure drop. That is why the vapor pressure drop is usually negligible in calculation of capillary limit.

Check the assumption of laminar flow ($Re_v < 2{,}300$) made. From Equations (4.26), the Reynolds number is

$$Re_v = \frac{4q}{\pi d_v \mu_v h_{fg}} = \frac{4\,(0.56)}{\pi\,(0.003)\,(0.91 \times 10^{-5})\,(888{,}600)} = 29.39 < 2{,}300 \quad (E4.1.15)$$

Hence, the assumption is justified. If not, you should repeat the calculation with turbulent flow using Equation (4.22).

Sonic Limit   Using Equation (4.35), the sonic limit will be

$$q_{s,\max} = 0.474 h_{fg} A_v \,(\rho_v P_v)^{\frac{1}{2}}$$

$$= 0.474\,(888{,}600)\,(7.069 \times 10^{-6})\,(0.38 \times 10{,}000)^{\frac{1}{2}}$$

$$= 183.5\,\text{W} \qquad (E4.1.16)$$

Entrainment Limit   Using Equation (4.38), the entrainment limit is

$$q_{e,\max} = A_v h_{fg} \left(\frac{\sigma \rho_v}{2 r_{h,w}}\right)^{\frac{1}{2}}$$

$$= (7.069 \times 10^{-6})\,(888{,}600)\left(\frac{(2.44 \times 10^{-2})\,(0.38)}{2\,(1.295 \times 10^{-4})}\right)^{\frac{1}{2}}$$

$$= 37.58\,\text{W} \qquad (E4.1.17)$$

where the hydraulic radius of the wick is obtained as

$$D_{h,w} = \frac{4A}{P} = \frac{4A_w}{\pi(d_i + d_v)} = \frac{4\left(1.273 \times 10^{-6}\right)}{\pi(0.003259 + 0.003)} = 2.59 \times 10^{-4}\,\text{m} \quad \text{(E4.1.18)}$$

$$r_{h,w} = \frac{D_{h,w}}{2} = 1.295 \times 10^{-4}\,\text{m} \quad \text{(E4.1.19)}$$

Boiling Limit   Using Equation (4.40), the boiling limit is

$$q_{b,\max} = \frac{4\pi L_e k_{eff} T_v \sigma}{h_{fg} \rho_v \ln\left(\dfrac{r_i}{r_v}\right)} \left(\frac{1}{r_n} - \frac{1}{r_{c,e}}\right) \quad \text{(E4.1.20)}$$

The effective thermal conductivity for wrapped screen using Equation (E4.1.11) is found in Table 4.3.

$$k_{eff} = \frac{k_l\left[(k_l + k_w) - (1 - \varepsilon)(k_l - k_w)\right]}{(k_l + k_w) + (1 - \varepsilon)(k_l - k_w)}$$

$$= \frac{(0.168)\left[(0.168 + 14.9) - (1 - 0.6495)(0.168 - 14.9)\right]}{(0.168 + 14.9) + (1 - 0.6495)(0.168 - 14.9)} = 0.3431 \quad \text{(E4.1.21)}$$

$$q_{b,\max} = \frac{4\pi(0.02)(0.3431)(30 + 273)(2.44 \times 10^{-2})}{(888{,}600)(0.38)\ln\left(\dfrac{0.003259/2}{0.003/2}\right)} \left(\frac{1}{2.54 \times 10^{-7}} - \frac{1}{2.54 \times 10^{-5}}\right)$$

$$= 88.8\,\text{W} \quad \text{(E4.1.22)}$$

Summary of Heat Transport Limits

Capillary limit =	0.56 W
Sonic limit =	183.5 W
Entertainment limit =	37.6 W
Boiling limit =	88.8 W

The dominant effect that limits the transport capacity of this heat pipe is the capillary limit. This limit could be increased by decreasing the pore radius.

4.4   HEAT PIPE THERMAL RESISTANCE

The temperature drop between the evaporator and condenser of a heat pipe is of particular interest to the designer of heat pipe thermal control system for electronics. Figure 4.9 illustrates the resistance for a typical heat pipe. The terms are defined as follows:

$R_{p,e}$: Radial conduction resistance of the heat pipe wall at the evaporator
$R_{p,c}$: Radial conduction resistance of the heat pipe wall at the condenser
$R_{w,e}$: Resistance of the liquid-wick combination at the evaporator

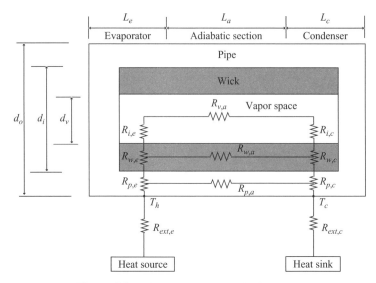

Figure 4.9 Thermal resistance of a heat pipe

$R_{w,c}$: Resistance of the liquid-wick combination at the condenser
$R_{i,e}$: Resistance of the liquid-vapor interface at the evaporator
$R_{i,c}$: Resistance of the liquid-vapor interface at the condenser
$R_{v,a}$: Resistance of the adiabatic vapor section
$R_{w,a}$: Resistance of the liquid-wick combination at the adiabatic section
$R_{p,a}$: Axial conduction resistance of the heat pipe wall at the adiabatic section
$R_{ext,e}$: Contact resistance between the heat source and the evaporator
$R_{ext,c}$: Contact resistance between the heat sink and the condenser

For a plate heat pipe, the conduction resistance at the evaporator is

$$R_{p,e} = \frac{\delta}{k_p A_e} \tag{4.43}$$

where δ is the plate thickness and A_e is the evaporator area.
  For a cylindrical heat pipe, the radial conduction resistance at the evaporator is

$$R_{p,e} = \frac{\ln\left(\dfrac{d_o}{d_i}\right)}{2\pi L_e k_p} \tag{4.44}$$

For a cylindrical wick, the resistance of the liquid-wick combination at the evaporator is

$$R_{w,e} = \frac{\ln\left(\dfrac{d_i}{d_v}\right)}{2\pi L_e k_{eff}} \tag{4.45}$$

Table 4.4   Comparative Values for Heat Pipe Resistances [11]

Resistance	°C/W
$R_{p,e}$ and $R_{p,c}$	10^{-1}
$R_{w,e}$ and $R_{w,c}$	10^{+1}
$R_{i,e}$ and $R_{i,c}$	10^{-5}
$R_{v,a}$	10^{-8}
$R_{p,a}$	10^{+2}
$R_{w,a}$	10^{+4}

The adiabatic vapor resistance can be found as

$$R_{v,a} = \frac{T_v \left(P_{v,e} - P_{v,c} \right)}{\rho_v h_{fg} q} \tag{4.46}$$

The effective thermal conductivity for a liquid-saturated wick can be found in Table 4.3. Expressions for the condenser would be similar.

In an electrical circuit in parallel, the resistances with large values may be treated as open circuit, while in a circuit in series the resistances with small values may be treated as short circuit.

In comparing the values in Table 4.4, it appears that the circuit in parallel can be simplified by omitting the two axial thermal resistances $R_{p,a}$ and $R_{w,a}$ due to the large values. And the circuit in series can be further simplified by making a short circuit of $R_{i,e}$, $R_{i,c}$, and $R_{v,a}$ due to the small values. Hence, the overall or total thermal resistance, excluding the external resistances, becomes the sum of a series of resistances:

$$R_{tot} = R_{p,e} + R_{w,e} + R_{w,c} + R_{p,c} \tag{4.47}$$

and

$$q = \frac{T_h - T_c}{R_{tot}} \tag{4.47a}$$

The original form of Equation (4.47) before simplification is

$$R_{tot} = \cfrac{1}{R_{p,e} + R_{w,e} + \cfrac{1}{\cfrac{1}{R_{v,a}} + \cfrac{1}{R_{p,a}} + \cfrac{1}{R_{w,a}}} + R_{w,c} + R_{p,c}} \tag{4.47b}$$

Two external resistances play a very important role in design of heat pipe for electric applications. These are the external or contact resistances occurring between the heat source and the heat sink. The contact resistance over an L-length cylindrical surface is

$$R_{ext,contact} = \frac{R''_{t,c}}{\pi d_o L} \tag{4.48}$$

where $R''_{t,c}$ is the contact resistance (m^{2}K/W).

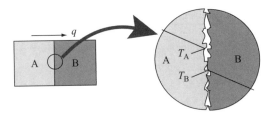

Figure 4.10 Temperature drop due to thermal contact resistance

The thermal resistance for an *L*-length cylindrical surface with a convection heat transfer coefficient *h* is

$$R_{ext,conv} = \frac{1}{h\pi d_o L} \tag{4.49}$$

4.4.1 Contact Resistance

It is important to recognize that, in composite systems, the temperature drop across the interface between materials may be appreciable, as shown in Figure 4.10. This temperature change is attributed to *thermal contact resistance*, $R_{t,c}$. Some thermal resistances are illustrated in Table 4.5. For a unit area of the interface, the thermal resistance is defined as

$$R''_{t,c} = \frac{T_A - T_B}{q''} \tag{4.50}$$

Example 4.2 Thermal Resistance A simple horizontal copper-water heat pipe is to be constructed from a 6-cm-long tube to cool an enclosed silicon casing, as shown in Figure E4.2. The inner and outer diameters of the heat pipe are 5 mm and 6 mm,

Table 4.5 Thermal Resistance of Respective Solid/Solid Interface

Interface	$R''_{t,c} \times 10^4 \ (\mathrm{m^2 K/W})$	Source
Silicon chip/lapped aluminum in air (27–500 kN/m²)	0.3–0.6	[13]
Aluminum/aluminum with indium foil filler (~100 kN/m²)	~0.07	[12,14]
Stainless/stainless with indium foil filler (~3500 kN/m²)	~0.04	[12,14]
Aluminum/aluminum with metallic (Pb) coating	0.01–0.1	[15]
Aluminum/aluminum with Dow Corning 340 grease (~100 kN/m²)	~0.07	[12,14]
Stainless/stainless with Dow Corning 340 grease (~3500 kN/m²)	~0.04	[12,14]
Silicon chip/aluminum with 0.02 mm epoxy	0.2–0.9	[16]
Brass/brass with 15 µm tin solder	0.025–0.14	[17]

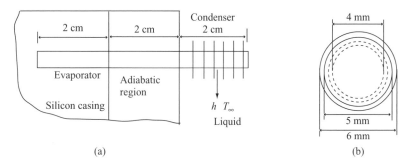

Figure E4.2 (a) A heat pipe arrangement on a silicon casing, (b) The cross-section of the heat pipe

respectively. The diameter of the vapor space is 4 mm, which is shown in the figure. The evaporator and condenser lengths of the heat pipe are 2 cm each and the wicking structure consists of two layers of #500-mesh copper screen with the wire diameter of 0.00085 in.

The evaporator of the heat pipe is embedded in a silicon casing. The contact resistance between the heat pipe and its casing is 2.0×10^{-5} m^2°C/W. The condenser is immerged in a stream of liquid, and the convection heat transfer coefficient at the condenser is 500 W/m^2°C. If the liquid is at 10°C and the heat pipe is operating at a heat transfer rate of 2.0 W, what is the temperature of the silicon casing?

Solution

Given dimensions:

$$d_o = 0.006 \text{ m}, \, d_i = 0.005 \text{ m}, \, d_v = 0.004 \text{ m}, \text{ and}$$

$$d_w = 0.00085 \text{ in.} = 2.159 \times 10^{-5} \text{ m}$$

$$L_e = L_a = L_c = 0.02 \text{ m}, \, L_{eff} = \frac{L_e}{2} + L_a + \frac{L_c}{2} = 0.04 \text{ m}$$

Properties:

$$\text{Water: } k_l = 0.603 \text{ W/m}^\circ\text{C}$$

$$\text{Copper: } k_p = k_w = 401 \text{ W/m}^\circ\text{C}$$

Assuming that the liquid-vapor interface resistances and the axial resistances are negligible, the total thermal resistance can be obtained using Equation (4.47):

$$R_{tot} = R_{p,e} + R_{w,e} + R_{w,c} + R_{p,c} \quad \text{(E4.2.1)}$$

The evaporator and condenser are physically identical, so the equation can be reduced to

$$R_{tot} = 2 \left(R_{p,e} + R_{w,e} \right) \quad \text{(E4.2.2)}$$

Using Equation (4.44), the radial conduction resistance is

$$R_{p,e} = \frac{\ln\left(\dfrac{d_o}{d_i}\right)}{2\pi L_e k_p} = \frac{\ln\left(\dfrac{0.006}{0.005}\right)}{2\pi(0.02)(401)} = 3.618 \times 10^{-3}\,^{\circ}\text{C/W} \qquad \text{(E4.2.3)}$$

$$R_{w,e} = \frac{\ln\left(\dfrac{d_i}{d_v}\right)}{2\pi L_e k_{eff}} = \frac{\ln\left(\dfrac{0.005}{0.004}\right)}{2\pi(0.02)(1.205)} = 1.474\,^{\circ}\text{C/W} \qquad \text{(E4.2.4)}$$

where

$$
\begin{aligned}
k_{eff} &= \frac{k_l\left[(k_l + k_w) - (1 - \varepsilon)\,(k_l - k_w)\right]}{(k_l + k_w) + (1 - \varepsilon)\,(k_l - k_w)} \\
&= \frac{(0.603)\left[(0.603 + 401) - (1 - 0.666)\,(0.603 - 401)\right]}{(0.603 + 401) + (1 - 0.666)\,(0.603 - 4.01)} \\
&= 1.205\,\text{W/m}^{\circ}\text{C} \qquad \text{(E4.2.5)}
\end{aligned}
$$

$$\varepsilon = 1 - \frac{\pi N d_w}{4} = 1 - \frac{\pi(500/\text{in.})(0.00085\,\text{in.})}{4} = 0.666 \qquad \text{(E4.2.6)}$$

$$R_{tot} = 2(0.003618 + 1.474) = 2.956\,^{\circ}\text{C/W}$$

Using Equations (4.48) and (4.49), the heat transfer rate is expressed as

$$q = \frac{T_{silicon} - T_\infty}{\dfrac{R''_{t,c}}{\pi d_o L_e} + R_{tot} + \dfrac{1}{h \pi d_o L_e}} \qquad \text{(E4.2.7)}$$

Solving for $T_{silicon}$ gives

$$
\begin{aligned}
T_{silicon} &= T_\infty + q\left(\frac{R''_{t,c}}{\pi d_o L_e} + R_{tot} + \frac{1}{h \pi d_o L_e}\right) \\
&= 10 + 2.0\left[\frac{2.0 \times 10^{-5}}{\pi(0.006)(0.02)} + 2.956 + \frac{1}{(500)\,\pi(0.006)(0.002)}\right] \\
&= 10 + 2.0\,[0.05 + 2.956 + 5.307] \\
&= 26.6\,^{\circ}\text{C} \qquad \text{(E4.2.8)}
\end{aligned}
$$

4.5 VARIABLE CONDUCTANCE HEAT PIPES (VCHP)

The variable conductance heat pipe, sometimes called the gas-controlled or gas–loaded heat pipe, has a unique feature that sets it apart from other types of heat pipe. This is its ability to maintain a device mounted at the evaporator at a near constant temperature, independent of the amount of power being generated by the device. The variable-conductance heat pipe offers temperature control within narrow limits, in addition to the simple heat transport function performed by basic heat pipes [21].

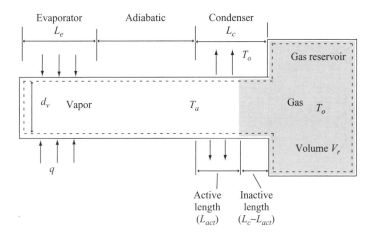

Figure 4.11 A gas-loaded heat pipe with a wicked reservoir

In Figure 4.11, the active condenser length varies in accordance with temperature changes in various parts of the system. An increase in the evaporator temperature causes an increase in vapor pressure of the working fluid, which causes the gas to compress into a smaller volume, releasing a large area of active condenser length for heat rejection. Conversely, a drop in the evaporator temperature results in a smaller area of active condenser length with a smaller amount of heat rejection. The net effect is to provide a passively controlled variable condenser area that increases or decreases heat transfer in response to the heat pipe vapor temperature.

4.5.1 Gas-Loaded Heat Pipes

As is the case with most phase cycles, the presence of noncondensible gases creates a problem, due to the partial blockage of the condensing area. Heat pipes are no exception. During normal operation, any noncondensible gases present are carried to the condenser and remain there, reducing the effective condenser area. This characteristic, although normally undesirable, can be used to control both the direction and amount of heat transfer [18, 19].

The important operating characteristic of such a heat pipe can be obtained from a simple model, based on five assumptions:

1. There is a flat front between the vapor and the gas that divides the condenser of length L_c into an active length L_{act} and inactive length $(L_c - L_{act})$.
2. The total pressure in the condenser P_c is consistent and equals the reservoir pressure.
3. Axial conduction along the wall and wick is negligible, so that there is a step change in temperature of the heat pipe and its contents at the vapor-gas front.
4. Since it is assumed that there is pure vapor in the active length, the active-length temperature T_a is the saturation temperature corresponding to pressure $P_c : T_a = T_{sat}(P_c)$.

5. Under steady operating conditions, the inactive condenser and reservoir are in thermal equilibrium with the heat rejection sink at temperature T_o. Locally, the vapor temperature at the vapor-gas interface at the condenser drops to T_o, which provides the partial pressure in the gas reservoir.

The vapor pressure in both the inactive condenser and the reservoir is P_{sat} (T_o), as mentioned before. Applying Dalton's law to the inactive condenser and the gas reservoir, we have

$$P_c = P_g + P_{sat} \ (T_o) \tag{4.51}$$

Applying the ideal gas law to the gas reservoir, we have

$$P_g V_{tot} = m_g R_g T_o \tag{4.52}$$

where R is the gas constant.

The total volume V_{tot} in the gas side consists of two parts, as shown in Figure 4.11:

$$V_{tot} = A_v \ (L_c - L_{act}) + V_r \tag{4.53}$$

Solving for the mass of gas in Equation (4.52) with Equations (4.51) and (4.52) gives

$$m_g = \frac{P_c - P_{sat} \ (T_o)}{R_g T_o} \ [A_v \ (L_c - L_{act}) + V_r] \tag{4.54}$$

We want to express Equation (4.54) for the active length in the condenser.

$$L_{act} = L_c + \frac{V_r}{A_v} - \frac{m_g R_g T_o}{[P_c - P_{sat} \ (T_o)] \ A_v} \tag{4.55}$$

The heat rejected at the condenser can be written in general as

$$q = U \pi d_v L_{act} \ (T_a - T_o) \tag{4.56}$$

where U is the overall heat transfer coefficient.

Inserting Equation (4.55) into Equation (4.56) gives

$$q = U \pi d_v (T_a - T_o) \left[L_c + \frac{V_r}{A_v} - \frac{m_g R_g T_o}{[P_c - P_{sat} \ (T_o)] \ A_v} \right] \tag{4.57}$$

When the condenser of the heat pipe is fully open, the heat rejected at the condenser will be

$$q = q_{max} \text{ and } T_a = T_{a,max} \tag{4.58}$$

Since the mass of gas is conservative, Equation (4.54) in terms of maximum values becomes

$$m_g = \frac{\left[P_{sat} \left(T_{a,max} \right) - P_{sat} \left(T_o \right) \right] V_r}{R_g T_o} \tag{4.59}$$

and Equation (4.56) becomes

$$q_{max} = U \pi d_v L_c \left(T_{a,max} - T_o\right) \tag{4.60}$$

Dividing Equation (4.57) by Equation (4.60) gives

$$\frac{q}{q_{max}} = \frac{T_a - T_o}{T_{a,max} - T_o} \left[1 + \frac{V_r}{L_c A_v} - \frac{m_g R_g T_o}{[P_c - P_{sat}(T_o)] L_c A_v}\right] \tag{4.61}$$

Denote the vapor core volume in the condenser region as

$$V_c = L_c A_v \tag{4.62}$$

Inserting Equation (4.59) into Equation (4.61) with (4.62) gives

$$\frac{q}{q_{max}} = \frac{T_a - T_o}{T_{a,max} - T_o} \left[\frac{1 + \dfrac{V_c}{V_r} - \dfrac{P_{sat}(T_{a,max}) - P_{sat}(T_o)}{P_{sat}(T_a) - P_{sat}(T_o)}}{\dfrac{V_c}{V_r}}\right] \tag{4.63}$$

The core-to-reservoir volume ratio V_c/V_r determines the sensitivity of the pipe.

An objective of gas loading is to maintain a nearly constant evaporator temperature when the heat load drops below the design value. The temperature drop across the evaporator and condenser can be maintained fairly constant, even though the evaporator heat flux may fluctuate.

An active controlled, gas-loaded heat pipe in which the gas volume at the reservoir end can be controlled externally. A temperature-sensing device at the evaporator signals the reservoir heater. This heater, when activated, can heat the gas contained in the reservoir, causing it to expand and thereby reduce the condensing area.

4.5.2   Clayepyron-Clausius Equation

The saturated pressure for a given temperature may be found in Table A.4 in Appendix A. However, we can use the Clapeyron-Clausius equation to estimate $P_{sat}(T)$.

$$\left(\frac{dP}{dT}\right)_{sat} = \frac{P h_{fg}}{R T^2} \tag{4.64}$$

where R is the gas constant and h_{fg} is the latent of vaporization. The gas constant has a relationship with the universal gas constant as

$$R = \frac{R_u}{M} \tag{4.65}$$

where R_u is the universal gas constant of 8.31447 kJ/kmol and M is the molar mass (kg/kmol), which is found in Table A.1 in Appendix A.

Equation (4.64) can be expressed as

$$\left(\frac{dP}{P}\right)_{sat} = \frac{h_{fg}}{R} \left(\frac{dT}{T^2}\right)_{sat} \tag{4.66}$$

For small temperature intervals, h_{fg} can be treated as a constant. Then, integrating this equation between two saturation states yields

$$\ln\left(\frac{P_2}{P_1}\right) \cong \frac{h_{fg}}{R}\left(\frac{1}{T_1} - \frac{1}{T_2}\right) \tag{4.67}$$

or

$$\frac{P_2}{P_1} \cong \exp\left[\frac{h_{fg}}{R}\left(\frac{1}{T_1} - \frac{1}{T_2}\right)\right] \tag{4.68}$$

This equation requires a set of saturated pressure (P_1) and temperature (T_1) as reference to determine a saturated pressure (P_2) for a given saturated temperature (T_2).

4.5.3   Applications

A VCHP was applied to sodium-sulphur batteries. These need to be kept above 300°C during operation, but if they overheat by more than 10°C, their life is reduced. The VCHP was shown to provide a satisfactory level of heat control [20].

Example 4.3 Variable Conductance Heat Pipe   A nitrogen-loaded methanol heat pipe is constructed from a 0.5-inch-O.D. stainless steel tube and is L-shaped, as shown in Figure E4.3.1. A fibrous stainless steel slab wick extends into the gas reservoir. The cross-sectional area of the vapor space in the pipe is 50.3 mm^2 and the reservoir volume is 5.28×10^4 mm^3. The condenser length L_c is 35 cm. When the sink temperature is 264 K, the heat pipe is fully open, for an adiabatic section temperature of 283 K and a heat load of 50 W. Determine the mass of gas in the heat pipe, and prepare a graph of heat load versus adiabatic section temperature. In addition, vary the reservoir volume to demonstrate its effect on performance. At a reference temperature $T_r = 5°C$, methanol has a vapor pressure of 5330 Pa and enthalpy of vaporization of 1.18×10^6 J/kg.

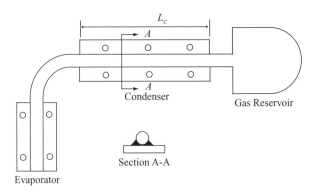

Figure E4.3.1   A L-shaped variable conductance heat pipe

Solution   Assumptions:

1. A flat front between the vapor and gas
2. Negligible pressure drop between adiabatic section and condenser

Given Information:

$$P_r = 5{,}330 \, \text{Pa}, \; T_r = (5 + 273.15) \, \text{K}, \; h_{fg} = 1.18 \times 10^6 \, \text{J/kg}$$

$$V_r = 5.28 \times 10^4 \, \text{mm}^3 \; \text{(reservoir volume)},$$

$$L_c = 35 \, \text{cm},$$

$$d_o = 0.5 \, \text{in}.$$

$$A_v = 50.3 \, \text{mm}^2.$$

When the heat pipe is fully open, we have

$$T_o = 264 \, \text{K}, \; T_{a,\max} = 283 \, \text{K}, \; q_{\max} = 50 \, \text{W} \; \text{(heat load)}$$

Properties (Table A.1):

Methanol: $M = 32.042, \; R = R_u/M = 8314.47/32.042 = 259.5 \, \text{J/kgK}$

Nitrogen: $R_g = 8314.47/28.013 = 296.8 \, \text{J/kgK}.$

The Clayepyron-Clausius equation may be used to determine $P_{sat}(T)$. Hence, using Equation (4.68)

$$\frac{P}{P_r} \cong \exp\left[\frac{h_{fg}}{R}\left(\frac{1}{T_r} - \frac{1}{T}\right)\right] \tag{E4.3.1}$$

$$P = 5{,}330 \exp\left[\frac{1.18 \times 10^6}{259.5}\left(\frac{1}{278.15} - \frac{1}{T}\right)\right] \tag{E4.3.2}$$

So, we find

$$P = P_{sat}(T_{a,\max}) = 7{,}048 \, \text{Pa for } T = T_{a,\max} = 283 \, \text{K} \tag{E4.3.3}$$

$$P = P_{sat}(T_o) = 2{,}217 \, \text{Pa for } T = T_o = 264 \, \text{K} \tag{E4.3.4}$$

From Equation (4.59),

$$m_g = \frac{\left[P_{sat}(T_{a,\max}) - P_{sat}(T_o)\right] V_r}{R_g T_o} = \frac{(7{,}048 - 2{,}217)(5.28 \times 10^4)(10^{-9})}{(296.8)(264)}$$

$$= 3.25 \times 10^{-6} \, \text{kg}$$

Using Equation (4.63) with Equation (E4.3.2) provides Figure E4.3.2:

$$\frac{q}{q_{\max}} = \frac{T_a - T_o}{T_{a,\max} - T_o}\left[\frac{V_r}{V_c} + 1 - \frac{V_r}{V_c}\frac{P_{sat}(T_{a,\max}) - P_{sat}(T_o)}{P_{sat}(T_a) - P_{sat}(T_o)}\right] \tag{E4.3.5}$$

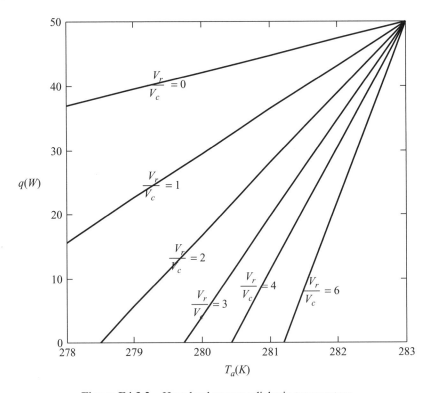

Figure E4.3.2   Heat load versus adiabatic temperature

$$V_c = L_c A_v = (0.35)\left(50.3 \times 10^{-6}\right) = 1.76 \times 10^{-5}\,\mathrm{m}^3$$

$$V_r/V_c = 5.28 \times 10^{-5}/1.76 \times 10^{-5} = 3$$

As the reservoir volume increases, the reservoir-to-core volume ratio increases and the adiabatic temperature becomes less sensitive to heat load.

4.6   LOOP HEAT PIPES

Loop heat pipes (LHPs) are two-phase heat transfer devices with capillary pumping of a working fluid, as shown in Figure 4.12. They are capable of transferring heat for distances up to several meters at any orientation in the gravity field, or several tens of meters in a horizontal position.

Consider operation at steady state. Note that State 1 is at vapor side adjacent to the vapor-liquid interface and state 9 is at liquid side adjacent to the interface, as shown in Figure 4.12(a). Evaporation takes place at the vapor-liquid interface as the evaporator receives heat from a heat source and, as a result of this, the wick menisci form on the wick surface of the vapor removal channel, creating a capillary pressure drop ΔP_c between state 1 and state 9, which should be the maximum pressure drop in the system and also be the driving force of capillary pumping, as shown in Figure 4.12(b) (see also Figure 4.7(a)). The vapor at state 1 thermodynamically constitutes the saturation

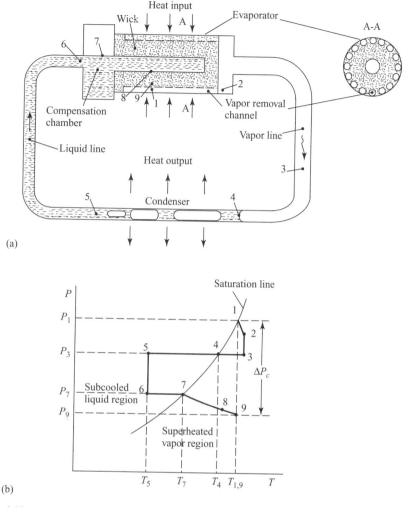

Figure 4.12   (a) Operating principle of a loop heat pipe, (b) Vapor-pressure curve for the working fluid (Revised from Maydanik (2005) [23])

temperature and the saturation pressure that is the highest pressure in the system, as shown in Figure 4.12(b).

However, the vapor generated at state 1 does not go to the wick side due to the wick meniscus. Instead, it passes through the vapor removal channel. As the vapor passes through the vapor removal channel, usually grooved as shown, it absorbs further heat from the evaporator shell surface and becomes a little superheated, as shown. The vapor isothermally advances to the vapor line from state 2 to state 3 and then reaches the condenser at state 4. We assume no heat losses in both the vapor line and the liquid line. Condensation at state 4 takes place as the condenser experiences heat rejection. Hence, state 4 returns to the saturation line, as shown. As the liquid condenses, it keeps rejecting heat through the condenser length, resulting in a lower temperature at state 5, as shown. Between state 5 and state 6, the temperature undergoes isothermal process as mentioned, although the pressure drops due to the friction in the liquid line. At state 7, phase change takes place; hence state 7 stay at the saturation line, as shown. The pressure difference between state 6 and state 7 will be negligible anyhow. The liquid reaches the wick at state 8 with slight pressure drop due to the friction but still penetrates into the wick up to state 9, since state 9 is the lowest pressure in the system created by the meniscus on the wick. Between state 8 and state 9, the liquid may be superheated by the heat flow from the evaporator. Therefore, boiling may or may not occur, depending on the power applied to the evaporator. However, boiling is not likely to occur because significant degrees of superheat, if the wick structure is well cleaned and the working fluid is clean, are usually required to boil the liquid, or equivalently, to form vapor bubbles on the wick elements, and also, the most superheated liquid will be consumed near the menisci on the surface of the vapor removal channel.

The LHP has the features of the classical heat pipe with the following additional advantages:

- High heat flux capability
- Capability to transport energy over long distance without restriction on the routing of the liquid and vapor lines
- Ability to operate over a range of gravity environments
- No wick within the transport lines
- Vapor and liquid flows separated; therefore, no entrainment

The structural elements of LHPs are typically stainless steel, but aluminum and copper are also used. Sintered nickel and titanium are commonly used in the manufacture of the wick. These materials are attractive because of their high strength and wide compatibility with working fluids, as well as their working performance.

Further details for loop heat pipes may be found in Reay and Kew (2006) [21] and Dunn and Reay (1976) [24].

4.7 MICRO HEAT PIPES

A micro heat pipe is a wickless, noncircular pipe with an approximate diameter of 0.1 to 1 mm and a length of about 10 to 60 mm. Cotter (1984) first introduced the concept of very small micro heat pipes incorporated into semiconductor devices to promote more uniform temperature and improve thermal control [24]. A micro heat pipe is defined as one that satisfies the condition that

$$\frac{r_c}{r_h} \geq 1 \tag{4.70}$$

where r_c is the capillary radius and r_h is the hydraulic radius (the hydraulic diameter is found in Equation (4.39)).

Figure 4.13 illustrates a micro heat pipe that utilizes the sharp-angled corner regions as liquid arteries working like capillary tubes. The fundamental operating principles of micro heat pipes are essentially the same as those occurring in large, more conventional heat pipes. Heat applied to one end of the heat pipe vaporizes the liquid in that region and forces it to move to the cooler end, where it condenses and gives up the latent heat of vaporization. This vaporization and condensation process causes the

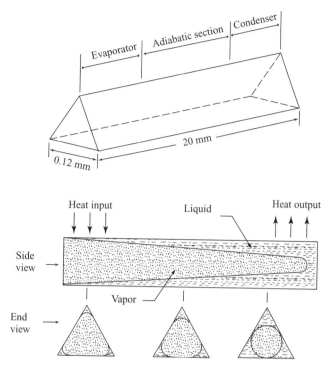

Figure 4.13   Micro-heat pipe operation

liquid-vapor interface in the liquid arteries to change continually along the pipe, as shown in Figure 4.13, and results in a capillary pressure difference between the evaporator and condenser regions. This capillary pressure difference promotes the flow of the working fluid from the condenser back to the evaporator through the triangular-shaped corner regions. These regions serve as liquid arteries; thus no wicking structure is required [1].

4.7.1   Steady-State Models

Many similarities exist between the micro heat pipes and the large, conventional heat pipes in the operation principles. One of the models developed by Babin and Peterson in 1990 [26] utilizes the early developed heat transfer limitations discussed in Section 4.3 for a trapezoidal heat pipe shown in Figure 4.14, showing a good agreement with the experimental results. Babin and Peterson also found that the maximum heat transport capacity of the micro heat pipe is primarily governed by the capillary pumping pressure.

4.7.1.1   Conventional Model

Rearranging the heat transfer limitations using the original equation numbers, we can rewrite the maximum capillary pressure limit.

$$\Delta P_{c,m} = \Delta P_v + \Delta P_l + \Delta P_{norm} + \Delta P_{axial} \tag{4.12}$$

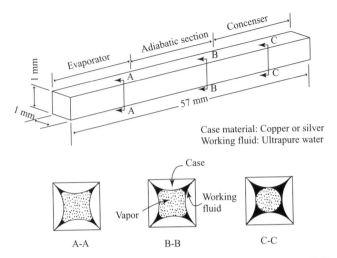

Figure 4.14   A trapezoidal micro-heat pipe (Redrawn from [26])

$$\Delta P_{c,m} = \frac{2\sigma}{r_{c,e}} \cos\theta \qquad (4.18)$$

where $\cos\theta$ is added to account the contact angle.

$$\Delta P_v = \frac{C\,(f_v \mathrm{Re}_v)\,\mu_v}{2r_v^2 A_v \rho_v h_{fg}} L_{eff}\, q \qquad (4.19)$$

$$\Delta P_l = \left(\frac{\mu_l}{KA_w h_{fg}\rho_l}\right) L_{eff}\, q \qquad (4.27)$$

$$\Delta P_{norm} = \rho_l g d_v \cos\psi \qquad (4.28)$$

$$\Delta P_{axial} = \rho_l g L \sin\psi \qquad (4.29)$$

The details for the equations—such as tables and definitions—are found in Section 4.3. A trapezoidal heat pipe is illustrated in Figure 4.14 and the cross-section for the trapezoidal heat pipe is detailed in Figure 4.15.

4.7.1.2 Cotter's Model

Cotter (1984) [25] first introduced an analytical model of a micro heat pipe with a basic assumption that the capillary pressure with the variation in radius of curvature along the pipe is the driving force for the liquid flow. He developed an expression for the maximum heat transport capacity considering the shape factors of both liquid and vapor as

$$q_{max} = \frac{0.16\beta\sqrt{K_l K_v}}{8\pi H} \frac{\sigma h_{fg}}{\nu_l} \sqrt{\frac{\nu_l}{\nu_v}} \frac{A^{\frac{3}{2}}}{L} \qquad (4.71)$$

where β is the dimensionless geometric constant, K_l and K_v are the dimensionless shape factors of liquid and vapor, respectively, H is the fraction of total heat transport, ν_l and ν_v are the kinematic viscosities of liquid and vapor, respectively, A is the total area that consists of the vapor and liquid areas, and L is the length of heat pipe.

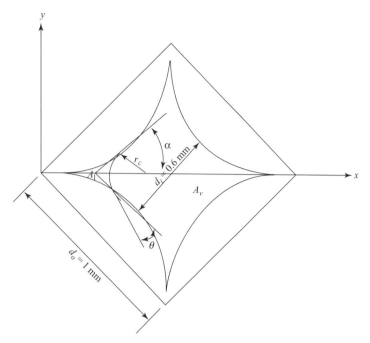

Figure 4.15   Cross-sectional dimensions of the micro heat pipe, where θ is the contact angle or liquid wetting angle and α is the half included angle [26]

The kinematic viscosity is defined by

$$\nu = \frac{\mu}{\rho} \qquad (4.72)$$

The value of shape factor K is 1 for circular disk and is less than 1 for any other shape. The shape factor H is an important parameter that should be studied analytically and experimentally for a universal constant rather than a specific geometry.

The dimensionless constants were developed mainly by Babin et al. (1990) [26] and Ha and Peterson (1998) [27], among other researchers, considering the contact angle and listed in Table 4.6.

The trapezoidal heat pipe is illustrated in Figure 4.14 with the detailed cross-sectional view in Figure 4.15.

Total Area (A) for Trapezoidal Heat Pipes   The total area, A, is the area within the heat pipe. The area ratio of the square area of the width d_i to the total area within

Table 4.6   Dimensionless Constants

Parameters	Triangular heat pipe [27]	Trapezoidal heat pipe [26]
H	0.666	0.5
K_l	0.86	0.975
K_v	0.135	0.6
β	1.433	2.044

the micro heat pipe is utilized to estimate the total area that consists of the liquid and vapor area.

From Figure 4.16(a), we have

$$x = \frac{d_o}{2}\left(1 - \frac{d_i}{d_o}\right) \tag{4.73}$$

From Figure 4.16(b),

$$C = \sqrt{\left(\frac{d_o}{2}\right)^2 + \left(\frac{d_o - d_i}{2}\right)^2} \tag{4.74}$$

From Figure 4.16(b),

$$r = \frac{d_o\sqrt{1 + \left(1 - \frac{d_i}{d_o}\right)^2}}{4\sin\left(\frac{\psi}{4}\right)} \tag{4.75}$$

From Figure 4.16(a),

$$r = \frac{d_o/2}{\sin\left(\frac{\psi}{2}\right)} \tag{4.76}$$

Combining Equations (4.75) and (4.76) yields

$$\frac{\sin\left(\frac{\psi}{2}\right)}{\sin\left(\frac{\psi}{4}\right)} = \frac{1}{2}\sqrt{1 + \left(1 - \frac{d_i}{d_o}\right)^2} \tag{4.77}$$

This can be simplified using trigonometry:

$$\psi = 4\cos^{-1}\left[\frac{1}{\sqrt{1 + \left(1 - \frac{d_i}{d_o}\right)^2}}\right] \tag{4.78}$$

The total area can be obtained:

$$A_{total} = d_o^2 - 2\psi r^2 + 2(r - x)d_o \tag{4.79}$$

$$A_{square,i} = d_i^2 \tag{4.80}$$

The area ratio of the d_i square in the heat pipe to the total area in the pipe is

$$\frac{A_{square,i}}{A_{total}} = \frac{d_i^2}{d_o^2 - 2\psi \cdot r^2 + 2(r - x)d_o} \tag{4.81}$$

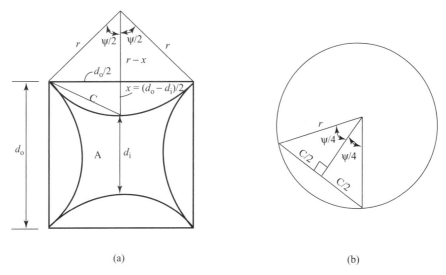

(a)   (b)

Figure 4.16   (a) Geometry for calculation of the total area, (b) Geometry for the circle with the radius r

Inserting Equations (4.73), (4.75), and (4.78) into (4.81) finally yields

$$\frac{A_{square,i}}{A_{total}} = \frac{(d_i/d_o)^2}{\dfrac{d_i}{d_o} - \dfrac{\psi\left[1+\left(1-\dfrac{d_i}{d_o}\right)^2\right]}{8\sin^2\left(\dfrac{\psi}{4}\right)} + \dfrac{\sqrt{1+\left(1-\dfrac{d_i}{d_o}\right)^2}}{2\sin\left(\dfrac{\psi}{4}\right)}} \tag{4.82}$$

Equation (4.82) is plotted in Figure 4.17, where there is no practical meaning when $\frac{d_i}{d_o}$ is less than 0.6, at which point the two faces of the circles contact each other tangentially, as shown in Figure 4.16. Figure 4.17 may be useful to determine the total area with the information of the width ratio $\frac{d_i}{d_o}$.

Example 4.4 Micro Heat Pipe   A trapezoidal micro heat pipe is designed to cool a semiconductor device as an element of the micro heat pipe array. The case material is silver and the working fluid is distilled deionized water. The heat pipe is operated at 60°C. The dimensions are given as follows:

Total length:	$L = 57\,\mathrm{mm}$
Evaporator length:	$L_e = 12.7\,\mathrm{mm}$
Condenser length:	$L_c = 12.7\,\mathrm{mm}$
Pipe width:	$d = 1\,\mathrm{mm}$
Inner vapor channel diameter:	$d_i = 0.6\,\mathrm{mm}$
Outer vapor channel diameter:	$d_o = 1.0\,\mathrm{mm}$
Charged liquid amount:	0.0032 g
Groove width:	$\omega = 0.0133\,\mathrm{mm}$
Half included angle α:	$\alpha = 0.5854\,\mathrm{rad}$

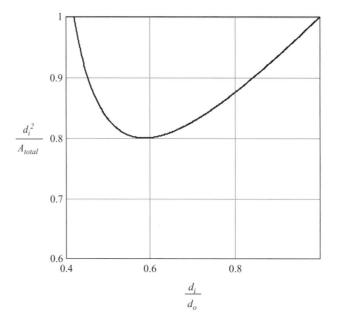

Figure 4.17 Area ratio of the inner square area to the total area

Determine the heat transport limitation for the micro heat pipe.

Solution: Properties of water at 60°C:

$$\sigma = 6.62 \times 10^{-2}\,\text{N/m}$$

$$h_{fg} = 2{,}359 \times 10^3\,\text{J/kg}$$

$$\rho_l = 983\,\text{kg/m}^3.$$

$$\rho_v = 0.13\,\text{kg/m}^3.$$

$$\mu_l = 0.47\,\text{cP} = 0.47 \times 10^{-3}\,\text{N} \cdot \text{s/m}^2.$$

$$\mu_v = 1.12 \times 10^{-2}\text{cP (centipoise)} = 1.12 \times 10^{-5}\,\text{N} \cdot \text{s/m}^2.$$

Using Equation (4.72)

$$v_l = \frac{\mu_l}{\rho_l} = \frac{0.47 \times 10^{-3}\,\text{N} \cdot \text{s/m}^2}{983\,\text{kg/m}^3} = 4.781 \times 10^{-7}\,\frac{\text{m}^2}{\text{s}} \qquad (\text{E4.4.1})$$

$$v_v = \frac{\mu_v}{\rho_v} = \frac{1.12 \times 10^{-5}\,\text{N} \cdot \text{s/m}^2}{0.13\,\text{kg/m}^3} = 8.615 \times 10^{-5}\,\frac{\text{m}^2}{\text{s}} \qquad (\text{E4.4.2})$$

The width ratio is

$$\frac{d_i}{d_o} = \frac{0.6\,\text{mm}}{1.0\,\text{mm}} = 0.6 \qquad (\text{E4.4.3})$$

The total area may be estimated using Figure 4.17:

$$\frac{d_i^2}{A_{total}} \approx 0.8 \tag{E4.4.4}$$

$$A_{total} = \frac{d_i^2}{0.8} = \frac{0.0006^2}{0.8} = 4.5 \times 10^{-7} \text{m}^2 \tag{E4.4.5}$$

From Table 4.6 for the trapezoidal micro heat pipe, the dimensionless constants are found

$$H = 0.5, \ K_l = 0.975, \ K_v = 0.6, \ \text{and} \ \beta = 2.044 \tag{E4.4.6}$$

Using Equation (4.71)

$$\begin{aligned}
q_{max} &= \frac{0.16 \beta \sqrt{K_l K_v}}{8 \pi H} \frac{\sigma h_{fg}}{v_l} \sqrt{\frac{v_l}{v_v}} \frac{A^{\frac{3}{2}}}{L} \\
&= \frac{0.16 \, (2.044) \, \sqrt{0.975 \times 0.6}}{8 \pi \, (0.5)} \frac{\left(6.62 \times 10^{-2}\right) (2359000)}{4.781 \times 10^{-7}} \\
&\qquad \times \sqrt{\frac{4.781 \times 10^{-7}}{8.615 \times 10^{-5}}} \frac{\left(4.5 \times 10^{-7}\right)^{\frac{3}{2}}}{0.057} \\
&= 2.565 \, \text{Watts} \tag{E4.4.7}
\end{aligned}$$

4.8   WORKING FLUID

The selection of a suitable working fluid is perhaps the most important aspect of the design and manufacturing process. The theoretical operating temperature range for a given heat pipe is typically between the critical temperature and triple state of the working fluid. Table 4.7 illustrates the typical operating temperature range for various working fluids.

4.8.1   Figure of Merit

Most heat pipe applications for electronic thermal control require the selection of a working fluid with boiling temperatures between 250 and 375 K. The liquid transport factor, or *figure of merit*, can be used to evaluate the effectiveness of various working fluids at specific operating temperatures. The figure of merit is defined as

$$N_l = \frac{\rho_l \sigma h_{fg}}{\mu_l} \tag{4.83}$$

Figure 4.18 illustrates the figure of merit for working fluids at boiling point. Sodium(Na) provides the high liquid figure of merit but is typically inappropriate at temperatures less than 500 K. In moderate temperature range, water appears to have the highest figure of merit. The presence of impurities in the working fluid can have significant detrimental effects on the operation of heat pipes and thermosyphons.

Table 4.7   Heat Pipe Working Fluids [21]

Medium	Melting point (°C)	Boiling point at atmos. press. (°C)	Useful range (°C)
Helium	−271	−261	−271 to −269
Nitrogen	−210	−196	−203 to −160
Ammonia	−78	−33	−60 to 100
Pentane	−130	28	−20 to 120
Acetone	−95	57	0 to 120
Methanol	−98	64	10 to 130
Flutec PP2[1]	−50	76	10 to 160
Ethanol	−112	78	0 to 130
Heptane	−90	98	0 to 150
Water	0	100	30 to 200
Toluene	−95	110	50 to 200
Flutec PP9[1]	−70	160	0 to 225
Thermex[2]	12	257	150 to 350
Mercury	−39	361	250 to 650
Caesium	29	670	450 to 900
Potassium	62	774	500 to 1000
Sodium	98	892	600 to 1200
Lithium	179	1340	1000 to 1800
Silver	960	2212	1800 to 2300

Note: The useful operating temperature range is indicative only. Full properties of most of the above are given in Appendix C.

[1]Included for cases where electrical insulation is a requirement.

[2]Also known as Dowtherm A, an eutectic mixture of diphenyl ether and diphenyl.

In heat pipe design, a high value of surface tension is desirable in order to enable the heat pipe to operate against gravity and to generate a high capillary driving force. In addition to high surface tension, it is necessary for the working fluid to wet the wick and container material. That is, the contact angle must be very small.

Working fluids used in heat pipes range from helium at 4 K up to lithium at 2,300 K. Figure 4.18 shows the superiority of water over the range 350–500 K, where the alternative organic fluids tend to have considerably lower values of the figure of merit. At the slightly lower temperatures, 270–350 K, ammonia(NH_3) is a desirable fluid, although it requires careful handling to retain high purity, and acetone and the alcohols are alternatives having lower vapor pressures. These fluids are commonly used in heat pipes for space applications. Water and methanol, both being compatible with copper, are often used for cooling electronic equipment.

More recently, the temperature range 400–700 K has received attention from NASA. The suggestion is that metallic halides might be used as working fluids within this temperature range. The halides are typically compounds of lithium, sodium, potassium, rubidium, and copper, with fluorine, iodine, bromine, and iodine.

For the temperature range 500–950 K, mercury has attractive thermodynamic properties. It is also liquid at room temperature, which facilitates handling, filling, and start-up of the heat pipe. Bienert [29], in proposing mercury/stainless steel heat pipes for solar energy concentrators, used Deverall's technique for wetting the wick in the evaporator section of heat pipe and achieved sufficient wetting for gravity-assisted

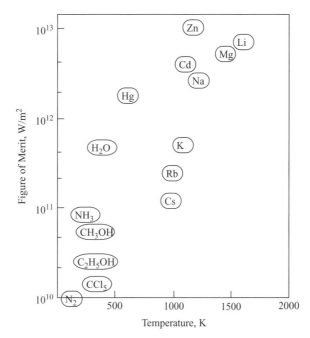

Figure 4.18 Figure of merit for selected working fluids for their boiling points (Revised from Reay and Kew (2006) [21])

operation. He argued that nonwetting in the condenser region of the heat pipe should enhance dropwise condensation, which would result in higher film coefficients, than those obtainable with film condensation.

For the higher-temperature range above 1400 K, lithium is generally a first choice as a working fluid but silver has also been used, with applications of liquid metal heat pipes for nuclear and space-related uses.

4.8.2 Compatibility

Compatibility is an important feature for the selection of the working fluid, wick, and containment vessel of the heat pipe. Two major results of incompatibility are corrosion and the generation of noncondensable gas. The environmental pressure must be also considered during the evaluation process of the container, particularly for spacecraft or satellite applications.

Table 4.8 illustrates the compatibility with working fluids, wicks, and case materials.

4.9 WICK STRUCTURES

The selection of the wick for a heat pipe depends on many factors, several of which are closely linked to the properties of the working fluid. The prime purpose of the wick is to generate capillary pressure to transport the working fluid from the condenser to the evaporator. It must also be able to distribute the liquid around the evaporator section to

Table 4.8   Heat Pipe Material Compatibility [28]

Wick material	Working fluids							
	H$_2$O	Acetone	NH$_3$	Methanol	Dow-A	Dow-E	Freon-11	Freon-13
Copper	RU	RU	NR	RU	RU	RU	RU	RU
Aluminum	GNC	RL	RU	NR	UK	NR	RU	RU
Stainless steel	GNT	PC	RU	GNT	RU	RU	RU	RU
Nickel	PC	PC	RU	RL	RU	RL	UK	UK
Refrasil fibre	RU	RU	RU	RU	RU	RU	UK	UK

RU Recommended by past successful use
RL Recommended by the literature
PC Probable compatible
NR Not recommended
UK Unknown
GNC Generation of noncondensable gas at all temperatures
GNT Generation of noncondensable gas at elevated temperature when oxide is present

any areas where heat is likely to be received by the heat pipe. The maximum capillary pressure difference generated by a wick increases with decreasing pore size. The wick permeability decreases with decreasing pore size. For homogeneous wicks, there is an optimum pore size, which is a compromise. In space, the constraints on size and the general high-power capability needs necessitate the use of nonhomogeneous or arterial wicks, aided by small pore structures for axial flow.

Figure 4.19 illustrates several common wicking structures in use, along with high-capacity concepts under development. Several improvements result from the separation of the liquid and vapor channels, handling the capillary pumping and axial fluid transport independency.

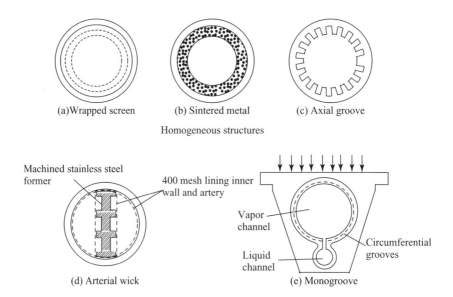

(a) Wrapped screen   (b) Sintered metal   (c) Axial groove

Homogeneous structures

(d) Arterial wick   (e) Monogroove

Figure 4.19   Different types of wicking structures (Reference [1] for (a), (b), and (c); Reference [21] for (d); Reference [30] for (e))

Another feature of the wick, which must be optimized, is its thickness. The heat transport capability of the heat pipe is raised by increasing the wick thickness. However, the increased radial thermal resistance of the wick created by this would work against increased capability and would lower the allowable maximum evaporator heat flux [21].

An arterial wick developed at IRD is shown in Figure 4.19(d). The bore of the heat pipe in this case was only 5.25 mm. This heat pipe, developed for European Space Organization (ESA), was designed to transport 15 W over a distance 1 m with an overall temperature drop not exceeding 6°C. The wall material was aluminum alloy and the working fluid acetone.

In the monogroove heat pipe presented by Alario et al. [30], the circumferential distribution and axial flow of the working fluid are accomplished through two separate mechanisms. The basic design of this high-capacity configuration consists of two large axial channels, one for vapor flow and the other for liquid flow. These two channels are separated by a small monogroove slot that creates a high capillary pressure difference and causes liquid to be pumped from the liquid channel to the fine circumferential grooves machined in the evaporator and condenser sections of the vapor channel.

4.10   DESIGN EXAMPLE

An electronic device in a satellite is cooled by a radiator. A heat pipe is required that will be capable of transferring a minimum of 25 W at a vapor temperature between 0°C and 80°C over a distance of 1 m in zero gravity. Restraints on the design are such that the evaporator and condenser sections are each 8 cm long, located at each end of the heat pipe. The maximum permissible temperature drop between the outside wall of the evaporator and the outside wall of the condenser is 14°C. Because of weight and volume limitations, the cross-sectional area of the vapor space should not exceed 0.197 cm². The heat pipe must also withstand bonding temperatures of 170°C. Design the heat pipe to meet the specification (Revised from [21]).

4.10.1   Selection of Material and Working Fluid

As this is an aerospace application, the mass is an important factor. Hence, aluminum alloy 6061 (HT30) ($k = 155.7$ W/m·K) is chosen for the wall and stainless steel for the wick. Working fluids compatible with these materials include acetone, ammonia, and Flutec PP9.

Acetone

Ammonia

Flutec PP9

Water must be dismissed at this stage, both on compatibility (see Table 4.8) and the operating range at 0°C with associated risk of freezing.

4.10.2   Working Fluid Properties

The properties of the three working fluids chosen are listed in Table 4.9 (from Table A.4 in Appendix A) to study each performance.

Table 4.9 Working Fluid Properties

	Acetone			Ammonia			Flutec PP9		
	0°C	40°C	80°C	0°C	40°C	80°C	0°C	30°C	90°C
Latent heat, kJ/kg	564	536	495	1263	1101	891	98.4	94.5	86.1
Liquid density, kg/m^3	812	768	719	638.6	579.5	505.7	2029	1960	1822
Vapor density, kg/m^3	0.26	1.05	4.30	3.48	12.0	34.13	0.01	0.12	1.93
Liquid thermal conductivity, W/m°C	0.813	0.175	0.160	0.298	0.272	0.235	0.059	0.057	0.054
Liquid viscosity, Ns/m^2 × 10^3	0.395	0.269	0.192	0.25	0.20	0.15	3.31	1.48	0.65
Vapor viscosity, Ns/m^2 × 10^5	0.78	0.86	0.95	0.92	1.16	1.40	0.90	1.06	1.21
Vapor pressure, bar	0.10	0.60	2.15	4.24	15.34	40.90	0.00	0.01	0.12
Vapor specific heat, kJ/kg °C	2.11	2.22	2.34	2.125	2.160	2.210	0.87	0.94	1.09
Liquid surface tension, N/m × 10^2	2.62	2.12	1.62	2.48	1.833	0.767	2.08	1.80	1.24

4.10.3   Estimation of Vapor Space Radius

If it is assumed that the heat pipe will be of a circular cross-section, the maximum vapor space area of $0.197\,cm^2$ given yields a radius of $2.5\,mm$. Hence, the vapor space area is

$$A_v = 0.197\,cm^2 \tag{4.84}$$

The vapor space radius is obtained as follows:

$$r_v = \sqrt{\frac{A_v}{\pi}} = \sqrt{\frac{0.197\,cm^2}{\pi}} = 0.25\,cm = 2.5\,mm \tag{4.85}$$

The vapor space diameter is

$$d_v = 5\,mm \tag{4.86}$$

4.10.4   Estimation of Operating Limits

4.10.4.1   Capillary Limits

The operating limits for each fluid must be examined. The capillary limit (or wicking limit), sonic limit, entrainment limit, and boiling limit are considered. The capillary limit at this stage requires more specification about the wick structure. However, the figure of merit may reveal the performance of the capillary pumping. Hence, using Equation (4.83) and Table 4.9, the figure of merit for the working fluids chosen can be plotted as a function of temperature, which is shown in Figure 4.20 using the equations for the curve fitting in Appendix D. We find that ammonia is superior to acetone and Flutec PP9.

4.10.4.2   Sonic Limits

The minimum axial heat transport due to the sonic limitation will occur at the minimum operating temperature, $0°C$ [21] and can be calculated from Equation (4.35) for acetone:

$$q_{s,max} = 0.474 h_{fg} A_v \left(\rho_v P_v \right)^{\frac{1}{2}} \tag{4.35}$$

$$q_{s,max} = (0.474)(564{,}000)\left(0.197 \times 10^{-4}\right)\left[(0.26)\left(0.10 \times 10^5\right)\right]^{\frac{1}{2}}$$

$$= 268.5\,W \tag{4.87}$$

Similar calculations may be carried out for the other working fluids, yielding

$$Acetone = 268.5\,W$$

$$Ammonia = 25670\,W$$

$$Flutec\ PP9 = 0.92\,W$$

Since the required axial heat flux is $25\,W$, Flutec PP9 is dismissed and the sonic limit would not be encountered for the other two fluids.

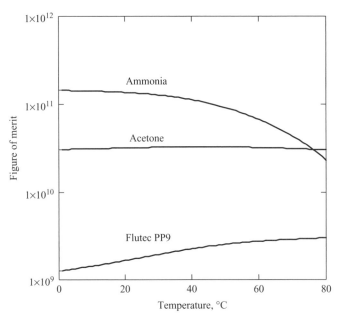

Figure 4.20   Figure of Merit for the working fluids

4.10.4.3 *Entrainment Limits*

The maximum entrainment limitation may be determined from Equation (4.38) with
(4.39)

$$q_{e,\max} = A_v h_{fg} \left(\frac{\sigma \rho_v}{2 r_{h,w}} \right)^{\frac{1}{2}} \tag{4.38}$$

where $r_{h,w}$ is the hydraulic radius of the wick. And the hydraulic diameter is defined

$$D_h = \frac{4 A_w}{P} \tag{4.39}$$

$$r_{h,w} = \frac{D_h}{2} \tag{4.88}$$

At this stage, it is assumed that the wick structure consist of two layers of #400
mesh of stainless steel and the wire thickness be the same order of the capillary radius.
Using Table 4.1 for the capillary radius,

$$d_w \approx r_{c,e} = \frac{1}{2N} = \frac{1}{2\,(400/\text{in.})\,(1\,\text{in.}/24.4\,\text{mm})} = 0.03175\,\text{mm} \tag{4.89}$$

The wick thickness for two layers of the screen mesh will be

$$t_w = 2 \times 2 d_w = 2 \times 2\,(0.03175) = 0.127\,\text{mm} \tag{4.89a}$$

The internal diameter of the heat pipe wall will be

$$d_i = d_v + 2 \times t_w = 5\,\text{mm} + 2 \times 0.127\,\text{mm} = 5.254\,\text{mm} \qquad (4.90)$$

The cross-sectional flow area of the wick structure will be

$$A_w = \frac{\pi\left(d_i^2 - d_v^2\right)}{4} = \frac{\pi\left(5.254^2 - 5^2\right)}{4} = 2.045\,\text{mm}^2 \qquad (4.91)$$

The wetted perimeter is

$$P = \pi\left(d_i + d_v\right) = \pi(5.254 + 5) = 32.21\,\text{mm} \qquad (4.92)$$

The hydraulic radius is

$$r_{h,w} = \frac{2A_w}{P} = \frac{2\left(2.045\,\text{mm}^2\right)}{32.21\,\text{mm}} = 0.127\,\text{mm} \qquad (4.93)$$

The entrainment limit is evaluated at the highest operating temperature, 80°C [21]. For acetone,

$$q_{e,\text{max}} = \left(0.197 \times 10^{-4}\right)(495{,}000)\left(\frac{\left(1.62 \times 10^{-2}\right)(4.30)}{2\left(0.127 \times 10^{-3}\right)}\right)^{\frac{1}{2}} = 161.49\,\text{W} \qquad (4.94)$$

The entrainment limits for the other two fluids may be calculated as in similar calculations:

$$\text{Acetone} = 161.49\,\text{W}$$

$$\text{Ammonia} = 564.13\,\text{W}$$

$$\text{Flutec PP9} = 16.48\,\text{W}$$

4.10.4.4   Boiling Limits

The minimum heat transport due to the boiling limit is obtained using Equation (4.40):

$$q_{b,\text{max}} = \frac{4\pi L_e k_{eff} T_v \sigma}{h_{fg}\rho_v \ln\left(\dfrac{r_i}{r_v}\right)}\left(\frac{1}{r_n} - \frac{1}{r_{c,e}}\right) \qquad (4.40)$$

At this stage, we keep assuming that stainless steel (AISI 316) will be used for the wick screen, the thermal conductivity is found in Table A.3 in Appendix A, and the thermal conductivity of the acetone is shown in Table 4.9.

$$k_w = 13.4\,\text{W/m·°C} \quad \text{and} \quad k_l = 0.16\,\text{W/m·°C} \qquad (4.95)$$

The effective thermal conductivity for the screen wick is obtained using Table 4.3.

$$k_{eff} = \frac{k_l\left[(k_l + k_w) - (1 - \varepsilon)(k_l - k_w)\right]}{(k_l + k_w) + (1 - \varepsilon)(k_l - k_w)} = 0.359\,\text{W/m·°C} \qquad (4.96)$$

The porosity for the screen wick is found in Table 4.2.

$$\varepsilon = 1 - \frac{\pi N d_w}{4} = 1 - \frac{\pi \, (400/25.4) \, (0.03175)}{4} = 0.607 \qquad (4.97)$$

$$L_e = 8 \, \text{cm} \qquad (4.98)$$

$$r_i = \frac{d_i}{2} = \frac{5.254 \, \text{mm}}{2} = 2.627 \, \text{mm} \qquad (4.99)$$

The nucleation site radius may be assumed to be

$$r_n = 2.54 \times 10^{-7} \, \text{m} \qquad (4.100)$$

The boiling limit using Equation (4.40) is obtained as follows:

$$q_{b,\text{max}} = \frac{4\pi \, \left(8 \times 10^{-2}\right) (0.359) \, (80 + 273) \left(1.62 \times 10^{-2}\right)}{495000 \, (4.30) \ln \left(\dfrac{2.627}{2.5}\right)}$$

$$\times \left(\frac{1}{2.54 \times 10^{-7}} - \frac{1}{0.03175 \times 10^{-3}}\right)$$

$$= 76.48 \, \text{W} \qquad (4.101)$$

Similar calculations may be carried out for the other fluids.

$$\text{Acetone} = 76.48 \, \text{W}$$

$$\text{Ammonia} = 3.68 \, \text{W}$$

$$\text{Flutec PP9} = 256.43 \, \text{W}$$

These results demand that ammonia is dismissed due to the occurrence of boiling. Hence, acetone is selected as the working fluid in spite of somewhat inferior thermal performance.

4.10.5 Wall Thickness

The bonding temperature of 170°C creates large vapor pressure that must be sustained in the heat pipe with a proper design. The vapor pressures for acetone and ammonia at this temperature are found to be 1.7 MPa and 11.3 MPa, respectively. Since the circumferential stress is twice of the longitudinal stress, the circumferential stress is only considered.

For equilibrium, we require

$$2\sigma_n \, (tdy) - P \, (2rdy) = 0 \qquad (4.102)$$

where σ_n is the normal stress and P is the vapor pressure at the bonding and t is the wall thickness, shown in Figure 4.21. Solving for t,

$$t = \frac{P \cdot r}{\sigma_n} \qquad (4.103)$$

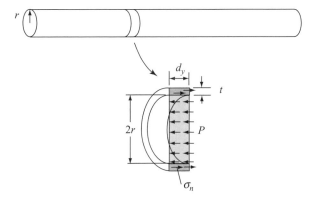

Figure 4.21   A free body diagram of the back segment along with the contained fluid

Taking the 10 percent of the material ultimate strength of 469 MPa for the aluminum 6061 chosen as the normal stress, the minimum wall thickness can be obtained.

For acetone,

$$t = \frac{1.7 \times 10^6 \, Pa \times 2.627 \times 10^{-3} \text{m}}{0.1 \times 469 \times 10^6 \, Pa} = 0.0952 \, \text{mm} \approx 0.1 \, \text{mm} \qquad (4.104)$$

For ammonia,

$$t = \frac{11.3 \times 10^6 \, \text{Pa} \times 2.627 \times 10^{-3} \, \text{m}}{0.1 \times 469 \times 10^6 \, \text{Pa}} = 0.0633 \, \text{mm} \qquad (4.104a)$$

Ammonia requires much thicker wall and, consequently, much more weight, so it is not desirable in space applications.

The heat pipe outer diameter will be

$$d_0 = d_i + 2t = 5.254 \, \text{mm} + 2 \times 0.1 \, \text{mm} = 5.454 \, \text{mm} \qquad (4.105)$$

4.10.6   Wick Selection

Two type of wick structure are proposed for this heat pipe, homogeneous and arterial wicks. A homogeneous wick may be a mesh, sintered, or grooved type, and arterial types normally incorporate a mesh to distribute liquid circumferentially.

Homogeneous meshes are easy to form but have inferior properties to arterial types. The first question is, therefore, will a homogeneous wick transport the required amount of fluid over 1 m to meet the heat transport specification?

To determine the maximum heat transport to compare with the requirement of 25 W, one can equate the maximum capillary pressure to the sum of the liquid, vapor, and gravitational pressure drops.

$$\Delta P_{c,m} = \Delta P_v + \Delta P_l + \Delta P_{norm} + \Delta P_{axial} \qquad (4.12)$$

where

$$\Delta P_{c,m} = \frac{2\sigma}{r_{c,e}} \tag{4.18}$$

$$\Delta P_v = \frac{C\,(f_v \mathrm{Re}_v)\,\mu_v}{2r_v^2 A_v \rho_v h_{fg}} L_{eff}\,q \tag{4.19}$$

$$\Delta P_l = \left(\frac{\mu_l}{K A_w h_{fg} \rho_l}\right) L_{eff}\,q \tag{4.27}$$

$$\Delta P_{norm} = \rho_l g d_v \cos\psi \tag{4.28}$$

$$\Delta P_{axial} = \rho_l g L \sin\psi \tag{4.29}$$

The gravitational effect is zero in this application, but is included to permit testing of the heat pipe on the ground. The elevation $L \sin\psi$ of end-to-end tilt during the tests is assumed to be 1 cm. The normal pressure drop is small and neglected. The vapor pressure drop is indeed small, but is included for information. The properties of acetone are evaluated at the maximum operating temperature of 80°C. The capillary radius is found to be 0.03175 mm in Equation (4.89). From Equation (4.18),

$$\Delta P_{c,m} = \frac{2\sigma}{r_{c,e}} = \frac{2\left(1.62 \times 10^{-2}\right)}{0.03175 \times 10^{-3}} = 1020\,\mathrm{Pa} \tag{4.106}$$

Assume laminar flow, using Equations (4.19) and (4.20):

$$\Delta P_v = \frac{16\left(0.95 \times 10^{-5}\right)}{2\left(2.5 \times 10^{-3}\right)^2 \left(0.197 \times 10^{-2}\right)(4.30)(495,000)}\,(1)\,q = 2.9 \times 10^{-3}q \tag{4.107}$$

The permeability for the screen wick is calculated using Table 4.2, the wire thickness in Equation (4.89), and the porosity in Equation (4.97)

$$K = \frac{d_w^2 \varepsilon^3}{122\,(1-\varepsilon)^2} = \frac{\left(0.03175 \times 10^{-3}\right)^2 (0.607)^3}{122\,(1 - 0.607)^2}$$

$$= 0.1196 \times 10^{-10}\,\mathrm{m}^2 \tag{4.108}$$

The liquid pressure drop in the wick over the pipe length, using Equation (4.27) with Equation (4.91), will be

$$\Delta P_l = \frac{\left(0.192 \times 10^{-3}\right)(1)\,q}{\left(0.1196 \times 10^{-10}\right)\left(2.045 \times 10^{-6}\right)(495,000)(719)}$$

$$= (22,056)q \tag{4.109}$$

The gravity pressure drop with $L \sin\psi = 1$ cm will be

$$\Delta P_{axial} = \rho_l g L \sin\psi = (719)(9.81)\left(1 \times 10^{-2}\right) = 70.5\,\mathrm{Pa} \tag{4.108}$$

Using Equation (4.12),

$$1020 = 2.9 \times 10^{-3}q + 22056q + 70.5 \tag{4.109}$$

Solving for q yields the heat transport for the maximum capillary pressure difference as

$$q = 0.043 \, \text{W} \tag{4.110}$$

Such a small amount of the heat transport 0.043 W results from the relatively long pipe length of 1 m. Since the required heat transport of 25 W is much greater than the calculated heat transport of 0.043 W, the homogeneous type wick is not acceptable. An arterial wick must be used. And it is noted that the vapor pressure drop appears negligible.

4.10.7   Maximum Arterial Depth

The maximum arterial depth may be determined considering Figure 4.22, where δ is the arterial depth and h is the elevation. The force balance between the surface tension with the contact angle and the elevated water weight may give approximately

$$(\rho_l - \rho_v) \, g \, (h + \delta) = \frac{2\sigma \cos \theta}{\delta} \tag{4.111}$$

Solving for δ yields

$$\delta = \frac{1}{2} \left[\sqrt{\left(h^2 + \frac{8\sigma \cos \theta}{(\rho_l - \rho_v) \, g} \right)} - h \right] \tag{4.112}$$

The maximum arterial depth δ has a significant physical meaning when a heat pipe starts its operation on the ground. The capillary force must overcome the gravity and allow the working fluid moving toward the evaporator from the condenser through the arterial channel. Once the arterial channel is filled with the working fluid, the capillary radius (either artery or wick) at the evaporator not the arterial depth, then governs the flow.

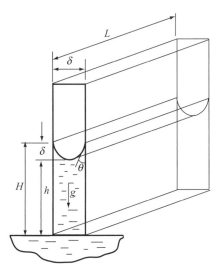

Figure 4.22   Rectangular artery

The maximum arterial depth δ in this work may be determined, assuming that the maximum end-to-end elevation h is taken as 1 cm to wet for the arteries to the top of the vapor space, the maximum contact angle is zero, and the properties are evaluated at an average operating temperature 40°C for convenience.

$$\delta = \frac{1}{2}\left[\sqrt{0.01^2 + \frac{8\,(0.0212)\cos 0}{(768 - 1.05)\,9.81}} - 0.01\right] = 0.534 \times 10^{-3}\,\text{m} \qquad (4.113)$$

Hence, the maximum permitted value is 0.534 mm. To allow for uncertainties in fluid properties and manufacturing tolerances, the maximum arterial depth is 0.5 mm.

4.10.8 Design of Arterial Wick

The arteries should transport the liquid with minimum pressure drop. The arteries should be away from the heat pipe wall to prevent nucleation (boiling) in the arteries. The arteries should be formed of low-conductivity material to prevent nucleation. The number of arteries should be determined based on the heat transport requirement. Considering all these features leads to an arterial wick shown in Figure 4.23. We want to calculate the heat transport capability starting with six arteries in the middle of the pipe as shown in Figure 4.23 covered by two layers of #400 mesh screen wicks.

We have already shown that the entrainment, sonic, and boiling limits meet the requirements. The heat pipe should also show that the capillary limitation meets the required heat transport capability.

4.10.9 Capillary Limitation

The capillary pressure difference should overcome the vapor pressure drop in the vapor space along with the pipe length, ΔP_v, the liquid pressure drop in the arteries along

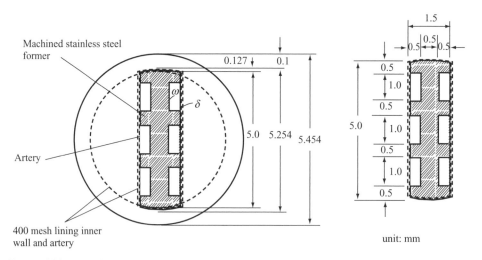

Figure 4.23 Arterial wick. The dimensions are modified from the one originally developed by IRD [21]

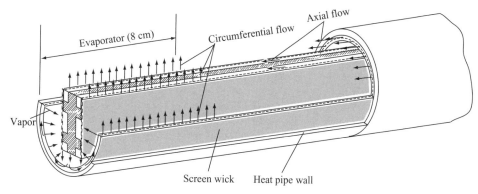

Figure 4.24 Liquid flows in the screen wick structure

with the pipe length, ΔP_{la}, the liquid pressure drop in the circumferential flow wick, ΔP_{lc}, the liquid pressure drop in the axial flow wick, ΔP_{lw} and the gravitational pressure drop, ΔP_{axial}. The circumferential and axial flows in the mesh are illustrated in Figure 4.24. It should be noted that all drops are connected in series except ΔP_v and ΔP_{lw}. Hence, the capillary pressure difference is expressed similar to the electrical circuit as

$$\Delta P_{c,m} = \Delta P_{la} + \Delta P_{lc} + \frac{1}{\dfrac{1}{\Delta P_v} + \dfrac{1}{\Delta P_{lw}}} + \Delta P_{norm} + \Delta P_{axial} \tag{4.114}$$

The axial liquid flow ΔP_{lw} in the wick along with the pipe length will have little effect, but we have included it in the evaluation for demonstration. The liquid pressure drop in the wick over the pipe length is the same as the one in Equation (4.109), as long as the same wick is used.

The normal gravitational pressure drop will be neglected. The capillary pressure drop has been calculated in Equation (4.106).

$$\Delta P_{c,m} = \frac{2\sigma}{r_{c,e}} = \frac{2\left(1.62 \times 10^{-2}\right)}{0.03175 \times 10^{-3}} = 1020 \, \text{Pa} \tag{4.106}$$

4.10.9.1 *Liquid Pressure Drop in the Arteries*

The liquid pressure drop in the arterial channels may be calculated using Equation (4.27) with Table 4.2 and the properties evaluated at 80°C.

$$\Delta P_l = \left(\frac{\mu_l}{KA_w h_{fg}\rho_l}\right) L_{eff} \, q \tag{4.27}$$

where, from Table 4.2, the hydraulic radius of the artery is

$$r_{h,l} = \frac{\delta\omega}{\delta + \omega} = \frac{0.5}{0.5 + 1.0} = 0.33 \, \text{mm} = 0.33 \times 10^{-3} \, \text{m} \tag{4.115}$$

Assuming laminar flow, the K value is

$$K = \frac{2r_{h,l}^2}{f_l Re_l} = \frac{2\left(0.33 \times 10^{-3}\right)^2}{16} = 1.386 \times 10^{-8}\,\text{m}^2 \tag{4.116}$$

The arterial cross-sectional area is

$$A_w = \delta\omega N = \left(0.5 \times 10^{-3}\right)\left(1 \times 10^{-3}\right)(6) = 3 \times 10^{-6}\,\text{m}^2 \tag{4.117}$$

where N is the number of arteries.

The liquid pressure drop in the arteries will be

$$\Delta P_{la} = \frac{0.192 \times 10^{-3}}{\left(1.386 \times 10^{-8}\right)\left(3 \times 10^{-6}\right)(495000)(719)}(1)\,q = 12.97q \tag{4.118}$$

4.10.9.2   *Liquid Pressure Drop in the Circumferential Wick*

The liquid pressure drop in the circumferential wick may be calculated using Equation (4.27). The porosity and permeability for the screen wick are already calculated in Equations (4.97) and (4.108), respectively. The properties are evaluated at 80°C.

$$\Delta P_{lc} = \left(\frac{\mu_l}{KA_w h_{fg}\rho_l}\right)L_{eff}\, q \tag{4.27}$$

For the circumferential flow area, we consider the arterial part, on which four flow areas are working, two upward and two downward as shown in Figure 4.24. Hence, the circumferential cross-sectional area A_w with Equation (4.89a) for t_w will be

$$A_w = 4L_e t_w = 4\left(8 \times 10^{-2}\right)\left(0.127 \times 10^{-3}\right) = 4.064 \times 10^{-5}\,\text{m}^2 \tag{4.120}$$

The effective length here would be approximately the sum of half of the inner diameter for the downward flow and one-eighth of the circumference for the upward flow, which may be visualized in Figure 4.24. Accordingly, the effective length indicates the maximum length.

$$L_{eff} = \frac{d_i}{2} + \frac{\pi d_i}{8} = \frac{5.254 \times 10^{-3}}{2} + \frac{\pi\left(5.254 \times 10^{-3}\right)}{8} = 4.69 \times 10^{-3}\,\text{m}$$

$$\Delta P_{lc} = \frac{0.192 \times 10^{-3}}{\left(0.1196 \times 10^{-10}\right)\left(4.064 \times 10^{-5}\right)(495,000)(719)}\left(4.69 \times 10^{-3}\right)q$$

$$= 5.205q \tag{4.121}$$

4.10.9.3   *Vapor Pressure Drop in the Vapor Space*

The vapor pressure drop occurs in two near-semicircular channels. The vapor pressure drop in the vapor space along with the pipe length may be calculated using Equation (4.19):

$$\Delta P_v = \frac{C\left(f_v Re_v\right)\mu_v}{2r_v^2 A_v \rho_v h_{fg}}L_{eff}\, q \tag{4.19}$$

Since there are two separated vapor channels, as shown in Figure 4.23, the hydraulic diameter concept is utilized to represent the noncircular vapor flow. Hence, the flow area for the one of the two vapor channels, considering the geometry of the channel, will be

$$A = \frac{1}{2}\left[\frac{\pi d_v^2}{4} - (1.5\,\text{mm})\,d_v\right] = \frac{1}{2}\left[\frac{\pi\,(5\,\text{mm})^2}{4} - (1.5\,\text{mm})\,(5\,\text{mm})\right]$$

$$= 6.067\,\text{mm}^2 \qquad (4.122)$$

The wetted perimeter will be

$$P = \left(\pi d_v - 2 \times \frac{1.5\,\text{mm}}{2}\right) + d_v = \left(\pi\,(5\,\text{mm}) - 2 \times \frac{1.5\,\text{mm}}{2}\right) + 5\,\text{mm}$$

$$= 19.208\,\text{mm} \qquad (4.123)$$

The hydraulic diameter of the vapor channel will be

$$d_v = \frac{4A}{P} = \frac{4\,(6.067\,\text{mm}^2)}{19.208\,\text{mm}} = 1.2634\,\text{mm} \qquad (4.124)$$

The hydraulic radius will be

$$r_v = \frac{d_v}{2} = \frac{1.2634}{2} = 0.6317\,\text{mm} \qquad (4.125)$$

The vapor flow area must be the total flow area; hence

$$A_v = 2A = 2 \times 6.067\,\text{mm}^2 = 12.134\,\text{mm}^2 \qquad (4.126)$$

Assuming turbulent flow at this moment, using Equation (4.22)

$$f_v Re_v = 0.038\left(\frac{2r_v q}{A_v \mu_v h_{fg}}\right)^{3/4}$$

$$= 0.038\left(\frac{2\,(0.6317 \times 10^{-3})\,q}{12.134 \times 10^{-6}\,(0.95 \times 10^{-5})\,(495000)}\right)^{3/4} = 0.387q^{3/4} \quad (4.126)$$

The vapor pressure drop in the vapor space along with the pipe length will be

$$\Delta P_v = \frac{(1)\,(0.387q^{3/4})\,(0.95 \times 10^{-5})}{2\,(0.6317 \times 10^{-3})^2\,(12.134 \times 10^{-6})\,(4.30)\,(495{,}000)}\,(1)\,q$$

$$= 0.178q^{\frac{7}{4}} \qquad (4.127)$$

The gravitational pressure drop for axial direction is the same as derived in Equation (4.108). Equation (4.114) with the evaluated terms is written as

$$1{,}020 = 12.97q + 5.205q + \frac{1}{\dfrac{1}{0.178q^{\frac{7}{4}}} + \dfrac{1}{22056q}} + 70.5 \qquad (4.128)$$

Solving for q gives the heat transport for the maximum capillary pressure difference as

$$q = 44.68 \, \text{W} \tag{4.129}$$

This meets the requirement of 25 W. However, the assumption of turbulent flow should be justified. Using Equations (4.25) and (4.26), the total mass flow rate is associated with two flow channels.

$$\dot{m} = \frac{q}{h_{fg}} = \frac{44.68 \, \text{W}}{495{,}000 \, \text{J/kg}} = 9.026 \times 10^{-5} \, \frac{\text{kg}}{\text{s}} \tag{4.130}$$

Using Equation (4.124) for d_v, the Reynolds number for one channel will be turbulent:

$$Re_v = \frac{4\dot{m}}{\pi d_v \mu_v} = \frac{4 \left(\dfrac{9.026 \times 10^{-5}}{2} \right)}{\pi \left(1.2634 \times 10^{-3} \right) \left(0.95 \times 10^{-5} \right)} = 4{,}787 > 2{,}300 \tag{4.131}$$

The assumption of the turbulent flow is justified since the Reynolds number is indeed greater than 2,300.

4.10.10   Performance Map

The heat transport for the maximum capillary pressure difference with respect to the operating temperature was plotted in Figure 4.25. The sonic, entrainment, and boiling limits for the specific arterial wick type heat pipe are also plotted. It is noted that the

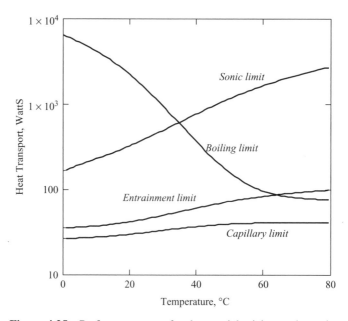

Figure 4.25   Performance map for the arterial wick type heat pipe

capillary limit decreases when the temperature decreases. The boiling limit only shows opposite behavior, decreasing with increasing temperature. The heat transport for the capillary pressure difference at zero temperature is obtained as 26.6 W, which is greater than 25 W. This graph clearly shows that the capillary limit governs the performance of the heat pipe.

4.10.11   Check the Temperature Drop

The temperature difference between the outside wall of the evaporator T_e and the outside wall of the condenser T_c is required to be 14°C. Considering the maximum heat transport for the capillary limit, we have

$$q = \frac{T_e - T_c}{R_{tot}} \tag{4.132}$$

The total resistance R_{tot} is found in Equation (4.47) neglecting the axial thermal resistances.

$$R_{tot} = R_{p,e} + R_{w,e} + R_{w,c} + R_{p,c} \tag{4.47}$$

The evaporator and condenser are physically identical, so the equation can be reduced

$$R_{tot} = 2\left(R_{p,e} + R_{w,e}\right) \tag{4.133}$$

Using Equations (4.44) and (4.45), for a cylindrical heat pipe, the radial conduction resistance at the evaporator is

$$R_{p,e} = \frac{\ln\left(\dfrac{d_o}{d_i}\right)}{2\pi L_e k_p} = \frac{\ln\left(\dfrac{5.454}{5.254}\right)}{2\pi(8 \times 10^{-2})(155.7)}$$

$$= 4.77 \times 10^{-4}°\mathrm{C/W} \tag{4.134}$$

For a cylindrical wick, the resistance of the liquid-wick combination at the evaporator is

$$R_{w,e} = \frac{\ln\left(\dfrac{d_i}{d_v}\right)}{2\pi L_e k_{eff}} = \frac{\ln\left(\dfrac{5.254}{5.0}\right)}{2\pi(8 \times 10^{-2})(0.359)} = 0.274°\mathrm{C/W} \tag{4.135}$$

Hence,

$$R_{tot} = 2\left(4.77 \times 10^{-4} + 0.274\right) = 0.549°\mathrm{C/W} \tag{4.136}$$

The temperature difference from Equation (4.132) with the heat transport requirement of 25 W will be

$$T_e - T_c = (25\,\mathrm{W})\left(0.549°\mathrm{C/W}\right) = 13.7°\mathrm{C} \tag{4.137}$$

Fortunately, the temperature drop of 13.7°C meets the temperature drop requirement of 14°C. If this does not meet the requirement, you should repeat the procedure with a smaller wick thickness. Take care to also satisfy the heat transport requirement. This section of design could be done at the earlier stages of designing to avoid the unnecessary repetitions.

REFERENCES

1. G.P. Peterson, *An Introduction to Heat Pipes—Modeling, Testing, and Applications.* New York: John Wiley and Sons, 1994.

2. Y.A. Cengel, *Heat Transfer—A Practical Approach.* New York: McGraw-Hill, 1998.

3. L.-T. Yeh and R. Chu, *Thermal Management of Microelectronic Equipment*. ASME Press, 2002.

4. W.M. Rohsenow, J.P. Hartnett, and Y.I. Cho, *Handbook of Heat Transfer*. New York: McGraw-Hill, 1998.

5. S.W. Chi, *Heat Pipe Theory and Practice*. New York: McGraw-Hill, 1976.

6. C.A. Busse, Pressure Drop in the Vapor Phase of Long Heat Pipes. *Proceedings of the IEEE International Thermionic Conversion Specialist Conference*, IEEE, New York, 1967.

7. F. Kreith, *The CRC Handbook of Mechanical Engineering*. New York: CRC Press, 1997.

8. J.E. Levy, Ultimate Heat Pipe Performance. *IEEE Transactions on Electron Devices*, 16 (1969): 717–723.

9. C.A. Busse, Theory of the Ultimate Heat Transfer of Cylindrical Heat Pipes. *International Journal of Heat Mass Transfer*, 16 (1973): 169.

10. T. P. Cotter, Heat Pipe Startup Dynamics. *Proc. SAE Thermionic Conversion Specialist Conference*, Palo Alto, CA, 1967.

11. G.A. Asselman and D.B. Green, Heat Pipes. *Phillips Technical Review*, 16, pp. 169–186.

12. E. Fried, Thermal Conduction Contribution to Heat Transfer at Contacts, in R. P. Tye, ed., *Thermal Conductivity*, vol. 2. London: Academic Press, 1969.

13. J.C. Eid and V.W. Antonetti, Small Scale Thermal Contact Resistance of Aluminum against Silicon, in C. L. Tien, V. P. Cary, and J. K. Ferrel, eds. *Heat Transfer*, vol. 2. New York: Hemisphere, 1986, pp. 659–664.

14. B. Snaith, P.W. O'Callaghan, and S.D. Probert, *Appl. Energy*, 16 (1984): 175.

15. M.M. Yovanovich, Theory and Application of Constriction and Spreading Resistance Concepts for Microelectronic Thermal Management. Presented at the International Symposium on Cooling Technology for Electronic Equipment, Honolulu, 1987.

16. G.P. Peterson and L.S. Fletcher, Thermal Contact Resistance of Silicon Chip Bonding Materials. *Proceedings of the International Symposium on Cooling Technology for Electronic Equipment* (Honolulu, 1987), pp. 438–448.

17. M.M. Yovanovich and M. Tuarze, *AIAA J. Spacecraft Rockets*, 6 (1969): 1013.

18. C.L. Tien and S.J. Chen, Noncondensible Gases in Heat Pipes. *Proc. 5th Int. Heat Transient* (Tsukuda, Japan, 1984), pp. 97–101.

19. A.F. Mills, *Heat Transfer*, 2nd ed. Upper Saddle River, NJ: Prentice Hall, 1999.

20. K. Watanabe, A. Kimura, K. Kawabata, T. Yanagida, and M. Yamauchi, Development of a variable-conductance heat-pipe for a sodium-sulphur (NAS) battery. *Furukawa Rev.*, no. 20 (2001): 71–76.

21. D.A. Reay and P. A. Kew, *Heat Pipes*, 5th ed. Butterworth-Heinemann, (Elsevier), 2006.

22. Anonymous, *Heat Pipes—Properties of Common Small-Pore Wicks*. Data items No. 79013, Engineering Sciences Data Unit, London, 1979.

23. Y.F. Maydanik, Loop Heat Pipes. *Applied Thermal Engineering*, 25, (2005): 635–657.

24. P.D. Dunn and D.A. Reay, *Heat Pipes*, 3rd ed. Oxford: Pergamon Press, 1982.

25. T.P. Cotter, Principles and Prospects of Micro Heat Pipes. *Proc. 5th International Heat Pipe Conference* (Tsukuba, Japan), pp. 328–335, 1984.

26. B.R. Babin, G.P. Peterson, and D. Wu, Steady-State Modeling and Testing of a Micro Heat Pipe. *Journal of Heat Transfer*, 112 (1990): 595–601.

27. J. M. Ha and G. P. Peterson, The Heat Transport Capacity of Micro Heat Pipes. *Journal of Heat Transfer*, 120 (1998): 1064–1071.

28. A. Basiulis and M. Filler, *Operating Characteristics and Long Life Capabilities of Organic Fluid Heat Pipes*, AIAA Paper 71-408, 6th AIAA Thermophy. Conference, Tullahoma, Tennessee, April 1971.

29. W. Bienert, *Heat pipes for solar energy collectors*. First International Heat Pipe Conference, Stuttgart, Paper 12-1, October 1973.

30. J. Alario, R. Brown, and P. Otterstadt, Space Constructable Radiation Prototype Test Program. Paper No. 84-1793, American Institute of Aeronautics and Astronautics, Washington, DC, 1984.

31. Y.Y. Hsu, On the Size Range of Active Nucleation Cavities on a Heating Surface. *J. Heat Transfer*, Trans. ASME, August 1962.

32. B.S. Petukhov, T.F. Ervine, and J.P. Hartnett, *Advances in Heat Transfer, Vol. 6*. New York: Academic Press, 1970.

PROBLEMS

4.1 A heat pipe for an electronics application uses water as the working fluid. The evaporator, condenser, and adiabatic section are 3 cm long each, as shown in Figure P4.1. The diameter of the vapor space is 5 mm, and the wicking structure consists of three layers of #400 mesh 304 stainless steel screen (wire diameter = 0.00098 in). If the heat pipe operates at 80°C in a horizontal position, determine the capillary, sonic, entrainment, and boiling limits.

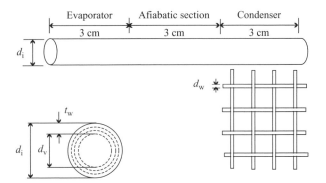

Figure P4.1   A heat pipe with the cross-sectional view and mesh

4.2 A simple horizontal copper-water heat pipe is to be constructed from a 3-cm-long tube to cool an enclosed silicon casing, as shown in Figure P4.2. The inner and outer diameters of the heat pipe are 4 mm and 4.5 mm, respectively. The diameter of the vapor space is 3.5 mm, also shown. The evaporator and condenser lengths of the heat pipe are 1 cm each and the wicking structure consists of two layers of #500-mesh copper screen with the wire diameter of 0.00085 in. The evaporator of the heat pipe is embedded in a silicon casing, and the contact resistance between the heat pipe and its casing is 1.5×10^{-5} m²°C/W. The condenser is immersed in a stream of liquid and the convection heat transfer coefficient at the condenser is 700 W/m^2°C. If the liquid is at 20°C, and the heat pipe is operating at a heat transfer rate of 1.5 W, what is the temperature of the silicon casing?

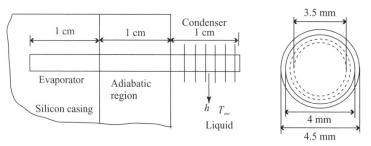

Figure P4.2 A heat pipe arrangement on a silicon casing

4.3 A nitrogen-loaded ethanol heat pipe is constructed from a 0.45-inch-O.D. stainless steel tube and is L-shaped, as shown in Figure P4.3. A fibrous stainless steel slab wick extends into the gas reservoir. The cross-sectional area of the vapor space in the pipe is 48.6 mm^2, and the reservoir volume is 6.2×10^4 mm^3. The condenser length L_c is 43 cm. When the sink temperature is 270 K, the heat pipe is fully open, for an adiabatic section temperature of 295 K and a heat load of 45 W. Determine the mass of gas in the heat pipe and prepare a graph of heat load versus adiabatic section temperature. In addition, vary the reservoir volume to demonstrate its effect on performance. At a reference temperature $T_r = 10$°C, ethanol has a vapor pressure of 3000 Pa and enthalpy of vaporization of 0.9048×10^6 J/kg.

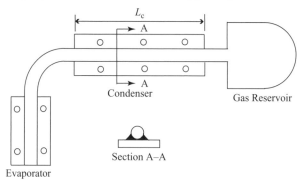

Figure P4.3 A L-shaped variable conductance heat pipe

4.4 A trapezoidal micro heat pipe is designed to cool a semiconductor device as an element of the micro heat pipe array. The case material is silver and the working fluid is distilled deionized water. The heat pipe is operated at 55°C. The dimensions are given as follows:

Total length:	$L = 52$ mm
Evaporator length:	$L_e = 13$ mm
Condenser length:	$L_c = 13$ mm
Pipe width:	$d = 1$ mm
Inner vapor channel diameter:	$d_i = 0.7$ mm
Outer vapor channel diameter:	$d_o = 0.9$ mm
Groove width:	$\omega = 0.014$ mm
Half included angle α:	$\alpha = 0.6032$ rad

Determine the heat transport limitation for the micro heat pipe.

4.5   A heat pipe for an electronics application uses methanol as the working fluid. The evaporator, condenser, and adiabatic sections are each 2 cm long. The diameter of the vapor space is 4 mm, and the wall is constructed of 1-mm-thick 2024-T6 aluminum with ten equally spaced 0.5 mm square axial grooves forming the wicking structure. The heat pipe operates at 30°C with an adverse tilt angle of 5 degrees. Determine the capillary, sonic, entrainment, and boiling limits for this heat pipe.

4.6   Estimate the liquid rate and heat transfer capacity of a simple water heat pipe operating at 100°C having a wick of two layers of 250 mesh against the inside wall. The heat pipe is 30 cm long and has a bore of 1 cm diameter. It is operating at an inclination to the horizontal of 30°C, with the evaporator above the condenser. It will be shown that the capability of the heat pipe is low. What improvement will be made if two layers of 100 mesh are added to the 250-mesh wick to increase liquid flow capability?

Design Problem

4.7   An electronic device in a satellite is cooled by a radiator. A heat pipe is required that will be capable of transferring a minimum of 30 W at a vapor temperature between 10°C and 80°C over a distance of 0.7 m in zero gravity. Restraints on the design are such that the evaporator and condenser sections are each 6 cm long, located at each end of the heat pipe and the maximum permissible temperature drop between the outside wall of the evaporator and the outside wall of the condenser is 15°C. Because of weight and volume limitations, the cross-sectional area of the vapor space should not exceed 0.205 cm^2. The heat pipe must also withstand bonding temperatures of 160°C. Design the heat pipe to meet the specification.

5

Compact Heat Exchangers

5.1 INTRODUCTION

A *heat exchanger* is a device to transfer thermal energy between two or more fluids, one comparatively hot and the other comparatively cold. Heat exchangers are typically classified according to flow arrangement. When the hot and cold fluids move in the same direction, it is called *parallel-flow* arrangement. When they are in opposite direction as shown in Figure 5.1(a), it is called *counterflow* arrangement. There is also *crossflow* arrangement, where the two fluids move in crossflow perpendicular each other. Double-pipe heat exchangers, shell-and-tube heat exchangers, and plate heat exchangers may have either parallel flow or counterflow arrangement. Finned-tube heat exchangers and plate-fin heat exchangers typically have a crossflow arrangement. These heat exchangers are depicted in Figure 5.1(a)–(e).

A special and important class of heat exchangers is used to achieve a very large heat transfer area per volume. Termed *compact heat exchangers*, these devices have dense arrays of finned tubes or plates and are typically used when at least one of the fluids is a gas, and is hence characterized by a small convection coefficient. Plate heat exchangers, finned-tube heat exchangers, and plate-fin heat exchangers are in the class of compact heat exchangers.

The *surface area density* β (m^2/m^3) which is defined as the ratio of the heat transfer area to the volume of the heat exchanger, is often used to describe the *compactness* of heat exchangers. The compactness of the various types of heat exchangers is shown in Figure 5.2, where the compact heat exchangers have a surface area density greater than about $600\,m^2/m^3$, or the hydraulic diameter is smaller than about 6 mm operating in a gas stream.

Double-pipe heat exchangers consist of two concentric pipes, as shown in Figure 5.1(a) and are perhaps the simplest heat exchanger. This heat exchanger is suitable where one or both fluids are at very high pressure. Double-pipe heat exchangers are generally used for small-capacity applications (less than $50\,m^2$ of total heat transfer surface area). Cleaning is done easily by disassembly. The exchanger with U tubes is referred to as a *hairpin exchanger*.

Shell-and-tube heat exchangers are generally built of a bundle of tubes mounted in a shell, as shown in Figure 5.1(b). The exchangers are custom designed for virtually any capacity and operating conditions, from high vacuum to high pressure over 100 MPa, from cryogenics to high temperature about $1{,}100^\circ$C. The surface area density β ranges from 60 to $500\,m^2/m^3$. Mechanical cleaning in the tubular side is done easily by disassembling the front and rear-end heads, while the shell side requires chemical

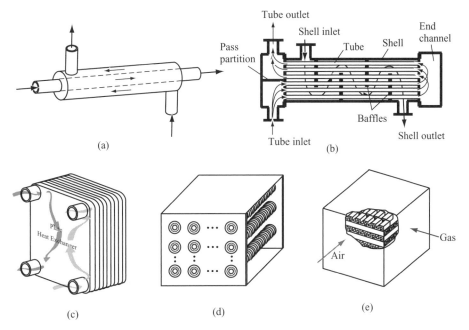

Figure 5.1   Typical heat exchangers: (a) double-pipe heat exchanger, (b) shell-and-tube heat exchanger, (c) brazed plate heat exchanger, (d) circular finned-tube heat exchanger, and (e) plate-fin heat exchanger (OSF)

cleaning. They are the most versatile exchangers, made from a variety of metals and nonmetals such as polymer to a supergiant surface area over $10^5 \, \text{m}^2$. They have held more than 65 percent of the market share in industry and design experience for about 100 years. The design codes and standards are available from the Tubular Exchanger Manufacturers Association (TEMA).

Plate heat exchangers (PHE) are one of the first compact heat exchangers, first built in 1923. They weight about 25 percent of the shell-and-tube heat exchangers of the same capacity. They are typically built of thin metal plates, which are either smooth or have some form of corrugation. Generally, these exchangers cannot accommodate very high pressures (up to 3 MPa) and temperatures (up to 260°C). The surface area density β typically ranges from 120 to 670 m^2/m^3. Plate heat exchangers are generally of two types. One is a plate-and-frame or gasketed plate heat exchanger and the other is a welded or brazed plate heat exchanger. The gasketed plate heat exchanger is designated to accommodate a mounted mechanical cleaning device. Hence, this type of exchanger is appropriate for situations requiring frequent cleaning, such as in food processing. One of the problems with the gasketed plate heat exchangers is the presence of gaskets, which limits operating temperatures and pressures. The welded or brazed plate heat exchangers overcome these limitations, which are shown in Figure 5.1(c).

Finned-tube heat exchangers are gas-to-liquid heat exchangers and have dense fins attached on the tubes of the air side, because the heat transfer coefficient on the air side is generally one order of magnitude less than that on the liquid side (Figure 5.1(d)). Circular finned-tube heat exchangers, as shown in Figure 5.3(a), are probably more rigid and practical in large heat exchangers such as in air conditioning and refrigerating

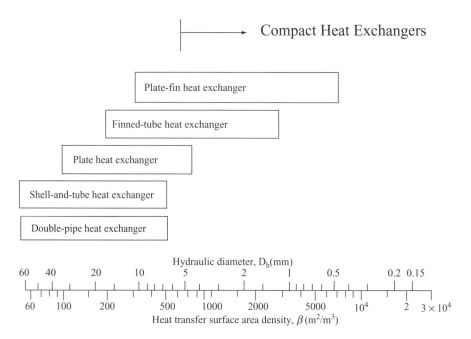

Compact Heat Exchangers

| Plate-fin heat exchanger |
| Finned-tube heat exchanger |
| Plate heat exchanger |
| Shell-and-tube heat exchanger |
| Double-pipe heat exchanger |

Hydraulic diameter, D_h(mm)

60   40   20   10   5   2   1   0.5   0.2 0.15

60   100   200   500   1000   2000   5000 10^4 2 3×10^4

Heat transfer surface area density, β (m²/m³)

Figure 5.2 Overview of the compactness of heat exchangers

industries. Flat-finned, flat-tube heat exchangers, as shown in Figure 5.3(b) are mostly used for automotive radiators. The circular finned-tube heat exchangers usually are less compact than the flat-finned, flat-tube heat exchangers, having a surface area density of about 3,300 m²/m³.

Plate-fin heat exchangers, as shown in Figure 5.1(e), are the most compact heat exchangers, commonly having triangular or rectangular cross sections. Plate-fin heat exchangers are generally designed for moderate pressures less than 700 kPa and temperatures up to about 840°C, with a surface area density of up to 5,900 m²/m³. These exchangers are widely used in electric power plants (gas turbine, nuclear, fuel cells, etc.). Recently, a condenser for an automotive air-conditioning system has been developed for operating pressures of 14 MPa.

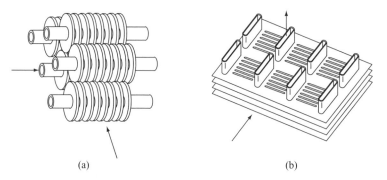

(a)   (b)

Figure 5.3 Typical components of finned-tube heat exchangers: (a) circular finned-tube heat exchanger, (b) louvered flat-finned flat-tube heat exchanger

There are other types of compact heat exchangers. *Printed-circuit heat exchangers* were developed for corrosive and reactive chemical processes, which do not tolerate dissimilar materials in fabrication. This exchanger is formed by diffusion bonding of a stack of plates, with fluid passages etched on one side of each plate using technology adapted from that used for electronic printed circuit boards—hence the name.

Polymer compact heat exchangers have become increasingly popular as an alternative to the use of exotic materials for combating corrosion in process duties involving strong acid solutions. Polymer compact heat exchangers also provide resistance to fouling. Most importantly, polymers offer substantial weight, volume, and cost savings. Polymer plate heat exchangers and polymer shell-and-tube heat exchangers are currently available in the market.

In the *thermal design* of heat exchangers, two of the most important problems involve *rating* and *sizing*. Determination of heat transfer and pressure drop is referred to as a rating problem. Determination of a physical size such as length, width, height, and surface areas on each side is referred to as a sizing problem. Before we discuss the thermal design of heat exchangers, we want to first develop the fundamentals of the heat exchangers for various flow arrangements.

5.2 FUNDAMENTALS OF HEAT EXCHANGERS

5.2.1 Counterflow and Parallel Flows

Consider two counterflow channels across a wall, as shown in Figure 5.4(a). The subscripts 1 and 2 denote the hot and cold fluids, respectively and the subscripts i and o indicate inlet and outlet, respectively. The mass flow rate is denoted as $\dot{m}$. The temperature distributions for the hot and cold fluids are shown in Figure 5.4(b), where the dotted line indicates the approximate wall temperatures.

Figure 5.4(c) shows parallel-flow channels across a wall. The hot fluid T_{1i} enters the lower channel and leaves at the decreased temperature T_{1o}, while the cold temperature T_{2i} enters the upper channel and leaves at the increased temperature T_{2o}. The temperature distributions are presented in Figure 5.4(d), where the dotted line indicates the wall temperatures along the length of the channel. Note that the wall temperatures in parallel flow are nearly constant compared to those changing in counterflow. We will discuss later the effectiveness of heat exchangers, but the effectiveness in counterflow surpasses that in parallel flow. Therefore, the counterflow heat exchanger is usually preferable. However, the nearly constant wall temperature is a characteristic of the parallel-flow heat exchanger, which is sometimes necessary (e.g., exhaust-gas heat exchangers usually require a constant wall temperature to avoid corrosion). The total heat transfer rate between the two fluids can be expressed considering an enthalpy flow that is the product of the mass flow rate and the specific heat and the temperature difference.

For the hot fluid, the heat transfer rate is

$$q = \dot{m}_1 c_{p1} \left(T_{1i} - T_{1o} \right) \tag{5.1}$$

where $\dot{m}_1$ is the mass flow rate for the hot fluid and c_{p1} is the specific heat for the hot fluid.

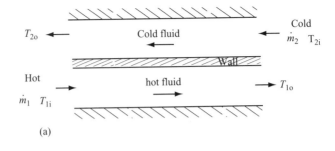

(a)

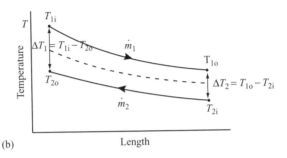

(b)

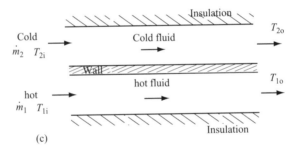

(c)

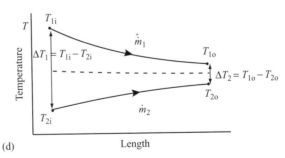

(d)

Figure 5.4 The counterflow arrangement: (a) Schematic for counterflow channels and (b) the temperature distributions. The parallel-flow arrangement: (c) Schematic for parallel-flow channels and (d) the temperature distributions

For the cold fluid, the same heat transfer is expressed as

$$q = \dot{m}_2 c_{p2} (T_{2o} - T_{2i})$$ (5.2)

where $\dot{m}_2$ is the mass flow rate for the cold fluid and c_{p2} is the specific heat for the cold fluid. The same heat transfer rate can be expressed in terms of the overall heat transfer coefficient,

$$q = UAF\Delta T_{lm}$$ (5.3)

where U is the overall heat transfer coefficient and A is the heat transfer surface area at the hot or cold side. F is the *correction factor*, depending on the flow arrangements. For example, $F = 1$ for counterflow or parallel flow such as the double-pipe heat exchangers, and usually $F \leq 1$ for other types of flow arrangements.

Note that

$$UA = U_1 A_1 = U_2 A_2$$ (5.4)

ΔT_{lm} is the *log mean temperature difference*—that is, it is defined (see Section 5.2.3 for the derivation) as

$$\Delta T_{lm} = \frac{\Delta T_1 - \Delta T_2}{\ln\left(\dfrac{\Delta T_1}{\Delta T_2}\right)}$$ (5.5)

where

$\Delta T_1 = T_{1i} - T_{2i}$ and $\Delta T_2 = T_{1o} - T_{2o}$ for parallel flow (see Figure 5.5).   (5.6)

$\Delta T_1 = T_{1i} - T_{2o}$ and $\Delta T_2 = T_{1o} - T_{2i}$ for counter flow (see Figure 5.4).   (5.7)

Equations (5.1), (5.2), and (5.3) are the basic equations for counterflow and parallel-flow heat exchangers. Hence, any combination of three unknowns among all parameters (T_1, T_2, t_1, t_2, A_o, and q) can be solved for. The heat transfer area is

$$A_1 = P_1 L$$ (5.8)

or

$$A_2 = P_2 L$$ (5.8a)

where P_1 and P_2 are perimeters of hot and cold fluid channels, respectively.

5.2.2   Overall Heat Transfer Coefficient

We construct a thermal circuit across a wall between hot and cold fluids, as shown in Figure 5.5. The temperature difference ($T_{1i-1o} - T_{2i-2o}$) seems complex, varying along the length, which can be represented by the log mean temperature difference ΔT_{lm} (the

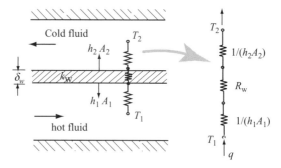

Figure 5.5   Thermal resistance and thermal circuit for a heat exchanger

derivation will be discussed in the next section). The heat transfer rate across the wall is:

$$q = \frac{T_{1i-1o} - T_{2i-2o}}{\dfrac{1}{UA}} = \frac{F\Delta T_{lm}}{\dfrac{1}{h_1 A_1} + R_w + \dfrac{1}{h_2 A_2}} \tag{5.9}$$

where h_1 and h_2 are the heat transfer coefficients for the hot and cold fluids, respectively, and A_1 and A_2 are the heat transfer surface areas for the hot and cold fluids, respectively, and R_w is the wall thermal resistance.

For flat walls, the wall thermal resistance is

$$R_w = \frac{\delta_w}{k_w A_w} \tag{5.10}$$

where δ_w is the thickness of the flat wall, k_w is the thermal conductivity of the wall, and A_w is the heat transfer area of the wall, which is the same as A_1 or A_2 in this case.

For concentric tubes (double-pipe heat exchanger), the wall thermal resistance is

$$R_w = \frac{\ln\left(\dfrac{d_o}{d_i}\right)}{2\pi k_w L} \tag{5.11}$$

where d_i and d_o are the inner and outer diameters of the circular wall and L is the tube length.

The overall heat transfer coefficient for the cold fluid with the heat transfer area A_2 is

$$U_2 = \frac{1/A_2}{\dfrac{1}{h_1 A_1} + R_w + \dfrac{1}{h_2 A_2}} \tag{5.12}$$

The UA value is defined using Equation (5.4) by

$$UA = \frac{1}{\dfrac{1}{h_1 A_1} + R_w + \dfrac{1}{h_2 A_2}} \tag{5.12a}$$

For a double-pipe heat exchanger (concentric pipes), neglecting the wall conduction, the overall heat transfer coefficient for the outer surface of the inner pipe is

$$U_o = \frac{1}{\dfrac{d_o}{h_i d_i} + \dfrac{1}{h_o}} \tag{5.12b}$$

5.2.3   Log Mean Temperature Difference (LMTD)

The *log mean temperature difference* in Equation (5.5) for parallel flow is derived in this section. We consider a control volume of a differential element for hot fluid, as shown in Figure 5.6(a). The energy (enthalpy) entering the left side of the element is given as a product of the mass flow rate, the specific heat, and the hot fluid temperature. The energy (enthalpy) leaving the right side of the element is supposed to have a change (dT) in the temperature.

Applying the heat balance to the control volume for the hot fluid at steady state provides

$$\dot{m}_1 c_{p1} T_1 - \dot{m}_1 c_{p1} (T_1 + dT_1) - dq = 0 \tag{5.13}$$

Rearranging this gives

$$\frac{dq}{\dot{m}_1 c_{p1}} = -dT_1 \tag{5.14}$$

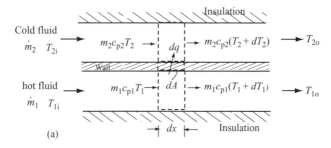

(a)

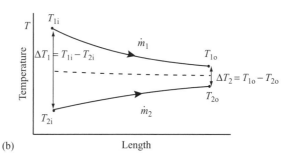

(b)

Figure 5.6   Parallel flow: (a) differential elements for parallel flow and (b) the temperature distributions

Also applying the heat balance to the control volume for the cold fluid and noting the direction of the differential heat transfer dq into the control volume provides

$$\dot{m}_2 c_{p2} T_2 - \dot{m}_1 c_{p1} (T_2 + dT_2) + dq = 0 \tag{5.15}$$

Rearranging Equation (5.15) gives

$$\frac{dq}{\dot{m}_2 c_{p2}} = dT_2 \tag{5.16}$$

Adding Equations (5.14) and (5.16) gives

$$dq \left(\frac{1}{\dot{m}_1 c_{p1}} + \frac{1}{\dot{m}_2 c_{p2}} \right) = -dT_1 + dT_2 = -d (T_1 - T_2) \tag{5.17}$$

From Figure 5.5, the local differential heat transfer can be formulated as

$$dq = \frac{T_1 - T_2}{\dfrac{1}{UdA}} = UdA (T_1 - T_2) \tag{5.18}$$

Inserting Equation (5.18) into Equation (5.17) yields

$$UdA(T_1 - T_2) \left(\frac{1}{\dot{m}_1 c_{p1}} + \frac{1}{\dot{m}_2 c_{p2}} \right) = -d (T_1 - T_2) \tag{5.19}$$

Rearranging this gives

$$\frac{-d (T_1 - T_2)}{T_1 - T_2} = U \left(\frac{1}{\dot{m}_1 c_{p1}} + \frac{1}{\dot{m}_2 c_{p2}} \right) dA \tag{5.20}$$

The inlet temperature difference is $T_{1i} - T_{2i}$ and the outlet temperature difference is $T_{1o} - T_{2o}$. Integrating the both sides of Equation (5.20) gives

$$\int_{T_{1i}-T_{2i}}^{T_{2o}-T_{1o}} \frac{-d (T_1 - T_2)}{T_1 - T_2} = U \left(\frac{1}{\dot{m}_1 c_{p1}} + \frac{1}{\dot{m}_2 c_{p2}} \right) \int_A dA \tag{5.21}$$

which yields

$$- \ln \left(\frac{T_{1o} - T_{2o}}{T_{1i} - T_{2i}} \right) = U \left(\frac{1}{\dot{m}_1 c_{p1}} + \frac{1}{\dot{m}_2 c_{p2}} \right) A \tag{5.22}$$

Equations (5.1) and (5.2) are rearranged for the inverse of the product of the mass flow rate and the specific heat, which are substituted into Equation (5.22).

$$- \ln \left(\frac{T_{1o} - T_{2o}}{T_{1i} - T_{2i}} \right) = UA \left(\frac{T_{1i} - T_{1o}}{q} + \frac{T_{2o} - T_{2i}}{q} \right)$$

$$= UA \left(\frac{(T_{1i} - T_{2i}) - (T_{1o} - T_{2o})}{q} \right) \tag{5.23}$$

Solving for q provides

$$q = UA\frac{\Delta T_1 - \Delta T_2}{\ln\left(\dfrac{\Delta T_1}{\Delta T_2}\right)} \tag{5.24}$$

where $\Delta T_1 = T_{1i} - T_{2i}$ and $\Delta T_2 = T_{1o} - T_{2o}$ for parallel flow (Figure 5.6). We can obtain Equation (5.24) in a similar way for counterflow. Hence, we generally define the log mean temperature difference as

$$\Delta T_{lm} = \frac{\Delta T_1 - \Delta T_2}{\ln\left(\dfrac{\Delta T_1}{\Delta T_2}\right)} \tag{5.25}$$

where

$$\Delta T_1 = T_{1i} - T_{2i} \text{ and } \Delta T_2 = T_{1o} - T_{2o} \text{ for parallel flow (Figure 5.6).}$$
$$\Delta T_1 = T_{1i} - T_{2o} \text{ and } \Delta T_2 = T_{1o} - T_{2i} \text{ for counter flow (Figure 5.4).}$$

Equation (5.24) is the same as Equation (5.3) if F = 1.

5.2.4   Flow Properties

The noncircular diameters in the flow channels are approximated using the hydraulic diameter D_h for the Reynolds number and the equivalent diameter D_e for the Nusselt number.

The *hydraulic diameter* is defined as

$$D_h = \frac{4A_c}{P_{wetted}} = \frac{4A_cL}{P_{wetted}L} = \frac{4A_cL}{A_t} \tag{5.26}$$

where P_{wetted} is the wetted perimeter, A_t the total heat transfer area, and L the length of the channel. The *mass velocity G* is defined as

$$G = \rho u_m \tag{5.27}$$

where ρ is the density of the fluid and u_m is the mean velocity of the fluid. The mass flow rate $\dot{m}$ is defined as

$$\dot{m} = \rho u_m A_c = G A_c \tag{5.28}$$

Then, the *Reynolds number* is expressed as

$$Re_D = \frac{\rho u_m D_h}{\mu} = \frac{\dot{m} D_h}{A_c \mu} = \frac{G D_h}{\mu} \tag{5.29}$$

where D_h is the hydraulic diameter, μ is the absolute viscosity, and A_c is the cross-sectional flow area. Note that the flow pattern is laminar when $Re_D < 2300$ and is turbulent when $Re_D > 2{,}300$. The *equivalent diameter*, which is often used for the heat transfer calculations, is defined as

$$D_e = \frac{4A_c}{P_{heated}} \tag{5.30}$$

where P_{heated} is the heated perimeter.

5.2.5 Nusselt Numbers

An empirical correlation was developed by Sieder and Tate [6] to predict the mean Nusselt number for laminar flow in a circular duct for the combined entry length with constant wall temperature. The average Nusselt number is

$$Nu_D = \frac{hD_e}{k_f} = 1.86 \left(\frac{D_h Re_D Pr}{L} \right)^{\frac{1}{3}} \left(\frac{\mu}{\mu_s} \right)^{0.14} \tag{5.31}$$

$$0.48 < Pr < 16{,}700$$

$$0.0044 < (\mu/\mu_s) < 9.75$$

Use $Nu_D = 3.66$ if $Nu_D < 3.66$

All the properties have been evaluated at the mean temperatures $T_{1m} = (T_{1i} + T_{1o})/2$ for a hot fluid or $T_{2m} = (T_{2i} + T_{2o})/2$ for a cold fluid except μ_s that is evaluated at the wall surface temperature.

Gnielinski [7] recommended the following correlation valid over a large Reynolds number range, including the transition region. The Nusselt number for turbulent is

$$Nu_D = \frac{hD_e}{k_f} = \frac{(f/2)(Re_D - 1000)Pr}{1 + 12.7(f/2)^{1/2}(Pr^{2/3} - 1)} \tag{5.32}$$

$$3000 < Re_D < 5 \times 10^6 \ [4]$$

$$0.5 \le Pr \le 2000$$

where the friction factor f is obtained assuming that the surface is smooth:

$$f = (1.58 \ln(Re_D) - 3.28)^{-2} \tag{5.33}$$

5.2.6 Effectiveness–NTU (ε-NTU) Method

When the heat transfer rate is not known or the outlet temperatures are not known, tedious iterations with the LMTD method are required. In an attempt to eliminate the iterations, Kays and London in 1955 developed a new method called the *ε-NTU* method. Current practice tends to favor the effectiveness approach because both effectiveness, ε, and the number of transfer units, *NTU*, have a unique physical significance for a given exchanger and given flow thermal capacities.

The *heat capacity rate* is defined as the product of mass flow rate and specific heat ($C_1 = \dot{m}_1 c_{p1}$). The *minimum capacity rate* is the one that has a lesser capacity rate. The *maximum capacity rate* is then the one that has a higher capacity rate. As shown in Figure 5.7, the minimum capacity curve always approaches the maximum capacity curve because the lower capacity fluid more quickly gains or loses thermal energy compared to the high capacity fluid. Considering both the maximum temperature difference $(T_{1i} - T_{2i})$ and the minimum heat capacity, the maximum possible heat transfer rate is defined by

$$q_{max} = \left(\dot{m} c_p \right)_{min} (T_{1i} - T_{2i}) \tag{5.34}$$

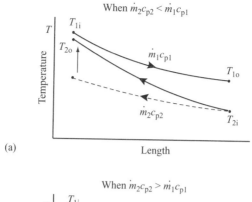

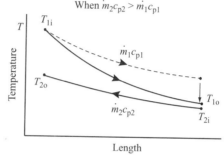

Figure 5.7 Maximum possible heat transfer rate, (a) when $\dot{m}_2 c_{p2} < \dot{m}_1 c_{p1}$, (b) $\dot{m}_2 c_{p2} > \dot{m}_1 c_{p1}$

The heat exchanger *effectiveness* ε is then given by

$$\varepsilon = \frac{\text{Actual heat transfer rate}}{\text{Maximum possible heat transfer rate}} = \frac{q}{q_{max}} \tag{5.35}$$

Using Equations (5.34) and (5.35), the actual heat transfer rate q is

$$q = \varepsilon \cdot (\dot{m} c_p)_{min} (T_{1i} - T_{2i}) \tag{5.35a}$$

The *heat transfer unit* (NTU) is defined as

$$NTU = \frac{UA}{(\dot{m} c_p)_{min}} \tag{5.36}$$

The *heat capacity* ratio C_r is defined as

$$C_r = \frac{(\dot{m} c_p)_{min}}{(\dot{m} c_p)_{max}} \tag{5.37}$$

Consider a parallel-flow heat exchanger for which $\dot{m}_2 c_{p2} > \dot{m}_1 c_{p1}$, or equivalently $(\dot{m} c_p)_{min} = \dot{m}_1 c_{p1}$. From Equation (5.37) with Equations (5.1) and (5.2), we obtain

$$C_r = \frac{(\dot{m} c_p)_{min}}{(\dot{m} c_p)_{max}} = \frac{T_{2o} - T_{2i}}{T_{1i} - T_{1o}} \tag{5.37a}$$

From Equation (5.35) with Equations (5.1) and (5.2), we can express

$$\varepsilon = \frac{q}{q_{max}} = \frac{(\dot{m}_1 c_{p1})(T_{1i} - T_{1o})}{(\dot{m}c_p)_{min}(T_{1i} - T_{2i})} = \frac{(\dot{m}_2 c_{p2})(T_{2o} - T_{2i})}{(\dot{m}c_p)_{min}(T_{1i} - T_{2i})} \tag{5.38}$$

$$\varepsilon = \frac{q}{q_{max}} = \frac{T_{1i} - T_{1o}}{T_{1i} - T_{2i}} \tag{5.38a}$$

5.2.6.1   Parallel Flow

Rearranging Equation (5.22) for *parallel flow* gives

$$\ln\left(\frac{T_{1o} - T_{2o}}{T_{1i} - T_{2i}}\right) = -UA\left(\frac{1}{(\dot{m}c_p)_{min}} + \frac{1}{(\dot{m}c_p)_{max}}\right)$$

$$= -\frac{UA}{(\dot{m}c_p)_{min}}\left(1 + \frac{(\dot{m}c_p)_{min}}{(\dot{m}c_p)_{max}}\right) \tag{5.39}$$

Using Equations (5.36) and (5.37), we have

$$\frac{T_{1o} - T_{2o}}{T_{1i} - T_{2i}} = \exp\left[-NTU(1 + C_r)\right] \tag{5.40}$$

Rearranging the left-hand side of Equation (5.40), we have

$$\frac{T_{1o} - T_{2o}}{T_{1i} - T_{2i}} = \frac{T_{1o} - T_{1i} + T_{1i} - T_{2o}}{T_{1i} - T_{2i}} \tag{5.41}$$

Using Equation (5.37a) and solving for T_{2o}, we have

$$T_{2o} = C_r(T_{1i} - T_{1o}) + T_{2i} \tag{5.42}$$

Inserting Equation (5.42) into Equation (5.41) gives

$$\frac{T_{1o} - T_{2o}}{T_{1i} - T_{2i}} = \frac{T_{1o} - T_{1i} + T_{1i} - C_r(T_{1i} - T_{1o}) - T_{2i}}{T_{1i} - T_{2i}}$$

$$= \frac{-(T_{1i} - T_{1o})(1 + C_r) + T_{1i} - T_{2i}}{T_{1i} - T_{2i}} \tag{5.43}$$

Inserting Equation (5.38a) into Equation (5.43) gives

$$\frac{T_{1o} - T_{2o}}{T_{1i} - T_{2i}} = -\varepsilon(1 + C_r) + 1 \tag{5.44}$$

Combining Equations (5.40) and (5.44), the heat exchanger effectiveness ε for parallel flow is obtained

$$\varepsilon = \frac{1 - \exp\left[-NTU(1 + C_r)\right]}{1 + C_r} \tag{5.45}$$

where *NTU* and C_r are referred to Equations (5.36) and (5.37).

Solving for *NTU* for parallel flow, we have

$$NTU = -\frac{1}{1+C_r} \ln\left[1 - \varepsilon\left(1 + C_r\right)\right] \tag{5.46}$$

Since the same result is obtained for $\dot{m}_2 c_{p2} < \dot{m}_1 c_{p1}$ or equivalently $\left(\dot{m}c_p\right)_{min} = \dot{m}_2 c_{p2}$, Equation (5.45) applies for any case of parallel-flow heat exchanger.

5.2.6.2   *Counterflow*

Based on a completely analogous analysis, the heat exchanger effectiveness for *counterflow* is obtained:

$$\varepsilon = \frac{1 - \exp\left[-NTU\left(1 - C_r\right)\right]}{1 - C_r \exp\left[-NTU\left(1 - C_r\right)\right]} \tag{5.47}$$

Solving for NTU for counterflow, we have

$$NTU = \frac{1}{1 - C_r} \ln\left(\frac{1 - \varepsilon C_r}{1 - \varepsilon}\right) \tag{5.48}$$

Using Equation (5.34) and (5.35), the actual heat transfer rate is expressed in terms of the effectiveness, inlet temperatures, and a minimum heat capacity rate as

$$q = \varepsilon \left(\dot{m}c_p\right)_{min} \left(T_{1i} - T_{2i}\right) \tag{5.49}$$

5.2.6.3   *Crossflow*

Consider a mixed-unmixed *crossflow* heat exchanger. This flow arrangement and the idealized temperature conditions are pictured schematically in Figure 5.8. The uniform hot fluid enters the exchanger, mixes, and leaves uniformly with an decreased temperature, as shown. The uniform cold fluid enters and leaves at nonuniform temperatures without mixing, as shown. We consider a differential element as a control volume, shown in the dotted lines.

We first define the heat capacity rates C_1 and C_2 for hot and cold fluids, respectively, as

$$C_1 = \dot{m}_1 c_{p1} \text{ and } C_2 = \dot{m}_2 c_{p2} \tag{5.50}$$

A uniform distribution of the heat transfer surface area A, the frontal area A_{fr}, and the heat capacity rate C_2 provides this relationship:

$$\frac{dA_{fr}}{A_{fr}} = \frac{dA}{A} = \frac{dC_2}{C_2} \tag{5.51}$$

The differential heat transfer rate for the element is given by an enthalpy flow as

$$dq = dC_2 \left(T_{2o} - T_{2i}\right) \tag{5.52}$$

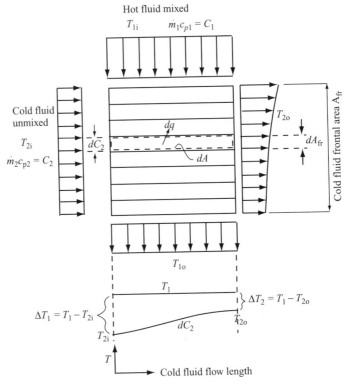

Figure 5.8 Temperature conditions for a crossflow exchanger, one fluid mixed, one unmixed [8]

The differential heat transfer rate for the element can also be expressed in terms of the overall heat transfer coefficient:

$$dq = U\,dA\frac{\Delta T_1 - \Delta T_2}{\ln\left(\dfrac{\Delta T_1}{\Delta T_2}\right)} = U\,dA\frac{(T_1 - T_{2i}) - (T_1 - T_{2o})}{\ln\left(\dfrac{T_1 - T_{2i}}{T_1 - T_{2o}}\right)} = U\,dA\frac{T_{2o} - T_{2i}}{\ln\left(\dfrac{T_1 - T_{2i}}{T_1 - T_{2o}}\right)}$$

(5.53)

Combining Equations (5.52) and (5.53) yields

$$\ln\left(\frac{T_1 - T_{2i}}{T_1 - T_{2o}}\right) = \frac{U\,dA}{dC_2}$$

(5.54)

Inverting the fraction in the left-hand side of Equation (5.54) leads to a minus sign in the right-hand side:

$$\ln\left(\frac{T_1 - T_{2o}}{T_1 - T_{2i}}\right) = -\frac{U\,dA}{dC_2}$$

(5.55)

Eliminating the logarithm gives

$$\frac{T_1 - T_{2o}}{T_1 - T_{2i}} = \exp\left(-\frac{U\,dA}{dC_2}\right)$$

(5.56)

Using the relationship in Equation (5.51), the temperature ratio is constant because U, A, and C_2 are constant. Denoting the ratio by Γ,

$$\frac{T_1 - T_{2o}}{T_1 - T_{2i}} = \exp\left(-\frac{UA}{C_2}\right) = \text{Const} = \Gamma \tag{5.57}$$

The differential heat transfer rate for the element can be written for the hot fluid as

$$dq = -dC_1 dT_1 \tag{5.58}$$

We rewrite the left-hand side of Equation (5.57) as

$$\frac{T_1 - T_{2o}}{T_1 - T_{2i}} = \frac{(T_1 - T_{2i}) - (T_{2o} - T_{2i})}{T_1 - T_{2i}} = 1 - \frac{T_{2o} - T_{2i}}{T_1 - T_{2i}} = \Gamma \tag{5.59}$$

and

$$T_{2o} - T_{2i} = (1 - \Gamma)(T_1 - T_{2i}) \tag{5.60}$$

From Equation (5.52), we now have

$$dq = (1 - \Gamma)(T_1 - T_{2i}) dC_2 \tag{5.61}$$

Combining Equations (5.58) and (5.61)

$$-C_1 dT_1 = (1 - \Gamma)(T_1 - T_{2i}) dC_2 \tag{5.62}$$

Rearranging this and using Equation (5.51) gives

$$\frac{dT_1}{(T_1 - T_{2i})} = -(1 - \Gamma)\frac{C_2}{C_1}\frac{1}{A_{fr}} dA_{fr} \tag{5.63}$$

Note that Γ, C_1, C_2, and A_{fr} are constant. Integration then yields

$$\int_{T_{1i}}^{T_{1o}} \frac{1}{T_1 - T_{2i}} dT_1 = -(1 - \Gamma)\frac{C_2}{C_1}\frac{1}{A_{fr}} \int_0^{A_{fr}} dA_{fr} \tag{5.64}$$

Knowing that T_1 is variable while T_{2i} is not,

$$dT_1 = d(T_1 - T_{2i}) \tag{5.65}$$

Equation (5.64) gives

$$\ln\left(\frac{T_{1o} - T_{2i}}{T_{1i} - T_{2i}}\right) = -(1 - \Gamma)\frac{C_2}{C_1} \tag{5.66}$$

Eliminating the logarithm gives

$$\frac{T_{1o} - T_{2i}}{T_{1i} - T_{2i}} = \exp\left(-(1 - \Gamma)\frac{C_2}{C_1}\right) \tag{5.67}$$

For $C_1 = C_{min}$ (*mixed*), Equation (5.37) becomes

$$C_r = \frac{C_{min}}{C_{max}} = \frac{C_1}{C_2} \tag{5.68}$$

From the definition of ε-*NTU* of Equation (5.38),

$$\varepsilon = \frac{q}{q_{max}} = \frac{C_1 (T_{1i} - T_{1o})}{C_{min} (T_{1i} - T_{2i})} = \frac{T_{1i} - T_{1o}}{T_{1i} - T_{2i}} \tag{5.69}$$

which is expended with Equation (5.67) as

$$\varepsilon = \frac{T_{1i} - T_{2i} - (T_{1o} - T_{2i})}{T_{1i} - T_{2i}} = 1 - \frac{T_{1o} - T_{2i}}{T_{1i} - T_{2i}} = 1 - \exp\left(-(1 - \Gamma)\frac{C_2}{C_1}\right) \tag{5.70}$$

Substituting Equation (5.57) gives

$$\varepsilon = 1 - \exp\left(-\left(1 - \exp\left(-\frac{UA}{C_2}\right)\right)\frac{C_2}{C_1}\right) \tag{5.71}$$

Using Equations (5.36) and (5.68),

$$NTU = \frac{UA}{C_{min}} = \frac{UA}{C_1} = \frac{UA}{C_2}\frac{C_2}{C_1} = \frac{UA}{C_2}\frac{1}{C_r} \tag{5.72}$$

The effectiveness for C_{min} (*mixed*) is finally expressed as

$$\varepsilon = 1 - \exp\left\{-\frac{1}{C_r}[1 - \exp(-C_r NTU)]\right\} \tag{5.73}$$

For $C_2 = C_{min}$ (*unmixed*), Equation (5.37) becomes

$$C_r = \frac{C_{min}}{C_{max}} = \frac{C_2}{C_1} \tag{5.74}$$

From the definition of ε-*NTU* of Equation (5.38),

$$\varepsilon = \frac{q}{q_{max}} = \frac{C_1 (T_{1i} - T_{1o})}{C_{min} (T_{1i} - T_{2i})} = \frac{C_1}{C_2}\frac{C_2}{C_{min}}\frac{T_{1i} - T_{1o}}{T_{1i} - T_{2i}} = \frac{1}{C_r}\frac{T_{1i} - T_{1o}}{T_{1i} - T_{2i}} \tag{5.75}$$

Using Equation (5.73), the effectiveness for C_{min} (*unmixed*) is

$$\varepsilon = \frac{1}{C_r}\left(1 - \exp\left\{-\frac{1}{C_r}[1 - \exp(-C_r NTU)]\right\}\right) \tag{5.76}$$

For *both fluids unmixed*, each fluid stream is assumed to have been divided into a large number of separate flow tubes for passage through the heat exchanger with no cross mixing. The numerical approaches provide an expression for the effectiveness based on Kays and London [8] and Mason [14]. The effectiveness for both fluids unmixed is

$$\varepsilon = 1 - \exp\left\{\left(\frac{1}{C_r}\right)NTU^{0.22}\left[\exp\left(-C_r \cdot NTU^{0.78}\right) - 1\right]\right\} \tag{5.77}$$

For a special case that $C_r = 0$, the crossflow effectiveness is indeterminate and the effectiveness for all exchangers with $C_r = 0$ is given as

$$\varepsilon = 1 - \exp(-NTU) \tag{5.78}$$

Note that for $C_r = 0$, as for an evaporator or condenser, the effectiveness is given by Equation (5.78) for all flow arrangements. Hence, for this special case, it follows that heat exchanger behavior is independent of flow arrangement.

Table 5.1 gives a summary for ε-NTU relationship for a large variety of configurations and Table 5.2 gives a summary of NTU-ε relations.

The heat exchanger effectiveness for parallel flow, counterflow, and crossflow is plotted in Figures 5.9, 5.10, and 5.11, respectively. For a given NTU and capacity ratio C_r, the counterflow heat exchanger shows higher effectiveness than the parallel-flow and crossflow heat exchangers. The effectiveness is independent of the capacity ratio C_r for NTU of less than 0.3. The effectiveness increases rapidly with NTU for small values up to 1.5 but rather slowly for larger values. Therefore, the use of a heat exchanger with an NTU of 3 or more and thus, a large size, cannot be justified economically.

5.2.7 Heat Exchanger Pressure Drop

The thermal design of a heat exchanger is aimed at calculating a surface area adequate to handle the thermal duty for the given specifications. Fluid friction effects in the heat exchanger are important because they determine the pressure drop of the fluids flowing in the system, and consequently the pumping power or fan work input necessary to maintain the flow. Heat transfer enhancement in heat exchangers is usually accompanied by increased pressure drop, and thus higher pumping power. Therefore,

Table 5.1 Heat Exchanger Effectiveness (ε)

Flow arrangement	Effectiveness	
Parallel flow	$\varepsilon = \dfrac{1 - \exp\left[-NTU\left(1 + C_r\right)\right]}{1 + C}$	(5.45)
Counterflow	$\varepsilon = \dfrac{1 - \exp\left[-NTU\left(1 - C_r\right)\right]}{1 - C_r \exp\left[-NTU\left(1 - C_r\right)\right]}$	(5.47)
Crossflow (single pass)		
Both fluids unmixed	$\varepsilon = 1 - \exp\left\{\left(\dfrac{1}{C_r}\right) NTU^{0.22}\left[\exp\left(-C_r \cdot NTU^{0.78}\right) - 1\right]\right\}$	(5.77)
$C_{\max}$ mixed, $C_{\min}$ unmixed	$\varepsilon = \dfrac{1}{C_r}\left(1 - \exp\left\{-C_r\left[1 - \exp\left(-NTU\right)\right]\right\}\right)$	(5.73)
$C_{\min}$ mixed, $C_{\max}$ unmixed	$\varepsilon = 1 - \exp\left\{-\dfrac{1}{C_r}\left[1 - \exp\left(-C_r \cdot NTU\right)\right]\right\}$	(5.76)
All exchangers ($C_r = 0$)	$\varepsilon = 1 - \exp\left(-NTU\right)$	(5.78)

Table 5.2 Heat Exchanger NTU

Flow arrangement	NTU	
Parallel flow	$NTU = -\dfrac{1}{1 + C_r} \ln\left[1 - \varepsilon\left(1 + C_r\right)\right]$	(5.79)
Counterflow	$NTU = \dfrac{1}{1 - C_r} \ln\left(\dfrac{1 - \varepsilon C_r}{1 - \varepsilon}\right)$	(5.80)
Crossflow (single pass)		
Both fluid unmixed		
$C_{\max}$ mixed, $C_{\min}$ unmixed	$NTU = -\ln\left[1 + \dfrac{1}{C_r} \ln\left(1 - \varepsilon C_r\right)\right]$	(5.81)
$C_{\min}$ mixed, $C_{\max}$ unmixed	$NTU = -\dfrac{1}{C_r} \ln\left[1 + C_r \cdot \ln\left(1 - \varepsilon\right)\right]$	(5.82)
All exchangers ($C_r = 0$)	$NTU = -\ln\left(1 - \varepsilon\right)$	(5.83)

any gain from the enhancement in heat transfer should be weighed against the cost of the accompanying pressure drop. Usually, the more viscous fluid is more suitable for the shell side (high passage area and thus lower pressure drop) and the fluid with the higher pressure is more suitable for the tube side.

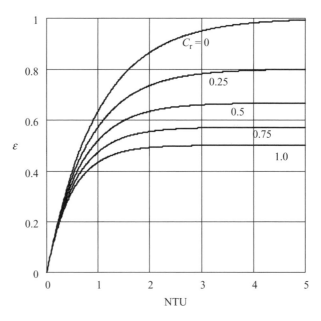

Figure 5.9 Effectiveness of a parallel-flow heat exchanger, Equation (5.45)

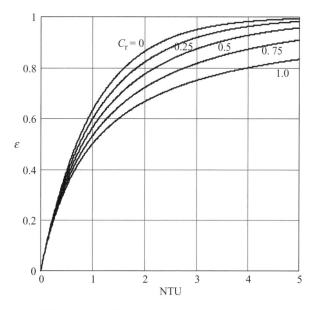

Figure 5.10   Effectiveness of a counterflow heat exchanger, Equation (5.47)

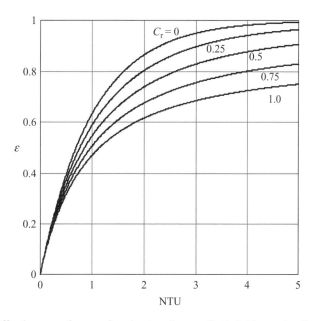

Figure 5.11   Effectiveness of a crossflow heat exchanger (both fluids unmixed), Equation (5.77)

The *power of a pump* or fan may be calculated by

$$\dot{W} = \frac{\dot{m}}{\rho} \Delta P \tag{5.84}$$

With the *pump efficiency* η_p, we have the actual pump power:

$$\dot{W}_{actual} = \frac{\dot{m}}{\eta_p \rho} \Delta P \tag{5.85}$$

Using the *Fanning friction factor* in a duct, we have

$$f_{Fanning} = \frac{\tau_w}{\frac{1}{2}\rho u_m^2} \tag{5.86}$$

Using the *Darcy friction factor*, we have

$$f_{Darcy} = \frac{4\tau_w}{\frac{1}{2}\rho u_m^2} \tag{5.86a}$$

We adopt herein the Fanning friction factor. For *laminar flow*, the friction factor is found analytically as

$$f = \frac{16}{Re_D} \tag{5.87}$$

where the Re_D was defined in Equation (5.29). The Fanning friction factor f is presented graphically in Figure 5.12 curve-fitted from the experimental data for fully developed flow, originally provided by Moody [16]. For smooth circular ducts for *turbulent flow*, the friction factor by Filonenko [15] for $10^4 < Re_D < 10^7$ is given by

$$f = (1.58 \ln (Re_D) - 3.28)^{-2} \tag{5.88}$$

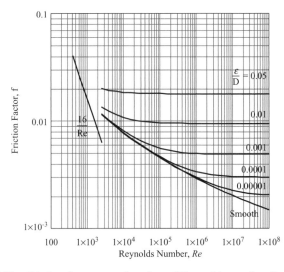

Figure 5.12 Friction factor as a function of Reynolds number for pipe flow

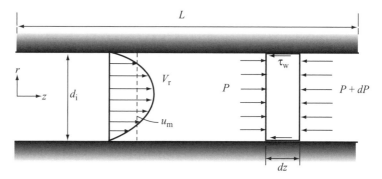

Figure 5.13 A fully developed flow in a duct

Consider the force balance for a small element in a circular duct assuming a fully developed flow as shown in Figure 5.13. Since the flow is fully developed, the sum of forces on the element is zero. Hence, we have

$$P\left(\frac{\pi d_i^2}{4}\right) - (P + dP)\left(\frac{\pi d_i^2}{4}\right) - \tau_w\,(\pi d_i)\,dz = 0 \qquad (5.89)$$

which reduces to

$$dP = \frac{4\tau_w}{d_i}dz \qquad (5.90)$$

Using Equation (5.86) and rearranging gives

$$dP = \frac{4f}{d_i}\frac{1}{2}\rho u_m^2 dz \qquad (5.91)$$

Integrating both sides of Equation (5.91) over the length L of the duct and rearranging gives the pressure drop along the duct.

$$\Delta P = \frac{4fL}{d_i}\frac{1}{2}\rho u_m^2 \qquad (5.92)$$

or for noncircular ducts using the hydraulic diameter D_h defined in Equation (5.26), we have a general form of the pressure drop for a circular or noncircular duct over the length L as

$$\Delta P = \frac{4fL}{D_h}\frac{1}{2}\rho u_m^2 = \frac{2fL}{D_h}\frac{G^2}{\rho} \qquad (5.93)$$

where G is the mass velocity.

5.2.8 Fouling Resistances (Fouling Factors)

When a heat exchanger is in service for a certain amount of time, scale and dirt will deposit on the surfaces of the tubes, as shown in Figure 5.14. These deposits reduce the

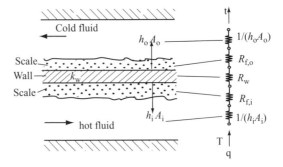

Figure 5.14 Thermal circuit with fouling for a heat exchanger

heat transfer rate and increase the pressure drop and pumping power as well. The heavy fouling fluid should be kept on the tube side for cleanability. Most often, the influence of fouling is included through an overdesign. In some applications, this overdesign accelerates fouling because of the low fluid velocity in the exchanger.

The fouling resistances on the inside and outside surfaces are denoted as $R_{f,i}$ and $R_{f,o}$. They affect the overall heat transfer coefficient defined earlier in Equation (5.12). The overall heat transfer coefficient with fouling is expressed as

$$U_o = \frac{1/A_o}{\dfrac{1}{h_i A_i} + \dfrac{R_{f,i}}{A_i} + R_w + \dfrac{R_{f,o}}{A_o} + \dfrac{1}{h_o A_o}} \tag{5.94}$$

Table 5.3 gives some representative values for fouling resistance per unit area. Clearly, the time–dependent nature of the fouling problem is such that it is very difficult to reliably estimate the overall heat transfer coefficient if fouling resistance is dominant. For high heat transfer applications, fouling may even dictate the design of the heat exchanger.

5.2.9   Overall Surface (Fin) Efficiency

Multiple fins are often used to increase the heat transfer area, as pictured in Figure 5.15.

Single fin efficiency presented in Chapter 2 is rewritten here for convenience. The *overall surface (fin) efficiency* is readily expressed in terms of the single fin efficiency, the fin, and primary (interfins) areas. The thermal analysis is then greatly simplified by the overall surface efficiency. We consider a multiple-finned plate (rectangular fin geometry) in both sides as shown in Figure 5.15(a). The single fin efficiency assuming adiabatic tips is given as

$$\eta_f = \frac{\tanh(mb)}{mb} \tag{5.95}$$

where b is the profile length and m is defined as

$$m = \sqrt{\frac{hP}{kA_c}} = \sqrt{\frac{h2(L+\delta)}{kL\delta}} = \sqrt{\frac{2h}{k\delta}\left(1 + \frac{\delta}{L}\right)} \cong \sqrt{\frac{2h}{k\delta}} \tag{5.96}$$

Table 5.3   Recommended Values of Fouling Resistances [9,10]

Fluid	Fouling resistance, $R_F \times 10^3 \, m^2 \cdot K/W$
Engine lube oil	0.176
Fuel oil	0.9
Vegetable oil	0.5
Gasoline	0.2
Kerosene	0.2
Refrigerant liquids	0.2
Refrigerant vapor (oil-bearing)	0.35
Engine exhaust gas	1.8
Steam	0.1
Compressed air	0.35
Sea water	0.1–0.2
Cooling tower water (treated)	0.2–0.35
Cooling tower water (untreated)	0.5–0.9
Ethylene glycol solutions	0.352
River water	0.2–0.7
Distilled water	0.1
Boiler water (treated)	0.1–0.2
City water or well water	0.2
Hard water	0.5
Methanol, ethanol, and ethylene glycol	0.4
Natural gas	0.2–0.4
Acid gas	0.4–0.5

where k is the thermal conductivity of the fin, h the convection coefficient, P the perimeter of the fin, δ the fin thickness, and A_c the cross-sectional area of the fin. It is assumed that $\delta \ll L$.

The single fin area A_f with the adiabatic tip is obtained as

$$A_f = 2(L + \delta)b \tag{5.97}$$

The total heat transfer area A_t is the sum of the fin area and the primary area.

$$A_t = n[2(L + \delta)b + Lz] \tag{5.98}$$

where n is the number of fins and z the fin spacing. The *overall surface efficiency* is given as

$$\eta_o = 1 - n\frac{A_f}{A_t}\left(1 - \eta_f\right) \tag{5.99}$$

The combined *thermal resistance* of the fin and primary surface area is given by

$$R_{t,o} = \frac{1}{\eta_o h A_t} \tag{5.100}$$

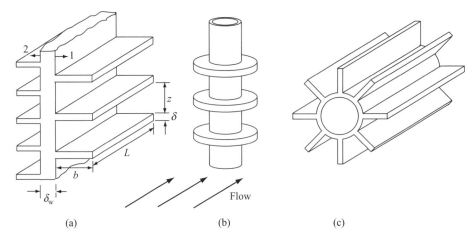

Figure 5.15   Extended fins: (a) plate-fin (rectangular fin), (b) circular finned-tube, and (c) longitudinal finned-tube

Considering the fin arrangement in Figure 5.15(a), the *overall heat transfer coefficient* based on the area A_{t1} is obtained as

$$U_{t1} = \cfrac{1/A_{t1}}{\cfrac{1}{\eta_{o1} h_1 A_{t1}} + R_w + \cfrac{1}{\eta_{o2} h_2 A_{t2}}} \tag{5.101}$$

Since the wall is flat, the wall resistance is given as

$$R_w = \frac{\delta_w}{k A_w} \tag{5.102}$$

where δ_w is the wall thickness and A_w the heat transfer area of the wall.

5.2.10   Reasonable Velocities of Various Fluids in Pipe Flow

With increasing fluid velocity in a pipe flow, the heat transfer rate usually increases, but the pressure drop also increases, increasing the cost of pumping. Therefore, an

Table 5.4   Reasonable velocities for various fluids in pipe flow

Fluid	Economic velocity range (m/s)	Fluid	Economic velocity range (m/s)
Acetone	1.5–3.0	Glycerin	0.43–0.86
Alcohol	1.5–3.0	Heptane	1.5–3.0
Benzene	1.4–2.8	Kerosene	1.4–2.8
Engine oil	0.5–1.0	Mercury	0.64–1.4
Ether	1.5–3.0	Propane	1.7–3.4
Ethylene glycol	1.2–2.4	Propylene glycol	1.4–2.8
R-11	1.2–2.4	Water	1.4–2.8

Sources adapted from Janna [5]

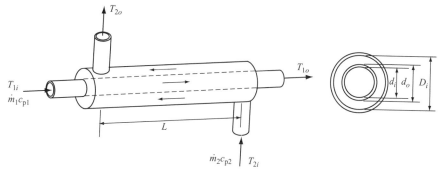

Figure 5.16   Schematic of a double pipe heat exchanger

optimum velocity exists. Furthermore, if the velocity is too high, it causes mechanical problems such as vibration and erosion. If the velocity is too low, it promotes fouling in the pipe. These facts lead to reasonable velocities for various fluids, which are shown in Table 5.4. The velocities in the table would be a good starting point for the design of a heat exchanger, but the velocities in a pipe flow are not strictly restricted or limited to these values.

5.3   DOUBLE-PIPE HEAT EXCHANGERS

A simple *double-pipe heat exchanger* consists of two concentric pipes as shown in Figure 5.16. One fluid flows in the inner pipe and the other fluid in the annulus between pipes in a counterflow direction for the ideal highest performance for the given surface area. However, if the application requires an almost constant wall temperature, the fluids may flow in a parallel direction. Double-pipe heat exchangers are typically suitable where one or both of the fluids are at very high pressure. Double-pipe exchangers are generally used for small–capacity applications where the total heat transfer surface required is 50 m² or less. One commercially available double-pipe heat exchanger is a hairpin exchanger shown in Figure 5.17, which can be stacked in series or series-parallel arrangements to meet the heat duty. The equations necessary for the rating and sizing problems are summarized in Table 5.5.

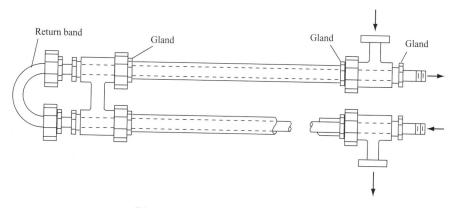

Figure 5.17   Hairpin heat exchanger

Table 5.5 Summary of Equations for a Double-pipe Heat Exchanger

Description	Equation	
Basic equations	$q = \dot{m}_1 c_{p1} (T_{1i} - T_{1o})$	(5.103)
	$q = \dot{m}_2 c_{p2} (T_{2o} - T_{2i})$	(5.104)
	$q = U_o A_o \Delta T_{lm}$	(5.105)
Log mean temperature difference	$\Delta T_{lm} = \dfrac{\Delta T_1 - \Delta T_2}{\ln\left(\dfrac{\Delta T_1}{\Delta T_2}\right)}$	(5.106)
For parallel flow	$\Delta T_1 = T_{1i} - T_{2i}$ and $\Delta T_2 = T_{1o} - T_{2o}$	(5.107)
For counterflow	$\Delta T_1 = T_{1i} - T_{2o}$ and $\Delta T_2 = T_{1o} - T_{2i}$	(5.108)
Heat transfer area (outer pipe)	$A_o = \pi \cdot d_o \cdot L$	(5.109)
Overall heat transfer coefficient	$U_o = \dfrac{1/A_o}{\dfrac{1}{h_i A_i} + \dfrac{\ln\left(\dfrac{d_o}{d_i}\right)}{2\pi k L} + \dfrac{1}{h_o A_o}}$	(5.110)
Reynolds number	$Re_D = \dfrac{\rho u_m D_h}{\mu} = \dfrac{\dot{m} D_h}{A_c \mu}$	(5.111)
Hydraulic diameter (annulus)	$D_h = \dfrac{4 A_c}{P_{wetted}} = \dfrac{4\pi (D_i^2 - d_o^2)/4}{\pi (D_i + d_o)} = D_i - d_o$	(5.112)
Equivalent diameter (annulus)	$D_e = \dfrac{4 A_c}{P_{heated}} = \dfrac{4\pi (D_i^2 - d_o^2)/4}{\pi d_o} = \dfrac{D_i^2 - d_o^2}{d_o}$	(5.113)
Laminar flow ($Re < 2{,}300$)	$Nu_D = \dfrac{h D_e}{k_f} = 1.86 \left(\dfrac{D_h Re_D Pr}{L}\right)^{\frac{1}{3}} \left(\dfrac{\mu}{\mu_s}\right)^{0.14}$ $0.48 < Pr < 16{,}700$ $0.0044 < (\mu/\mu_s) < 9.75$ Use $Nu_D = 3.66$ if $Nu_D < 3.66$	(5.114)
Turbulent flow ($Re > 2{,}300$)	$Nu_D = \dfrac{h D_e}{k_f} = \dfrac{(f/2)(Re_D - 1000) Pr}{1 + 12.7(f/2)^{1/2} (Pr^{2/3} - 1)}$ $3{,}000 < Re_D < 5 \times 10^6$ [4] $0.5 \le Pr \le 2000$	(5.115)

Table 5.5   (*continued*)

Description	Equation
Friction factor	$f = (1.58 \ln (Re_D) - 3.28)^{-2}$ turbulent

$$(5.116)$$

$$f = 16/Re_D \text{ laminar}$$

ε-NTU Method

Heat transfer unit (NTU)	$NTU = \dfrac{U_o A_o}{(\dot{m} c_p)_{min}}$	(5.117)
Heat capacity ratio	$C_r = \dfrac{(\dot{m} c_p)_{min}}{(\dot{m} c_p)_{max}}$	(5.118)
Effectiveness $\varepsilon = f(NTU, C)$	Parallel flow $\varepsilon = \dfrac{1 - \exp\left[-NTU(1 + C_r)\right]}{1 + C_r}$	(5.119)
	Counterflow $\varepsilon = \dfrac{1 - \exp\left[-NTU(1 - C_r)\right]}{1 - C_r \exp\left[-NTU(1 - C_r)\right]}$	(5.120)
$NTU = f(\varepsilon, C)$	Parallel flow $NTU = -\dfrac{1}{1 + C_r} \ln\left[1 - \varepsilon(1 + C_r)\right]$	(5.121)
	Counterflow $NTU = \dfrac{1}{1 - C_r} \ln\left(\dfrac{1 - \varepsilon C_r}{1 - \varepsilon}\right)$	(5.122)
Effectiveness ε	$\varepsilon = \dfrac{q}{q_{max}} = \dfrac{(\dot{m}_1 c_{p1})(T_{1i} - T_{1o})}{(\dot{m} c_p)_{min}(T_{1i} - T_{2i})} = \dfrac{(\dot{m}_2 c_{p2})(T_{2o} - T_{2i})}{(\dot{m} c_p)_{min}(T_{1i} - T_{2i})}$	

$$(5.123)$$

Actual heat transfer rate	$q = \varepsilon (\dot{m} c_p)_{min}(T_{1i} - T_{2i})$	(5.124)

Pressure Drop

Pressure drop	$\Delta P = \dfrac{4fL}{D_h} \dfrac{1}{2} \rho u_m^2$	(5.125)
Laminar flow	$f = 16/Re_D$	(5.126)
Turbulent flow	$f = (1.58 \ln (Re_D) - 3.28)^{-2}$	(5.127)

Example 5.3.1 Double-Pipe Heat Exchanger   A   counterflow   double-pipe   heat exchanger is used to cool the engine oil for a large engine, as shown in Figure E5.3.1. Oil at a flow rate of 0.82 kg/s is required to be cooled from 95°C to 90°C using water at a flow rate of 1.2 kg/s and at 25°C. A 7-m long carbon-steel hairpin is to be used (see Figure 5.17). The inner and outer pipes are 1 1/4 and 2 inches nominal schedule 40, respectively. The engine oil flows through the inner tube. How many hairpins

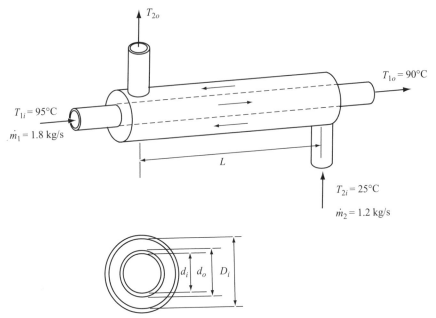

Figure E5.3.1 Double pipe heat exchanger (counterflow) and the cross-section with dimensions

will be required? When the heat exchanger is initially in service (no fouling) with the hairpins, determine the outlet temperature, the heat transfer rate, and the pressure drop for the exchanger.

MathCAD Format Solution Two methods are typically available to solve this problem, the LMTD method and ε-NTU method. However, we use MathCAD minimizing the approximations in calculations, so we would prefer to use the ε-NTU method, which includes the important parameters such as the effectiveness and NTU.

The properties of oil and water are obtained from Table A.12 in Appendix A with the average temperatures estimated assuming the water outlet temperature to be $30°C$.

$$T_{oil} := \frac{95°C + 90°C}{2} = 92.5 \cdot °C \qquad T_{water} := \frac{25°C + 30°C}{2} = 27.5 \cdot °C \quad (E5.3.1)$$

Engine oil (subscript 1)

$$\rho_1 := 848 \frac{kg}{m^3}$$

$$c_{p1} := 2161 \frac{J}{kg \cdot K}$$

$$k_1 := 0.137 \frac{W}{m \cdot K}$$

$$\mu_1 := 2.52 \cdot 10^{-2} \frac{N \cdot s}{m^2}$$

$$Pr_1 := 395$$

Water (subscript 2)

$$\rho_2 := 995 \frac{kg}{m^3}$$

$$c_{p2} := 4178 \frac{J}{kg \cdot K}$$

$$k_2 := 0.62 \frac{W}{m \cdot K}$$

$$\mu_2 := 769 \cdot 10^{-6} \frac{N \cdot s}{m^2}$$

$$Pr_2 := 5.2$$

(E5.3.2)

The mass flow rates given are defined

$$mdot_1 := 0.82 \, \frac{kg}{s} \qquad\qquad mdot_2 := 1.2 \, \frac{kg}{s} \qquad (E5.3.3)$$

The inlet and outlet temperatures given are defined

$$T_{1i} := 95°C \qquad T_{1o} := 90°C \qquad T_{2i} := 25°C \qquad (E5.3.4)$$

From Equations (5.103) and (5.104), the heat transfer rate and the water outlet temperature are readily calculated. The actual outlet temperature will be recalculated with a final number of hairpins (tube length).

$$T_{2o} := T_{2i} + \frac{q}{mdot_2 \cdot c_{p2}} \qquad T_{2o} = 26.767 \cdot °C \qquad (E5.3.5)$$

The pipe dimensions for the hairpin heat exchanger are obtained in Table C.1 in Appendix C.

1 1/4 nominal schedule 40 $d_i := 35.05 \, mm$ $d_o := 42.16 \, mm$
2 nominal schedule 40 $D_i := 52.50 \, mm$ $(E5.3.6)$

Since the pipe is made of carbon steel, the thermal conductivity is obtained in Table A.3 in Appendix A using an average wall temperature by

$$T_w := \frac{1}{2} \cdot \left(\left(\frac{95°C + 90°C}{2} + \frac{25°C + 30°C}{2} \right) \right) = 333.15 \, K \qquad (E5.3.7)$$

$$k_w := 56.7 \, \frac{W}{m \cdot K} \qquad (E5.3.8)$$

Initially assume the tube length L_t for iteration, starting with $L_t = 7$ m (one hairpin) and increasing the number of hairpin until T_{1o} meets 90°C, or slightly less.

$$L_t := 21 \, m \qquad (E5.3.9)$$

Calculate the cross-sectional areas for the tube and annulus.

$$A_{c1} := \frac{\pi \cdot d_i^2}{4} \qquad A_{c1} = 9.643 \times 10^{-4} \, m^2$$
$$A_{c2} := \frac{\pi}{4} \cdot \left(D_i^2 - d_o^2 \right) \qquad A_{c2} = 7.704 \times 10^{-4} \, m^2 \qquad (E5.3.10)$$

From Equations (5.112) and (5.113), the hydraulic diameter and the equivalent diameter for the annulus are calculated. The equivalent diameter will be used in calculation of the Nusselt number.

$$D_h := D_i - d_o \qquad D_h = 1.036 \, cm \qquad (E5.3.11)$$

$$D_e := \frac{D_i^2 - d_o^2}{d_o} \qquad D_e = 2.327 \, cm \qquad (E5.3.12)$$

From Equation (5.111), the Reynolds numbers are calculated, indicating that the oil flow is laminar while the water flow is turbulent, since the critical Reynolds number is 2,300.

$$Re_1 := \frac{mdot_1 \cdot d_i}{A_{c1} \cdot \mu_1} \qquad Re_1 = 1.182 \times 10^3$$

$$Re_2 := \frac{mdot_2 \cdot D_h}{A_{c2} \cdot \mu_2} \qquad Re_2 = 2.098 \times 10^4 \tag{E5.3.13}$$

The velocities can be calculated. Note that, for proper design, the velocities should not be too low to avoid fouling, nor too high to avoid vibration (typically less than 3 m/s for light viscous liquids; refer to Table 5.4).

$$v_1 := \frac{mdot_1}{\rho_1 \cdot A_{c1}} \qquad v_1 = 1.003 \frac{m}{s}$$

$$v_2 := \frac{mdot_2}{\rho_2 \cdot A_{c2}} \qquad v_2 = 1.565 \frac{m}{s} \tag{E5.3.14}$$

The friction factors are programmed to take into account either laminar or turbulent flow, using Equation (5.116).

$$f(Re_D) := \begin{vmatrix} (1.58 \cdot \ln(Re_D) - 3.28)^{-2} & \text{if } Re_D > 2300 \\ \dfrac{16}{Re_D} & \text{otherwise} \end{vmatrix} \tag{E5.3.15}$$

The Nusselt number is programmed for either turbulent or laminar flow using Equation (5.114) and (5.115), assuming μ changes moderately with temperature.

$$Nu_D(D_h, L_t, Re_D, Pr) := \begin{vmatrix} \left(\dfrac{f(Re_D)}{2}\right) \cdot \dfrac{(Re_D - 1000) \cdot Pr}{1 + 12.7 \cdot \left(\dfrac{f(Re_D)}{2}\right)^{0.5} \cdot \left(Pr^{\frac{2}{3}} - 1\right)} \\ \text{if } Re_D > 2300 \\ 1.86 \cdot \left(\dfrac{D_h \cdot Re_D \cdot Pr}{L_t}\right)^{\frac{1}{3}} \text{ otherwise} \end{vmatrix} \tag{E5.3.16}$$

The heat transfer coefficients are obtained as

$$h_1 := Nu_D(d_i, L_t, Re_1, Pr_1) \cdot \frac{k_1}{d_i} \qquad h_1 = 66.922 \cdot \frac{W}{m^2 \cdot K}$$

$$h_2 := Nu_D(D_h, L_t, Re_2, Pr_2) \cdot \frac{k_2}{D_e} \qquad h_2 = 3.658 \times 10^3 \cdot \frac{W}{m^2 \cdot K} \tag{E5.3.17}$$

The heat transfer coefficient in the oil side is an order smaller than that in the water side. The heat transfer areas are calculated as

$$A_i := \pi \cdot d_i \cdot L_t \qquad A_i = 2.312 \, m^2$$

$$A_o := \pi \cdot d_o \cdot L_t \qquad A_o = 2.781 \, m^2 \tag{E5.3.18}$$

The overall heat transfer coefficient is calculated using Equation (5.110)

$$UA_o := \frac{1}{\dfrac{1}{h_1 \cdot A_i} + \dfrac{\ln\left(\dfrac{d_o}{d_i}\right)}{2 \cdot \pi \cdot k_w \cdot L_t} + \dfrac{1}{h_2 \cdot A_o}} \qquad UA_o = 151.822 \cdot \dfrac{W}{K} \qquad \text{(E5.3.19)}$$

Note that the oil-side heat-transfer coefficient, h_1, is an order smaller than the water-side coefficient, h_2, and it dominates the overall UA value. This can be considerably improved with an extended heat transfer area such as fins. The ε-NTU method is used to determine the outlet temperatures. Define the heat capacities for oil and water flows.

$$C_1 := mdot_1 \cdot c_{p1} \qquad C_1 = 1.772 \times 10^3 \cdot \frac{W}{K}$$

$$C_2 := mdot_2 \cdot c_{p2} \qquad C_2 = 5.014 \times 10^3 \cdot \frac{W}{K} \qquad \text{(E5.3.20)}$$

Define the minimum and maximum heat capacities C_1 and C_2 for the ε-NTU method using the MathCAD functions. And define the heat capacity ratio C_r.

$$C_{min} := \min(C_1, C_2) \qquad C_{max} := \max(C_1, C_2) \qquad \text{(E5.3.21)}$$

$$C_r := \frac{C_{min}}{C_{max}} \qquad \text{(E5.3.22)}$$

Define the number of heat transfer unit NTU.

$$NTU := \frac{UA_o}{C_{min}} \qquad NTU = 0.086 \qquad \text{(E5.3.23)}$$

The effectiveness of the double-pipe heat exchanger for counterflow is calculated using Equation (5.120)

$$\varepsilon_{hx} := \frac{1 - \exp[-NTU \cdot (1 - C_r)]}{1 - C_r \cdot \exp[-NTU \cdot (1 - C_r)]} \qquad \varepsilon_{hx} = 0.081 \qquad \text{(E5.3.24)}$$

Using Equation (5.123), the effectiveness is given by

$$\varepsilon_{hx} = \frac{q}{q_{max}} = \frac{C_1 \cdot (T_{1i} - T_{1o})}{C_{min} \cdot (T_{1i} - T_{2i})} = \frac{C_2 \cdot (T_{2o} - T_{2i})}{T_{1i} - T_{2i}} \qquad \text{(E5.3.25)}$$

The actual outlet temperatures are calculated as

$$T_{1o} := T_{1i} - \varepsilon_{hx} \cdot \frac{C_{min}}{C_1} \cdot (T_{1i} - T_{2i}) \qquad T_{1o} = 89.333°C$$

$$T_{2o} := T_{2i} + \varepsilon_{hx} \cdot \frac{C_{min}}{C_2} \cdot (T_{1i} - T_{2i}) \qquad T_{2o} = 27.000°C \qquad \text{(E5.3.26)}$$

The heat transfer rate is

$$q := \varepsilon_{hx} \cdot C_{min} \cdot (T_{1i} - T_{2i}) \qquad q = 1.004 \times 10^4 \, W \qquad (E5.3.27)$$

The iteration between Equations (E5.3.9) and (E5.3.26) with increased tube length L_t (and number of hairpins) continues until the engine-oil outlet temperature reaches $T_{1o} = 90°C$ or slightly less.

The inlet temperatures are rewritten for comparing the outlet temperatures.

$$T_{1i} = 95°C \quad T_{2i} = 25°C$$

Once the oil outlet temperature is satisfied, the pressure drops for both fluids, as calculated using Equation (5.125). The allowable pressure drops depend on the types of fluids and the types of heat exchangers. For liquids, an allowance in the range of 50 to 140 kPa (7–20 psi) is commonly used. For gases, a value in the range of 7 to 30 kPa (1–5 psi) is often specified. An allowance of 70 kPa (10 psi) is widely used for a double-pipe heat exchanger.

$$\Delta P_1 := \frac{4 \cdot f(Re_1) \cdot L_t}{d_i} \cdot \frac{1}{2} \cdot \rho_1 \cdot v_1^2 \qquad\qquad \Delta P_1 = 13.831 \, kPa$$

$$(E5.3.28)$$

$$\Delta P_2 := \frac{4 \cdot f(Re_2) \cdot L_t}{D_h} \cdot \frac{1}{2} \cdot \rho_2 \cdot v_2^2 \qquad\qquad \Delta P_2 = 63.846 \, kPa$$

The number of hairpins for the requirement of oil outlet temperature less than $90°C$ is now found to be three (3).

$$L_t = 21 \, m$$

$$N_{hairpin} := \frac{L_t}{7 \, m} = 3$$

Comments: This example was to find the rating of an exchanger without fouling. In order to see the fouling effect after years of service, the fouling factors should be included in Equation (E5.3.19).

5.4 SHELL-AND-TUBE HEAT EXCHANGERS

The most common type of heat exchanger in industrial applications is the shell-and-tube heat exchanger. These exchangers hold more than 65 percent of the market share. Shell-and-tube heat exchangers typically provide a surface area density ranging from 50 to 500 m^2/m^3 and are easily cleaned. The design codes and standards are available from the Tubular Exchanger Manufacturers Association (TEMA). A simple exchanger, which involves one shell and one pass, is shown in Figure 5.18.

Table 5.6 provides a summary of equations used with shell-and-tube heat exchangers.

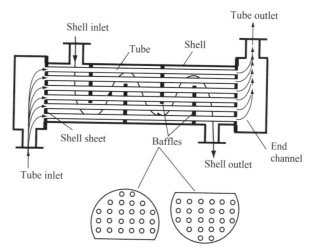

Figure 5.18   Schematic of one-shell one-pass (1-1) shell-and-tube heat exchanger

5.4.1   Baffles

In Figure 5.18, baffles are placed within the shell of the heat exchanger first to support the tubes, preventing tube vibration and sagging, and second to direct the flow to have a higher heat transfer coefficient. The distance between two baffles is called the *baffle spacing*.

5.4.2   Multiple Passes

Shell-and-tube heat exchangers can have multiple passes, such as 1-1, 1-2, 1-4, 1-6, and 1-8 exchangers, where the first number denotes the number of the shells and the second number denotes the number of passes. An odd number of tube passes is seldom used except the 1-1 exchanger. A 1-2 shell-and-tube heat exchanger is illustrated in Figure 5.19.

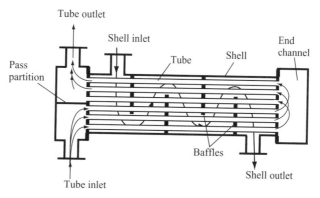

Figure 5.19   Schematic of one-shell two-pass (1-2) shell-and-tube heat exchanger

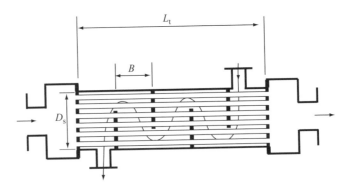

Figure 5.20 Dimensions of 1-1 shell-and-tube heat exchanger

5.4.3 Dimensions of Shell-and-Tube Heat Exchanger

Some of the following dimensions are pictured in Figure 5.20.

$$L_t = \text{tube length}$$
$$N_t = \text{number of tube}$$
$$N_p = \text{number of passes}$$
$$D_s = \text{Shell inside diameter}$$
$$N_b = \text{number of baffles}$$
$$B = \text{baffle spacing}$$

The baffle spacing is obtained as

$$B = \frac{L_t}{N_b + 1} \tag{5.128}$$

5.4.4 Shell-side Tube Layout

Figure 5.21 shows a cross-section of both a square and a triangular pitch layout. The tube pitch P_t and the clearance C_t between adjacent tubes are both defined. Equation

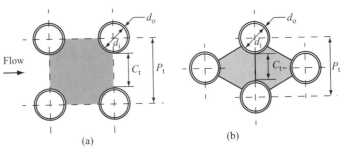

Figure 5.21 (a) Square-pitch layout, (b) triangular-pitch layout

(5.30) for the equivalent diameter is rewritten here for convenience:

$$D_e = \frac{4A_c}{P_{heated}} \tag{5.129}$$

From Figure 5.21(a), the *equivalent diameter* for the square pitch layout is

$$D_e = \frac{4\left(P_t^2 - \pi d_o{}^2/4\right)}{\pi d_o} \tag{5.130a}$$

From Figure 5.21(b), the *equivalent diameter* for the triangular pitch layout is

$$D_e = \frac{4\left(\dfrac{\sqrt{3}P_t^2}{4} - \dfrac{\pi d_o^2}{8}\right)}{\pi d_o/2} \tag{5.130b}$$

The cross-flow area of the shell A_c is defined as

$$A_c = \frac{D_s C_t B}{P_t} \tag{5.131}$$

The diameter ratio d_r is defined by

$$d_r = \frac{d_o}{d_t} \tag{5.132}$$

Some diameter ratios for nominal pipe sizes are illustrated in Table C.1 in Appendix C. The tube pitch ratio P_r is defined by

$$P_r = \frac{P_t}{d_o} \tag{5.133}$$

The tube clearance C_t is obtained from Figure 5.21.

$$C_t = P_t - d_o \tag{5.134}$$

The number of tubes N_t can be predicted in fair approximation with the shell inside diameter D_s.

$$N_t = (CTP)\frac{\pi D_s^2/4}{\text{Shaded area}} \tag{5.135}$$

where *CTP* is the *tube count constant* that accounts for the incomplete coverage of the shell diameter by the tubes, due to necessary clearance between the shell and the outer tube circle and tube omissions due to tube pass lanes for multiple pass design [1].

$$
\begin{aligned}
CTP &= 0.93 \quad \text{for one-pass exchanger} \\
CTP &= 0.90 \quad \text{for two-pass exchanger} \\
CTP &= 0.85 \quad \text{for three-pass exchanger}
\end{aligned}
\tag{5.136}
$$

$$\text{Shaded Area} = CL \cdot P_t^2 \tag{5.137}$$

Table 5.6 Summary of Shell-and-Tube Heat Exchangers

Description	Equation	
Basic equations	$q = \dot{m}_1 c_{p1}\left(T_{1i} - T_{1o}\right)$	(5.140)
	$q = \dot{m}_2 c_{p2}\left(T_{2o} - T_{2i}\right)$	(5.141)
Heat transfer areas of inner and outer surfaces of an inner pipe	$A_i = \pi d_i N_t L$	(5.142a)
	$A_o = \pi d_o N_t L$	(5.142b)
Overall heat transfer coefficient	$U_o = \dfrac{1/A_o}{\dfrac{1}{h_i A_i} + \dfrac{\ln\left(\dfrac{d_o}{d_i}\right)}{2\pi k L} + \dfrac{1}{h_o A_o}}$	(5.143)

Tube side

Reynolds number	$Re_D = \dfrac{\rho u_m d_i}{\mu} = \dfrac{\dot{m} d_i}{A_c \mu}$	(5.144)
	$A_c = \dfrac{\pi d_i^2}{4}\dfrac{N_t}{N_p}$	(5.144a)
Laminar flow ($Re < 2,300$)	$Nu_D = \dfrac{h d_i}{k_f} = 1.86\left(\dfrac{d_i\,Re\,Pr}{L}\right)^{\frac{1}{3}}\left(\dfrac{\mu}{\mu_s}\right)^{0.14}$	(5.145)
	$0.48 < Pr < 16{,}700$ $0.0044 < (\mu/\mu_s) < 9.75$ Use $Nu_D = 3.66$ if $Nu_D < 3.66$	
Turbulent flow ($Re > 2,300$)	$Nu_D = \dfrac{h d_i}{k_f} = \dfrac{(f/2)\,(Re_D - 1000)\,Pr}{1 + 12.7\,(f/2)^{1/2}\left(Pr^{2/3} - 1\right)}$	(5.146)
	$300 < Re_D < 5\times10^6$ [4] $0.5 \le Pr \le 2000$	
Friction factor	$f = (1.58\ln(Re_D) - 3.28)^{-2}$	(5.147)

Shell side

Square pitch layout (Figure 5.21)	$D_e = \dfrac{4\left(P_t^2 - \pi d_o^2/4\right)}{\pi d_o}$	(5.148a)

Table 5.6   (*continued*)

Description	Equation
Triangular pitch layout (Figure 5.21)	$$D_e = \frac{4\left(\frac{\sqrt{3}P_t^2}{4} - \frac{\pi d_o^2}{8}\right)}{\pi d_o/2} \qquad (5.148\mathrm{b})$$
Cross-flow area	$$A_c = \frac{D_s C_t B}{P_t} \qquad (5.149)$$
Reynolds number	$$Re_D = \frac{\rho u_m D_e}{\mu} = \frac{\dot{m} D_e}{A_c \mu} \qquad (5.150)$$
Nusselt number	$$Nu = \frac{h_o D_e}{k_f} = 0.36 Re^{0.55} Pr^{1/3} \left(\frac{\mu}{\mu_s}\right)^{0.14} \qquad (5.151)$$ $$2{,}000 < Re < 1 \times 10^6$$

ε-NTU Method

Description	Equation
Heat transfer unit (NTU)	$$NTU = \frac{U_o A_o}{\left(\dot{m} c_p\right)_{\min}} \qquad (5.152)$$
Capacity ratio	$$C_r = \frac{\left(\dot{m} c_p\right)_{\min}}{\left(\dot{m} c_p\right)_{\max}} \qquad (5.153)$$
Effectiveness One shell (2,4,..passes)	$$\varepsilon = 2\left\{1 + C_r + \left(1 + C_r^2\right)^{1/2} \frac{1 + \exp\left[-NTU_1\left(1 + C_r^2\right)^{1/2}\right]}{1 - \exp\left[-NTU\left(1 + C_r^2\right)^{1/2}\right]}\right\}^{-1}$$ $$NTU_1 = NTU/N_p$$ $$(5.154)$$
Heat transfer unit (NTU)	$$NTU = -\left(1 + C_r^2\right)^{-1/2} \ln\left(\frac{E-1}{E+1}\right) \qquad (5.155)$$ where $E = \dfrac{2/\varepsilon - (1 + C_r)}{\left(1 + C_r^2\right)^{1/2}}$
Effectiveness	$$\varepsilon = \frac{q}{q_{\max}} = \frac{\left(\dot{m}_1 c_{p1}\right)\left(T_{1i} - T_{1o}\right)}{\left(\dot{m} c_p\right)_{\min}\left(T_{1i} - R_{2i}\right)} = \frac{\left(\dot{m}_2 c_{p2}\right)\left(T_{2o} - T_{2i}\right)}{\left(\dot{m} c_p\right)_{\min}\left(T_{1i} - T_{2i}\right)} \qquad (5.156)$$
Heat transfer rate	$$q = \varepsilon \left(\dot{m} c_p\right)_{\min}\left(T_{1i} - T_{2i}\right) \qquad (5.157)$$

(*continues*)

Table 5.6   (*continued*)

Description	Equation
Tube Side **Pressure Drop** Pressure drop	$$\Delta P = 4 \left(\frac{f \cdot L_t}{d_i} + 1 \right) N_p \frac{1}{2} \rho \cdot v^2 \qquad (5.158)$$
Laminar flow	$$f = 16/Re_D \qquad (5.159)$$
Turbulent flow	$$f = (1.58 \ln (Re_D) - 3.28)^{-2} \qquad (5.160)$$
Shell-side **Pressure Drop**	$$\Delta P = f \frac{D_s}{D_e} (N_b + 1) \frac{1}{2} \rho \cdot v^2 \qquad (5.161)$$ $$f = \exp (0.576 - 0.19 \ln (Re_s)) \qquad (5.162)$$

where *CL* is the tube layout constant.

$$CL = 1 \qquad \text{for square-pitch layout} \qquad (5.138)$$

$$CL = \sin(60^\circ) = 0.866 \quad \text{for triangular-pitch layout}$$

Plugging Equation (5.137) into (5.135) gives

$$N_t = \frac{\pi}{4} \left(\frac{CTP}{CL} \right) \frac{D_s^2}{P_t^2} = \frac{\pi}{4} \left(\frac{CTP}{CL} \right) \frac{D_s^2}{P_r^2 d_o^2} \qquad (5.139)$$

Example 5.4.1 Miniature Shell-and-Tube Heat Exchanger   A miniature shell-and-tube heat exchanger is designed to cool engine oil with the engine coolant (50 percent ethylene glycol) (see Figure E5.4.1). The engine oil at a flow rate of 0.23 kg/s enters the exchanger at 120°C and leaves at 115°C. The 50 percent ethylene glycol at a rate of 0.47 kg/s enters at 90°C. The tube material is Cr alloy ($k_w = 42.7$ W/mK). Fouling factors of 0.176×10^{-3} m^{2}K/W for engine oil and 0.353×10^{-3} m^{2}K/W for 50 percent ethylene glycol are specified. Route the engine oil through the tubes. The maximum permissible pressure drop for each fluid is 10 kPa. The volume of the exchanger is required to be minimized. Since the exchanger is custom designed, the tube size can be smaller than NPS 1/8 (DN 6 mm)—that is, the smallest size in Table C.1 in Appendix C, wherein the tube pitch ratio of 1.25 and the diameter ratio of 1.3 can be used. Design the shell-and-tube heat exchanger.

MathCAD Format Solution   The design concept is to develop a MathCAD model for a miniature shell-and-tube heat exchanger and then seek the solution by iterating the calculations by varying the parameters to satisfy the design requirements. Remember that the design requirements are the engine oil outlet temperature less than 115°C and the pressure drop less than 10 kPa in each of the fluids.

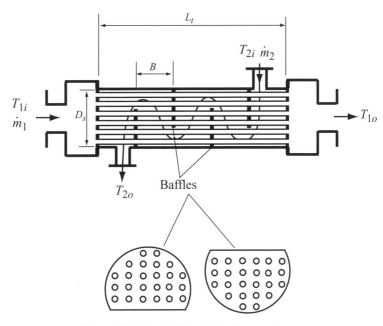

Figure E5.4.1   Shell-and-tube heat exchanger

The properties of engine oil and ethylene glycol are obtained using the average temperatures from Table A.12 in Appendix A.

$$T_{oil} := \frac{(120°C + 115°C)}{2} = 117.5 \cdot °C \quad T_{cool} := \frac{(90°C + 100°C)}{2} = 95 \cdot °C$$
$$\text{(E5.4.1)}$$

Engine oil (subscript 1) - tube side   50% Ethylene glycol (subscript 2) - shell side

$$\rho_1 := 828 \frac{kg}{m^3} \qquad\qquad \rho_2 := 1020 \frac{kg}{m^3}$$

$$c_{p1} := 2307 \frac{J}{kg \cdot K} \qquad\qquad c_{p2} := 3650 \frac{J}{kg \cdot K}$$

$$k_1 := 0.135 \frac{W}{m \cdot K} \qquad\qquad k_2 := 0.442 \frac{W}{m \cdot K}$$

$$\mu_1 := 1.027 \cdot 10^{-2} \frac{N \cdot s}{m^2} \qquad\qquad \mu_2 := 0.08 \cdot 10^{-2} \frac{N \cdot s}{m^2}$$

$$Pr_1 := 175 \qquad\qquad Pr_2 := 6.6$$
$$\text{(E5.4.2)}$$

The thermal conductivity for the tube material (Chromium alloy) is given by

$$k_w := 42.7 \frac{W}{m \cdot K} \qquad\qquad \text{(E5.4.3)}$$

Given information   The inlet temperatures are given as

$$T_{1i} := 120°C \qquad T_{2i} := 90°C \tag{E5.4.4}$$

The mass flow rates are given as

$$\text{mdot}_1 := 0.23 \frac{kg}{s} \qquad \text{mdot}_2 := 0.47 \frac{kg}{s} \tag{E5.4.5}$$

The fouling factors for engine oil and 50 percent ethylene glycol are given as

$$R_{fi} := 0.176 \cdot 10^{-3} \frac{m^2 \cdot K}{W} \qquad R_{fo} := 0.353 \cdot 10^{-3} \frac{m^2 \cdot K}{W} \tag{E5.4.6}$$

Design requirement   The engine oil outlet temperature must be less than or equal to 115°C.

$$T_{1o} \le 115°C \tag{E5.4.7}$$

The pressure drop on each side must be

$$\Delta P \le 10\,\text{kPa} \tag{E5.4.8}$$

Design Parameters Sought by Iterations   Initially, estimate the following boxed parameters and then iterate the calculations with different values in order to satisfy the design requirements.

$D_s := 2.0\,\text{in}$	Shell inside diameter		$D_s = 50.8\,\text{mm}$	(E5.4.9)
$L_t := 10\,\text{in}$	Tube length		$L_t = 254\,\text{mm}$	(E5.4.10)
$d_o := \dfrac{1}{8}\,\text{in}$	Tube outside diameter		$d_o = 3.175\,\text{mm}$	(E5.4.11)

The diameter ratio ($d_r = d_o/d_i$) is given as suggested in the problem description.

$$d_r := 1.3 \qquad d_i := \frac{1}{d_r} \cdot d_o \qquad d_i = 2.442 \cdot \text{mm} \tag{E5.4.12}$$

The tube pitch ratio ($P_r = P_t/d_o$) is given as suggested in the problem description.

$$P_r := 1.25 \tag{E5.4.13}$$

The tube pitch is then obtained from Equation (5.133).

$$P_t := P_r \cdot d_o \tag{E5.4.14}$$

The baffle spacing is assumed and may be iterated, and the baffle number from Equation (5.128) is defined.

$$B := \frac{8}{8}\,\text{in} \qquad B = 25.4\,\text{mm} \tag{E5.4.15}$$

$$N_b := \frac{L_t}{B} - 1 \qquad N_b = 9 \tag{E5.4.16}$$

The number of passes is defined by

$$N_p := 1 \tag{E5.4.17}$$

The tube clearance C_t is obtained from Figure 5.21 as

$$C_t := P_t - d_o \qquad C_t = 0.794 \, \text{mm} \tag{E5.4.18}$$

From Equation (5.136), the tube count calculation constants (CTP) up to three-passes are

$$CTP := \begin{vmatrix} 0.93 \text{ if } N_p = 1 \\ 0.9 \text{ if } N_p = 2 \\ 0.85 \text{ otherwise} \end{vmatrix} \tag{E5.4.19}$$

From Equation (5.138), the tube layout constant (*CL*) for a triangular-pitch layout is given by

$$CL := 0.866 \tag{E5.4.20}$$

The number of tubes N_t is estimated using Equation (5.139) and rounded off in practice. Note that the number of tubes in the shell indicates the compactness of a miniature exchanger. A 253-tube exchanger is commercially available for a 2-inch shell diameter.

$$N_{tube}(D_s, d_o, P_r) := \frac{\pi}{4} \cdot \left(\frac{CTP}{CL} \right) \cdot \frac{D_s^2}{P_r^2 \cdot d_o^2} \qquad N_{tube}(D_s, d_o, P_r) = 138.189 \tag{E5.4.21}$$

$$N_t := \text{round}(N_{tube}(D_s, d_o, P_r)) \qquad N_t = 138 \tag{E5.4.22}$$

Tube Side (Engine Oil) The cross-flow area, velocity, and Reynolds number are defined as follows:

$$A_{c1} := \frac{\pi \cdot d_i^2}{4} \cdot \frac{N_t}{N_p} \qquad A_{c1} = 6.465 \times 10^{-4} \, \text{m}^2 \tag{E5.4.23}$$

$$v_1 := \frac{mdot_1}{\rho_1 \cdot A_{c1}} \qquad v_1 = 0.43 \, \frac{\text{m}}{\text{s}} \tag{E5.4.24}$$

$$Re_1 := \frac{\rho_1 \cdot v_1 \cdot d_i}{\mu_1} \qquad Re_1 = 84.603 \tag{E5.4.25}$$

The Reynolds number indicates very laminar flow. The velocity in the tubes appears acceptable when considering a reasonable range of 0.5 to 1.0 m/s in Table 5.4 for the engine oil. The friction factor is programmed for either laminar or turbulent flow as

$$f(Re_D) := \begin{vmatrix} (1.58 \cdot \ln(Re_D) - 3.28)^{-2} \text{ if } Re_D > 2300 \\ \dfrac{16}{Re_D} \text{ otherwise} \end{vmatrix} \tag{E5.4.26}$$

The Nusselt number for turbulent or laminar flow is defined using Equations (5.146) and (5.147), assuming that μ changes moderately with temperature. The convection heat transfer coefficient is then obtained.

$$\text{Nu}_D\,(D_h, L_t, \text{Re}_D, \text{Pr}) := \begin{vmatrix} \left(\dfrac{f\,(\text{Re}_D)}{2}\right) \cdot \dfrac{(\text{Re}_D - 1000)\cdot \text{Pr}}{1 + 12.7 \cdot \left(\dfrac{f(\text{Re}_D)}{2}\right)^{0.5} \cdot \left(\text{Pr}^{\frac{2}{3}} - 1\right)} \\ \qquad \text{if } \text{Re}_D > 2300 \\[6pt] 1.86 \left(\dfrac{D_h \cdot \text{Re}_D \cdot \text{Pr}}{L_t}\right)^{\frac{1}{3}} \quad \text{otherwise} \end{vmatrix}$$

(E5.4.27)

$$\text{Nu}_1 := \text{Nu}_D\,(d_i, L_t, \text{Re}_1, \text{Pr}_1) \qquad \text{Nu}_1 = 9.712 \qquad \text{(E5.4.28)}$$

$$h_1 := \frac{\text{Nu}_1 \cdot k_1}{d_i} \qquad h_1 = 536.839 \,\frac{W}{m^2 \cdot k} \qquad \text{(E5.4.29)}$$

Shell Side (50 % Ethylene Glycol)   The free-flow area is obtained using Equation (5.131) and the velocity in the shell is also calculated

$$A_{c2} := \frac{D_s \cdot C_t \cdot B}{P_t} \qquad A_{c2} = 2.581 \times 10^{-4}\,m^2 \qquad \text{(E5.4.30)}$$

$$v_2 := \frac{\text{mdot}_2}{\rho_2 \cdot A_{c2}} \qquad v_2 = 1.786 \,\frac{m}{s} \qquad \text{(E5.4.31)}$$

The velocity of 1.786 m/s in the shell is acceptable because the reasonable range of 1.2 to 2.4 m/s for the similar fluid shows in Table 5.4. The equivalent diameter for a triangular pitch is given in Equation (5.148b):

$$D_e := 4 \left[\frac{\dfrac{P_t^2 \cdot \sqrt{3}}{4} - \dfrac{\pi \cdot d_o^2}{8}}{\left(\dfrac{\pi \cdot d_o}{2}\right)} \right] \qquad D_e = 2.295 \cdot mm \qquad \text{(E5.4.32)}$$

$$\text{Re}_2 := \frac{\rho_2 \cdot v_2 \cdot D_e}{\mu_2} \qquad \text{Re}_2 = 5.225 \times 10^3 \qquad \text{(E5.4.33)}$$

The Nusselt number is given in Equation (5.151) and the heat transfer coefficient is obtained.

$$\text{Nu}_2 := 0.36 \cdot \text{Re}_2^{0.55} \cdot \text{Pr}_2^{\frac{1}{3}} \qquad \text{(E5.4.34)}$$

$$h_2 := \frac{\text{Nu}_2 \cdot k_2}{D_e} \qquad h_2 = 1.442 \times 10^4 \cdot \frac{W}{m^2 \cdot K} \qquad \text{(E5.4.35)}$$

The total heat transfer areas for both fluids are obtained as follows:

$$A_i := \pi \cdot d_i \cdot L_t \cdot N_t \qquad A_i = 0.269\,\mathrm{m}^2 \tag{E5.4.36}$$

$$A_o := \pi \cdot d_o \cdot L_t \cdot N_t \qquad A_o = 0.35 \cdot \mathrm{m}^2 \tag{E5.4.37}$$

The overall heat transfer coefficient is calculated using Equation (5.143), with the fouling factors as

$$UA_o := \cfrac{1}{\cfrac{1}{h_1 \cdot A_i} + \cfrac{R_{fi}}{A_i} + \cfrac{\ln\left(\cfrac{d_o}{d_i}\right)}{2 \cdot \pi \cdot k_w \cdot L_t \cdot N_t} + \cfrac{R_{fo}}{A_o} + \cfrac{1}{h_2 \cdot A_o}} \tag{E5.4.38}$$

$$UA_o = 113.425\,\frac{\mathrm{W}}{\mathrm{K}}$$

ε -*NTU Method*   The heat capacities for both fluids are defined and then the minimum and maximum heat capacities are obtained using the MathCAD built-in functions:

$$C_1 := mdot_1 \cdot c_{p1} \qquad C_1 = 530.61 \cdot \frac{\mathrm{W}}{\mathrm{K}} \tag{E5.4.39}$$

$$C_2 := mdot_2 \cdot c_{p2} \qquad C_2 = 1.716 \times 10^3 \cdot \frac{\mathrm{W}}{\mathrm{K}} \tag{E5.4.40}$$

$$C_{min} := \min(C_1 \cdot C_2) \qquad C_{max} := \max(C_1, C_2) \tag{E5.4.41}$$

The heat capacity ratio is defined as

$$C_r := \frac{C_{min}}{C_{max}} \qquad C_r = 0.309 \tag{E5.4.42}$$

The number of transfer units is defined as

$$NTU := \frac{UA_o}{C_{min}} \qquad NTU = 0.214 \tag{E5.4.43}$$

The effectiveness for a shell-and-tube heat exchanger is given using Equation (5.154) as

$$NTU_1 := \frac{NTU}{N_p} \tag{E5.4.44a}$$

$$\varepsilon_{hx} := 2 \cdot \left[1 + C_r + (1 + C_r^2)^{0.5} \cdot \frac{1 + \exp\left[-NTU_1 \cdot (1 + C_r^2)^{0.5}\right]}{1 - \exp\left[-NTU_1 \cdot (1 + C_r^2)^{0.5}\right]}\right]^{-1} \tag{E5.4.44b}$$

$$\varepsilon_{hx} = 0.187$$

Using Equation (5.156), the effectiveness is expressed as

$$\varepsilon_{hx} = \frac{q}{q_{max}} = \frac{C_1 \cdot (T_{1i} - T_{1o})}{C_{min} \cdot (T_{1i} - T_{2i})} = \frac{C_2 \cdot (T_{2o} - T_{2i})}{C_{min} \cdot (T_{1i} - T_{2i})} \tag{E5.4.45}$$

The inlet temperatures are rewritten for comparison with the outlet temperatures.

$$T_{1i} = 120 \cdot {}^\circ C \qquad\qquad T_{2i} = 90 \cdot {}^\circ C$$

$$T_{1o} := T_{1i} - \varepsilon_{hx} \cdot \frac{C_{min}}{C_1} \cdot (T_{1i} - T_{2i}) \qquad T_{1o} = 114.395^\circ C \qquad (E5.4.46)$$

$$T_{2o} := T_{2i} + \varepsilon_{hx} \cdot \frac{C_{min}}{C_2} \cdot (T_{1i} - T_{2i}) \quad T_{2o} = 91.734^\circ C \qquad (E5.4.47)$$

The engine oil outlet temperature of 114.544°C satisfies the requirement of 115°C. The heat transfer rate is obtained

$$q := \varepsilon_{hx} \cdot C_{min} \cdot (T_{1i} - T_{2i}) \qquad q = 2.974 \times 10^3 \text{ W} \qquad (E5.4.48)$$

The pressure drops for both fluids are obtained using Equations (5.158) and (5.161) as

$$\Delta P_1 := 4 \cdot \left(\frac{f(Re_1) \cdot L_t}{d_i} + 1 \right) \cdot N_p \cdot \frac{1}{2} \cdot \rho_1 \cdot v_1^2 \quad \Delta P_1 = 6.319 \cdot kPa \qquad (E5.4.49)$$

$$\Delta P_2 := f(Re_2) \cdot \frac{D_s}{D_e} (N_b + 1) \cdot \frac{1}{2} \cdot \rho_2 \cdot v_2^2 \qquad \Delta P_2 = 3.427 \cdot kPa \qquad (E5.4.50)$$

Both the pressure drops calculated are less than the requirement of 10 kPa. The iteration between Equations (E5.4.9) and (E5.4.46) is terminated. The surface density β for the engine oil side is obtained using the relationship of the heat transfer area over the volume of the exchanger.

$$\beta := \frac{A_o + A_i}{\left(\frac{\pi \cdot D_s^2}{4} \right) \cdot L_t} \qquad \beta = 1202 \cdot \frac{m^2}{m^3} \qquad (E5.4.51)$$

Summary of the Design of the Miniature Shell-and-Tube Heat Exchanger   Given information:

$T_{1i} = 120 \cdot {}^\circ C$	Engine oil inlet temperature
$T_{2i} = 90 \cdot {}^\circ C$	50 percent ethylene glycol inlet temperature
$mdot_1 = 0.23 \frac{kg}{s}$	Mass flow rate of engine oil
$mdot_2 = 0.47 \frac{kg}{s}$	Mass flow rate of 50% ethylene glycol
$R_{fi} = 1.76 \times 10^{-4} \cdot m^2 \cdot \frac{K}{W}$	Fouling factor of engine oil
$R_{fo} = 3.53 \times 10^{-4} \cdot m^2 \cdot \frac{K}{W}$	Fouling factor of 50% ethylene giycol

Requirements for the exchanger:

$T_{1o} \leq 115^\circ C$	Engine outlet temperature
$\Delta P_1 \leq 10 \, kPa$	Pressure drop on both sides
$\Delta P_2 \leq 10 \, kPa$	Pressure drop on coolant sides

Design obtained:

$N_p = 1$	Number of passes	
$D_s = 50.8 \cdot mm$	Shell inside diameter	$D_s = 2 in$
$d_o = 3.175 \cdot mm$	Tube outer diameter	
$d_i = 2.442 \cdot mm$	Tube inner diameter	
$L_t = 254 \cdot mm$	Tube length	$L_t = 10 \cdot in$
$N_t = 138$	Number of tube	
$C_t = 0.794 \cdot mm$	Tube clearance	
$B = 25.4 \cdot mm$	Baffle spacing	$B = 1 in$
$N_b = 9$	Number of baffles	
$T_{1o} = 114.395°C$	Engine oil outlet temperature	
$T_{2o} = 91.734°C$	50 percent ethylene glycol outlet temperature	
$q = 2.974 \cdot kW$	Heat transfer rate	
$\beta = 1202 \cdot \dfrac{m^2}{m^3}$	Surface area density	
$\Delta P_1 = 6.319 \cdot kPa$	Pressure drop for engine oil	
$\Delta P_2 = 3.427 \cdot kPa$	Pressure drop for 50 percent ethylene glycol	
$UA_o = 113.425 \cdot \dfrac{W}{K}$	UA value	

The design satisfies the requirements.

5.5 PLATE HEAT EXCHANGERS (PHE)

There are two types of plate heat exchangers. One is the *plate-and-frame* or *gasketed plate heat exchanger*, which was developed in the 1930s for the food industries because of easy cleaning and quality control (small temperature differences). The other is the welded or brazed plate heat exchanger, which usually allows higher pressures and temperatures, being widely used as a compact heat exchanger. However, since this exchanger cannot be opened, applications are limited to negligible fouling cases. The weights of the brazed plate heat exchangers are about 25 percent of the shell-and-tube heat exchangers for the same heat duty. The brazed plate heat exchanger consists of a number of thin rectangular metal plates. The flow arrangement is pictured in Figure 5.22.

5.5.1 Flow Pass Arrangements

A group of channels for which the flow is in the same direction is referred to as a *pass*. Figure 5.23(a) shows a single *pass arrangement* for each fluid, which is denoted as a 1-pass/1-pass flow system. Figure 5.23(b) indicates a two-pass arrangement for each fluid, which is denoted as a 2-pass/2-pass flow system. Finally, Figure 5.23(c) is two-pass for one fluid (line) and one-pass for the other fluid (dashed), which is referred to as a 2-pass/1-pass flow system. However, only a single pass is illustrated here, with one example problem.

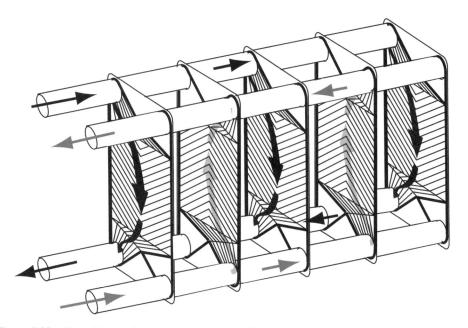

Figure 5.22   Flow diagram in a single-pass counter flow arrangement for a plate heat exchanger

5.5.2   Geometric Properties

Each plate is made by stamping or embossing a corrugated or wavy surface pattern on sheet metal. Typical plate geometries (chevron patterns) are shown in Figure 5.24. The chevron pattern is the most common in use today. Alternate plates are assembled

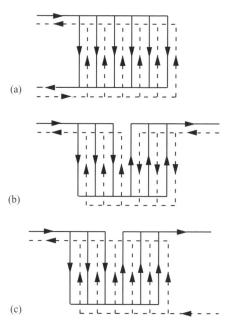

Figure 5.23 Flow pass arrangements: (a) 1-pass/1-pass flow system, (b) 2-pass/2-pass flow system, and (c) 2-pass/1-pass flow system

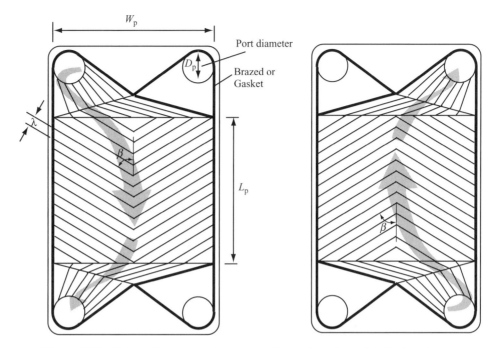

Figure 5.24   Plates with chevron-type corrugation pattern for a plate heat exchanger

such that the corrugations on successive plates contact or cross each other to provide mechanical support to the plate pack through a large number of contact points. The resulting flow passages are narrow, highly interrupted, and tortuous. They enhance the heat transfer rate and decrease fouling resistance by increasing the shear stress, producing secondary flow and increasing the level of turbulence. The corrugations also improve the rigidity of the plates and form the desired plate spacing [17]. Each plate has four corner ports that provide access to the flow passages on either side of the plate. Corrugation inclination angle (chevron angle) β can be between $0°$ and $90°$, typically with $30°$, $45°$, or $60°$. The high β provides the *hard channel* due to the turbulence, and the low β provides the *soft channel*.

One problem associated with the PHE design is the precise matching of both the thermal and hydrodynamic loads. It is difficult to accommodate the required thermal duty and at the same time fully utilize the available pressure drop, as the minimum cross-flow area and the surface area in a PHE are interdependent, unlike in other types of compact heat exchangers [19]. This limits the exchanger designed to be either pressure drop or heat transfer.

A brazed plate heat exchanger and the inter-plate cross-corrugated flow are illustrated in Figures 5.24 and 5.25. The wavelength λ of the chevron pattern is the corrugation pitch as shown in Figure 5.25(b). The amplitude of the corrugation is denoted as $2a$, where a is the amplitude of the sinusoidal corrugation with the plate thickness δ. The number of wavelength per plate N_λ is calculated by dividing the width

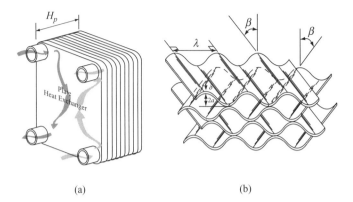

Figure 5.25   (a) Brazed plate heat exchanger and (b) inter-plate cross-corrugated flow channels

W_p by the wavelength λ.

$$N_\lambda = \frac{W_p}{\lambda} \tag{5.163}$$

Considering the front and end covers, we can calculate the number of channels per each fluid N_c by dividing the total even number of channels $(N_t + 1)$ by two for the odd number of plates $(1, 3, 5...)$.

$$N_c = \frac{N_t + 1}{2N_p} \tag{5.164}$$

where N_t is the total number of plates and N_p the number of channels per pass. The amplitude a can be expressed in terms of the PHE height H_p, the total number of plates N_t and the plate thickness δ.

$$a = \frac{1}{2}\left(\frac{H_p}{N_t + 1} - \delta\right) \tag{5.165}$$

The surface waviness can be essentially represented by two dimensionless parameters, namely, the *corrugation aspect ratio* γ and the *surface enlargement factor* ϕ. The corrugation aspect ratio γ is defined as the ratio of the double of corrugation amplitude (channel amplitude), $4a$, to the *corrugation pitch* or *wavelength,* λ. The corrugation aspect ratio is

$$\gamma = \frac{4a}{\lambda} \tag{5.166}$$

It is noted that when $\gamma = 0$, a flat-parallel plate is obtained. Thus, $\gamma > 0$ is indicative of surface area enlargement as well. Increasing γ enlarges the surface area, but high γ may induce vortexes at the top and bottom of the channel which can trap the fluids locally and reduce the heat transfer. Therefore, the PHE design is commonly limited to $\gamma < 1$, depending on the Reynolds number. The optimum surface compactness may be obtained with $0.2 < \gamma < 0.6$ [26].

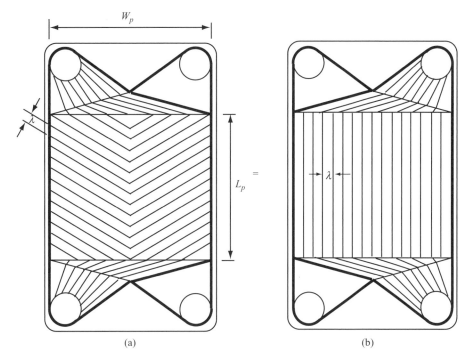

Figure 5.26   Surface area equivalence between (a) $\beta > 0$ and (b) $\beta = 0$

The surface enlargement factor ϕ is given by

$$\phi = \frac{\text{Corrugated area}}{\text{Projected area}} \tag{5.167}$$

The corrugated area is obtained using the path distance of the sinusoidal corrugation. Note that the corrugated area is independent of the chevron angle β, as shown in Figure 5.26. Therefore, the *enlarged length per wavelength* for the plate in Figure 5.26(b) is mathematically obtained by

$$L_\lambda = \int_0^\lambda \sqrt{1 + \left(\frac{2\pi a}{\lambda}\right)^2 \cos\left(\frac{2\pi x}{\lambda}\right)^2}\, dx \tag{5.168}$$

The corrugated area (heat transfer area) A_t for each fluid is then calculated by

$$A_t = 2L_\lambda N_\lambda L_p N_c \tag{5.169}$$

The free-flow area A_c in a channel that is equivalent to a rectangular area for taking half of each fluid is calculated by

$$A_c = 2a W_p N_c \tag{5.170}$$

Finally, the surface enlargement factor ϕ is expressed as

$$\phi = \frac{2L_\lambda N_\lambda L_p N_c}{2W_p L_p N_c} = \frac{L_\lambda N_\lambda}{W_p} \tag{5.171}$$

The surface enlargement factor can be calculated approximately for sinusoidal corrugation by

$$\phi \approx \frac{1}{6} \left(1 + \sqrt{1 + X^2} + 4\sqrt{1 + \frac{X^2}{2}} \right) \tag{5.172}$$

where $X = \frac{2\pi a}{\lambda}$

The hydraulic diameter of the plate heat exchanger is obtained as follows:

$$D_h = \frac{4 A_c L_p}{P_{wet} L_p} = \frac{4 A_c L_p}{A_t} = \frac{4 (2a W_p N_c) L_p}{2 L_\lambda N_\lambda N_p N_c} = \frac{4a}{\phi} \tag{5.173}$$

5.5.3 Friction Factor

One of the correlations for the friction factors of a plate heat exchanger with chevron pattern is provided by Martin [19]. The Fanning friction factor is

$$f = \left[\frac{\cos \beta}{(0.045 \tan \beta + 0.09 \sin \beta + f_0/\cos \beta)^{1/2}} + \frac{1 - \cos \beta}{\sqrt{3.8 f_1}} \right]^{-96} \tag{5.174}$$

[handwritten: $^{-96}$ & still neg]

where

$$f_0 = \begin{cases} \dfrac{16}{Re} & \text{for } Re < 2{,}000 \\[2mm] (1.56 \ln Re - 3.0)^{-2} & \text{for } Re > 2{,}000 \end{cases} \tag{5.174a}$$

$$f_1 = \begin{cases} \dfrac{149.25}{Re} + 0.9625 & \text{for } Re < 2{,}000 \\[2mm] \dfrac{9.75}{Re^{0.289}} & \text{for } Re > 2{,}000 \end{cases} \tag{5.174b}$$

5.5.4 Nusselt Number

Martin [19] also provided the Nusselt number correlation for a plate heat exchanger with chevron pattern as

$$Nu = \frac{h D_h}{k_f} = 0.205 \, Pr^{\frac{1}{3}} \left(f Re^2 \sin 2\beta \right)^{0.374} \left(\frac{\mu}{\mu_s} \right)^{1/6} \tag{5.175}$$

where $10° < \beta < 80°$ and k_f is the thermal conductivity of the fluid and μ_s the dynamic viscosity at the wall temperature. $\mu/\mu_s = 1$ may be used with the assumption that μ changes moderately with temperature.

5.5.5 Pressure Drops

The total pressure drop in a plate heat exchanger is composed of the frictional pressure drop of the channels ΔP_f and the port pressure drop ΔP_p. We assumed that the pressure drop due to the elevation (gravity) change is negligible. The frictional pressure drop is calculated using Equation (5.93):

$$\Delta P_f = \frac{2fL}{D_h} \frac{G^2}{\rho} N_p \tag{5.176}$$

The mass velocity for each fluid at the port of the plate heat exchanger is defined by

$$G_p = \frac{4\dot{m}}{\pi D_p^2} \tag{5.177}$$

where D_p is the port diameter. The port pressure drop for each fluid is then calculated by

$$\Delta P_p = \frac{1.5 N_p G_p^2}{2\rho} \tag{5.178}$$

The total pressure drop is the sum of Equations (5.176) and (5.178).

$$\Delta P_t = \Delta P_f + \Delta P_p \tag{5.179}$$

The port diameter is designed such that the port pressure drop is usually less than 10 percent of the total pressure drop, but it may be as high as 25 to 30 percent in some designs.

Example 5.5.1 Plate Heat Exchanger (PHE)   Cold water will be heated by wastewater using a plate heat exchanger. The cold water with a flow rate of 130 kg/s enters the plate heat exchanger at 22°C, and it will be heated to 42°C. The hot wastewater enters at a flow rate of 140 kg/s and 65°C. The maximum permissible pressure drop for each fluid is 70 kPa (10 psi). Using single-pass chevron plates (stainless steel AISI 304) with $\beta = 30°$, determine the rating (T, q, ε, and ΔP) and sizing (W_p, L_p, and H_p) of the plate heat exchanger. The following data list is used for the calculations.

Description	value
Number of passes N_p	1
Chevron angle β	30°
Total number of plates N_t	109
Plate thickness δ	0.6 mm
Corrugation pitch λ	9 mm
Port diameter D_p	200 mm
Thermal conductivity k_w	14.9 W/mK

MathCAD Format Solution   The design concept is to develop a MathCAD model as a function of the sizing (W_p, L_p, and H_p) and then to seek the solution for the sizing in order to satisfy the design requirements.

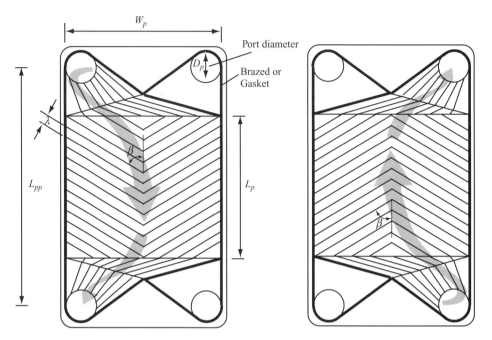

Figure E5.5.1   Plates with chevron-type corrugation pattern for a plate heat exchanger

The hot and cold water properties at the average fluid temperatures are obtained from Table A.12 in Appendix A.

$$T_{hot\_water} := \frac{65°C + 45°C}{2} = 55 \cdot °C \quad T_{cold\_water} := \frac{22°C + 42°C}{2} = 32 \cdot °C$$

(E5.5.1)

Hot (waste) water (subscript 1)   Cold water(subscript 2)

$$\rho_1 := 985 \frac{kg}{m^3} \qquad\qquad \rho_2 := 997 \frac{kg}{m^3}$$

$$c_{p1} := 4184 \frac{J}{kg \cdot K} \qquad\qquad c_{p2} := 4179 \frac{J}{kg \cdot K}$$

$$k_1 := 0.651 \frac{W}{m \cdot K} \qquad\qquad k_2 := 0.6125 \frac{W}{m \cdot K}$$

(E5.5.2)

$$\mu_1 := 471 \cdot 10^{-6} \frac{N \cdot s}{m^2} \qquad \mu_2 := 880 \cdot 10^{-6} \frac{N \cdot s}{m^2}$$

$$Pr_1 := 3.02 \qquad\qquad Pr_2 := 5.68$$

The thermal conductivity of the plate (Stainless Steel AISI 304) given is defined.

$$k_w := 14.9 \frac{W}{m \cdot K}$$

(E5.5.3)

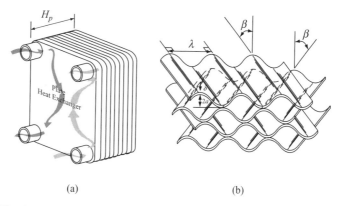

(a)   (b)

Figure E5.5.2   (a) Brazed plate heat exchanger and (b) the inter-plate cross-corrugated flow

Given information: The inlet temperatures for the hot and cold fluids given are defined by

$$T_{1i} := 65°C \qquad T_{2i} := 22°C \qquad\qquad (E5.5.4)$$

The mass flow rates given are defined as

$$mdot_1 := 140\,\frac{kg}{s} \qquad mdot_2 := 130\,\frac{kg}{s} \qquad\qquad (E5.5.5)$$

Design Requirements: The cold water outlet temperature is required to be

$$T_{2o} := 42°C \qquad\qquad (E5.5.6)$$

The pressure drop for each fluid is required to be

$$\Delta P \le 70\,kPa \qquad\qquad (E5.5.7)$$

Geometric parameters:

$N_p := 1$	Number of passes	(E5.5.8)
$\beta := 30\,deg$	Corrugation inclination angle (chevron angle)	(E5.5.9)
$N_t := 109$	Total number of plates	(E5.5.10)
$D_p := 0.2\,m$	Port diameter	(E5.5.11)
$\delta := 0.6\,mm$	Thickness of the plates	(E5.5.12)
$\lambda := 9\,mm$	Corrugation Pitch (=wavelength)	(E5.5.13)

Initially, we will guess the following boxed values of the sizing and compute the quantities. This will be helpful for understanding the procedure and for finally updating the calculations with the final values.

$$W_p := 0.524 \, \text{m} \qquad L_p := 1.022 \, \text{m} \qquad H_p := 0.5 \, \text{m} \qquad \text{(E5.5.14)}$$

The number of wavelengths per single plate is obtained by dividing the PHE width W_p by the corrugation pitch λ.

$$N_\lambda (W_p) := \frac{W_p}{\lambda} \qquad N_\lambda (W_p) = 58.222 \qquad \text{(E5.5.15)}$$

The number of channels for each fluid is calculated by dividing the total number of channels $(N_t + 1)$ by two and the number of passes N_p, as shown in Equation (5.164).

$$N_{c1} := \frac{N_t + 1}{2 \cdot N_p} \qquad \text{(E5.5.16)}$$

The number of channels for one fluid is the same as for the other fluid.

$$N_{c2} := N_{c1} \qquad N_{c2} = 55 \qquad \text{(E5.5.17)}$$

The corrugation amplitude is given by Equation (5.165). The amplitude is expressed as a function of the PHE height, H_p.

$$a (H_p) := \frac{1}{2} \cdot \left(\frac{H_p}{N_t + 1} - \delta \right) \qquad a (H_p) = 1973 \cdot \text{mm} \qquad \text{(E5.5.18)}$$

The corrugation ratio γ is defined as in Equation (5.166):

$$\gamma(H_p) := 4 \cdot \frac{a (H_p)}{\lambda} \qquad \gamma(H_p) = 0.877 \qquad \text{(E5.5.19)}$$

The PHE design is commonly limited to $\gamma < 1$. The corrugation length is given using Equation (5.168) by

$$L_\lambda (H_p) := \int_0^\lambda \sqrt{1 + \left(\frac{2 \cdot \pi \cdot a (H_p)}{\lambda} \right)^2 \cdot \cos \left(\frac{2 \cdot \pi}{\lambda} \cdot x \right)^2} \, dx \qquad \text{(E5.5.20)}$$

$$L_\lambda (H_p) = 12.365 \, \text{mm}$$

The heat transfer area for each fluid is calculated considering two surfaces per channel.

$$A_1 (W_p, L_p, H_p) := 2 \cdot L_\lambda (H_p) \cdot N_\lambda (W_p) \cdot L_p \cdot N_{c1} \qquad \text{(E5.5.21)}$$

$$A_2 (W_p, L_p, H_p) := A_1 (W_p, L_p, H_p) \qquad A_2 (W_p, L_p, H_p) = 80.933 \, \text{m}^2 \qquad \text{(E5.5.22)}$$

The projected area for the plate is calculated by

$$A_{p1} (W_p, L_p) := 2 \cdot W_p \cdot L_p \cdot N_{c1} \qquad A_{p2} (W_p, L_p) := A_{p1} (W_p, L_p)$$

$$A_{p2} (W_p, L_p) = 58.908 \, \text{m}^2 \qquad \text{(E5.5.23)}$$

Surface Enlargement Factor   Equation (5.171) gives the surface enlargement factor. The numerical value of the right-hand side depends on the sizing, which will not be known until the final sizing is updated in Equation (E5.5.14).

$$\Phi\left(W_p, H_p\right) := \frac{L_\lambda\left(H_p\right) \cdot N_\lambda\left(W_p\right)}{W_p} \qquad \Phi\left(W_p, H_p\right) = 1.374 \qquad \text{(E5.5.24)}$$

The hydraulic diameter is given from Equation (5.173):

$$D_h\left(W_p, H_p\right) := \frac{4 \cdot a\left(H_p\right)}{\Phi\left(W_p, H_p\right)} \qquad D_h\left(W_p, H_p\right) = 5.743\,\text{mm} \qquad \text{(E5.5.25)}$$

The free-flow area A_c for each fluid is obtained using Equation (5.170):

$$A_{c1}\left(W_p, H_p\right) := 2 \cdot a\left(H_p\right) \cdot W_p \cdot N_{c1} \qquad\qquad\qquad \text{(E5.5.26)}$$

$$A_{c2}\left(W_p, H_p\right) := A_{c1}\left(W_p, H_p\right) \qquad A_{c2}\left(W_p, H_p\right) = 0.114\,\text{m}^2 \qquad \text{(E5.5.27)}$$

The mass velocity, velocity, and Reynolds number are defined using the general definitions as

$$G_1\left(W_p, H_p\right) := \frac{mdot_1}{A_{c1}\left(W_p, H_p\right)} \qquad v_1 := \frac{G_1\left(W_p, H_p\right)}{\rho_1} = 1.25\,\frac{m}{s} \qquad \text{(E5.5.28)}$$

$$Re_1\left(W_p, H_p\right) := \frac{G_1\left(W_p, H_p\right) \cdot D_h\left(W_p, H_p\right)}{\mu_1} \qquad Re_1\left(W_p, H_p\right) = 1.501 \times 10^4$$
$$\text{(E5.5.29)}$$

After updating the final sizing, the velocities may be checked against the reasonable values in Table 5.4 (this is not a strict rule, but a good guideline). The velocity of 1.25 m/s is acceptable. The Reynolds number of 1.5×10^4 indicates a fully turbulent flow for the enhancement of heat transfer.

$$G_2\left(W_p, H_p\right) := \frac{mdot_2}{A_{c2}\left(W_p, H_p\right)} \qquad v_2 := \frac{G_2\left(W_p, H_p\right)}{\rho_2} = 1.147\,\frac{m}{s} \qquad \text{(E5.5.30)}$$

$$Re_2\left(W_p, H_p\right) := \frac{G_2\left(W_p, H_p\right) \cdot D_h\left(W_p, H_p\right)}{\mu_2} \qquad Re_2\left(W_p, H_p\right) = 7.642 \times 10^3$$
$$\text{(E5.5.31)}$$

Friction Factors   Martin's correlation (1996) developed with the Darcy friction factor is modified with the Fanning friction factor. Equation (5.174) provides

$$f_{o1}\left(W_p, H_p\right) := \begin{vmatrix} \left(1.56 \ln\left(Re_1\left(W_p, H_p\right)\right) - 3.0\right)^{-2} & \text{if } Re_1\left(W_p, H_p\right) \geq 2000 \\[2mm] \dfrac{16}{Re_1\left(W_p, H_p\right)} & \text{otherwise} \end{vmatrix}$$
$$\text{(E5.5.32)}$$

$$f_{o2}\left(W_p, H_p\right) := \left| \begin{array}{l} \left(1.56\ln\left(Re_2\left(W_p, H_p\right)\right) - 3.0\right)^{-2} \text{ if } Re_2\left(W_p, H_p\right) \geq 2000 \\[2ex] \dfrac{16}{Re_2\left(W_p, H_p\right)} \text{ otherwise} \end{array} \right.$$

$$(E5.5.33)$$

$$f_{m1}\left(W_p, H_p\right) := \left| \begin{array}{l} \dfrac{9.75}{Re_1\left(W_p, H_p\right)^{0.289}} \text{ if } Re_1\left(W_p, H_p\right) \geq 2000 \\[3ex] \dfrac{149.25}{Re_1\left(W_p, H_p\right)} + 0.9625 \text{ otherwise} \end{array} \right.$$

$$(E5.5.34)$$

$$f_{m2}\left(W_p, H_p\right) := \left| \begin{array}{l} \dfrac{9.75}{Re_2\left(W_p, H_p\right)^{0.289}} \text{ if } Re_2\left(W_p, H_p\right) \geq 2000 \\[3ex] \dfrac{149.25}{Re_2\left(W_p, H_p\right)} + 0.9625 \text{ otherwise} \end{array} \right.$$

$$(E5.5.35)$$

The Fanning friction factor for each fluid is given using Equation (5.174) by

$$f_1\left(W_p, H_p\right) := \left[\frac{\cos(\beta)}{\left(0.045 \cdot \tan(\beta) + 0.09 \cdot \sin(\beta) + \dfrac{f_{o1}\left(W_p, H_p\right)}{\cos(\beta)}\right)^{0.5}} \right.$$
$$\left. + \left(\frac{1 - \cos(\beta)}{\sqrt{3.8 \cdot f_{m1}\left(W_p, H_p\right)}}\right) \right]^{-2}$$

$$(E5.5.36)$$

$$f_1\left(W_p, H_p\right) = 0.1$$

$$f_2\left(W_p, H_p\right) := \left[\frac{\cos(\beta)}{\left(0.045 \cdot \tan(\beta) + 0.09 \cdot \sin(\beta) + \dfrac{f_{o2}\left(W_p, H_p\right)}{\cos(\beta)}\right)^{0.5}} \right.$$
$$\left. + \left(\frac{1 - \cos(\beta)}{\sqrt{3.8 \cdot f_{m2}\left(W_p, H_p\right)}}\right) \right]^{-2}$$

$$(E5.5.37)$$

$$f_2\left(W_p, H_p\right) = 0.102$$

Using Equation (5.175), the heat transfer coefficients are given by

$$h_1\left(W_p, H_p\right) := \frac{k_1}{D_h\left(W_p, H_p\right)} \cdot \left[0.205 \cdot Pr_1^{\frac{1}{3}} 1^{\frac{1}{6}} \cdot \left(f_1\left(W_p, H_p\right) \cdot Re_1\left(W_p, H_p\right)^2\right.\right.$$
$$\left.\left.\cdot \sin(2 \cdot \beta)\right)^{0.374}\right] \tag{E5.5.38}$$

$$h_1\left(W_p, H_p\right) = 1.787 \times 10^4 \cdot \frac{W}{m^2 \cdot K}$$

$$h_2\left(W_p, H_p\right) := \frac{k_2}{D_h\left(W_p, H_p\right)} \cdot \left[0.205 \cdot Pr_2^{\frac{1}{3}} 1^{\frac{1}{6}} \cdot \left(f_2\left(W_p, H_p\right) \cdot Re_2\left(W_p, H_p\right)^2\right.\right.$$
$$\left.\left.\cdot \sin(2 \cdot \beta)\right)^{0.374}\right] \tag{E5.5.39}$$

$$h_2\left(W_p, H_p\right) = 1.242 \times 10^4 \cdot \frac{W}{m^2 \cdot K}$$

The overall heat transfer coefficient is obtained using Equation (5.12) by

$$UA\left(W_p, L_p, H_p\right) := \frac{1}{\left[\dfrac{1}{h_1\left(W_p, H_p\right) \cdot A_1\left(W_p, L_p, H_p\right)} + \dfrac{\delta}{k_w \cdot A_1\left(W_p, L_p, H_p\right)} + \dfrac{1}{h_2\left(W_p, H_p\right) \cdot A_2\left(W_p, L_p, H_p\right)}\right]} \tag{E5.5.40}$$

$$UA\left(W_p, L_p, H_p\right) = 4.579 \times 10^5 \frac{W}{K}$$

ε-NTU Method   The heat capacities for both fluids are defined, and then the minimum and maximum heat capacities are obtained using the MathCAD built-in functions.

$$C_1 := mdot_1 \cdot c_{p1} \qquad C_1 = 5.858 \times 10^5 \frac{m^2 \cdot kg}{K \cdot s^3} \tag{E5.5.41}$$

$$C_2 := mdot_2 \cdot c_{p2} \qquad C_2 = 5.433 \times 10^5 \frac{m^2 \cdot kg}{K \cdot s^3} \tag{E5.5.42}$$

$$C_{min} := min\left(C_1, C_2\right) \qquad C_{min} = 5.433 \times 10^5 \frac{m^2 \cdot kg}{K \cdot s^3} \tag{E5.5.43}$$

$$C_{max} := max\left(C_1, C_2\right) \qquad C_{max} = 5.858 \times 10^5 \frac{m^2 \cdot kg}{K \cdot s^3} \tag{E5.5.44}$$

The heat capacity ratio C_r is defined as

$$C_r := \frac{C_{min}}{C_{max}} \qquad C_r = 0.927 \tag{E5.5.45}$$

The number of transfer units is defined as

$$\mathrm{NTU}\left(W_p, L_p, H_p\right) := \frac{UA\left(W_p, L_p, H_p\right)}{C_{min}} \qquad \mathrm{NTU}\left(W_p, L_p, H_p\right) = 0.843 \quad \text{(E5.5.46)}$$

The effectiveness of the PHE for counterflow is obtained using Equation (5.47):

$$\varepsilon_{PHE}\left(W_p, L_p, H_p\right) := \frac{1 - \exp\left[-\mathrm{NTU}\left(W_p, L_p, H_p\right) \cdot (1 - C_r)\right]}{1 - C_r \cdot \exp\left[-\mathrm{NTU}\left(W_p, L_p, H_p\right) \cdot (1 - C_r)\right]} \qquad \text{(E5.5.47)}$$

$$\varepsilon_{PHE}\left(W_p, L_p, H_p\right) = 0.465$$

The heat transfer rate for the PHE is obtained

$$q\left(W_p, L_p, H_p\right) := \varepsilon_{PHE}\left(W_p, L_p, H_p\right) \cdot C_{min} \cdot (T_{1i} - T_{2i}) \qquad \text{(E5.5.48)}$$

$$q\left(W_p, L_p, H_p\right) = 1.086 \times 10^7 \, W$$

The outlet temperatures for the hot and cold water are calculated using Equation (5.123) by

$$T_{1o}\left(W_p, L_p, H_p\right) := T_{1i} - \varepsilon_{PHE}\left(W_p, L_p, H_p\right) \cdot \frac{C_{min}}{C_1} \cdot (T_{1i} - T_{2i}) \qquad \text{(E5.5.49)}$$

$$T_{1o}\left(W_p, L_p, H_p\right) = 46.456^\circ C$$

$$T_{2o}\left(W_p, L_p, H_p\right) := T_{2i} - \varepsilon_{PHE}\left(W_p, L_p, H_p\right) \cdot \frac{C_{min}}{C_2} \cdot (T_{1i} - T_{2i}) \qquad \text{(E5.5.50)}$$

$$T_{2o}\left(W_p, L_p, H_p\right) = 41.995^\circ C$$

The outlet temperature of the cold fluid is close to $42^\circ C$ and so satisfies the requirement.

Pressure Drops   The frictional channel pressure drops for both fluids are obtained using Equation (5.176).

$$\Delta P_{f1}\left(W_p, L_p, H_p\right) := \frac{2 \cdot f_1\left(W_p, H_p\right) \cdot L_p}{D_h\left(W_p, H_p\right)} \cdot \frac{G_1\left(W_p, H_p\right)^2}{\rho_1} \cdot N_p$$

$$\Delta P_{f1}\left(W_p, L_p, H_p\right) = 54.518 \cdot kPa \qquad \text{(E5.5.51)}$$

$$\Delta P_{f2}\left(W_p, L_p, H_p\right) := \frac{4 \cdot f_2\left(W_p, H_p\right) \cdot L_p}{D_h\left(W_p, H_p\right)} \cdot \frac{G_2\left(W_p, H_p\right)^2}{2 \cdot \rho_2} \cdot N_p$$

$$\Delta P_{f2}\left(W_p, L_p, H_p\right) = 47.661 \cdot kPa \qquad \text{(E5.5.52)}$$

The connection and port pressure drops are obtained using Equations (5.177) and (5.178)

$$G_{p1} := \frac{4 \cdot \mathrm{mdot}_1}{\pi \cdot D_p^2} \qquad G_{p2} := \frac{4 \cdot \mathrm{mdot}_2}{\pi \cdot D_p^2} \qquad \text{(E5.5.53)}$$

$$\Delta P_{p1} := 1.5 \cdot N_p \cdot \frac{G_{p1}^2}{2 \cdot \rho_1} = 15.121 \cdot kPa$$

$$\Delta P_{p2} := 1.5 \cdot N_p \cdot \frac{G_{p2}^2}{2 \cdot \rho_2} = 12.881 \cdot kPa$$

(E5.5.54)

The total pressure drop for each fluid is the sum of the frictional pressure drop and the port pressure drop.

$$\Delta P_1 \left(W_p, L_p, H_p \right) := \Delta P_{f1} \left(W_p, L_p, H_p \right) + \Delta P_{p1}$$

$$\Delta P_1 \left(W_p, L_p, H_p \right) = 69.639 \, kPa$$

(E5.5.55)

$$\Delta P_2 \left(W_p, L_p, H_p \right) := \Delta P_{f2} \left(W_p, L_p, H_p \right) + \Delta P_{p2}$$

$$\Delta P_2 \left(W_p, L_p, H_p \right) = 60.542 \, kPa$$

(E5.5.56)

Find the Dimensions (W_p, L_p, and H_p) Using a MathCAD Find *Function* The sizing problem usually involves time-consuming calculations with a number of iterations. This can be simplified using MathCAD. In a plate heat exchanger, the thermal load (heat duty) and the hydrodynamic load (pressure drop) are interdependent, and also the pressure drops for both fluids are interdependent. Some combinations of the requirements for the heat duty and the pressure drops may not provide a convergence in the solution because they are interdependent. But, in the present case, the heat duty and one of the pressure drops appear sufficient to determine the size (W_p, L_p, and H_p). The present combination of the two requirements, (E5.5.58) and (E5.5.59), was chosen because ΔP_2 is not geometrically independent of ΔP_1. This combination allows ΔP_2 a degree of freedom for the third requirement; consequently allowing a quick convergence in the solution. Initial guess values are required for the *solve block* in MathCAD. Note that the word *Given* is a MathCAD command, not text (entering *space* on a keyboard after typing a word converts to a *text* region).

Initial guesses W_p : 0.5m L_p : 0.5m H_p : 0.5m   (E5.5.57)

Given

$$T_{2o} \left(W_p, L_p, H_p \right) = 42^\circ C \qquad \text{Requirements} \qquad \text{(E5.5.58)}$$

$$\Delta P_1 \left(W_p, L_p, H_p, \right) = 70 \, kPa \qquad \text{Requirements} \qquad \text{(E5.5.59)}$$

$$\begin{bmatrix} W_p \\ L_p \\ H_p \end{bmatrix} := \text{Find} \left(W_p, L_p, H_p \right)$$

(E5.5.60)

The results for the dimensions are

$$W_p = 0.524 \, m \qquad L_p = 1.022 \, m \qquad H_p = 0.499 \, m \qquad \text{(E5.5.61)}$$

Now we want to check the result of ΔP_2 from Equation (E5.5.75), whether it satisfies the requirement ($\Delta P_2 \leq 70$ kPa). After confirming the satisfaction, we now return to Equation (E5.5.14) and update the boxed dimensions with the results obtained here. The surface area density, which is the ratio of the heat transfer area to the volume between plates, is obtained as

$$\beta_1 := \frac{A_1\left(W_p, L_p, H_p\right)}{W_p \cdot \left(2 \cdot a\left(H_p\right)\right) \cdot L_p \cdot N_{c1}} = 698 \cdot \frac{m^2}{m^3} \qquad (E5.5.62)$$

$$\beta_2 := \frac{A_2\left(W_p, L_p, H_p\right)}{W_p \cdot \left(2 \cdot a\left(H_p\right)\right) \cdot L_p \cdot N_{c2}} = 698 \cdot \frac{m^2}{m^3} \qquad (E5.5.63)$$

Summary of Results and Geometry
Given information:

$T_{1i} = 65°C$	Hot wastewater inlet temperature	(E5.5.64)
$T_{2i} = 22°C$	Cold water inlet temperature	(E5.5.65)
$mdot_1 = 140\frac{kg}{s}$	Mass flow rate at the hot wastewaterside	(E5.5.66)
$mdot_2 = 130\frac{kg}{s}$	Mass flow rate at the cold water side	(E5.5.67)

Requirements:

$T_{2o} = 42°C$	Cold water outlet temperature	(E5.5.68)
$\Delta P_1 \leq 70\,kPa$	Pressure drop for cold water	(E5.5.69)
$\Delta P_2 \leq 70\,kPa$	Pressure drop for hot wastewater	(E5.5.70)

Dimensions for the sizing:

$$W_p = 0.524\,m \qquad L_p = 1.022\,m \qquad H_p = 0.499\,m \qquad (E5.5.71)$$

Outlet temperatures and pressure drops:

$T_{2o}\left(W_p, L_p, H_p\right) = 42°C$	Cold water outlet temperature	(E5.5.72)
$T_{1o}\left(W_p, L_p, H_p\right) = 46.451°C$	Hot wastewater outlet temperature	(E5.5.73)
$\Delta P_1\left(W_p, L_p, H_p\right) = 70\,kPa$	Pressure drop for hot wastewater	(E5.5.74)
$\Delta P_2\left(W_p, L_p, H_p\right) = 60.857\,kPa$	Pressure drop for cold water	(E5.5.75)

Geometry:

$N_p = 1$	Number of passes	(E5.5.76)
$N_t = 109$	Number of plates	(E5.5.77)
$\lambda = 9 \cdot mm$	Corrugation pitch (wavelength)	(E5.5.78)
$\delta = 0.6 \cdot mm$	Plate thickness	(E5.5.79)
$\beta = 30 \cdot deg$	Corrugation inclination angle (chevron angle)	(E5.5.80)

$$D_p = 0.2 \, \text{m} \qquad\qquad \text{Port diameter} \qquad\qquad (E5.5.81)$$

$$\gamma\left(H_p\right) = 0.874 \qquad\qquad \text{Corrugation aspect ratio less than 1} \qquad (E5.5.82)$$

$$\Phi\left(W_p, H_p\right) = 1.372 \qquad\quad \text{Surface enlargement factor} \qquad\quad (E5.5.83)$$

$$\beta_1 = 697.612 \frac{\text{m}^2}{\text{m}^3} \qquad\qquad \text{Surface area density} \qquad\qquad (E5.5.84)$$

$$A_2\left(W_p, L_p, H_p\right) = 80.821 \, \text{m}^2 \quad \text{Heat transfer area} \, (A_1 = A_2) \qquad (E5.5.85)$$

Thermal and hydraulic quantities:

$$D_h\left(W_p, H_p\right) = 5.734 \text{mm} \qquad\qquad \text{Hydraulic diameter} \qquad\qquad (E5.5.86)$$

$$\varepsilon_{PHE}\left(W_p, L_p, H_p\right) = 0.465 \qquad\qquad \text{Effectiveness} \qquad\qquad (E5.5.87)$$

$$q\left(W_p, L_p, H_p,\right) = 1.087 \times 10^7 \text{W} \qquad \text{Heat transferrate} \qquad\qquad (E5.5.88)$$

$$UA\left(W_p, L_p, H_p\right) = 4.581 \times 10^5 \frac{\text{W}}{\text{K}} \qquad \text{UA value} \qquad\qquad (E5.5.89)$$

The cold water outlet temperature and the pressure drops for each fluid satisfy the design requirements.

Comments: This problem can be further developed by looking for an unknown corrugation pitch λ and an unknown number of plates N_t under the same conditions of the requirements. A new MathCAD model was developed (not shown here) as a function of N_t and λ in addition to the dimensions of the exchanger. The results are shown here. Note that Figures E5.5.3, E5.5.4, and E5.5.5 were obtained with satisfaction of the same requirements shown in Equations (E5.5.58) and (E5.5.59). Figure E5.5.3 shows the surface area density β and the corrugation aspect ratio γ with respect to the corrugation pitch (wavelength) λ. The surface area density β increases with decreasing the corrugation pitch λ. However, the corrugation pitch λ is limited to a constraint of the corrugation aspect ratio ($\gamma < 1$ see Equation (5.166)). As a result of this, λ was taken to be 9 mm in Example 5.5.1.

Figure E5.5.4 shows the surface area density β and the corrugation aspect ratio γ with respect to the number of plates N_t. With increasing N_t, β increases while γ decreases. Figure E5.5.5 indicates that the volume of the exchanger decreases with increasing the surface area density β. Accordingly, we can reduce the volume of the exchanger (or increase the compactness) by increasing the number of plates N_t. In Example 5.5.1, $N_t = 109$ was used, but the number of plates N_t may be increased further until other constraints meet (such as manufacturability or hydraulic diameters). The new model was calculated using the guess values of $W_p = 0.5$ m, $L_p = 0.6$ m, and $H_p = 0.5$ m). The solution negligibly varied depending on the guess values. This is probably due to the fact that the heat transfer area and the free-flow area are not independent.

A *solve block* for the new MathCAD model is illustrated here. Note that the dimensions of the exchanger are now a function of N_t and λ.

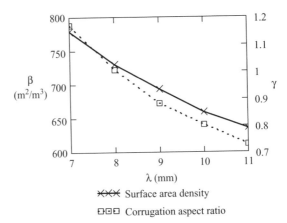

Figure E5.5.3   Surface area density β and corrugation aspect ratio γ versus corrugation pitch λ

Initial guess $W_p := 0.5\,\text{m}$ $L_p := 0.6\,\text{m}$ $H_p := 0.5\,\text{m}$ (E5.5.90)

Given   (E5.5.91)

$T_{2o}\left(W_p, L_p, H_p, N_t, \lambda\right) = 42\,^{\circ}\text{C}$ Requirement   (E5.5.92)

$\Delta P_1\left(W_p, L_p, H_p, N_t, \lambda\right) = 70\,\text{kPa}$ Requirement   (E5.5.93)

$$\begin{pmatrix} W_p\,(N_t, \lambda) \\ L_p\,(N_t, \lambda) \\ H_p\,(N_t, \lambda) \end{pmatrix} := \text{Find}\left(W_p, L_p, H_p\right) \qquad\qquad \text{(E5.5.94)}$$

The solution for the sizing is

$$W_p\,(N_t, \lambda) = 0.52\,\text{m} \quad L_p\,(N_t, \lambda) = 1.028\,\text{m} \quad H_p\,(N_t, \lambda) = 0.503\,\text{m} \qquad \text{(E5.5.95)}$$

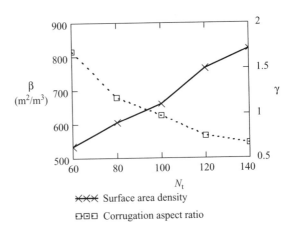

Figure E5.5.4   Surface area density β and corrugation aspect ratio γ versus number of plates N_t

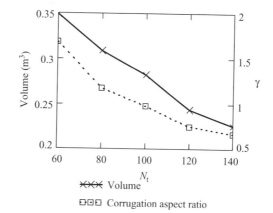

✕✕✕ Volume
◻◻◻ Corrugation aspect ratio

Figure E5.5.5 Volume and corrugation aspect ratio γ versus number of plates N_t

The volume of the heat exchanger is defined by

$$V_{PHE}(N_t, \lambda) := W_p(N_t, \lambda) \cdot L_p(N_t, \lambda) \cdot H_p(N_t, \lambda) \qquad (E5.5.96)$$

The surface area density is defined by

$$\beta_1(N_t, \lambda) := \frac{A_1\left(W_p(N_t, \lambda), L_p(N_t, \lambda), H_p(N_t, \lambda), N_t, \lambda\right)}{W_p(N_t, \lambda) \cdot \left(2 \cdot a\left(H_p(N_t, \lambda), N_t\right)\right) \cdot L_p(N_t, \lambda) \cdot N_{c1}(N_t)} \qquad (E5.5.97)$$

These expressions allow us to provide three plots for analysis of the corrugation pitch λ and the number of plates N_t, shown in Figures E5.5.3, E5,5.4, and E5.5.5.

5.6 PRESSURE DROPS IN COMPACT HEAT EXCHANGERS

The pumping power is an important part of the system design and the operating cost. The pumping power is proportional to the fluid pressure drop, which is associated with the fluid friction and the others (see Section 5.6.2). The typical design values of pressure drop are 70 kPa for water (a typical value in shell-and-tube heat exchanger) and 250 Pa for air (for compact heat exchangers with airflow near ambient pressures). A core flow in a heat exchanger is shown in Figure 5.27, wherein the contraction at the entrance and expansion at the exit are illustrated.

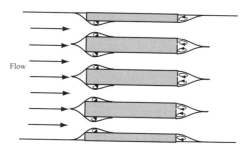

Flow

Figure 5.27 Core flow in a heat exchanger

5.6.1   Fundamentals of Core Pressure Drop

We want to derive a theoretical pressure drop in a unit-cell flow, as shown in Figure 5.28(a). The pressure drop within the core consists of two contributions: (1) the pressure loss caused by fluid friction; and (2) the pressure change due to momentum rate change in the core. The friction losses take into account both skin friction and drag effects. We consider first the core flow between section 2 and 3. The entrance and exit effects will be considered later. Consider a differential element as a control volume, as shown in Figure 5.28(b). Newton's second law of motion is applied to a control volume of fluid:

$$\sum F = ma = \underbrace{\left.\frac{d\,(mV)}{dt}\right|_{sys}}_{\substack{\text{Total rate of change of momentum}\\\text{within system}}} = \underbrace{\left.\frac{d\,(mV)}{dt}\right|_{CV}}_{\substack{\text{Rate of momentum stored}\\\text{in control volume}}} + \underbrace{\iint_{CS} V\rho V\, dA}_{\substack{\text{Rate of momentum out minus}\\\text{in across control surface}}} \qquad (5.180)$$

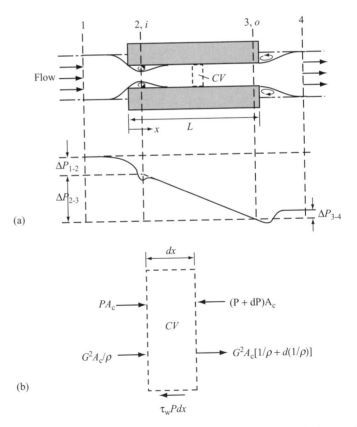

Figure 5.28 Pressure drops at the entrance and exit and the differential element (control volume), modified from Shah and Sekulic [17]

The forces acting on the control volume are the pressures on both sides and the shear stress on the surface.

$$\sum F = P A_c - (P + dP)\, A_c - \tau_w \mathrm{P} dx \tag{5.181}$$

where P is the pressure, P the perimeter, A_c the free-flow area, τ_w the shear stress, and P the wetted perimeter. The first term of the right-hand side of Equation (5.180) vanishes at steady state. The second term depicts the momentum entering or exiting on the control surface. The time rate of change of momentum is expressed in terms of mass velocity as

$$\dot{m} V = \rho V A_c V = \frac{G^2 A_c}{\rho} \tag{5.182}$$

The momentum entering $(-)$ and exiting $(+)$ on the control surface in Figure 5.28(b) is written by

$$\iint\limits_{CS} V \rho V\, dA = G^2 A_c \left(\frac{1}{\rho} + d\left(\frac{1}{\rho}\right) \right) - \frac{G^2 A_c}{\rho} \tag{5.183}$$

Inserting Equation (5.181) and (5.183) into Equation (5.180) gives

$$P A_c - (P + dP)\, A_c - \tau_w \mathrm{P} dx = G^2 A_c \left(\frac{1}{\rho} + d\left(\frac{1}{\rho}\right) \right) - \frac{G^2 A_c}{\rho} \tag{5.184}$$

Dividing by $A_c dx$ and rearranging it gives

$$-\frac{dP}{dx} = G^2 \frac{d}{dx}\left(\frac{1}{\rho}\right) + \tau_w \frac{\mathrm{P}}{A_c} \tag{5.185}$$

The friction factor defined in Equation (5.86) is rewritten here in terms of G.

$$f = \frac{\tau_w}{\frac{1}{2}\rho u_m^2} = \frac{2\rho \tau_w}{G^2} \tag{5.186}$$

The hydraulic diameter defined in Equation (5.26) is rewritten here

$$D_h = \frac{4 A_c}{\mathrm{P}} \tag{5.187}$$

Using Equations (5.186) and (5.187) and eliminating τ_w and P from Equation (5.185) gives

$$-\frac{dP}{dx} = G^2 \frac{d}{dx}\left(\frac{1}{\rho}\right) + \frac{4 f G}{2\rho D_h} = \frac{G^2}{2}\left(2\frac{d}{dx}\left(\frac{1}{\rho}\right) + \frac{4f}{\rho D_h}\right) \tag{5.188}$$

Integration of this from $x = 0$ to $x = L$ will provide the expression for the core pressure drop.

$$-\int\limits_2^3 dP = \frac{G^2}{2}\left(2 \int\limits_{x=0}^{x=L} \frac{d}{dx}\left(\frac{1}{\rho}\right) dx + \int\limits_{x=0}^{x=L} \frac{4f}{\rho D_h} dx \right) \tag{5.189}$$

Performing the integral from $\rho = \rho_i$ to $\rho = \rho_o$ gives

$$P_2 - P_3 = \frac{G^2}{2} \left(2 \int_{\rho_i}^{\rho_o} d\left(\frac{1}{\rho}\right) + \frac{4f}{D_h} \int_{x=0}^{x=L} \frac{1}{\rho} dx \right) \tag{5.190}$$

$$\Delta P_{2-3} = \frac{G^2}{2} \left[2\left(\frac{1}{\rho_o} - \frac{1}{\rho_i}\right) + \frac{4fL}{D_h} \frac{1}{L} \int_{x=0}^{x=L} \frac{1}{\rho} dx \right] \tag{5.191}$$

Finally, the pressure drop between section 2 and 3 is expressed as

$$\Delta P_{2-3} = \frac{G^2}{2\rho_i} \left[2\left(\frac{\rho_i}{\rho_o} - 1\right) + \frac{4fL}{D_h} \rho_i \left(\frac{1}{\rho}\right)_m \right] \tag{5.192}$$

where the mean specific volume is defined by

$$\left(\frac{1}{\rho}\right)_m = \frac{1}{L} \int_{x=0}^{x=L} \frac{1}{\rho} dx \tag{5.193}$$

The mean specific volume can be expressed [17] by

$$\left(\frac{1}{\rho}\right)_m = \frac{1}{\rho_m} = \frac{1}{2}\left(\frac{1}{\rho_i} + \frac{1}{\rho_o}\right) \tag{5.193a}$$

5.6.2   Core Entrance and Exit Pressure Drops

The core entrance pressure drop is composed of two parts: (1) the pressure drop due to the flow area change, and (2) the pressure losses that follow sudden contraction. Consider a flow between section 1 and 2. The first part of the pressure drop can be developed from the Bernoulli equation (frictionless and incompressible flow) as

$$\frac{P}{\rho} + \frac{V^2}{2} + gz = \text{constant} \tag{5.194}$$

The pressure drop between section 1 and 2 in Figure 5.28(a) is then expressed by

$$P_1 - P_2 = \frac{1}{2}\rho_i \left(u_2^2 - u_1^2\right) = \frac{\rho_i u_2^2}{2} \left(1 - \left(\frac{u_1}{u_2}\right)^2\right) \tag{5.195}$$

where ρ_i is the fluid density at the core entrance and it is assumed that $\rho_i = \rho_1 = \rho_2$ in Figure 5.28(a). The continuity equation gives

$$\rho_i u_1 A_{c1} = \rho_i u_2 A_{c2} = \rho_i u_3 A_{c3} = \rho_i u_4 A_{c4} \tag{5.196}$$

The velocity ratio becomes the porosity σ, which is the ratio of free-flow area A_c to frontal area A_{fr}.

$$\frac{u_1}{u_2} = \frac{A_{c2}}{A_{c1}} = \frac{A_{c3}}{A_{c4}} = \frac{A_c}{A_{fr}} = \sigma \tag{5.197}$$

Equation (5.195) can be expressed in terms of G.

$$\Delta P_{1-2} = \frac{1}{2}\rho_i \left(u_2^2 - u_1^2\right) = \frac{G^2}{2\rho_i}\left(1 - \sigma^2\right) \tag{5.198}$$

Since this is derived from the Bernoulli equation, it is based on the assumption of a frictionless and incompressible flow. Friction losses due to the contraction must be added to this equation. The total pressure drop at the core entrance is then given as

$$\Delta P_{1-2} = \frac{G^2}{2\rho_i}\left(1 - \sigma^2 + K_c\right) \tag{5.199}$$

where K_c is the contraction loss coefficient. The core exit pressure drop is similarly developed as

$$\Delta P_{3-4} = \frac{G^2}{2\rho_o}\left(1 - \sigma^2 - K_e\right) \tag{5.200}$$

where K_e is the expansion loss coefficient.

The total pressure drop is the sum of those in Equations (5.192), (5.199), and (5.200) as shown in Figure 5.28(a).

$$\Delta P = \Delta P_{1-2} + \Delta P_{2-3} - \Delta P_{3-4} \tag{5.201}$$

The core pressure drop is obtained by

$$\Delta P = \frac{G^2}{2\rho_i}\left[\underbrace{1 - \sigma^2 + K_c}_{\text{Entrance effect}} + \underbrace{2\left(\frac{\rho_i}{\rho_o} - 1\right)}_{\text{Momentum effect}} + \underbrace{\frac{4fL}{D_h}\rho_i\left(\frac{1}{\rho}\right)_m}_{\text{Core friction}} - \underbrace{\left(1 - \sigma^2 - K_e\right)\frac{\rho_i}{\rho_o}}_{\text{Exit effect}}\right] \tag{5.202}$$

Generally, the frictional pressure drop is the dominate term, about 90 percent or more of ΔP for gas in many compact heat exchangers. The entrance effect represents the pressure loss, and the exit effect in many cases represents a pressure rise; thus, the net effect of entrance and exit losses is usually compensating [17].

5.6.3 Contraction and Expansion Loss Coefficients

The loss coefficients at the entrance and exit for four different flow passages were studied by Kays [23] in 1950, wherein four graphs on both the contraction loss coefficient K_c and the expansion coefficient K_e were provided in Figures 5.30 to 5.33. The four graphs have been widely used for design purposes. In our design method, we have a problem that the graphs do not work with MathCAD. We need the equations, not the graphs. Efforts were made by the author to provide the mathematical expressions rather than the graphs by tracking the literature related to this problem. Most of the equations were found in Kays [23] and some in Rouse [24] and also by curve fitting from graphs. Finally, the equations for the four graphs were obtained and the contraction loss coefficient K_c and the expansion coefficient K_e for four different flow passages were presented in Figures 5.30 to 5.33.

The contraction loss coefficient K_c is given by

$$K_c = \frac{1 - 2C_c + C_c^2(2K_d - 1)}{C_c^2} \tag{5.203}$$

where $C_c(\sigma)$ is the *jet contraction ratios* (a function of porosity σ) and K_d the velocity-distribution coefficient (a function of Re). They are usually dependent of the flow geometry and will be discussed later in this section.

The expansion loss coefficient K_e is given by

$$K_e = 1 - 2K_a\sigma + \sigma^2 \tag{5.204}$$

The friction factor f for all the loss coefficient calculations for a turbulent flow is given by

$$f = 0.049 Re^{-0.2} \tag{5.205}$$

For a laminar flow, the friction factor used in the calculations is given by

$$f = \frac{16}{Re} \tag{5.206}$$

These equations are commonly used in the following four flow geometries.

5.6.3.1 Circular-Tube Core

For multiple-circular-tube core, the velocity-distribution coefficient K_d is given by

$$K_{d\_tube} = \begin{cases} 1.09068\left[4f(Re)\right] + 0.05884\sqrt{4f(Re)} + 1 & \text{if } Re \geq 2{,}300 \\ 1.33 & \text{if } Re < 2{,}300 \end{cases} \tag{5.207}$$

The jet contraction ratio, $C_c = A_2/A_3$, is a function of porosity σ as shown in Figure 5.29. The tabulated data of jet contraction ratio for circular-tube and flat-tube geometries were found in Rouse [24]. Curve fitting was used to produce an equation for circular-tube core:

$$C_c = 4.374 \times 10^{-4} \exp\left(6.737\sqrt{\sigma}\right) + 0.621 \tag{5.208}$$

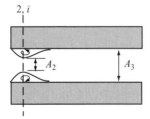

Figure 5.29 Jet contraction ratio, related to Figure 5.28

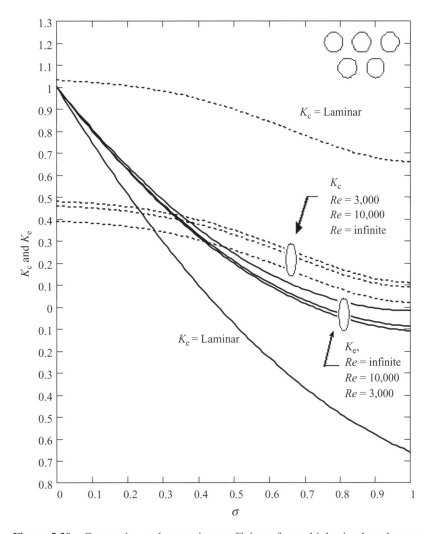

Figure 5.30   Contraction and expansion coefficients for multiple circular-tube core

5.6.3.2   *Square-Tube Core*

For multiple-square-tube core, the velocity-distribution coefficient K_d is given by

$$K_{d\_square} = \begin{cases} 1 + 1.17\left(K_{d\_tube} - 1\right) & \text{if } Re \geq 2{,}300 \\ 1.39 & \text{if } Re < 2{,}300 \end{cases} \tag{5.209}$$

The jet contraction ratio C_c for the square-tube geometry was assumed to be the same as the circular-tube geometry. The jet contraction ratio C_c for the square-tube geometry is given by

$$C_c = 4.374 \times 10^{-4} \exp\left(6.737\sqrt{\sigma}\right) + 0.621 \tag{5.210}$$

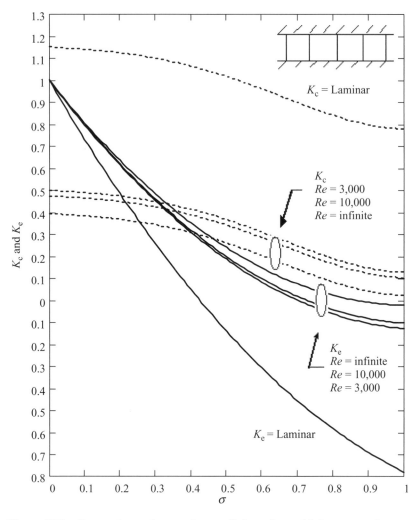

Figure 5.31   Contraction and expansion coefficients for multiple square-tube core

5.6.3.3   *Flat-Tube Core*

For multiple flat-tube core, the velocity-distribution coefficient K_d is given by

$$K_{d\_flat} = \begin{cases} 0.750\left[4f\,(Re)\right] + 0.024\sqrt{4f\,(Re)} + 1 & \text{if } Re \geq 2{,}300 \\ 1.2 & \text{if } Re < 2{,}300 \end{cases} \tag{5.211}$$

The tabulated data of the jet contraction ratio for flat-tube geometries were found in Rouse [24]. Curve fitting was used to produce an equation for the flat-tube geometry, which is given by

$$C_c = 4.374 \times 10^{-4} \exp\left(6.737\sigma\right) + 0.621 \tag{5.212}$$

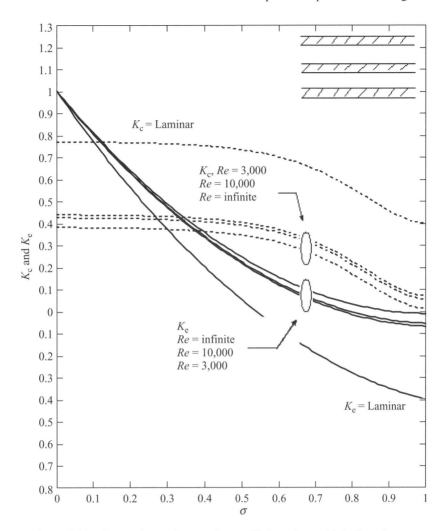

Figure 5.32   Contraction and expansion coefficients for multiple flat-tube core

5.6.3.4   *Triangular-Tube Core*

For multiple-triangular-tube core, the velocity-distribution coefficient K_d is given by

$$K_{d\_tri} = \begin{cases} 1 + 1.29\left(K_{d\_tube} - 1\right) & \text{if } Re \geq 2{,}300 \\ 1.43 & \text{if } Re < 2{,}300 \end{cases} \tag{5.213}$$

The jet contraction ratio for triangular-tube geometries is the same as the circular-tube geometry, which is given by

$$C_c = 4.374 \times 10^{-4} \exp\left(6.737\sqrt{\sigma}\right) + 0.621 \tag{5.214}$$

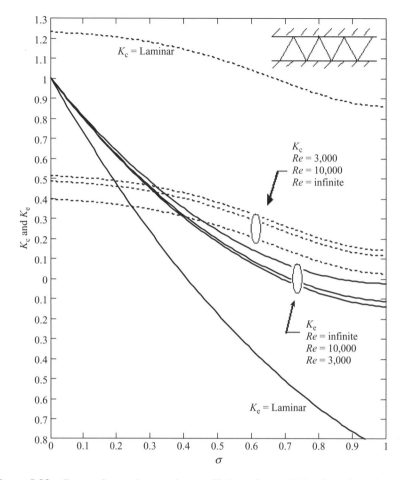

Figure 5.33 Contraction and expansion coefficients for multiple triangular-tube core

5.7 FINNED-TUBE HEAT EXCHANGERS

Finned-tube heat exchangers have been used for heat exchange between gases and liquids for many years. In a gas-to-liquid exchanger, the heat transfer coefficient on the liquid side is generally one order of magnitude higher than that on the gas side. Hence, to have balanced thermal conductances on both sides for a minimum-size heat exchanger, fins are used on the gas side to increase the surface area. Figure 5.34 shows three important finned-tube heat exchanger construction types. Figure 5.34(a) shows circular finned-tube geometry, Figure 5.34(b) shows the plate finned-tube geometry, and Figure 5.34(c) shows the plate-fin flat-tube geometry. These exchangers are widely used in the air conditioning, refrigerator, and automotive industry. Some examples are cooling towers, evaporators, condensers, and radiators.

Finned-tube exchangers can withstand high pressure on the tube side. The highest temperature is again limited by the type of bonding, materials employed, and material thickness. Finned-tube exchangers usually are less compact than plate-fin exchangers. Finned-tube exchangers with a surface area density of about 3,300 m^2/m^3 are available

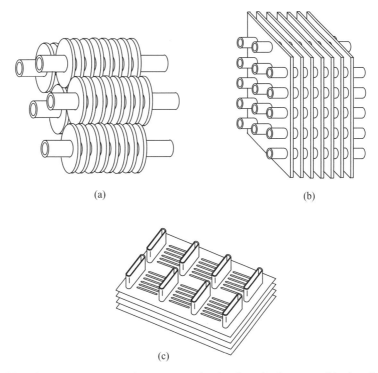

Figure 5.34   Finned-tube heat exchangers (a) circular finned-tube type, (b) plate finned-tube type, and (c) louvered plate-fin flat-tube type

commercially. On the fin side, the desired surface area can be achieved through the proper fin density and fin geometry. Typical fin densities for plate fins vary from 250 to 800 fins per meter (6 to 20 fins per inch), fin thicknesses vary from 0.08 to 0.25 mm, and fin flow lengths vary from 25 to 250 mm. A plate finned-tube exchanger with 400 fins per meter has a surface area density of about $720 \, m^2/m^3$ [17]. Since the liquid-side quantities were discussed previously in Section 5.4, Shell-and-Tube Heat Exchangers, we focus here on the gas-side quantities.

5.7.1   Geometrical Characteristics

The basic core geometry for an idealized single-pass cross-flow circular finned-tube exchanger with a staggered tube arrangement is shown in Figure 5.35. We assume that the staggered tube arrangement has an equilateral triangular shape.

The geometrical characteristics are derived for the staggered arrangement in Figures 5.35 and 5.36. The total number of tubes can be calculated in terms of tube pitches P_t and P_c and the dimensions of the exchanger, where P_t is the transverse tube pitch and P_c the longitudinal tube pitch. Note that, in Figure 5.35, the number of tubes in the first row perpendicular to the flow direction is L_3/P_t and the number of tubes in the second row is $(L_3/P_t - 1)$. Also, note that the number of rows with the same pattern as the first row is $(L_2/P_c+1)/2$ and the number of rows with the same

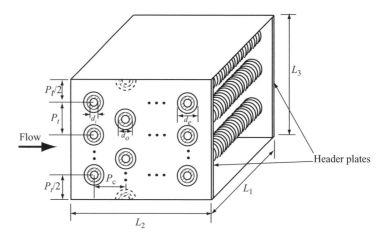

Figure 5.35 Circular finned-tube heat exchanger

pattern as the second row is $(L_2/P_c-1)/2$. Therefore, the total number of tubes is

$$N_t = \frac{L_3}{P_t}\frac{\dfrac{L_2}{P_c}+1}{2} + \left(\frac{L_3}{P_t}-1\right)\frac{\dfrac{L_2}{P_c}-1}{2} \tag{5.215}$$

The total heat transfer area A_t is composed of the primary surface area A_p and the fin surface area A_f. The primary surface area consists of the tube surface area, except the fin base area (Figure 5.36(b)) and two header plate surface areas except the tube outer diameters (Figure 5.35). N_f is the number of fins per unit length, as shown in Figure 5.36(b). The primary surface area is

$$A_p = \pi d_o \left(L_1 - \delta N_f L_1\right) N_t + 2 \left(L_2 L_3 - \frac{\pi d_o^2}{4} N_t\right) \tag{5.216}$$

The fin surface area is

$$A_f = \left(\frac{2\pi \left(d_e^2 - d_o^2\right)}{4} + \pi d_e \delta\right) N_f L_1 N_t \tag{5.217}$$

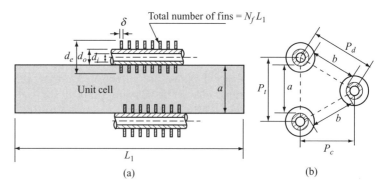

Figure 5.36 Unit cell of a staggered finned-tube arrangement

The total heat transfer area is then

$$A_t = A_p + A_f \tag{5.218}$$

We consider a unit cell of a staggered finned-tube exchanger, as shown in Figure 5.36. The total minimum free-flow area can be then calculated by multiplying the unit cell area by the number of unit cells (L_3/P_t). The minimum free-flow area occurs either at a plane through a or at two planes through b in Figure 5.36(b) whichever the smaller area takes the minimum free-flow area. However, since two planes through b are always greater than one plane through a for the equilateral triangular arrangement, the minimum free-flow area must occur at the plane through a. In other words, the flow through a is not restricted by the diagonal planes through b because $2b > a$. Noting that $a = (P_t - d_o)$, the minimum free-flow area is calculated using the dark area in Figure 5.36(a) as

$$A_c = \frac{L_3}{P_t} \left[(P_t - d_o) \, L_1 - (d_e - d_o) \, \delta N_f L_1 \right] \tag{5.219}$$

5.7.2   Flow Properties

The hydraulic diameter is given using Equation (5.26) by

$$D_h = \frac{4A_c L_2}{A_t} \tag{5.220}$$

The mass velocity G is given by

$$G = \frac{\dot{m}}{A_c} \tag{5.221}$$

The Reynolds number is given by

$$\mathrm{Re} = \frac{\rho v D_h}{\mu} = \frac{G D_h}{\mu} \tag{5.222}$$

The pin pitch is defined by

$$P_f = \frac{1}{N_f} \tag{5.223}$$

5.7.3   Thermal Properties

In compact heat exchangers, the *Stanton number St* is often used, which is defined by

$$St = \frac{Nu}{Re \cdot Pr} \tag{5.224}$$

The *Nusselt number* is given by

$$Nu = \frac{h D_h}{k} \tag{5.225}$$

The *Reynolds number* is given using Equation (5.29) by

$$Re = \frac{GD_h}{\mu} \tag{5.226}$$

With the kinematic viscosity ($v = \mu/\rho$), and the *thermal diffusivity* ($\alpha = k/\rho c_p$), the *Prandtl number, Pr*, can be written as

$$Pr = \frac{v}{\alpha} = \frac{\dfrac{\mu}{\rho}}{\dfrac{k}{\rho c_p}} = \frac{c_p \mu}{k} \tag{5.227}$$

The *Stanton number, St*, is then expressed using Equations (5.224) to (5.227) as

$$St = \frac{Nu}{Re \cdot Pr} = \frac{\dfrac{hD_h}{k}}{\dfrac{GD_h}{\mu} \dfrac{c_p \mu}{k}} = \frac{h}{Gc_p} \tag{5.228}$$

It is customary to use the *Colburn factor, j*, to represent the thermal characteristics of the compact heat exchangers. The Colburn factor j is defined and expressed using Equation (5.228) as

$$j = St \cdot Pr^{\frac{2}{3}} = \frac{h}{Gc_p} Pr^{\frac{2}{3}} \tag{5.229}$$

This gives the convection heat transfer coefficient h in terms of the Colburn factor j as

$$h = j \frac{Gc_p}{Pr^{\frac{2}{3}}} \tag{5.230}$$

5.7.4 Correlations for Circular Finned-Tube Geometry

The fin geometry, helically wrapped (or extruded), circular fins on a tube similar to the one shown in Figure 5.34(a), are commonly used in process and waste heat industry. Briggs and Young [21] recommended a correlation for the Colburn factor j.

$$j = 0.134 Re^{-0.319} \left(\frac{P_f - \delta}{d_e - d_o} \right)^{0.2} \left(\frac{P_f - \delta}{\delta} \right)^{0.11} \tag{5.231}$$

where P_f is the fin pitch and the symbols are found in Figures 5.35 and 5.36. From the Colburn factor j, the convection heat transfer coefficient h can be calculated.

For friction factors, Robinson and Briggs [20] recommended the following correlation:

$$f_{RB} = 9.465 Re^{-316} \left(\frac{P_t}{d_o} \right)^{-0.927} \left(\frac{P_t}{P_d} \right)^{0.515} \tag{5.232}$$

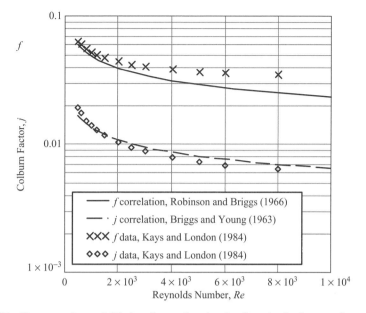

Figure 5.37 Heat transfer and friction factor for circular finned-tube heat exchangers. Graph shows correlations of Briggs and Young (1963) and Robinson and Briggs (1966) and experimental data for Surface CF8.72C, Kays and London (1984)

The friction factor f_{RB} used by Robinson and Briggs was given by

$$f_{RB} = \frac{\Delta P \rho}{N_r G^2} = \frac{\Delta P \rho}{\frac{L_2}{P_c} G^2} = \frac{\Delta P \rho}{G^2} \frac{P_c}{L_2} \tag{5.233}$$

where N_r is the number of rows that is obtained by dividing the flow length L_2 by the longitudinal pitch P_c in Figure 5.35. The general friction factor f was defined in Equation (5.93) and expressed in terms of ΔP for the geometry shown in Figure 5.35 by

$$f = \frac{\Delta P \rho}{G^2} \frac{D_h}{L_2} \tag{5.234}$$

The relationship between f and f_{RB} is given by

$$f = \frac{1}{2} \frac{D_h}{P_c} f_{RB} \tag{5.235}$$

The correlations for the Colburn factor j and the friction factor f were compared with the experimental data of CF8.72C by Kays and London [8] (Figure 5.37).

5.7.5 Pressure Drop

For a circular finned-tube heat exchanger, the tube outside flow in each tube row experiences a contraction and an expansion. Thus, the pressure losses associated with a tube row within the core are of the same order of magnitude as those at the entrance with the first tube row and those at the exit with the last tube row. Consequently, the entrance and exit pressure drops are not calculated separately, but they are generally lumped into the friction factor (which is generally derived experimentally) for individually finned tubes. By eliminating the entrance and exit terms from Equation (5.202), the total pressure drop associated with the core becomes [17]

$$\Delta P = \frac{G^2}{2\rho_i} \left[2 \left(\frac{\rho_i}{\rho_o} - 1 \right) + \frac{4fL}{D_h} \rho_i \left(\frac{1}{\rho} \right)_m \right] \tag{5.236}$$

where $(1/\rho)_m$ was defined in Equation (5.193a) and is rewritten here for convenience.

$$\left(\frac{1}{\rho} \right)_m = \frac{1}{\rho_m} = \frac{1}{2} \left(\frac{1}{\rho_i} + \frac{1}{\rho_o} \right) \tag{5.237}$$

It should be emphasized that the friction factor f in Equation (5.236) is based on the hydraulic diameter.

5.7.6 Correlations for Louvered Plate-Fin Flat-Tube Geometry

Figure 5.38 shows a widely used automotive radiator geometry, having louvered plate fins on flat tubes. Achaichia and Cowell [22] developed a correlation for the inline louver fin geometry. Their test data span $1.7 \le P_f \le 3.44$ mm, $8 \le P_t \le 14$ mm, $1.7 \le$

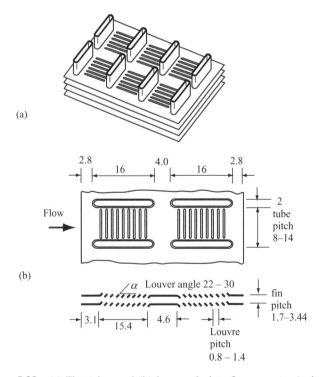

Figure 5.38 (a) Flat tubes and (b) louvered plate fin geometry (unit: mm)

$P_f \leq 3.44$ mm and $22 \leq \alpha \leq 30°$ [25]. The Reynolds number limits are $150 \leq Re \leq 3,000$.

The Colburn factor j is given by

$$j = 1.234\gamma \, Re^{-0.59} \left(\frac{P_t}{P_L}\right)^{-0.09} \left(\frac{P_f}{P_L}\right)^{-0.04} \tag{5.238}$$

where

$$\gamma = \frac{1}{\alpha}\left(0.936 - \frac{243}{Re} - 1.76\frac{P_f}{P_L} + 0.995\alpha\right) \tag{5.239}$$

where P_L is the louver pitch, P_t the tube pitch, P_f the fin pitch, and α the Louver angle (must be in degrees).

The friction factor f is given by

$$f = 834 \cdot CF \cdot P_f^{-0.22} P_L^{0.25} P_t^{0.26} H_L^{0.33} \left[Re^{(0.318\log_{10} Re - 2.25)}\right]^{1.07} \tag{5.240}$$

where CF is the correction factor (see Equation (5.242)). H_L is the louver height and can be calculated by

$$H_L = P_L \sin(\alpha \deg) \tag{5.241}$$

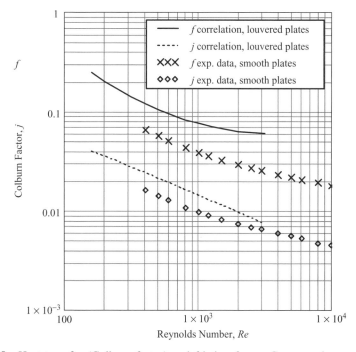

Figure 5.39 Heat transfer (Colburn factor) and friction factor. Compare the correlations for louvered plate-fin flat-tube geometry developed by Achaichia and Cowell (1988) and the experimental data from a smooth plate-fin flat-tube geometry obtained by Kays and London (1984)

A conversion factor (CF) is inserted into Equation (5.240) in order to correct the units in dimensions because, in the original equations, all dimensions were in millimeters. Hence, we can enter meters in Equation (5.240). Meters is the default metric unit in MathCAD.

$$CF = \left(\frac{1000}{m}\right)^{(-0.22+0.25+0.26+0.33)} \qquad (5.242)$$

where m is the meter. The correlations, Equations (5.238) and (5.240), are compared with the experimental data of Kays and London [8] with smooth plate-fin flat tube geometry without louvers. The correlations show an enhanced heat transfer, but a higher pressure drop with louvered fins. See Figure 5.39.

Example 5.7.1 Circular Finned-Tube Heat Exchanger Cold air is heated by hot wastewater using a circular finned-tube heat exchanger. The cold air enters at 5°C with a flow rate of 124 kg/s and will leave at 30°C. The hot water enters at a flow rate of 50 kg/s and 92°C. The allowable air pressure loss is 250 Pa and the allowable water pressure loss is 3 kPa. The thermal conductivities of the aluminum fins and the copper tubes are 200 W/mK and 385 W/mK, respectively. The tubes of the exchanger have an equilateral triangular pitch arrangement. The tube pitch is 4.976 cm (1.959 in.) and the fin pitch is 3.46 fins/cm (8.8 fins/in.). The fin thickness is 0.0305 cm (0.012 in.). The three diameters are given as $d_i = 2.2098$ cm for the tube inside diameter, $d_o = 2.601$ cm for the tube outside diameter, and $d_e = 4.412$ cm for the fin outside diameter. Design the circular finned-tube heat exchanger by determining the dimensions of the exchanger, outlet temperatures, effectiveness, and pressure drops. See Figures E5.7.1 to E5.7.3.

MathCAD Format Solution The design concept is to develop a MathCAD model as a function of dimensions (L_1, L_2, and L_3) in Figure E5.7.2 and solve for the dimensions.

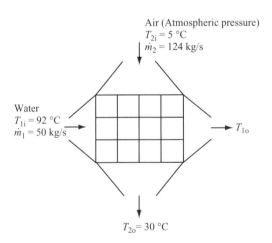

Figure E5.7.1   Cross-flow heat exchanger

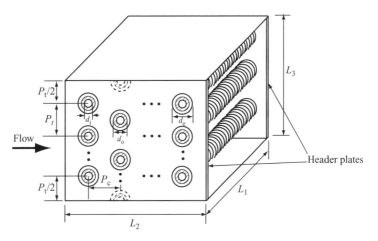

Figure E5.7.2   Circular finned-tube heat exchanger

Properties: The thermophysical properties of cold air and wastewater are obtained from Tables A.2 and A.12 in Appendix A, assuming the wastewater outlet temperature of 76°C.

$$T_{water} := \frac{92°C + 76°C}{2} = 84 \cdot °C \quad T_{air} := \frac{5°C + 30°C}{2} = 17.5 \cdot °C \qquad (E5.7.1)$$

Water (subscript 1) - tube side   Air (subscript 2) - finned - tube side

$\rho_1 := 974 \, \dfrac{kg}{m^3}$ ρ_2 : defined later

$c_{p1} := 4196 \, \dfrac{J}{kg \cdot K}$ $c_{p2} := 1007 \, \dfrac{J}{kg \cdot K}$

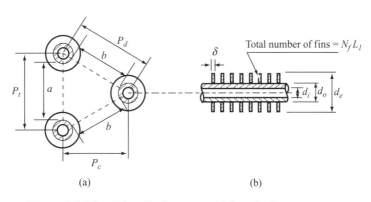

Figure E5.7.3   Unit cell of a staggered finned-tube arrangement

$$k_1 := 0.668 \frac{W}{m \cdot K} \qquad\qquad k_2 := 0.0255 \frac{W}{m \cdot K}$$

$$\mu_1 := 355 \cdot 10^{-6} N \cdot \frac{s}{m^2} \qquad \mu_2 := 1.798 \cdot 10^{-5} N \cdot \frac{s}{m^2} \qquad\text{(E5.7.2)}$$

$$Pr_1 := 2.22 \qquad\qquad Pr_2 := 0.710$$

The thermal conductivities for copper tubes and aluminum fins are given

$$k_{co} := 385 \frac{W}{m \cdot K} \quad k_{al} := 200 \frac{W}{m \cdot K} \qquad\text{(E5.7.3)}$$

Given information: The inlet temperatures are given as

$$
\begin{array}{lll}
T_{1i} := 92°C & \text{Hot water inlet temperature} & \\
T_{2i} := 5°C & \text{Cold air inlet temperature} & \text{(E5.7.4)}
\end{array}
$$

The mass flow rates for both fluids are defined as

$$md_1 := 50 \frac{kg}{s} \quad md_2 := 124 \frac{kg}{s} \qquad\text{(E5.7.5)}$$

Design requirement:

$$
\begin{array}{lll}
T_{2o} = 30°C & \text{Air outlet temperature} & \text{(E5.7.6)} \\
\Delta P_2 \le 250\,\text{Pa} & \text{For air side} & \text{(E5.7.7)} \\
\Delta P_1 \le 3\,\text{kPa} & \text{For water side} & \text{(E5.7.8)}
\end{array}
$$

Geometric information:

$$
\begin{array}{lll}
d_i := 2.2098\,\text{cm} & \text{Tube inside diameter} & \text{(E5.7.9)} \\
d_o := 2.601\,\text{cm} & \text{Tube outside diameter} & \text{(E5.7.10)} \\
d_e := 4.412\,\text{cm} & \text{Fin outside diameter} & \text{(E5.7.11)} \\
N_f := 3.46\,\text{cm}^{-1} & \text{Number of fins per unit length} & \text{(E5.7.12)} \\
P_t := 4.976\,\text{cm} & \text{Transverse tube pitch} & \text{(E5.7.13)} \\
P_c := P_t \cdot \sin\left(\frac{\pi}{3}\right) = 4.309\,\text{cm} & \text{Longitudinal pitch} & \text{(E5.7.14)} \\
\delta := 0.0305\,\text{cm} & \text{Fin thickness} & \text{(E5.7.15)}
\end{array}
$$

Assume the Sizing of the Heat Exchanger   Initially, assume the dimensions of the heat exchanger to have the following numerical values and update the dimensions when the final size is obtained.

$$L_1 := 4.451\,\text{m} \quad L_2 := 7.559\,\text{cm} \quad L_3 := 3.006\,\text{m} \qquad\text{(E5.7.16)}$$

The total number of tubes is obtained using Equation (5.215) as

$$N_t(L_2, L_3) := \frac{L_3}{P_t} \cdot \frac{\frac{L_2}{P_c} + 1}{2} + \left(\frac{L_3}{P_t} - 1\right) \cdot \frac{\frac{L_2}{P_c} - 1}{2} \tag{E5.7.17}$$

$$N_t(L_2, L_3) = 105.588$$

Note that N_t is expressed as a function of L_2 and L_3 (which is the design concept). The primary surface area A_p is given in Equation (5.216) as

$$A_p(L_1, L_2, L_3) := \pi \cdot d_o \cdot (L_1 - \delta \cdot N_f \cdot L_1) \cdot N_t(L_2, L_3)$$

$$+ 2 \cdot \left(L_2 \cdot L_3 - \frac{\pi \cdot d_o^2}{4} \cdot N_t(L_2, L_3)\right) \tag{E5.7.18}$$

$$A_p(L_1, L_2, L_3) = 34.692 \, \text{m}^2$$

The fin surface area A_f is given in Equation (5.217) by

$$A_f(L_1, L_2, L_3) := \left[\frac{2 \cdot \pi \cdot (d_e^2 - d_o^2)}{4} + \pi \cdot d_e \cdot \delta\right] \cdot N_f \cdot L_1 \cdot N_t(L_2, L_3) \tag{E5.7.19}$$

$$A_f(L_1, L_2, L_3) = 331.281 \, \text{m}^2$$

The factor of 2 on the right-hand side of Equation (E5.7.19) is due to two sides of a fin. The total heat transfer surface area A_2 on the air side is then

$$A_2(L_1, L_2, L_3) := A_p(L_1, L_2, L_3) + A_f(L_1, L_2, L_3)$$
$$A_2(L_1, L_2, L_3) = 366 \, \text{m}^2 \tag{E5.7.20}$$

The dimension b must be the same as a in the equilateral triangular tube arrangement.

$$a := (P_t - d_o) \quad a = 2.375 \, \text{cm} \tag{E5.7.21}$$

$$b := a \quad b = 2.375 \, \text{cm} \tag{E5.7.22}$$

It is obvious for the equilateral triangular tube arrangement that $2b > a$ in Figure E5.7.3. For the staggered tube arrangement, the minimum free-flow area could occur either through the front row (area a) or through the diagonals (area $2b$), depending on whichever is larger. The area blocked by the circular fins must be included when computing the minimum free-flow area. The minimum free-flow area A_{c2} on the air side is given in Equation (5.219) as

$$A_{c2}(L_1, L_3) := \frac{L_3}{P_t} \left[(P_t - d_o) L_1 - (d_e - d_o) \delta \cdot N_f \cdot L_1\right] \tag{E5.7.23}$$

$$A_{c2}(L_1, L_3) = 5.872 \, \text{m}^2$$

The porosity on the air side is calculated by

$$\sigma_2 (L_1, L_3) := \frac{A_{c2} (L_1, L_3)}{L_1 \cdot L_3} \tag{E5.7.24}$$

$$\sigma_2 (L_1, L_3) = 0.439 \tag{E5.7.25}$$

The hydraulic diameter of the air side is calculated by

$$D_{h2} (L_1, L_2, L_3) := \frac{4 \cdot A_{c2} (L_1, L_3) \cdot L_2}{A_2 (L_1, L_2, L_3)} \tag{E5.7.26}$$

$$D_{h2} (L_1, L_2, L_3) = 4.851 \cdot mm$$

The air mass velocity is

$$G_2 (L_1, L_3) := \frac{md_2}{A_{c2} (L_1, L_3)} \quad G_2 (L_1, L_3) = 21.117 \frac{kg}{m^2 \cdot s} \tag{E5.7.27}$$

The Reynolds number on the air side is calculated by

$$Re_2 (L_1, L_2, L_3) := \frac{G_2 (L_1, L_3) \cdot D_{h2} (L_1, L_2, L_3)}{\mu_2} \tag{E5.7.28}$$

$$Re_2 (L_1, L_2, L_3) = 5698$$

The fin pitch P_f is obtained by dividing the number of fins per unit length.

$$p_f := \frac{1}{N_f} = 0.289 \, cm \tag{E5.7.29}$$

Correlation for the Heat Transfer   The Colburn factor j on the air side is obtained using Equation (5.231):

$$j_2 (L_1, L_2, L_3) := 0.134 \cdot Re_2 (L_1, L_2, L_3)^{-0.319} \cdot \left(\frac{p_f - \delta}{d_e - d_o} \right)^{0.2} \cdot \left(\frac{p_f - \delta}{\delta} \right)^{0.11} \tag{E5.7.30}$$

The heat transfer coefficient h on the air side is obtained using Equation (5.230).

$$h_2 (L_1, L_2, L_3) := \frac{j_2 (L_1, L_2, L_3) \cdot G_2 (L_1, L_3) \cdot c_{p2}}{Pr_2^{\frac{2}{3}}} \tag{E5.7.31}$$

$$h_2 (L_1, L_2, L_3) = 194.476 \frac{W}{m^2 \cdot K}$$

In order to calculate the overall surface (fin) efficiency η_o, the single fin efficiency η_f is first calculated. The quantity m_f is found using Equation (5.96).

$$m_f (L_1, L_2, L_3) := \sqrt{\frac{2 \cdot h_2 (L_1, L_2, L_3)}{k_{al} \cdot \delta}} \tag{E5.7.32}$$

Converting the diameters to radii,

$$r_o := \frac{d_o}{2} \qquad r_e := \frac{d_e}{2} \qquad (E5.7.33)$$

$$mr_1\,(L_1, L_2, L_3) := m_f\,(L_1, L_2, L_3) \cdot r_o \quad mr_2\,(L_1, L_2, L_3) := m_f\,(L_1, L_2, L_3) \cdot r_e$$
$$(E5.7.34)$$

$$C_a\,(L_1, L_2, L_3) := \frac{2 \cdot r_o}{m_f\,(L_1, L_2, L_3)} \cdot \frac{1}{r_e^2 - r_o^2} \qquad (E5.7.35)$$

The single fin efficiency η_f for a circular finned tube is obtained using Equation 2.143 in Table 2.1 (consisting of Bessel functions)

$$\eta_f\,(L_1, L_2, L_3) := C_a\,(L_1, L_2, L_3) \cdot \frac{\begin{aligned}&K1\,(mr_1\,(L_1, L_2, L_3)) \cdot I1\,(mr_2\,(L_1, L_2, L_3))\\&- I1\,(mr_1\,(L_1, L_2, L_3)) \cdot K1\,(mr_2\,(L_1, L_2, L_3))\end{aligned}}{\begin{aligned}&I0\,(mr_1\,(L_1, L_2, L_3)) \cdot K1\,(mr_2\,(L_1, L_2, L_3))\\&+ K0\,(mr_1\,(L_1, L_2, L_3)) \cdot I1\,(mr_2\,(L_1, L_2, L_3))\end{aligned}}$$
$$(E5.7.36)$$

$$\eta_f\,(L_1, L_2, L_3) = 0.819$$

The overall surface (fin) efficiency η_o is readily obtained using Equation (5.99).

$$\eta_o\,(L_1, L_2, L_3) := 1 - \frac{A_f\,(L_1, L_2, L_3)}{A_2\,(L_1, L_2, L_3)} \cdot (1 - \eta_f\,(L_1, L_2, L_3))$$
$$(E5.7.37)$$

$$\eta_o\,(L_1, L_2, L_3) = 0.837$$

Water-side Reynolds Number and Mass Velocity The Reynolds number on the water side is calculated. The flow area for the water side is

$$A_{c1}\,(L_2, L_3) := N_t\,(L_2, L_3) \cdot \frac{\pi \cdot d_i^2}{4} \qquad G_1\,(L_2, L_3) := \frac{md_1}{A_{c1}\,(L_2, L_3)} \qquad (E5.7.38)$$

$$v_1\,(L_2, L_3) := \frac{G_1\,(L_2, L_3)}{\rho_1} \qquad v_1\,(L_2, L_3) = 1.268\,\frac{m}{s} \qquad (E5.7.39)$$

$$Re_1\,(L_2, L_3) := \frac{G_1\,(L_2, L_3) \cdot d_i}{\mu_1} \qquad Re_1\,(L_2, L_3) = 7.686 \times 10^4 \qquad (E5.7.40)$$

Water-side Friction Factor and Heat Transfer Coefficient Using Equations (5.33) and (5.32), obtain the friction factor and the Nusselt number for a turbulent flow.

$$f_1\,(L_2, L_3) := (1.58 \cdot \ln\,(Re_1\,(L_2, L_3)) - 3.28)^{-2} \qquad f_1\,(L_2, L_3) = 0.00476$$
$$(E5.7.41)$$

$$Nu_1\,(L_2, L_3) := \frac{f_1\,(L_2, L_3)}{2} \cdot \frac{(Re_1\,(L_2, L_3) - 1000) \cdot Pr_1}{1 + 12.7 \cdot \left(\frac{f_1\,(L_2, L_3)}{2}\right)^{0.5} \cdot \left(Pr_1^{\frac{2}{3}} - 1\right)} \qquad (E5.7.42)$$

$$h_1(L_2, L_3) := \frac{Nu_1(L_2, L_3) \cdot k_1}{d_i} \qquad h_1(L_2, L_3) = 8444 \cdot \frac{W}{m^2 \cdot K} \qquad \text{(E5.7.43)}$$

The wall conduction resistance across the tube wall is obtained as follows:

$$R_W(L_1, L_2, L_3) := \frac{\ln\left(\frac{d_o}{d_i}\right)}{2 \cdot \pi \cdot k_{co} \cdot L_1 \cdot N_t(L_2, L_3)} \qquad \text{(E5.7.44)}$$

$$R_W(L_1, L_2, L_3) = 1.434 \times 10^{-7} \cdot \frac{K}{W}$$

The heat transfer area on the water side is

$$A_1(L_1, L_2, L_3) := N_t(L_2, L_3) \cdot \pi \cdot d_i \cdot L_1 \quad A_1(L_1, L_2, L_3) = 32.627 \, m^2 \quad \text{(E5.7.45)}$$

The overall heat transfer coefficient U is calculated as follows:

$$R_{o1}(L_1, L_2, L_3) := \frac{1}{h_1(L_2, L_3) \cdot A_1(L_1, L_2, L_3)}$$

$$R_{o1}(L_1, L_2, L_3) = 3.63 \times 10^{-6} \cdot \frac{K}{W} \qquad \text{(E5.7.46a)}$$

$$R_{o2}(L_1, L_2, L_3) := \frac{1}{\eta_o(L_1, L_2, L_3) \cdot h_2(L_1, L_2, L_3) \cdot A_2(L_1, L_2, L_3)}$$

$$R_{o2}(L_1, L_2, L_3) = 1.68 \times 10^{-5} \cdot \frac{K}{W} \qquad \text{(E5.7.46b)}$$

$$UA(L_1, L_2, L_3) := \frac{1}{R_{o1}(L_1, L_2, L_3) + R_W(L_1, L_2, L_3) + R_{o2}(L_1, L_2, L_3)} \quad \text{(E5.7.46c)}$$

$$UA(L_1, L_2, L_3) = 4.862 \times 10^4 \cdot \frac{W}{K} \qquad \text{(E5.7.46a)}$$

Note that the wall conduction resistance in Equation (E5.7.44) is small compared to Equations (E5.7.46 a, b). Sometimes the resistance is neglected.

ε -NTU Analysis   The heat capacities are defined as

$$C_1 := md_1 \cdot c_{p1} \qquad C_1 = 2.098 \times 10^5 \cdot \frac{W}{K} \qquad \text{(E5.7.47)}$$

$$C_2 := md_2 \cdot c_{p2} \qquad C_2 = 1.249 \times 10^5 \cdot \frac{W}{K} \qquad \text{(E5.7.48)}$$

The minimum and maximum heat capacities are found using the MathCAD built-in functions.

$$C_{min} := \min(C_1, C_2) \qquad C_{min} = 1.249 \times 10^5 \frac{W}{K} \qquad \text{(E5.7.49)}$$

$$C_{max} := \max(C_1, C_2) \qquad C_{max} = 2.098 \times 10^5 \frac{W}{K} \qquad \text{(E5.7.50)}$$

The heat capacity ratio is defined by

$$C_r := \frac{C_{min}}{C_{max}} \qquad C_r = 0.595 \qquad\qquad \text{(E5.7.51)}$$

The number of transfer unit *NTU* is defined by

$$NTU\,(L_1, L_2, L_3) := \frac{UA\,(L_1, L_2, L_3)}{C_{min}} \quad NTU\,(L_1, L_2, L_3) = 0.389 \qquad \text{(E5.7.52)}$$

The effectiveness for both fluids unmixed cross-flow is given in Equation (5.77).

$$\varepsilon_{hx}\,(L_1, L_2, L_3) := 1 - \exp\left[\frac{1}{C_r} \cdot NTU\,(L_1, L_2, L_3)^{0.22}\right.$$

$$\left. \times \left(\exp\left(-C_r \cdot NTU\,(L_1, L_2, L_3)^{0.78}\right) - 1\right)\right] \qquad \text{(E5.7.53)}$$

$$\varepsilon_{hx}\,(L_1, L_2, L_3) = 0.287$$

The heat transfer rate is calculated by

$$q\,(L_1, L_2, L_3) := \varepsilon_{hx}\,(L_1, L_2, L_3) \cdot C_{min} \cdot (T_{1i} - T_{2i})$$

$$q\,(L_1, L_2, L_3) = 3.121 \times 10^6 \, W \qquad\qquad \text{(E5.7.54)}$$

The outlet temperatures are calculated by

$$T_{1o}\,(L_1, L_2, L_3) := T_{1i} - \varepsilon_{hx}\,(L_1, L_2, L_3) \cdot \frac{C_{min}}{C_1} \cdot (T_{1i} - T_{2i})$$

$$T_{1o}\,(L_1, L_2, L_3) = 77.122 \cdot {}^\circ C \qquad\qquad \text{(E5.7.55)}$$

$$T_{2o}\,(L_1, L_2, L_3) := T_{2i} + \varepsilon_{hx}\,(L_1, L_2, L_3) \cdot \frac{C_{min}}{C_2} \cdot (T_{1i} - T_{2i})$$

$$T_{2o}\,(L_1, L_2, L_3) = 29.998 \cdot {}^\circ C \qquad\qquad \text{(E5.7.56)}$$

Pressure Drop on the Air Side   The tube diagonal pitch p_d is the same as P_t, as shown in Figure E5.7.3.

$$P_d := P_t \qquad \text{For equilateral triangular tubes} \qquad \text{(E5.7.57)}$$

The friction factor on the air side is obtained using Equations (5.232) and (5.235).

$$f_{RB}\,(L_1, L_2, L_3) := 9.465 \, Re_2\,(L_1, L_2, L_3)^{-0.316} \left(\frac{P_t}{d_o}\right)^{-0.927} \cdot \left(\frac{P_t}{P_d}\right)^{0.515} \qquad \text{(E5.7.58)}$$

$$f_2\,(L_1, L_2, L_3) := f_{RB}\,(L_1, L_2, L_3) \cdot \frac{D_{h2}\,(L_1, L_2, L_3)}{2 \cdot P_c} \quad f_2\,(L_1, L_2, L_3) = 0.019$$

$$\text{(E5.7.59)}$$

The inlet air density is calculated using the ideal-gas law.

$$R_a := 287 \frac{J}{kg \cdot K} \qquad \text{Gas constant for air} \qquad (E5.7.60)$$

$$P_{2i} := 101.33 \, kPa \qquad \text{Cold air inlet pressure (atmospheric pressure)} \qquad (E5.7.61)$$

$$\rho_{2i} := \frac{P_{2i}}{R_a \cdot T_{2i}} \qquad \rho_{2i} = 1.269 \, \frac{kg}{m^3} \qquad (E5.7.62)$$

The outlet air density is calculated using the ideal-gas law.

$$\rho_{2o}(L_1, L_2, L_3, P_{2o}) := \frac{P_{2o}}{R_a \cdot T_{2o}(L_1, L_2, L_3)} \qquad (E5.7.63)$$

The mean specific volume $(1/\rho)_m$ is defined using Equation (5.237) as

$$\rho_m(L_1, L_2, L_3, P_{2o}) := \left[\frac{1}{2} \cdot \left(\frac{1}{\rho_{2i}} + \frac{1}{\rho_{2o}(L_1, L_2, L_3, P_{2o})} \right) \right]^{-1} \qquad (E5.7.64)$$

The outlet air density ρ_{2o} depends on both the outlet air pressure P_{2o} and the outlet temperature, and the pressure can be determined from Equation (5.236), solving implicitly using a MathCAD function.

Guess a value for the 'Find' built-in function:

$$P_{2o} := 100 \, kPa \qquad (E5.7.65)$$

Given

$$P_{2i} - P_{2o} = \frac{G_2(L_1, L_3)^2}{2 \cdot \rho_{2i}} \cdot \left[2 \left(\frac{\rho_{2i}}{\rho_{2o}(L_1, L_2, L_3, P_{2o})} - 1 \right) \right.$$
$$\left. + \frac{4 \cdot f_2(L_1, L_2, L_3) \cdot L_2}{D_{h2}(L_1, L_2, L_3)} \cdot \frac{\rho_{2i}}{\rho_m(L_1, L_2, L_3, P_{2o})} \right] \qquad (E5.7.66)$$

The outlet air pressure is found using the MathCAD Find function.

$$\underset{\wedge\wedge\wedge\wedge}{P_{2o}}(L_1, L_2, L_3) := \text{Find}(P_{2o}) \qquad P_{2o}(L_1, L_2, L_3) = 1.011 \times 10^5 \, Pa \qquad (E5.7.67)$$

The pressure drop on the air side is obtained from

$$\Delta P_2(L_1, L_2, L_3) := P_{2i} - P_{2o}(L_1, L_2, L_3) \qquad \Delta P_2(L_1, L_2, L_3) = 250.051 \cdot Pa$$
$$(E5.7.68)$$

In order to calculate the pressure drop on the air side, we considered the air density change due to the temperature change and the momentum, which leads to an improvement of about 10 percent in pressure drop (the computation was not presented here).

Finally, the outlet air density is calculated as

$$\rho_{2o}(L_1, L_2, L_3, P_{2o}(L_1, L_2, L_3)) = 1.162 \, \frac{kg}{m^3} \qquad (E5.7.69)$$

When we compare this with the inlet air density in Equation (E5.7.62), the inlet air density is reduced about 8 percent due to the pressure drop and the temperature change.

The air velocity at the exit is

$$v_2\,(L_1, L_2, L_3) := \frac{G_2\,(L_1, L_3)}{\rho_{2o}\,(L_1, L_2, L_3, P_{2o}\,(L_1, L_2, L_3))}$$

(E5.7.70)

$$v_2\,(L_1, L_2, L_3) = 18.176\frac{m}{s}$$

Pressure Drop on the Water Side The pressure drop inside the tubes is determined using Equation (5.93). Water is incompressible (no density change) and the net pressure drop for the contraction and expansion at the entrance and exit is negligible compared to the core friction pressure drop. Therefore, the pressure drop on the water side is given by

$$\Delta P_1\,(L_1, L_2, L_3) := \frac{4f_1\,(L_2, L_3)\cdot G_1\,(L_2, L_3)^2\cdot L_1}{2\cdot\rho_1\cdot d_i}$$

(E5.7.70)

$$\Delta P_1\,(L_1, L_2, L_3) = 3.001\times 10^3\,Pa$$

Determination of the Dimensions With the equations obtained up to now, we try to solve exact dimensions that will satisfy the design requirements. Three unknowns with three requirements may be mathematically solvable. The most important criterion is the air outlet temperature, as shown in Equation (E5.7.73). The second-most important criterion is the airside pressure drop in Equation (E5.7.74). The third-most important criterion is the water pressure drop in Equation (E5.7.75). If the computation takes too long or does not show the convergence, you may try decreasing the value until the convergence occurs.

The initial guesses for the following "Find" function are sometimes important, depending on the problem. The "Given" is a command for the MathCAD Find function. Try to give initial guesses close to the solution by updating the initial guesses if it takes too long to compute. Sometimes you get the solution in a few seconds or in several minutes.

Initial guesses:

$$L_1 := 3m \quad L_2 := 0.1m \quad L_3 := 3m$$

(E5.7.71)

Given

(E5.7.72)

$$T_{2o}\,(L_1, L_2, L_3) = 30°C$$

(E5.7.73)

$$\Delta P_2\,(L_1, L_2, L_3) = 250\,Pa$$

(E5.7.74)

$$\Delta P_1\,(L_1, L_2, L_3) = 3kPa$$

(E5.7.75)

$$\begin{pmatrix} L_1 \\ L_2 \\ L_3 \end{pmatrix} := Find\,(L_1, L_2, L_3)$$

(E5.7.76)

The final dimensions are obtained as

$$L_1 = 4.451\,m \quad L_2 = 7.559\,cm \quad L_3 = 3.006\,m$$

(E5.7.77)

You should update the numerical values in Equation (E5.7.16) with these values in Equation (E5.7.77). Then the numerical values after each equation given up to now will be correct. The solution is very sensitive to the design requirements. There may be no solution for the given criteria from Equations (5.7.73) to (5.75). In this case, you may try with the different set of criteria.

The total heat transfer area is the sum of each heat transfer area by

$$A_t\,(L_1, L_2, L_3) := A_1\,(L_1, L_2, L_3) + A_2\,(L_1, L_2, L_3) \qquad \text{(E5.7.78)}$$

The surface area density β for a circular finned-tube heat exchanger is generally defined by

$$\beta\,(L_1, L_2, L_3) := \frac{A_t\,(L_1, L_2, L_3)}{L_1 \cdot L_2 \cdot L_3} \qquad \beta\,(L_1, L_2, L_3) = 394.118\,\frac{m^2}{m^3} \qquad \text{(E5.7.79)}$$

The fan power can be calculated using Equation (5.85).

$$\text{Assuming} \quad \eta_f := 0.8 \qquad \text{(E5.7.80)}$$

$$\text{Power}_{fan}\,(L_1, L_2, L_3) := \frac{md_2}{\eta_f \cdot \rho_{2i}} \cdot \Delta P_2\,(L_1, L_2, L_3)$$

$$\text{(E5.7.81)}$$

$$\text{Power}_{fan}\,(L_1, L_2, L_3) = 40.938\,\text{hp}$$

The pump power can be calculated by assuming

$$\eta_p := 0.8 \qquad \text{(E5.7.82)}$$

$$\text{Power}_{pump}\,(L_1, L_2, L_3) := \frac{md_1}{\eta_p \cdot \rho_1} \cdot \Delta P_1\,(L_1, L_2, L_3)$$

$$\text{(E5.7.83)}$$

$$\text{Power}_{pump}\,(L_1, L_2, L_3) = 0.258\,\text{hp}$$

Summary of the Results and Geometry

Given information:

$T_{1i} = 92 \cdot^\circ C$	Water inlet temperature	(E5.7.84)
$T_{2i} = 5 \cdot^\circ C$	Air inlet temperature	(E5.7.85)
$md_1 = 50\,\dfrac{kg}{s}$	Mass flow rate at the water side	(E5.7.86)
$md_2 = 124\,\dfrac{kg}{s}$	Mass flow rate at the air side	(E5.7.87)

Requirements:

$T_{2o} = 30^\circ C$	Outlet air temperature	(E5.7.88)
$\Delta P_2 \le 250\,\text{kPa}$	Pressure drop at air side	(E5.7.89)
$\Delta P_1 \le 3\,\text{kPa}$	Pressure drop at water side	(E5.7.90)

Dimensions of the sizing obtained:

$$L_1 = 4.451 \, \text{m} \quad L_2 = 7.559 \, \text{cm} \quad L_3 = 3.006 \, \text{m} \qquad \text{(E5.7.91)}$$

Outlet temperatures, pressure drops, and fan and pump powers:

$T_{2o}(L_1, L_2, L_3) = 30 \cdot{}^\circ\text{C}$	Air outlet temperature	(E5.7.92)
$T_{1o}(L_1, L_2, L_3) = 77.1 \cdot{}^\circ\text{C}$	Water outlet temperature	(E5.7.93)
$\Delta P_1(L_1, L_2, L_3) = 3 \cdot \text{kPa}$	Pressure drop for water flow	(E5.7.94)
$\Delta P_2(L_1, L_2, L_3) = 250 \, \text{Pa}$	Pressure drop for air flow	(E5.7.95)
$\text{Power}_{\text{pump}}(L_1, L_2, L_3) = 0.258 \, \text{hp}$	Pump power	(E5.7.96)
$\text{Power}_{\text{fan}}(L_1, L_2, L_3) = 40.938 \, \text{hp}$	Fan power	(E5.7.97)

Note that the fan power of 40.938 hp in Equation (5.7.97) is significant compared to the pump power of 0.258 hp.

Thermal quantities:

$\text{NTU}(L_1, L_2, L_3) = 0.389$	Number of transfer unit	(E5.7.98)
$\varepsilon_{\text{hx}}(L_1, L_2, L_3) = 0.287$	Effectiveness	(E5.7.99)
$q(L_1, L_2, L_3) = 3.122 \times 10^6 \text{W}$	Heat transfer rate	(E5.7.100)
$h_1(L_2, L_3) = 8442 \dfrac{\text{W}}{\text{m}^2 \cdot \text{K}}$ $A_1(L_1, L_2, L_3) = 33 \, \text{m}^2$		(E5.7.101)
$h_2(L_1, L_2, L_3) = 194.456 \dfrac{\text{W}}{\text{m}^2 \cdot \text{K}}$ $A_2(L_1, L_2, L_3) = 366 \, \text{m}^2$		(E5.7.102)
$\text{UA}(L_1, L_2, L_3) = 4.862 \times 10^4 \cdot \dfrac{\text{W}}{\text{K}}$ UA value		(E5.7.103)

Note that the low heat transfer coefficient h_2 in Equation (E5.7.102) is compensated for by the increased heat transfer area A_2.

Hydraulic quantities:

$D_{h2}(L_1, L_2, L_3) = 4.851 \cdot \text{mm}$	Hydraulic diameter	(E5.7.104)
$\text{Re}_1(L_2, L_3) = 7.684 \times 10^4$	Reynolds number	(E5.7.105)
$\text{Re}_2(L_1, L_2, L_3) = 5.697 \times 10^3$	Reynolds number	(E5.7.106)
$v_2(L_1, L_2, L_3) = 18.173 \dfrac{\text{m}}{\text{s}}$	Air velocity at the exit	(E5.7.107)
$v_1(L_2, L_3) = 1.267 \dfrac{\text{m}}{\text{s}}$	Water velovity	(E5.7.108)

Note that the hydraulic diameter on the air side in Equation (E5.7.104) is very small compared to a wide open frontal area (5.4 m × 4 m). The Reynolds numbers indicate the fully turbulent flows. The velocity on the water side is reasonable.

Geometric quantities, given:

$d_i = 2.21 \, \text{cm}$	Tube inside diameter	(E5.7.109)
$d_o = 2.601 \, \text{cm}$	Tube outside diameter	(E5.7.110)
$d_e = 44.12 \, \text{mm}$	Fin outside diameter	(E5.7.111)
$P_t = 4.976 \, \text{cm}$	Transverse tube pitch	(E5.7.112)
$P_c = 4.309 \, \text{cm}$	Longitudinal tube pitch	(E5.7.113)
$\delta = 0.305 \, \text{mm}$	Finthickness	(E5.7.114)
$N_f = 3.46 \, \dfrac{1}{\text{cm}}$	Number of fins per centimeter	(E5.7.115)

Obtained quantities:

$N_t \,(L_2, L_3) = 105.6$	Number of tubes	(E5.7.116)
$A_2 \,(L_1, L_2, L_3) = 366 \, \text{m}^2$	Total heat transfer area atair side	(E5.7.117)
$\sigma_2 \,(L_1, L_3) = 0.439$	Porosity	(E5.7.118)
$\beta \,(L_1, L_2, L_3) = 394.118 \, \dfrac{\text{m}^2}{\text{m}^3}$	Surface area density	(E5.7.119)

The dimensions given in Equation (E5.7.91) satisfy the design requirements shown in Equations (E5.7.88) to (E5.7.90), indicating the need for a large heat exchanger. Since the surface area density ($394.118 \, \text{m}^2/\text{m}^3$) obtained is much less than the criterion (greater than $600 \, \text{m}^2/\text{m}^3$) of the compact heat exchanger, this heat exchanger is not really very compact.

5.8   PLATE-FIN HEAT EXCHANGERS

Plate-fin exchangers have various geometries in order to compensate for the high thermal resistance, particularly if one of the fluids is a gas. This type of exchanger has corrugated fins sandwiched between parallel plates or formed tubes. The plate fins are categorized as triangular fin, rectangular fin, wavy fin, *offset strip fin (OSF)*, louvered fin, and perforated fin. One of the most widely used enhanced fin geometries is the offset strip fin type, which is shown in Figure 5.40.

Plate-fin heat exchangers have been used since the 1910s in the auto industry and since the 1940s in the aerospace industry. They are now widely used in many applications: cryogenics, gas turbines, nuclear, and fuel cells.

Plate-fin heat exchangers are generally designed for moderate operating pressures less than 700 kPa (gauge pressure) and have been built with a surface area density

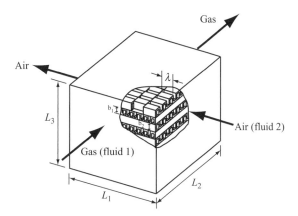

Figure 5.40   Plate-fin heat exchanger, employing offset strip fin

of up to $5,900\,\text{m}^2/\text{m}^3$. Common fin thickness ranges between 0.05 and 0.25 mm. Fin heights may range from 2 to 25 mm. Although typical fin densities are 120 to 700 fins/m, applications exist for as many as 2,100 fins/m [17].

5.8.1   Geometric Characteristics

A schematic of a single-pass crossflow plate-fin heat exchanger, employing offset strip fins, is shown in Figure 5.40. The idealized fin geometry is shown in Figure 5.41. Formulating the total heat transfer area for each fluid is an important task in the analysis. The total heat transfer area consists of the primary area and the fin area. The primary area consists of the plate area except the fin base area, multipassage side walls, and multipassage front and back walls. Usually, the numbers of passages for the hot fluid side and the cold fluid side are N_p and $N_p + 1$ to minimize the heat loss to the ambient. The top and bottom passages in Figure 5.40 are designated to be cold fluid. The number of passages (don't confuse with the number of passes) can be obtained

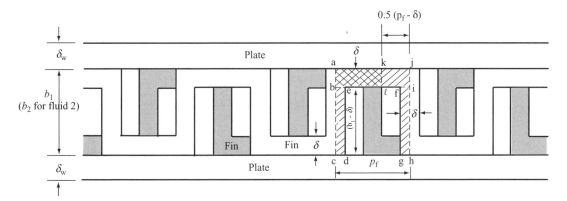

Figure 5.41   Schematic of offset strip fin (OSF) geometry

from an expression for L_3 as

$$L_3 = \underbrace{N_p b_1}_{\text{Length of fluid 1}} + \underbrace{(N_p + 1) b_2}_{\text{Length of fluid2}} + \underbrace{2(N_p + 1) \delta_w}_{\text{Thickness of total plates}} \tag{5.243}$$

Solving for N_p gives the number of passages for hot fluid

$$N_p = \frac{L_3 - b_2 - 2\delta_w}{b_1 + b_2 + 2\delta_w} \tag{5.244}$$

By definition, the number of passages counts the number based on one flow passage between two plates, not all the individual channels between the plates. The total number of fins for fluid 1 (hot) is calculated by

$$n_{f1} = \frac{L_1}{p_{f1}} N_p \tag{5.245}$$

For fluid 2 (cold),

$$n_{f2} = \frac{L_2}{p_{f2}} (N_p + 1) \tag{5.245a}$$

where p_{f1} is the fin pitch that is usually obtained by taking the inverse of the fin density. The total number of fins n_{f1} is based on the hatched area (a-c-d-e-f-g-h-j-a in Figure 5.41) counting as a unit fin. Since total primary area = Total plate areas − Fin base areas + Passage side wall areas + Passage front and back wall areas, the primary area for fluid 1 is expressed by

$$A_{p1} = \underbrace{2L_1 L_2 N_p}_{\text{Total plate area}} - \underbrace{2\delta L_2 n_{f1}}_{\text{Fin base area}} + \underbrace{2b_1 L_2 N_p}_{\text{Passage side wall area}} + \underbrace{2(b_2 + 2\delta_w) L_1 (N_p + 1)}_{\text{Passage front and back wall area}} \tag{5.246}$$

The number of offset strip fins n_{off1} per the number of fins is obtained from

$$n_{off2} = \frac{L_1}{\lambda_2} \tag{5.247}$$

$$n_{off1} = \frac{L_2}{\lambda_1} \tag{5.247a}$$

where $\lambda_{1,2}$ is the offset strip fin length for fluid 1 and 2. The total fin area A_{f1} consists of the fin area and offset-strip edge areas.

$$A_{f1} = \underbrace{2(b_1 - \delta) L_2 n_{f1}}_{\substack{\text{Fin surface areas} \\ \text{(d - e and g - f)}}} + \underbrace{2(b_1 - \delta) \delta n_{off1} n_{f1}}_{\substack{\text{Offset - strip edge areas} \\ 2(b - c - d - e)}}$$

$$+ \underbrace{(p_{f1} - \delta) \delta (n_{off1} - 1) n_{f1}}_{\substack{\text{Internal offset - strip edge area} \\ 2(i-j-k-l-i)}} + \underbrace{2 p_{f1} \delta n_{f1}}_{\substack{\text{First \& last offset - strip edge area,} \\ 2(a - b - i - j - a)}} \tag{5.248}$$

Note that the cross-hatched area (a-b-l-k-a) was not included in the offset-strip edge area because the area is blocked by the next strip fin, as shown in Figure 5.41 so that no heat transfer in the area is expected. The total heat transfer area A_{t1} is the sum of the primary area and the fin area.

$$A_{t1} = A_{p1} + A_{f1} \tag{5.249}$$

The free-flow (cross sectional) area A_{c1} is obtained from

$$A_{c1} = (b_1 - \delta)\left(p_{f1} - \delta\right)n_{f1} \tag{5.250}$$

It is assumed in Equation (5.250) that there exists a small gap between the offset strip fins, whereby the next strip fin shown in the unit fin in Figure 5.41 is not considered as an obstructing structure to the free flow. The frontal area A_{fr1} for fluid 1 where fluid 1 is entering is defined by

$$A_{fr1} = L_1 L_3 \tag{5.251}$$

The hydraulic diameter for fluid 1 is generally defined by

$$D_{h1} = \frac{4A_{c1}L_2}{A_{t1}} \tag{5.252}$$

For the fin efficiency η_f of the offset strip fin, it is assumed that the heat flow from both plates is uniform and that the adiabatic plane occurs at the middle of the plate spacing b_1. Hence, the fin profile length L_{f1} is defined by

$$L_{f1} = \frac{b_1}{2} - \delta \tag{5.253}$$

The m value is obtained using Equation (5.96) as

$$m_1 = \sqrt{\frac{2h}{k_f \delta}\left(1 + \frac{\delta}{\lambda_1}\right)} \tag{5.254}$$

The single fin efficiency η_f is obtained using Equation (5.95) as

$$\eta_f = \frac{\tanh\left(m_1 L_{f1}\right)}{m_1 L_{f1}} \tag{5.255}$$

The overall surface (fin) efficiency η_o is then obtained using Equations (5.99), (5.248), and (5.249) as

$$\eta_{o1} = 1 - \frac{A_{f1}}{A_{t1}}\left(1 - \eta_f\right) \tag{5.256}$$

5.8.2 Correlations for Offset Strip Fin (OSF) Geometry

This geometry has one of the highest heat transfer performances relative to the friction factors. Extensive analytical, numerical and experimental investigations have been conducted over the last 50 years. The most comprehensive correlations for j and f factors for the laminar, transition, and turbulent regions are provided by Manglik and Bergles [27] as follows.

$$j = 0.6522 \text{Re}^{-0.5403} \left(\frac{p_f - \delta}{b - \delta} \right)^{-0.1541} \left(\frac{\delta}{\lambda} \right)^{0.1499} \left(\frac{\delta}{p_f - \delta} \right)^{-0.0678}$$

$$\times \left[1 + 5.269 \times 10^{-5} \text{Re}^{1.34} \left(\frac{p_f - \delta}{b - \delta} \right)^{0.504} \left(\frac{\delta}{\lambda} \right)^{0.456} \left(\frac{\delta}{p_f - \delta} \right)^{-1.055} \right]^{0.1}$$

(5.257)

$$f = 9.6243 \text{Re}^{-0.7422} \left(\frac{p_f - \delta}{b - \delta} \right)^{-0.1856} \left(\frac{\delta}{\lambda} \right)^{0.3053} \left(\frac{\delta}{p_f - \delta} \right)^{-0.2659}$$

$$\times \left[1 + 7.669 \times 10^{-8} \text{Re}^{4.429} \left(\frac{p_f - \delta}{b - \delta} \right)^{0.92} \left(\frac{\delta}{\lambda} \right)^{3.767} \left(\frac{\delta}{p_f - \delta} \right)^{0.236} \right]^{0.1}$$

(5.258)

where the hydraulic diameter was defined by them as

$$D_{h\_MB} = \frac{4 \left(p_f - \delta \right) (b - \delta) \lambda}{2 \left[\left(p_f - \delta \right) \lambda + (b - \delta) \lambda + (b - \delta) \delta \right] + \left(p_f - \delta \right) \delta}$$

(5.258a)

Equation (5.258) is an approximation of Equation (5.252). However, Equation (5.258a) is in good agreement with Equation (5.258a), which indicates that the general definition of hydraulic diameter, Equation (5.252), can be used in place of Equation (5.258a). The correlations of Manglik and Bergles [27] were compared with the experiments (surface 1/8-19.86) reported by Kays and London [8] in Figure 5.42. The comparison for both j and f shows good agreement. Note that the comparison covers both laminar through turbulent regions.

Example 5.8.1 Plate-Fin Heat Exchanger   A gas-to-air single-pass crossflow heat exchanger is designed for heat recovery from the exhaust gas to preheat incoming air in a solid oxide fuel cell (SOFC) cogeneration system. Offset strip fins of the same geometry are employed on the gas and air sides; the geometrical properties and surface characteristics are provided in Figures E5.8.1 and E5.8.2. Both fins and plates (parting sheets) are made from Inconel 625 with $k = 18$ W/mK. The anode gas flows into the heat exchanger at 3.494 m³/s and 900°C. The cathode air on the other fluid side flows in at 1.358 m³/s and 200°C. The inlet pressure of the gas is at 160 kPa absolute, whereas that of air is at 200 kPa absolute. Both the gas and air pressure drops are limited to 10 kPa. It is desirable to have an equal length for L_1 and L_2. Design a gas-to-air single-pass cross-flow heat exchanger operating at $\varepsilon = 0.824$. Determine the core dimensions of this exchanger. Then, determine the heat transfer rate, outlet fluid temperatures, and pressure drops on each fluid. Use the properties of air for the gas. Also use the following geometric information.

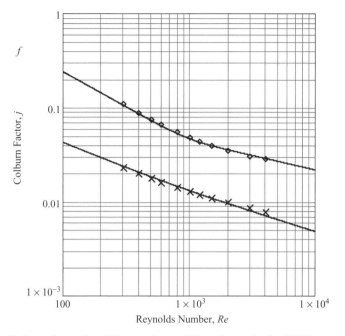

Figure 5.42 Colburn factor j and friction factor f for offset-strip-fin (OSF) type plate-fin heat exchangers. The correlation of Manglik and Bergles [27] were compared with the experiments (1/8-19.86 surface) of Kays and London [8]

Description	Value
Fin thickness δ	0.102 mm
Plate thickness δ_w	0.5 mm
Fin density N_f	782 m^{-1}
Spacing between plates b_1 and b_2	2.49 mm
Offset strip lengths λ_1 and λ_2	3.175 mm

MathCAD Format Solution The design concept is to develop a MathCAD model as a function of the dimensions and then solve for them while satisfying the design requirements.

Properties: We will use the arithmetic average as the appropriate mean temperature on each side by assuming the outlet temperatures at this time.

$$T_{gas} := \frac{900^\circ C + 400^\circ C}{2} = 923.15\,K \quad T_{air} := \frac{200^\circ C + 600^\circ C}{2} = 673.15\,K$$

(E5.8.1)

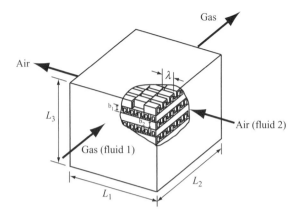

Figure E5.8.1   Plate-fin heat exchanger, employing offset strip fin

Gas (subscript1)

$\rho_1 = \text{defined\_later}$

$c_{p1} := 1126 \dfrac{\text{J}}{\text{kg} \cdot \text{K}}$

$k_1 := 0.063 \dfrac{\text{W}}{\text{m} \cdot \text{K}}$

$\mu_1 := 40.1 \cdot 10^{-6} \dfrac{\text{N} \cdot \text{s}}{\text{m}^2}$

$\text{Pr}_1 := 0.731$

Air (subscript2)

$\rho_2 = \text{defined\_later}$

$c_{p2} := 1073 \dfrac{\text{J}}{\text{kg} \cdot \text{K}}$

$k_2 := 0.0524 \dfrac{\text{W}}{\text{m} \cdot \text{K}}$

$\mu_2 := 33.6 \cdot 10^{-6} \dfrac{\text{N} \cdot \text{s}}{\text{m}^2}$

$\text{Pr}_2 := 0.694$

(E5.8.2)

The thermal conductivities of the fins and walls are given as

$$k_f := 18 \frac{\text{W}}{\text{m} \cdot \text{K}} \quad k_w := 18 \frac{\text{W}}{\text{m} \cdot \text{K}}$$

(E5.8.3)

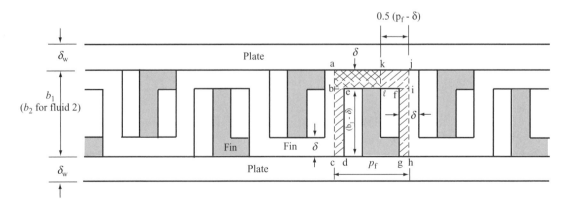

Figure E5.8.2   Schematic of offset strip fin (OSF) geometry

Given information:

$$T_{1i} := 900°C \qquad \text{Gas inlet temperature} \qquad (E5.8.4)$$

$$T_{2i} := 200°C \qquad \text{Air inlet temperature} \qquad (E5.8.5)$$

$$P_{1i} := 160\,\text{kPa} \qquad \text{Gas inlet pressure} \qquad (E5.8.6)$$

$$P_{2i} := 200\,\text{kPa} \qquad \text{Air inlet pressure} \qquad (E5.8.7)$$

$$Q_1 := 3.494\,\frac{m^3}{s} \qquad \text{Volume flow rate at gas side} \qquad (E5.8.8)$$

$$Q_2 := 1.358\,\frac{m^3}{s} \qquad \text{Volume flow rate at air side} \qquad (E5.8.9)$$

Air and gas densities (inlet):   The gas constant for air is known to be

$$R_2 := 287.04\,\frac{J}{kg \cdot K} \qquad (E5.8.10)$$

We calculate the air and gas inlet densities using the ideal gas law with an assumption that the gas constants for the air and gas are equal.

$$\text{Assuming} \qquad R_1 := R_2 \qquad (E5.8.11)$$

$$\rho_{1i} := \frac{P_{1i}}{R_1 \cdot T_{1i}} \qquad \rho_{1i} = 0.475\,\frac{kg}{m^3} \qquad (E5.8.12)$$

$$\rho_{2i} := \frac{P_{2i}}{R_2 \cdot T_{2i}} \qquad \rho_{2i} = 1.473\,\frac{kg}{m^3} \qquad (E5.8.13)$$

Hence, the mass flow rate for each fluid is calculated as

$$\text{mdot}_1 := \rho_{1i} \cdot Q_1 \qquad \text{mdot}_1 = 1.66\,\frac{kg}{s} \qquad (E5.8.14)$$

$$\text{mdot}_2 := \rho_{2i} \cdot Q_2 \qquad \text{mdot}_2 = 2\,\frac{kg}{s} \qquad (E5.8.15)$$

Design requirements: In this design problem, effectiveness is a major concern for heat recovery. The designers want to derive the relationship between the effectiveness and the size. The frontal widths for both fluids are required to be the same ($L_1 = L_2$).

$$\varepsilon = 0.824 \qquad \text{Effectiveness} \qquad (E5.8.16)$$

$$\Delta P_1 \leq 10\,\text{kPa} \qquad \text{Pressure drop at gas side} \qquad (E5.8.17)$$

$$\Delta P_2 \leq 10\,\text{kPa} \qquad \text{Pressure drop at air side} \qquad (E5.8.18)$$

$$L_1 = L_2 \qquad \text{Desirable} \qquad (E5.8.19)$$

Geometric information:

$$\delta_{\substack{w}} := 0.102 \, \text{mm} \qquad \text{Fin thickness} \qquad\qquad \text{(E5.8.20)}$$

$$\delta_w := 0.5 \, \text{mm} \qquad \text{Plate thickness} \qquad\qquad \text{(E5.8.21)}$$

$$N_f := 782 \, \text{m}^{-1} \qquad \text{Fin density} \qquad\qquad\quad \text{(E5.8.22)}$$

$$p_{f1} := \frac{1}{N_f} \qquad\qquad \text{Fin pitch} \qquad\qquad\qquad \text{(E5.8.22a)}$$

$$p_{f2} := p_{f1} \qquad\qquad p_{f1} = 1.2788 \, \text{mm} \qquad \text{(E5.8.23)}$$

$$b_1 := 2.49 \, \text{mm} \qquad \text{Plate distance} \qquad\qquad \text{(E5.8.24)}$$

$$b_2 := 2.49 \, \text{mm} \qquad\qquad\qquad\qquad\qquad\qquad \text{(E5.8.25)}$$

$$\lambda_1 := 3.175 \, \text{mm} \qquad \text{Offset strip length} \qquad \text{(E5.8.26)}$$

$$\lambda_2 := 3.175 \, \text{mm} \qquad\qquad\qquad\qquad\qquad\qquad \text{(E5.8.27)}$$

Assume the Dimensions of the Heat Exchanger   Initially, assume the dimensions to have the following values and update the dimensions later when the final size is determined.

$$L_1 := 0.294 \, \text{m} \qquad L_2 := 0.294 \, \text{m} \qquad L_3 := 0.982 \, \text{m} \qquad \text{(E5.8.28)}$$

It is assumed that the number of passages for the gas side and the air side are N_p and $N_p + 1$, respectively, to minimize the heat loss to the ambient. Using Equation (5.243), L_3 is given by

$$L_3 = N_p \cdot b_1 + \left(N_p + 1\right) \cdot b_2 + 2 \cdot \left(N_p + 1\right) \cdot \delta_w \qquad \text{(E5.8.29)}$$

Thus, the total number of passages N_p for the gas side can be obtained using Equation (5.244).

$$N_p \, (L_3) := \frac{L_3 - b_2 - 2 \cdot \delta_w}{b_1 + b_2 + 2 \cdot \delta_w} \quad N_p \, (L_3) = 163.63 \qquad \text{(E5.8.30)}$$

Note that the total number of passages is a function of L_3. The total number of fins for each fluid is

$$n_{f1} \, (L_1, L_3) := \frac{L_1}{p_{f1}} \cdot N_p \, (L_3) \qquad n_{f1} \, (L_1, L_3) = 3.762 \times 10^4 \qquad \text{(E5.8.31)}$$

$$n_{f2} \, (L_2, L_3) := \frac{L_2}{p_{f2}} \cdot \left(N_p \, (L_3) + 1\right) \quad n_{f2} \, (L_2, L_3) = 3.785 \times 10^4 \qquad \text{(E5.8.32)}$$

The total primary area A_p for each fluid is calculated using Equation (5.246).

$$A_{p1} \, (L_1, L_2, L_3) := 2 \cdot L_1 \cdot L_2 \cdot N_p \, (L_3) - 2 \cdot \delta \cdot L_2 \cdot n_{f1} \, (L_1, L_3) + 2 \cdot b_1 \cdot L_2 \cdot N_p \, (L_3)$$
$$+ \, 2 \cdot (b_2 + 2 \cdot \delta_w) \cdot L_1 \cdot \left(N_p \, (L_3) + 1\right) \qquad \text{(E5.8.33)}$$

$$A_{p2}(L_1, L_2, L_3) := 2 \cdot L_1 \cdot L_2 \cdot (N_p(L_3) + 1) - 2 \cdot \delta \cdot L_1 \cdot n_{f2}(L_2, L_3)$$
$$+ 2 \cdot b_2 \cdot L_1 \cdot (N_p(L_3) + 1)$$
$$+ 2 \cdot (b_1 + 2 \cdot \delta_w) \cdot L_2 \cdot N_p(L_3) \tag{E5.8.34}$$

$$A_{p1}(L_1, L_2, L_3) = 26.608\,\mathrm{m}^2 \quad A_{p2}(L_1, L_2, L_3) = 26.767\,\mathrm{m}^2 \tag{E5.8.35}$$

The number of offset strip fins for each fluid (per the number of fins) is

$$n_{off1}(L_2) := \frac{L_2}{\lambda_1} \quad n_{off2}(L_1) := \frac{L_1}{\lambda_2} \tag{E5.8.36}$$

The total fin area A_f for each fluid is obtained using Equation (5.248).

$$A_{f1}(L_1, L_2, L_3) := 2 \cdot (b_1 - \delta) \cdot L_2 \cdot n_{f1}(L_1, L_3)$$
$$+ 2 \cdot (b_1 - \delta) \cdot \delta \cdot n_{off1}(L_2) \cdot n_{f1}(L_1, L_3)$$
$$+ (p_{f1} - \delta) \cdot \delta \cdot (n_{off1}(L_2) - 1) \cdot n_{f1}(L_1, L_3)$$
$$+ 2 \cdot p_{f1} \cdot \delta \cdot n_{f1}(L_1, L_3) \tag{E5.8.37}$$

$$A_{f2}(L_1, L_2, L_3) := 2 \cdot (b_2 - \delta) \cdot L_1 \cdot n_{f2}(L_2, L_3)$$
$$+ 2 \cdot (b_2 - \delta) \cdot \delta \cdot n_{off2}(L_1) \cdot n_{f2}(L_2, L_3)$$
$$+ (p_{f2} - \delta) \cdot \delta \cdot (n_{off2}(L_1) - 1) \cdot n_{f2}(L_2, L_3)$$
$$+ 2 \cdot p_{f2} \cdot \delta \cdot n_{f2}(L_2, L_3) \tag{E5.8.38}$$

$$A_{f1}(L_1, L_2, L_3) = 54.944\,\mathrm{m}^2 \quad A_{f2}(L_1, L_2, L_3) = 55.28\,\mathrm{m}^2 \tag{E5.8.39}$$

The total surface area A_t for each fluid is the sum of the fin area A_f and the primary area A_p.

$$A_{t1}(L_1, L_2, L_3) := A_{f1}(L_1, L_2, L_3) + A_{p1}(L_1, L_2, L_3) \tag{E5.8.40}$$

$$A_{t2}(L_1, L_2, L_3) := A_{f2}(L_1, L_2, L_3) + A_{p2}(L_1, L_2, L_3) \tag{E5.8.41}$$

$$A_{t1}(L_1, L_2, L_3) = 81.553\,\mathrm{m}^2 \quad A_{t2}(L_1, L_2, L_3) = 82.047\,\mathrm{m}^2 \tag{E5.8.42}$$

The free-flow area A_c for each side is obtained using Equation (5.250).

$$A_{c1}(L_1, L_3) := (b_1 - \delta) \cdot (p_{f1} - \delta) \cdot n_{f1}(L_1, L_3) \quad A_{c1}(L_1, L_3) = 0.1057\,\mathrm{m}^2 \tag{E5.8.43}$$

$$A_{c2}(L_2, L_3) := (b_2 - \delta) \cdot (p_{f2} - \delta) \cdot n_{f2}(L_2, L_3) \quad A_{c2}(L_2, L_3) = 0.1064\,\mathrm{m}^2 \tag{E5.8.44}$$

The frontal area for each fluid is found using Equation (5.251).

$$A_{fr1}(L_1, L_3) := L_1 \cdot L_3 \quad A_{fr1}(L_1, L_3) = 0.289\,\mathrm{m}^2 \tag{E5.8.45}$$

$$A_{fr2}(L_2, L_3) := L_2 \cdot L_3 \quad A_{fr2}(L_2, L_3) = 0.289\,\mathrm{m}^2 \tag{E5.8.46}$$

The hydraulic diameter is found using Equation (5.252).

$$D_{h1}(L_1, L_2, L_3) := \frac{4 \cdot A_{c1}(L_1, L_3) \cdot L_2}{A_{t1}(L_1, L_2, L_3)} \quad D_{h1}(L_1, L_2, L_3) = 1.524\, \text{mm} \quad \text{(E5.8.47)}$$

$$D_{h2}(L_1, L_2, L_3) := \frac{4 \cdot A_{c2}(L_2, L_3) \cdot L_1}{A_{t2}(L_1, L_2, L_3)} \quad D_{h2}(L_1, L_2, L_3) = 1.525\, \text{mm} \quad \text{(E5.8.48)}$$

The hydraulic diameter of Equation (5.258a) given by Manglik and Bergles [27] is calculated in Equation (E5.8.49) and compared with the general definition in Equation (E5.8.47). We find that they are in agreement. Therefore, we use Equations (E5.8.47) and (E5.8.48) in the rest of the calculations. The reason of the preference of the general hydraulic diameter rather than the Manglik and Bergles's expression for the correlations is to reduce any loss of the information of the dimensions (L_1, L_2 and L_3) with the approximation of Equation (E5.8.49) for modeling purposes. The discrepancy between Equation (E5.8.49) and (E5.8.47) is calculated to be about 1.5 percent.

$$D_{h\_MB} := \frac{4 \cdot (p_{f1} - \delta) \cdot (b_1 - \delta) \cdot \lambda_1}{2 \cdot \left[(p_{f1} - \delta) \cdot \lambda_1 + (b_1 - \delta) \cdot \lambda_1 + (b_1 - \delta) \cdot \delta \right] + (p_{f1} - \delta) \cdot \delta}$$

$$D_{h\_MB} = 1.535\, \text{mm} \quad \text{(E5.8.49)}$$

The porosity σ for each fluid is calculated as

$$\sigma_1(L_1, L_3) := \frac{A_{c1}(L_1, L_3)}{A_{fr1}(L_1, L_3)} \quad \sigma_1(L_1, L_3) = 0.366 \quad \text{(E5.8.50)}$$

$$\sigma_2(L_2, L_3) := \frac{A_{c2}(L_2, L_3)}{A_{fr2}(L_2, L_3)} \quad \sigma_2(L_2, L_3) = 0.368 \quad \text{(E5.8.51)}$$

The volume of the exchanger for each fluid is calculated as

$$V_{p1}(L_1, L_2, L_3) := b_1 \cdot L_1 \cdot L_2 \cdot N_p(L_3) \quad V_{p1}(L_1, L_2, L_3) = 0.0352\, \text{m}^3 \quad \text{(E5.8.52)}$$

$$V_{p2}(L_1, L_2, L_3) := b_2 \cdot L_1 \cdot L_2 \cdot (N_p(L_3) + 1) \quad V_{p2}(L_1, L_2, L_3) = 0.0354\, \text{m}^3 \quad \text{(E5.8.53)}$$

The surface area density β for each side is calculated as

$$\beta_1(L_1, L_2, L_3) := \frac{A_{t1}(L_1, L_2, L_3)}{V_{p1}(L_1, L_2, L_3)} \quad \beta_1(L_1, L_2, L_3) = 2316\, \frac{\text{m}^2}{\text{m}^3} \quad \text{(E5.8.54)}$$

$$\beta_2(L_1, L_2, L_3) := \frac{A_{t2}(L_1, L_2, L_3)}{V_{p2}(L_1, L_2, L_3)} \quad \beta_2(L_1, L_2, L_3) = 2316\, \frac{\text{m}^2}{\text{m}^3} \quad \text{(E5.8.55)}$$

The mass velocities, velocities, Reynolds Numbers are calculated as

$$G_1(L_1, L_3) := \frac{\dot{m}_1}{A_{c1}(L_1, L_3)} \quad G_1(L_1, L_3) = 15.704\, \frac{\text{kg}}{\text{m}^2 \cdot \text{s}} \quad \text{(E5.8.56)}$$

$$G_2(L_2, L_3) := \frac{\dot{m}_2}{A_{c2}(L_2, L_3)} \quad G_2(L_2, L_3) = 18.802\, \frac{\text{kg}}{\text{m}^2 \cdot \text{s}} \quad \text{(E5.8.57)}$$

$$\text{Re}_1\,(L_1, L_2, L_3) := \frac{G_1\,(L_1, L_3) \cdot D_{h1}\,(L_1, L_2, L_3)}{\mu_1} \qquad \text{Re}_1\,(L_1, L_2, L_3) = 596.998$$

$$(\text{E5.8.58})$$

$$\text{Re}_2\,(L_1, L_2, L_3) := \frac{G_2\,(L_2, L_3) \cdot D_{h2}\,(L_1, L_2, L_3)}{\mu_2} \qquad \text{Re}_2\,(L_1, L_2, L_3) = 853.09$$

$$(\text{E5.8.59})$$

Colburn Factors and Friction Factors   The Reynolds numbers indicate laminar flows at both the gas and air sides. j and f factors are calculated using Equations (5.257) and (5.258) as

$$j_{1a}\,(L_1, L_2, L_3) := 0.6522 \cdot \text{Re}_1\,(L_1, L_2, L_3)^{-0.5403} \cdot \left(\frac{p_{f1} - \delta}{b_1 - \delta}\right)^{-0.1541}$$

$$\cdot \left(\frac{\delta}{\lambda_1}\right)^{0.1499} \cdot \left(\frac{\delta}{p_{f1} - \delta}\right)^{-0.0678} \qquad (\text{E5.8.60})$$

$$j_{1b}\,(L_1, L_2, L_3) := \left[1 + 5.269 \cdot 10^{-5} \cdot \text{Re}_1\,(L_1, L_2, L_3)^{1.34} \cdot \left(\frac{p_{f1} - \delta}{b_1 - \delta}\right)^{0.504}\right.$$

$$\left. \cdot \left(\frac{\delta}{\lambda_1}\right)^{0.456} \cdot \left(\frac{\delta}{p_{f1} - \delta}\right)^{-1.055}\right]^{0.1} \qquad (\text{E5.8.61})$$

The Colburn factor j_1 on the gas side is calculated by

$$j_1\,(L_1, L_2, L_3) := j_{1a}\,(L_1, L_2, L_3) \cdot j_{1b}\,(L_1, L_2, L_3) \qquad j_1\,(L_1, L_2, L_3) = 0.017$$

$$(\text{E5.8.62})$$

$$f_{1a}\,(L_1, L_2, L_3) := 9.6243\,\text{Re}_1\,(L_1, L_2, L_3)^{-0.7422} \cdot \left(\frac{p_{f1} - \delta}{b_1 - \delta}\right)^{-0.1856}$$

$$\cdot \left(\frac{\delta}{\lambda_1}\right)^{0.3053} \cdot \left(\frac{\delta}{p_{f1} - \delta}\right)^{-0.2659} \qquad (\text{E5.8.63})$$

$$f_{1b}\,(L_1, L_2, L_3) := \left[1 + 7.669 \cdot 10^{-8} \cdot \text{Re}_1\,(L_1, L_2, L_3)^{4.429} \cdot \left(\frac{p_{f1} - \delta}{b_1 - \delta}\right)^{0.92}\right.$$

$$\left. \cdot \left(\frac{\delta}{\lambda_1}\right)^{3.767} \cdot \left(\frac{\delta}{p_{f1} - \delta}\right)^{0.236}\right]^{0.1} \qquad (\text{E5.8.64})$$

The friction factor f_1 on the gas side is

$$f_1\,(L_1, L_2, L_3) := f_{1a}\,(L_1, L_2, L_3) \cdot f_{1b}\,(L_1, L_2, L_3) \qquad f_1\,(L_1, L_2, L_3) = 0.065$$

$$(\text{E5.8.65})$$

$$j_{2a}\,(L_1, L_2, L_3) := 0.6522 \cdot Re_2\,(L_1, L_2, L_3)^{-0.5403} \cdot \left(\frac{p_{f2} - \delta}{b_2 - \delta}\right)^{-0.1541}$$

$$\cdot \left(\frac{\delta}{\lambda_2}\right)^{0.1499} \cdot \left(\frac{\delta}{p_{f2} - \delta}\right)^{-0.0678} \tag{E5.8.66}$$

$$j_{2b}\,(L_1, L_2, L_3) := \left[1 + 5.269 \cdot 10^{-5} \cdot Re_2\,(L_1, L_2, L_3)^{1.34} \cdot \left(\frac{p_{f2} - \delta}{b_2 - \delta}\right)^{0.504}\right.$$

$$\left. \cdot \left(\frac{\delta}{\lambda_2}\right)^{0.456} \cdot \left(\frac{\delta}{p_{f2} - \delta}\right)^{-1.055}\right]^{0.1} \tag{E5.8.67}$$

The Colburn factor j_2 on the air side is

$$j_2\,(L_1, L_2, L_3) := j_{2a}\,(L_1, L_2, L_3) \cdot j_{2b}\,(L_1, L_2, L_3) \qquad j_2\,(L_1, L_2, L_3) = 0.014$$

$$\tag{E5.8.68}$$

$$f_{2a}\,(L_1, L_2, L_3) := 9.6243 \cdot Re_2\,(L_1, L_2, L_3)^{-0.7422} \cdot \left(\frac{p_{f2} - \delta}{b_2 - \delta}\right)^{-0.1856}$$

$$\cdot \left(\frac{\delta}{\lambda_2}\right)^{0.3053} \cdot \left(\frac{\delta}{p_{f2} - \delta}\right)^{-0.2659} \tag{E5.8.69}$$

$$f_{2b}\,(L_1, L_2, L_3) := \left[1 + 7.669 \cdot 10^{-8} \cdot Re_2\,(L_1, L_2, L_3)^{4.429} \cdot \left(\frac{p_{f2} - \delta}{b_2 - \delta}\right)^{0.92}\right.$$

$$\left. \cdot \left(\frac{\delta}{\lambda_2}\right)^{3.767} \cdot \left(\frac{\delta}{p_{f2} - \delta}\right)^{0.236}\right]^{0.1} \tag{E5.8.70}$$

The friction factor f_2 on the air side is

$$f_2\,(L_1, L_2, L_3) := f_{2a}\,(L_1, L_2, L_3) \cdot f_{2b}\,(L_1, L_2, L_3) \qquad f_2\,(L_1, L_2, L_3) = 0.051$$

$$\tag{E5.8.71}$$

Heat Transfer Coefficients   The heat transfer coefficients in terms of the j and f factors are calculated using Equation (5.230) as

$$h_1\,(L_1, L_2, L_3) := \frac{j_1\,(L_1, L_2, L_3) \cdot G_1\,(L_1, L_3) \cdot c_{p1}}{Pr_1^{\frac{2}{3}}}$$

$$h_1\,(L_1, L_2, L_3) = 368.869 \frac{W}{m^2 \cdot K} \tag{E5.8.72}$$

$$h_2\,(L_1, L_2, L_3) := \frac{j_2\,(L_1, L_2, L_3) \cdot G_2\,(L_1, L_3) \cdot c_{p2}}{Pr_2^{\frac{2}{3}}}$$

$$h_2\,(L_1, L_2, L_3) = 366.268 \frac{W}{m^2 \cdot K} \tag{E5.8.73}$$

Overall Surface (Fin) Efficiency   Since the offset strip fins are used on both the gas and air sides, we will use the multiple fin analysis. *m* values are obtained using Equation (5.254) as

$$m_1\,(L_1, L_2, L_3) := \left[\frac{2 \cdot h_1\,(L_1, L_2, L_3)}{k_f \cdot \delta} \cdot \left(1 + \frac{\delta}{\lambda_1}\right)\right]^{0.5} \tag{E5.8.74}$$

$$m_2\,(L_1, L_2, L_3) := \left[\frac{2 \cdot h_2\,(L_1, L_2, L_3)}{k_f \cdot \delta} \cdot \left(1 + \frac{\delta}{\lambda_2}\right)\right]^{0.5} \tag{E5.8.75}$$

Using Equation (5.253), the fin length L_f for each fluid will be

$$L_{f1} := \frac{b_1}{2} - \delta \qquad L_{f2} := \frac{b_2}{2} - \delta \tag{E5.8.76}$$

The single fin efficiency η_f for each fluid is calculated using Equation (5.255) as

$$\eta_{f1}\,(L_1, L_2, L_3) := \frac{\tanh\,(m_1\,(L_1, L_2, L_3) \cdot L_{f1})}{m_1\,(L_1, L_2, L_3) \cdot L_{f1}}$$
$$\eta_{f2}\,(L_1, L_2, L_3) := \frac{\tanh\,(m_2\,(L_1, L_2, L_3) \cdot L_{f2})}{m_2\,(L_1, L_2, L_3) \cdot L_{f2}} \tag{E5.8.77}$$

The overall surface (fin) efficiencies η_o are obtained using Equation (E5.456)

$$\eta_{o1}\,(L_1, L_2, L_3) := 1 - (1 - \eta_{f1}\,(L_1, L_2, L_3)) \cdot \frac{A_{f1}\,(L_1, L_2, L_3)}{A_{t1}\,(L_1, L_2, L_3)} \tag{E5.8.78}$$

$$\eta_{o1}\,(L_1, L_2, L_3) = 0.9$$

$$\eta_{o2}\,(L_1, L_2, L_3) := 1 - (1 - \eta_{f2}\,(L_1, L_2, L_3)) \cdot \frac{A_{f2}\,(L_1, L_2, L_3)}{A_{t2}\,(L_1, L_2, L_3)} \tag{E5.8.79}$$

$$\eta_{o2}\,(L_1, L_2, L_3) = 0.901$$

The conduction area for the wall thermal resistance is given by

$$A_w\,(L_1, L_2, L_3) := 2 \cdot L_1 \cdot L_2 \cdot (N_p\,(L_3) + 1) \quad A_w\,(L_1, L_2, L_3) = 28.46\,\text{m}^2 \tag{E5.8.80}$$

The thermal resistances are given by

$$R_w\,(L_1, L_2, L_3) := \frac{\delta_w}{k_w \cdot A_w\,(L_1, L_2, L_3)} \tag{E5.8.81}$$

$$R_{o1}\,(L_1, L_2, L_3) := \frac{1}{\eta_{o1}\,(L_1, L_2, L_3) \cdot h_1\,(L_1, L_2, L_3) \cdot A_{t1}\,(L_1, L_2, L_3)} \tag{E5.8.82}$$

$$R_{o2}\,(L_1, L_2, L_3) := \frac{1}{\eta_{o2}\,(L_1, L_2, L_3) \cdot h_2\,(L_1, L_2, L_3) \cdot A_{t2}\,(L_1, L_2, L_3)} \tag{E5.8.83}$$

$$UA\,(L_1, L_2, L_3) := \frac{1}{R_{o1}\,(L_1, L_2, L_3) + R_w\,(L_1, L_2, L_3) + R_{o2}\,(L_1, L_2, L_3)} \tag{E5.8.84}$$

The overall heat transfer coefficient times the heat transfer area is given by

$$\text{UA}(L_1, L_2, L_3) = 13357 \cdot \frac{W}{K} \qquad \text{(E5.8.84a)}$$

ε-NTU Method   The heat capacity rates are given as

$$C_1 := \text{mdot}_1 \cdot c_{p1} \qquad C_1 = 1.869 \times 10^3 \cdot \frac{W}{K} \qquad \text{(E5.8.85)}$$

$$C_2 := \text{mdot}_2 \cdot c_{p2} \qquad C_2 = 2.146 \times 10^3 \cdot \frac{W}{K} \qquad \text{(E5.8.86)}$$

$$C_{min} := \min(C_1, C_2) \qquad C_{min} = 1.869 \times 10^3 \cdot \frac{W}{K} \qquad \text{(E5.8.87)}$$

$$C_{max} := \max(C_1, C_2) \qquad C_{max} = 2.146 \times 10^3 \cdot \frac{W}{K} \qquad \text{(E5.8.88)}$$

$$C_r := \frac{C_{min}}{C_{max}} \qquad C_r = 0.871 \qquad \text{(E5.8.89)}$$

The number of transfer unit (NTU) is calculated using Equation (5.77) by

$$\text{NTU}(L_1, L_2, L_3) := \frac{\text{UA}(L_1, L_2, L_3)}{C_{min}} \quad \text{NTU}(L_1, L_2, L_3) = 7.145 \qquad \text{(E5.8.90)}$$

The effectiveness of the plate-fin heat exchanger is calculated using Equation (5.38) as

$$\varepsilon_{hx}(L_1, L_2, L_3) := 1 - \exp\left[\frac{1}{C_r} \cdot \text{NTU}(L_1, L_2, L_3)^{0.22}\right.$$

$$\left. \cdot \left(\exp\left(-C_r \cdot \text{NTU}(L_1, L_2, L_3)^{0.78}\right) - 1\right)\right] \qquad \text{(E5.8.91)}$$

$$\varepsilon_{hx}(L_1, L_2, L_3) = 0.8241$$

The heat transfer rate is calculated as

$$q(L_1, L_2, L_3) := \varepsilon_{hx}(L_1, L_2, L_3) \cdot C_{min} \cdot (T_{1i} - T_{2i})$$

$$q(L_1, L_2, L_3) = 1.078 \times 10^6 \, W \qquad \text{(E5.8.92)}$$

The air outlet temperatures are calculated as

$$T_{1o}(L_1, L_2, L_3) := T_{1i} - \varepsilon_{hx}(L_1, L_2, L_3) \cdot \frac{C_{min}}{C_1}(T_{1i} - T_{2i})$$

$$T_{1o}(L_1, L_2, L_3) = 323.101 \cdot {}^{\circ}C \qquad \text{(E5.8.93)}$$

$$T_{2o}(L_1, L_2, L_3) := T_{2i} + \varepsilon_{hx}(L_1, L_2, L_3) \cdot \frac{C_{min}}{C_2}(T_{1i} - T_{2i})$$

$$T_{2o}(L_1, L_2, L_3) = 702.57 \cdot {}^{\circ}C \qquad \text{(E5.8.94)}$$

Pressure Drops   The pressure drop for the plate-fin heat exchanger is expressed using Equation (5.202) as

$$\Delta P = \frac{G^2}{2 \cdot \rho_i} \left[\left(1 - \sigma^2 + K_c\right) + 2 \cdot \left(\frac{\rho_i}{\rho_o} - 1\right) + \frac{4f \cdot L}{D_h} \cdot \left(\frac{\rho_i}{\rho_m}\right) \right.$$
$$\left. - \left(1 - \sigma^2 - K_e\right) \cdot \frac{\rho_i}{\rho_o} \right] \tag{E5.8.95}$$

Contraction and Expansion Coefficients, K_c and K_e   We develop equations for Figure 5.31. The jet contraction ratio for the square-tube geometry is given using Equation (5.210) by

$$C_{c\_tube}(\sigma) := 4.374 \cdot 10^{-4} \cdot e^{6.737 \cdot \sqrt{\sigma}} + 0.621 \tag{E5.8.96}$$

The friction factor used for this calculation is given in Equation (5.205).

$$f_d(Re) := \begin{vmatrix} 0.049\,Re^{-0.2} & \text{if } Re \geq 2300 \\ \dfrac{16}{Re} & \text{otherwise} \end{vmatrix} \tag{E5.8.97}$$

The velocity-distribution coefficient for circular tubes is given in Equation (5.207).

$$K_{d\_tube}(Re) := \begin{vmatrix} 1.09068\,(4 \cdot f_d(Re)) + 0.05884\sqrt{4 \cdot f_d(Re)} \\ \quad + 1 \text{ if } Re \geq 2300 \\ 1.33 \text{ otherwise} \end{vmatrix} \tag{E5.8.98}$$

which is converted to the square tubes as shown in Equation (5.209).

$$K_{d\_square}(Re) := \begin{vmatrix} 1 + 1.17 \cdot \left(K_{d\_tube}(Re) - 1\right) & \text{if } Re \geq 2300 \\ 1.39 \text{ otherwise} \end{vmatrix} \tag{E5.8.99}$$

The contraction coefficient is given in Equation (5.203).

$$K_{c\_sqaure}(\sigma, Re) := \frac{1 - 2 \cdot C_{c\_tube}(\sigma) + C_{c\_tube}(\sigma)^2 \cdot \left(2 \cdot K_{d\_square}(Re) - 1\right)}{C_{c\_tube}(\sigma)^2}$$
$$\tag{E5.8.100}$$

The expansion coefficient is given in Equation (5.204).

$$K_{e\_square}(\sigma, Re) := 1 - 2 \cdot K_{d\_square}(Re) \cdot \sigma + \sigma^2 \tag{E5.8.101}$$

K_c and K_e for each fluid are determined with the porosity and the Reynolds number where the Reynolds number is assumed to be fully turbulent ($Re = 10^7$) because of the frequent boundary layer interruptions due to the offset strip fins. The values may also be obtained graphically from Figure 5.31.

$$K_{c1}(L_1, L_3) := K_{c\_square}\left(\sigma_1(L_1, L_3), 10^7\right) \quad K_{c1}(L_1, L_3) = 0.33 \tag{E5.8.102}$$
$$K_{c2}(L_2, L_3) := K_{c\_square}\left(\sigma_2(L_2, L_3), 10^7\right) \quad K_{c2}(L_2, L_3) = 0.329 \tag{E5.8.103}$$

$$K_{e1}(L_1, L_3) := K_{e\_square}\left(\sigma_1(L_1, L_3), 10^7\right) \quad K_{e1}(L_1, L_3) = 0.39 \qquad \text{(E5.8.104)}$$

$$K_{e2}(L_2, L_3) := K_{e\_square}\left(\sigma_2(L_2, L_3), 10^7\right) \quad K_{e2}(L_2, L_3) = 0.387 \qquad \text{(E5.8.105)}$$

Gas and Air Densities   The gas and air outlet densities are a function of both the outlet temperatures and the outlet pressures that are unknown. The outlet densities are expressed using the ideal gas law as

$$\rho_{1o}(P_{1o}, L_1, L_2, L_3) := \frac{P_{1o}}{R_1 \cdot T_{1o}(L_1, L_2, L_3)} \qquad \text{(E5.8.106)}$$

$$\rho_{2o}(P_{2o}, L_1, L_2, L_3) := \frac{P_{2o}}{R_2 \cdot T_{2o}(L_1, L_2, L_3)} \qquad \text{(E5.8.107)}$$

Since ρ_m is the mean density and defined in Equation (5.193a),

$$\frac{1}{\rho_m} = \frac{1}{2} \cdot \left(\frac{1}{\rho_i} + \frac{1}{\rho_o}\right) \qquad \text{(E5.8.108)}$$

Thus, for each fluid,

$$\rho_{1m}(P_{1o}, L_1, L_2, L_3) := \left[\frac{1}{2} \cdot \left(\frac{1}{\rho_{1i}} + \frac{1}{\rho_{1o}(P_{1o}, L_1, L_2, L_3)}\right)\right]^{-1} \qquad \text{(E5.8.109)}$$

$$\rho_{2m}(P_{2o}, L_1, L_2, L_3) := \left[\frac{1}{2} \cdot \left(\frac{1}{\rho_{2i}} + \frac{1}{\rho_{2o}(P_{2o}, L_1, L_2, L_3)}\right)\right]^{-1} \qquad \text{(E5.8.110)}$$

Pressure Drops   We divide Equation (E5.8.95) into several short equations to fit the present page.

$$\Delta P1_{12}(L_1, L_3) := \frac{G_1(L_1, L_3)^2}{2 \cdot \rho_{1i}} \cdot \left(1 - \sigma_1(L_1, L_3)^2 + K_{c1}(L_1, L_3)\right) \qquad \text{(E5.8.111)}$$

$$K1_{23}(P_{1o}, L_1, L_2, L_3) := 2 \cdot \left(\frac{\rho_{1i}}{\rho_{1o}(P_{1o}, L_1, L_2, L_3)} - 1\right)$$
$$+ \frac{4f_1(L_1, L_2, L_3) \cdot L_2}{D_{h1}(L_1, L_2, L_3)} \cdot \left(\frac{\rho_{1i}}{\rho_{1m}(P_{1o}, L_1, L_2, L_3)}\right) \qquad \text{(E5.8.112)}$$

$$\Delta P1_{23}(P_{1o}, L_1, L_2, L_3) := \frac{G_1(L_1, L_3)^2}{2 \cdot \rho_{1i}} \cdot K1_{23}(P_{1o}, L_1, L_2, L_3) \qquad \text{(E5.8.113)}$$

$$\Delta P1_{34}(P_{1o}, L_1, L_2, L_3) := \frac{G_1(L_1, L_3)^2}{2 \cdot \rho_{1i}} \cdot \left[(1 - \sigma_1(L_1, L_3)^2\right.$$
$$\left. - K_{e1}(L_1, L_3)) \cdot \frac{\rho_{1i}}{\rho_{1o}(P_{1o}, L_1, L_2, L_3)}\right] \qquad \text{(E5.8.114)}$$

The pressure drop on the gas side, which is Equation (E5.8.95), is the sum of these three pressure drops, Equations (E5.8.111), (E5.8.113) and (E5.8.114).

$$\Delta P1\,(P_{1o}, L_1, L_2, L_3) := \Delta P1_{12}\,(L_1, L_3) + \Delta P1_{23}\,(P_{1o}, L_1, L_2, L_3)$$
$$- \Delta P1_{34}\,(P_{1o}, L_1, L_2, L_3) \tag{E5.8.115}$$

Now the gas outlet pressure can be obtained using the MathCAD's "root" function.

An initial guess that would be a closest value to the solution is needed for the root function as

$$P_{1o} := 160\,\text{kPa} \tag{E5.8.116}$$

Solving for the pressure outlet on the gas side as

$$\underset{\sim}{P_{1o}}\,(L_1, L_2, L_3) := \text{root}\,[[(P_{1i} - P_{1o}) - \Delta P1\,(P_{1o}, L_1, L_2, L_3)]\,, P_{1o}] \tag{E5.8.117}$$

The gas outlet pressure is finally obtained by

$$P_{1o}\,(L_1, L_2, L_3) = 150.006\,\text{kPa} \tag{E5.8.118}$$

The pressure drop on the gas side is

$$\Delta P_1\,(L_1, L_2, L_3) := P_{1i} - P_{1o}\,(L_1, L_2, L_3) \quad \Delta P_1\,(L_1, L_2, L_3) = 9.994\,\text{kPa} \tag{E5.8.119}$$

The air outlet pressure can be obtained in a similar way.

$$\Delta P2_{12}\,(L_2, L_3) := \frac{G_2\,(L_2, L_3)^2}{2 \cdot \rho_{2i}} \cdot \left(1 - \sigma_2\,(L_2, L_3)^2 + K_{c2}\,(L_2, L_3)\right) \tag{E5.8.120}$$

$$K2_{23}\,(P_{2o}, L_1, L_2, L_3) := 2 \cdot \left(\frac{\rho_{2i}}{\rho_{2o}\,(P_{2o}, L_1, L_2, L_3)} - 1\right)$$
$$+ \frac{4f_2\,(L_1, L_2, L_3) \cdot L_1}{D_{h2}\,(L_1, L_2, L_3)} \cdot \left(\frac{\rho_{2i}}{\rho_{2m}\,(P_{2o}, L_1, L_2, L_3)}\right) \tag{E5.8.121}$$

$$\Delta P2_{23}\,(P_{2o}, L_1, L_2, L_3) := \frac{G_2\,(L_2, L_3)^2}{2 \cdot \rho_{2i}} \cdot K2_{23}\,(P_{2o}, L_1, L_2, L_3) \tag{E5.8.122}$$

$$\Delta P2_{34}\,(P_{2o}, L_1, L_2, L_3) := \frac{G_2\,(L_2, L_3)^2}{2 \cdot \rho_{2i}}$$
$$\cdot \left(1 - \sigma_2\,(L_2, L_3)^2 - K_{e2}\,(L_2, L_3)\right) \cdot \frac{\rho_{2i}}{\rho_{2o}\,(P_{2o}, L_1, L_2, L_3)} \tag{E5.8.123}$$

$$\Delta P2\,(P_{2o}, L_1, L_2, L_3) := \Delta P2_{12}\,(L_2, L_3) + \Delta P2_{23}\,(P_{2o}, L_1, L_2, L_3)$$
$$- \Delta P2_{34}\,(P_{2o}, L_1, L_2, L_3) \tag{E5.8.124}$$

An initial guess is needed for MathCAD.

$$P_{2o} := 200\,\text{kPa} \tag{E5.8.125}$$

$$\underset{\sim\sim\sim}{P_{2o}}\,(L_1, L_2, L_3) := \text{root}\,[[(P_{2i} - P_{2o}) - \Delta P2\,(P_{2o}, L_1, L_2, L_3)], P_{2o}] \tag{E5.8.126}$$

$$P_{2o}\,(L_1, L_2, L_3) = 192.246\,\text{kPa} \tag{E5.8.127}$$

The pressure drop on the air side is

$$\Delta P_2\,(L_1, L_2, L_3) := P_{2i} - P_{2o}\,(L_1, L_2, L_3) \qquad \Delta P_2\,(L_1, L_2, L_3) = 7.754\,\text{kPa} \tag{E5.8.128}$$

Determine Dimensions of the Plate-Fin Heat Exchanger   The calculations for the size usually require a number of iterations, which is time-consuming work. This can be easily performed by MathCAD. The three unknowns L_1, L_2, and L_3, are found using a combination of three requirements. A higher pressure drop for fluid 1 (gas) is expected due to the higher volume flow rate. The following combination of the three requirements will be sufficient for the solution. Initial guess may be updated and iterated with the new values to have a convergence.

Initial guesses

$$\underset{\sim\sim\sim}{L_1} := 1\,\text{m} \qquad \underset{\sim\sim\sim}{L_2} := 1\,\text{m} \qquad \underset{\sim\sim\sim}{L_3} := 1\,\text{m} \tag{E5.8.129}$$

$$\text{Given} \tag{E5.8.130}$$

$$\varepsilon_{\text{hx}}\,(L_1, L_2, L_3) = 0.824 \tag{E5.8.131}$$

$$\Delta P_1\,(L_1, L_2, L_3) = 10\,\text{kPa} \tag{E5.8.132}$$

$$L_1 = L_2 \tag{E5.8.133}$$

$$\begin{pmatrix} \underset{\sim\sim\sim}{L_1} \\ \underset{\sim\sim\sim}{L_2} \\ \underset{\sim\sim\sim}{L_3} \end{pmatrix} := \text{Find}\,(L_1, L_2, L_3) \tag{E5.8.134}$$

The dimensions with the requirements are finally found as

$$L_1 = 0.294\,\text{m} \qquad L_2 = 0.294\,\text{m} \qquad L_3 = 0.982\,\text{m} \tag{E5.8.135}$$

Now return to Equation (E5.8.28) and update the values for all parameters.

Summary of the Results and Information

Given information:

$T_{1i} = 900^\circ\text{C}$	Gas inlet temperature	(E5.8.136)
$T_{2i} = 200^\circ\text{C}$	Air inlet temperature	(E5.8.137)
$P_{1i} = 160\,\text{kPa}$	Gas inlet pressure	(E5.8.138)
$P_{2i} = 200\,\text{kPa}$	Air inlet pressure	(E5.8.139)
$Q_1 = 3.494\,\text{m}^3/\text{s}$	Volume flow rate at gas side	(E5.8.140)
$Q_2 = 1.358\,\text{m}^3/\text{s}$	Volume flow rate at air side	(E5.8.141)

Requirements:

$$e_{\text{hx}} = 0.824 \qquad \text{Effectiveness} \qquad \text{(E5.8.142)}$$

$$\Delta P_1 \leq 10\,\text{kPa} \qquad \text{Pressure drop at gas side} \qquad \text{(E5.8.143)}$$

$$\Delta P_2 \leq 10\,\text{kPa} \qquad \text{Pressure drop at air side} \qquad \text{(E5.8.144)}$$

$$L_1 = L_2 \qquad \text{Desirable} \qquad \text{(E5.8.145)}$$

Dimensions of the sizing:

$$L_1 = 0.294\,\text{m} \quad L_2 = 0.294\,\text{m} \quad L_3 = 0.982\,\text{m} \qquad \text{(E5.8.146)}$$

Outlet temperatures and pressure drops:

$$T_{1\text{o}}(L_1, L_2, L_3) = 323.2 \cdot {}^\circ\text{C} \qquad \text{Gas outlet temperature} \qquad \text{(E5.8.147)}$$

$$T_{2\text{o}}(L_1, L_2, L_3) = 702.484^\circ\text{C} \qquad \text{Air outlet temperature} \qquad \text{(E5.8.148)}$$

$$\Delta P_1(L_1, L_2, L_3) = 10 \cdot \text{kPa} \qquad \text{Pressure drop at gas side} \qquad \text{(E5.8.149)}$$

$$\Delta P_2(L_1, L_2, L_3) = 7.759 \cdot \text{kPa} \qquad \text{Pressure drop at air side} \qquad \text{(E5.8.150)}$$

Densities of gas and air at inlet and outlet

$$\rho_{1\text{i}} = 0.475\,\text{kg/m}^3 \quad \rho_{1\text{o}}(P_{1\text{o}}(L_1, L_2, L_3), L_1, L_2, L_3) = 0.876\,\text{kg/m}^3 \quad \text{(E5.8.151)}$$

$$\rho_{2\text{i}} = 1.473\,\text{kg/m}^3 \quad \rho_{2\text{o}}(P_{2\text{o}}(L_1, L_2, L_3), L_1, L_2, L_3) = 0.686\,\text{kg/m}^3 \quad \text{(E5.8.152)}$$

Thermal quantities:

$$NTU(L_1, L_2, L_3) = 7.134 \qquad \text{Number of transfer unit} \qquad \text{(E5.8.153)}$$

$$\varepsilon_{\text{hx}}(L_1, L_2, L_3) = 0.824 \qquad \text{Effectiveness} \qquad \text{(E5.8.154)}$$

$$q(L_1, L_2, L_3) = 1.078 \times 10^6\,\text{W} \qquad \text{Heat transfer rate} \qquad \text{(E5.8.155)}$$

$$h_1(L_1, L_2, L_3) = 369.088\,\text{W/m}^2\text{K} \qquad \text{Heat transfer coefficient} \qquad \text{(E5.8.156)}$$

$$h_2(L_1, L_2, L_3) = 366.492\,\text{W/m}^2\text{K} \qquad \text{(E5.8.157)}$$

$$UA(L_1, L_2, L_3) = 13335\,\text{W/K} \qquad \text{UA value} \qquad \text{(E5.8.157a)}$$

Hydraulic quantities:

$$D_{\text{h1}}(L_1, L_2, L_3) = 1.524\,\text{mm} \qquad \text{Hydraulic diameter} \qquad \text{(E5.8.158)}$$

$$D_{\text{h2}}(L_1, L_2, L_3) = 1.525\,\text{mm} \qquad \text{(E5.8.159)}$$

$$G_1(L_1, L_3) = 15.722\,\text{kg/m}^2\,\text{s} \qquad \text{Mass velocity} \qquad \text{(E5.8.160)}$$

$$G_2(L_2, L_3) = 18.824\,\text{kg/m}^2\,\text{s} \qquad \text{(E5.8.161)}$$

$$Re_1(L_1, L_2, L_3) = 597.69 \qquad \text{Reynolds number} \qquad \text{(E5.8.162)}$$

$$Re_2(L_1, L_2, L_3) = 854.08 \qquad \text{(E5.8.163)}$$

Geometric quantities:

$$b_1 = 2.49 \, \text{mm} \qquad\qquad b_2 = 2.49 \, \text{mm} \qquad\qquad\qquad \text{(E5.8.164)}$$

$$p_{f1} = 1.279 \, \text{mm} \qquad\qquad p_{f2} = 1.279 \, \text{mm} \qquad\qquad \text{(E5.8.165)}$$

$$\lambda_1 = 3.175 \, \text{mm} \qquad\qquad \lambda_2 = 3.175 \, \text{mm} \qquad\qquad \text{(E5.8.166)}$$

$$\delta = 0.102 \, \text{mm} \qquad\qquad \text{Fin thickness} \qquad\qquad\qquad\quad \text{(E5.8.167)}$$

$$\delta_w = 0.5 \, \text{mm} \qquad\qquad \text{Wall thickness} \qquad\qquad\qquad\quad \text{(E5.8.168)}$$

$$N_p(L_3) = 163.611 \qquad\quad \text{Number of passages at gas side} \qquad \text{(E5.8.169)}$$

$$N_p(L_3) + 1 = 164.611 \qquad \text{Number of passages at air side} \qquad \text{(E5.8.170)}$$

$$\sigma_1(L_1, L_3) = 0.366 \qquad\quad \text{Porosity} \qquad\qquad\qquad\qquad\qquad \text{(E5.8.171)}$$

$$\sigma_2(L_2, L_3) = 0.368 \qquad\qquad\qquad\qquad\qquad\qquad\qquad\qquad\quad \text{(E5.8.172)}$$

$$\beta_1(L_1, L_2, L_3) = 2316 \, \text{m}^2/\text{m}^3 \quad \text{Surface area density} \qquad\quad \text{(E5.8.173)}$$

$$\beta_2(L_1, L_2, L_3) = 2316 \, \text{m}^2/\text{m}^3 \qquad\qquad\qquad\qquad\qquad\qquad\quad \text{(E5.8.174)}$$

$$A_{t1}(L_1, L_2, L_3) = 81.372 \, \text{m}^2 \quad \text{Total heat transfer area} \qquad \text{(E5.8.175)}$$

$$A_{t2}(L_1, L_2, L_3) = 81.866 \, \text{m}^2 \qquad\qquad\qquad\qquad\qquad\qquad\quad \text{(E5.8.176)}$$

The dimensions given in Equation (E5.8.146) satisfy the design requirements shown in Equations (E5.8.142) to (E5.8.145). The surface area density ($2316 \, \text{m}^2/\text{m}^3$) in Equation (E5.8.173) is much greater than the criterion (greater than $600 \, \text{m}^2/\text{m}^3$) of the compact heat exchanger in Section 5.1, indicating that this heat exchanger is indeed very compact. This exchanger is one of the most compact heat exchangers. Note that this sizing problem actually becomes the rating problem with given dimensions of a plate-fin heat exchanger if we eliminate the equations after Equation (E5.8.129).

5.9   LOUVER-FIN-TYPE FLAT-TUBE PLATE-FIN HEAT EXCHANGERS

Louver surface is the standard geometry for automotive radiators and also is often used in aircraft heat exchangers. A core of the louver-fin-type plate-fin heat exchanger is shown in Figure 5.43. The louver-fin geometry provides heat transfer coefficients comparable to those of the offset-strip-fin geometry (OSF).

5.9.1   Geometric Characteristics

A louver fin geometry is illustrated in Figure 5.44. For convenience, coolant and air are referred to as fluid 1 and fluid 2, respectively. We consider first the air side.

The number of passages N_{pg} on the air side is defined as the air flow passages between the flat tubes (don't confuse with the number of passes N_p). The core width L_1 is expressed in terms of the number of passages N_{pg} as

$$L_1 = N_{pg}b + (N_{pg} + 1) H_t \qquad\qquad (5.260)$$

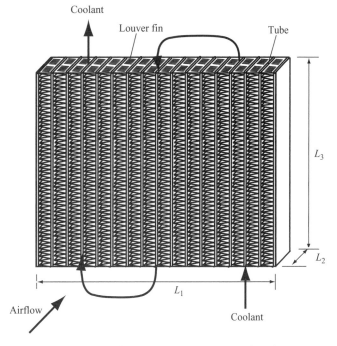

Figure 5.43   Louver fin type plate-fin heat exchanger (1-pass for air and 3-pass for coolant)

where b is the tube spacing and H_t the tube height. Solving for the number of passages N_{pg} gives

$$N_{pg} = \frac{L_1 - H_t}{b + H_t} \tag{5.261}$$

The total number of fins n_f is

$$n_f = \frac{L_3}{p_f} N_{pg} \tag{5.262}$$

where p_f is the fin pitch as shown in Figure 5.44(b) and L_3 the core height. The total heat transfer area A_t is generally obtained from the sum of the primary area A_p and the fin area A_f. The primary area A_p is calculated by subtracting the fin base areas from the tube outer surface areas, considering the circular front and end of the tubes (Figure 5.44(c)).

$$A_p = \underbrace{[2 \left(L_2 - H_t \right) + \pi H_t] \, L_3 \left(N_{pg} + 1 \right)}_{\text{Tube outside surface area}} - \underbrace{2\delta L_2 n_f}_{\text{Fin base area}} \tag{5.263}$$

The total number of louvers in the core is obtained from Figure 5.44(d) as

$$n_{louv} = \left(\frac{L_f}{l_p} - 1 \right) n_f \tag{5.264}$$

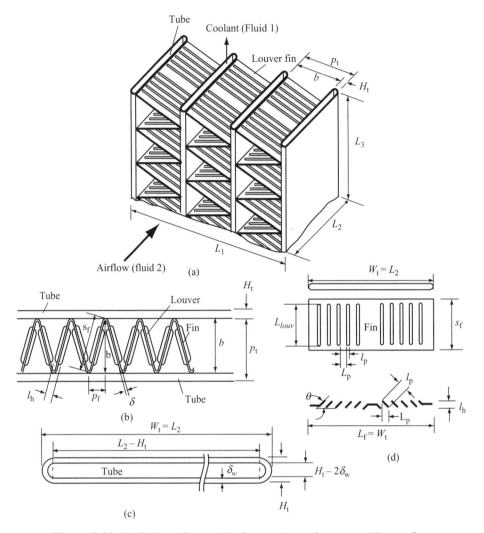

Figure 5.44   Definition of geometrical parameters of corrugated louver fins

where L_f is the fin length in the flow direction and l_p the louver pitch. Note that when the louvers change the direction of opening we allow one louver missing in the middle of the fin. The total fin area A_f is the sum of the fin area and the louver edge area as

$$A_f = \underbrace{2\left(s_f L_2 + s_f \delta\right) n_f}_{\text{Fin area}} + \underbrace{2 L_{louv} \delta n_{louv}}_{\text{Louver edge area}} \qquad (5.265)$$

where s_f is the fin width and L_{louv} the louver length. The total heat transfer area A_t is obtained by

$$A_{t2} = A_p + A_f \qquad (5.266)$$

The minimum free-flow area A_c is expressed by

$$A_{c2} = \underbrace{bL_3N_{pg}}_{\text{Spacing between tubes}} - \underbrace{\left[\delta\left(\delta_f - L_{louv}\right) + L_{louv}l_h\right]n_f}_{\text{Fin and louver edge area}} \tag{5.267}$$

Now we consider coolant-side geometry in Figure 5.44(a)–(c). The total number of tubes N_t is obtained by

$$N_t = N_{pg} + 1 \tag{5.268}$$

Considering the circular shapes at both ends, the total heat transfer area A_{t1} on the coolant side is obtained by

$$A_{t1} = \left[\underbrace{2\left(L_2 - H_t\right)}_{\text{Tube straight length}} + \underbrace{\pi\left(H_t - 2\delta_w\right)}_{\text{Circular shape at the ends}} \right] L_3 N_t \tag{5.269}$$

The free-flow area A_{c1} on the coolant side is obtained by

$$A_{c1} = \left[\underbrace{\left(L_2 - H_t\right)\left(H_t - 2\delta_w\right)}_{\text{Rectangle of tube}} + \underbrace{\frac{\pi}{4}\left(H_t - 2\delta_w\right)^2}_{\text{Circular parts at the ends}} \right] \frac{N_t}{N_p} \tag{5.270}$$

5.9.2   Correlations for Louver Fin Geometry

Louver fins are used extensively in the auto industry due to their mass production manufacturability and, hence, lower cost. They have generally higher j and f factors than those for the offset strip fin geometry, and also the increase in the friction factors is, in general, higher than the increase in j factors [17]. Davenport [28] in 1983 tested 32 one-row louver fin geometries. Davenport systematically varied the louver dimensions for two louver heights (12.7 and 7.8 mm) and developed correlations for j and f versus Re. Chang and Wang [29] in 1997 provided a correlation for the Colburn factor j based on an extensive database for airflow over louver fins.

$$j = Re^{-0.49}\left(\frac{\theta}{90}\right)^{0.27}\left(\frac{p_f}{l_p}\right)^{-0.14}\left(\frac{b}{l_p}\right)^{-0.29}$$
$$\times \left(\frac{W_t}{l_p}\right)^{-0.23}\left(\frac{L_{louv}}{l_p}\right)^{0.68}\left(\frac{p_t}{l_p}\right)^{-0.28}\left(\frac{\delta}{l_p}\right)^{-0.05} \tag{5.271}$$

where l_p is the louver pitch, θ the louver angle, p_f the fin pitch, p_t the tube pitch, b the tube spacing, W_t the tube outside width, L_{louv} the louver cut length, and δ the fin thickness. Figure 5.43 shows the dimensions of the geometrical parameters of louver fins.

Equation (5.271) is valid for the following ranges of the parameters: $0.82 \le D_h \le 5.02$ mm, $0.51 \le p_f \le 3.33$ mm, $0.5 \le l_p \le 3$ mm, $2.84 \le b \le 20$ mm, $15.6 \le W_t \le 57.4$ mm, $2.13 \le L_l \le 18.5$ mm, $7.51 \le p_t \le 25$ mm, $0.0254 \le \delta \le$

0.16 mm, $1 \leq N_r \leq 2$, and $8.4 \leq \theta \leq 35°$. This correlation predicts j factors with in ± 15 percent for $30 < Re < 5,000$. The correlation for the Fanning friction factor based on the same database by Chang [30] is

$$f = f_a f_b f_c \tag{5.272}$$

where

$$f_a = \begin{cases} 14.39 Re^{-0.805 p_f/b} \left[\ln \left(1.0 + p_f/l_p \right) \right]^{3.04} & Re < 150 \\ 4.97 Re^{\left(0.6049 - 1.064/\theta^{0.2} \right)} \left\{ \ln \left[\left(\delta/p_f \right)^{0.5} + 0.9 \right] \right\}^{-0.527} & 150 \leq Re < 5,000 \end{cases}$$
$$\tag{5.273}$$

$$f_b = \begin{cases} \left\{ \ln \left[\left(\delta/p_f \right)^{0.48} + 0.9 \right] \right\}^{-1.453} \left(D_h/l_p \right)^{-3.01} \left[\ln \left(0.5 \, Re \right) \right]^{-3.01} & Re < 150 \\ \left[\left(D_h/l_p \right) \ln \left(0.3 \, Re \right) \right]^{-2.966} \left(p_f/L_{louv} \right)^{-0.7931 p_f/b} & 150 < Re < 5,000 \end{cases}$$
$$\tag{5.274}$$

$$f_c = \begin{cases} \left(p_f/L_{louv} \right)^{-0.308} \left(L_f/L_{louv} \right)^{-0.308} \left[\exp \left(-0.1167 p_t/H_t \right) \right] \theta^{0.35} & Re < 150 \\ \left(p_t/H_t \right)^{-0.0446} \left\{ \ln \left[1.2 + \left(l_p/p_f \right)^{1.4} \right] \right\}^{-3.553} \theta^{-0.477} & 150 < Re < 5,000 \end{cases}$$
$$\tag{5.275}$$

where H_t is the tube outside height and L_f the fin length.

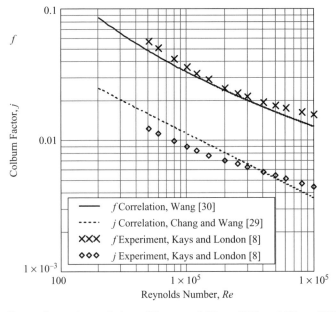

Figure 5.45 Comparison of correlations (Chang and Wang [29] and Wang [30]) and experiments of Surface $^3/_4$(b)-11.1 (Kays and London [17])

The correlations in Equations (5.271) and (5.272) were compared with the experiments of Surface $^3/_4$ (b)-11.1 reported by Kays and London [8], as shown in Figure 5.45. The detailed information for the louver fin geometry of the experiments was not provided. For some unknown parameters, the middle values of the valid ranges given in the previous paragraph were used in the calculations. However, the comparison between them shows fair agreement (Figure 5.45), indicating that the j and f factors are not affected substantially by the louver fin geometry. Note that the Reynolds number may extend up to 10,000 without significant errors.

Example 5.9.1 Louver-Fin-Type Plate-Fin Heat Exchanger   A coolant-to-air cross-flow heat exchanger is designed to cool the coolant (50 percent ethylene glycol with water). Louver fin geometry is employed on the air side. The geometrical properties and surface characteristics are provided in Figures E5.9.1 and E5.9.2. Both fins and tubes are made from aluminum alloy with $k = 117 \, \text{W/mK}$. The coolant flows in the flat tubes at $1.65 \times 10^{-3} \, \text{m}^3/\text{s}$ and $95°C$, and shall leave at $90°C$. Air enters at $1.05 \, \text{m}^3/\text{s}$ and $25°C$. The inlet pressure of the air is at $100 \, \text{kPa}$ absolute and the inlet pressure of the coolant is at $200 \, \text{kPa}$ absolute. The air pressure drop is required to be less than $250 \, \text{Pa}$. The coolant pressure drop is recommended to be less than $70 \, \text{kPa}$. Design a coolant-to-air cross flow heat exchanger. Determine the core dimensions of this exchanger. Then, determine the heat transfer rate, the outlet fluid temperature, and the pressure drop on each fluid. Use the following information in your calculations.

Description	value
Fin thickness δ	0.152 mm
Tube wall thickness δ_w	0.3 mm
Fin density N_f	437 m^{-1}
Tube spacing b	6.35 mm
Louver pitch l_p	1.0 mm
Louver angle θ	20 degree
Tube outside height H_t	2 mm

MathCAD Format Solution   The design concept is to develop a MathCAD model as a function of the dimensions in order to determine the dimensions of the heat exchanger.

Properties: We will use the arithmetic average as the approximate mean temperature at this time on each side by assuming the air outlet temperature. Define the coolant and the air as fluid 1 and fluid 2, respectively.

$$T_{\text{coolant}} := \frac{95°C + 90°C}{2} = 92.5 \cdot °C \quad T_{\text{air}} := \frac{25°C + 35°C}{2} = 303.15 \cdot K \quad (E5.9.1)$$

50% Ethylene Glycol (subscript 1)   Air (subscript 2)

$$\rho_1 := 1020 \, \frac{\text{kg}}{\text{m}^3} \qquad \rho_2 = \text{defined\_later}$$

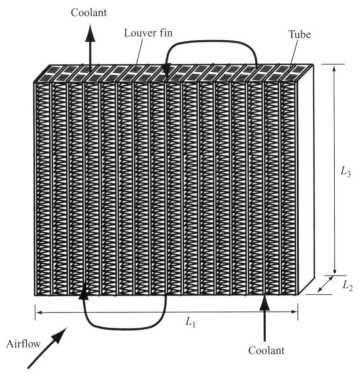

Figure E5.9.1   Louver-fin-type plate-fin heat exchanger (1-pass for air and 3-pass for coolant)

$$c_{p1} := 3650 \frac{J}{kg \cdot K} \qquad c_{p2} := 1007 \frac{J}{kg \cdot K} \qquad (E5.9.2)$$

$$k_1 := 0.442 \frac{W}{m \cdot K} \qquad k_2 := 0.026 \frac{W}{m \cdot K}$$

$$\mu_1 := 0.08 \cdot 10^{-2} \frac{N \cdot s}{m^2} \qquad \mu_2 := 18.46 \cdot 10^{-6} \frac{N \cdot s}{m^2}$$

$$Pr_1 := 6.6 \qquad Pr_2 := 0.707$$

The thermal conductivities of fins and walls (aluminum) are given as

$$k_f := 117 \frac{W}{m \cdot K} \qquad k_W := 117 \frac{W}{m \cdot K} \qquad (E5.9.3)$$

Given information:

$$T_{1i} := 95°C \qquad \text{Coolant inlet temperature} \qquad (E5.9.4)$$

$$T_{2i} := 25°C \qquad \text{Air inlet temperature} \qquad (E5.9.5)$$

$$P_{1i} := 500 \, kPa \qquad \text{Coolant inlet pressure} \qquad (E5.9.6)$$

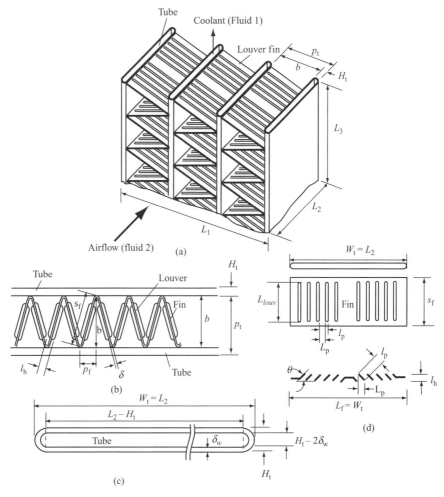

Figure E5.9.2   Definition of geometrical parameters of corrugated louver fins

$P_{2i} := 100\,\text{kPa}$ Air inlet pressure   (E5.9.7)

$Q_1 := 1.65 \cdot 10^{-3}\,\dfrac{\text{m}^3}{\text{s}}$ Volume flow rate at coolant side (26.15 gall/min)   (E5.9.8)

$Q_2 := 1.05\,\dfrac{\text{m}^3}{\text{s}}$ Volume flow rate at air side   (E5.9.9)

Design requirements:

$T_{1o} = 90^\circ\text{C}$ Coolent outlet temperature   (E5.9.10)

$\Delta P_1 \leq 70\,\text{kPa}$ Pressure drop at coolent side   (E5.9.11)

$\Delta P_2 \leq 250\,\text{Pa}$ Pressure drop at air side   (E5.9.12)

Inlet air density:

$$R_g := 287.04 \, \frac{J}{kg \cdot K} \qquad \text{Gas constant for the air} \qquad \text{(E5.9.13)}$$

$$\rho_{2i} := \frac{P_{2i}}{R_g \cdot T_{2i}} \qquad \rho_{2i} = 1.168 \, \frac{kg}{m^3} \qquad \text{(E5.9.14)}$$

Then, the mass flow rate for each fluid is calculated as

$$mdot_1 := \rho_1 \cdot Q_1 \qquad mdot_1 = 1.683 \, \frac{kg}{s} \qquad \text{(E5.9.15)}$$

$$mdot_2 := \rho_{2i} \cdot Q_2 \qquad mdot_2 = 1.227 \, \frac{kg}{s} \qquad \text{(E5.9.16)}$$

Assume the Sizing of the Heat Exchanger Initially, assume the sizing to have the instant values and update later the dimensions when the final sizing is determined.

$$L_1 := 50.596 \, cm \quad L_2 := 3.33 \, cm \quad L_3 := 30.213 \, cm \qquad \text{(E5.9.17)}$$

Geometric information:

$N_p := 3$	Number of passes at coolant side		(E5.9.18)
$\delta := 0.152 \, mm$	Fin thickness		(E5.9.19)
$\delta_w := 0.3 \, mm$	Tube wall thickness		(E5.9.20)
$N_f := 437 \, m^{-1}$	Fin density		(E5.9.21)
$p_f := \dfrac{1}{N_f}$	Fin pitch	$p_f = 2.2883 \, mm$	(E5.9.22)
$b := 6.35 \, mm$	Plate distance		(E5.9.23)
$s_f := \sqrt{b^2 + p_f^2}$	Fin height	$s_f = 6.75 \cdot mm$	(E5.9.24)
$l_p := 1 \, mm$	Louver pitch		(E5.9.25)
$L_{louv} := 0.85 \cdot s_f$	Louver cut length	$L_{louv} = 5.737 \cdot mm$	(E5.9.26)
$\theta := 20$	Louver angle		(E5.9.27)
$H_t := 2 \, mm$	Tube outside height		(E5.9.28)
$p_t := b + H_t$	Tube pitch	$p_t = 8.35 \cdot mm$	(E5.9.29)
$l_h := l_p \cdot \sin(20deg)$	Louver height	$l_h = 0.342 \cdot mm$	(E5.9.30)
$W_t(L_2) := L_2$	Tube outside width	$W_t(L_2) = 33.3 \cdot mm$	(E5.9.31)
$L_f(L_2) := W_t(L_2)$	Fin length	$L_f(L_2) := 33.3 \cdot mm$	(E5.9.32)

The core width L_1 is obtained using Equation (5.260) and Figure E5.9.2(a) as

$$L_1 = N_{pg} \cdot b + (N_{pg} + 1) \cdot H_t \qquad \text{(E5.9.33)}$$

The total number of passages N_{pg} on the air side is obtained using Equation (5.261) by

$$N_{pg}(L_1) := \frac{L_1 - H_t}{b + H_t} \qquad N_{pg}(L_1) = 60.354 \qquad \text{(E5.9.34)}$$

The total number of fins for the air side is obtained by

$$n_f(L_1, L_3) := \frac{L_3}{p_f} \cdot N_{pg}(L_1) \qquad n_f(L_1, L_3) = 7969 \qquad \text{(E5.9.35)}$$

The primary area on the air side is given in Equation (5.263) by

$$A_p(L_1, L_2, L_3) := [2 \cdot (L_2 - H_t) + \pi \cdot H_t] \cdot L_3$$
$$\cdot (N_{pg}(L_1) + 1) - 2 \cdot \delta \cdot L_2 \cdot n_f(L_1, L_3) \qquad \text{(E5.9.36)}$$
$$A_p(L_1, L_2, L_3) = 1.196\,\text{m}^2$$

The total number of louvers is calculated using Equation (5.264) by

$$n_{louv}(L_1, L_2, L_3) := \left(\frac{L_f(L_2)}{l_p} - 1\right) \cdot n_f(L_1, L_3) \qquad \text{(E5.9.37)}$$

The total fin area is given using Equation (5.265) by

$$A_f(L_1, L_2, L_3) := 2 \cdot (s_f \cdot L_2 + s_f \cdot \delta) \cdot n_f(L_1, L_3)$$
$$+ 2 \cdot L_{louv} \cdot \delta \cdot n_{louv}(L_1, L_2, L_3) \qquad \text{(E5.9.38)}$$
$$A_f(L_1, L_2, L_3) = 4.047\,\text{m}^2$$

The total surface area is the sum of the fin area and the primary area.

$$A_{t2}(L_1, L_2, L_3) := A_f(L_1, L_2, L_3) + A_p(L_1, L_2, L_3)$$
$$A_{t2}(L_1, L_2, L_3) = 5.244\,\text{m}^2 \qquad \text{(E5.9.39)}$$

The minimum free-flow area on the air side is given in Equation (5.267) by

$$A_{c2}(L_1, L_3) := b \cdot L_3 \cdot N_{pg}(L_1) - [\delta \cdot (s_f - L_{louv}) + L_{louv} \cdot l_h] \cdot n_f(L_1, L_3) \qquad \text{(E5.9.40)}$$
$$A_{c2}(L_1, L_3) = 0.0989\,\text{m}^2$$

The frontal area on the air side is

$$A_{fr2}(L_1, L_3) := L_1 \cdot L_3 \qquad A_{fr2}(L_1, L_3) = 0.153\,\text{m}^2 \qquad \text{(E5.9.41)}$$

The hydraulic diameter on the air side is calculated by

$$D_{h2}(L_1, L_2, L_3) := \frac{4 \cdot A_{c2}(L_1, L_3) \cdot L_2}{A_{t2}(L_1, L_2, L_3)} \qquad D_{h2}(L_1, L_2, L_3) = 2.513 \, \text{mm} \qquad \text{(E5.9.42)}$$

The porosity on the air side is then obtained by

$$\sigma_2(L_1, L_3) := \frac{A_{c2}(L_1, L_3)}{A_{fr2}(L_1, L_3)} \qquad \sigma_2(L_1, L_3) = 0.647 \qquad \text{(E5.9.43)}$$

The volume of the exchanger on the air side is

$$V_{hx2}(L_1, L_2, L_3) := L_2 \cdot L_3 \cdot b \cdot N_{pg}(L_1)$$
$$V_{hx2}(L_1, L_2, L_3) = 3.8559 \times 10^{-3} \cdot \text{m}^3 \qquad \text{(E5.9.44)}$$

The surface area density on the air side is

$$\beta_2(L_1, L_2, L_3) := \frac{A_{t2}(L_1, L_2, L_3)}{V_{hx2}(L_1, L_2, L_3)} \qquad \beta_2(L_1, L_2, L_3) = 1360 \, \frac{\text{m}^2}{\text{m}^3} \qquad \text{(E5.9.45)}$$

Mass Velocities, Reynolds Numbers, and j and f Factors

$$G_2(L_1, L_3) := \frac{mdot_2}{A_{c2}(L_1, L_3)} \qquad G_2(L_1, L_3) = 12.402 \, \frac{\text{kg}}{\text{m}^2 \cdot \text{s}} \qquad \text{(E5.9.46)}$$

$$v_2(L_1, L_3) := \frac{G_2(L_1, L_3)}{\rho_{2i}} \qquad v_2(L_1, L_3) = 10.614 \, \frac{\text{m}}{\text{s}} \qquad \text{(E5.9.47)}$$

$$Re_2(L_1, L_2, L_3) := \frac{G_2(L_1, L_3) \cdot D_{h2}(L_1, L_2, L_3)}{\mu_2}$$
$$Re_2(L_1, L_2, L_3) = 1688 \qquad \text{(E5.9.48)}$$

The Reynolds number on the air side indicates a laminar flow pattern. The Colburn factor *j* is given in Equation (5.271):

$$j(Re) := Re^{-0.49} \cdot \left(\frac{\theta}{90}\right)^{0.27} \left(\frac{p_f}{l_p}\right)^{-0.14} \left(\frac{b}{l_p}\right)^{-0.29}$$
$$\times \left(\frac{W_t(L_2)}{l_p}\right)^{-0.23} \left(\frac{L_{louv}}{l_p}\right)^{0.68} \left(\frac{p_t}{l_p}\right)^{-0.28} \left(\frac{\delta}{l_p}\right)^{-0.05} \qquad \text{(E5.9.49)}$$

$$j_2(L_1, L_2, L_3) := j(Re_2(L_1, L_2, L_3)) \qquad j_2(L_1, L_2, L_3) = 0.00808 \qquad \text{(E5.9.50)}$$

The Fanning friction factor *f* is given in Equation (5.272) as follows:

$$f_a(Re) := \begin{vmatrix} 14.39 \cdot Re^{-0.805 \cdot \frac{p_f}{b}} \cdot \left(\ln\left(1.0 + \frac{p_f}{l_p}\right)\right) & \text{if } Re < 150 \\ 4.97 \cdot Re^{\left(0.6049 - \frac{1.064}{\theta^{0.2}}\right)} \cdot \left[\ln\left[\left(\frac{\delta}{p_f}\right)^{0.5} + 0.9\right]\right]^{-0.527} & \text{otherwise} \end{vmatrix}$$

$$\text{(E5.9.51)}$$

$$f_b(\text{Re}) := \begin{vmatrix} \left[\ln\left[\left(\dfrac{\delta}{p_f}\right)^{0.48} + 0.9\right]\right]^{-1.453} \cdot \left(\dfrac{D_{h2}(L_1, L_2, L_3)}{l_p}\right)^{-3.01} \\[2em] \quad \times (\ln(0.5 \cdot \text{Re}))^{-3.01} \text{ if Re} < 150 \\[2em] \left[\left(\dfrac{D_{h2}(L_1, L_2, L_3)}{l_p}\right) \cdot \ln(0.3 \cdot \text{Re})\right]^{-2.966} \cdot \left(\dfrac{p_f}{L_{louv}}\right)^{-0.7931 \cdot \frac{p_t}{b}} \text{ otherwise} \end{vmatrix}$$

$$\text{(E5.9.52)}$$

$$f_c(\text{Re}) := \begin{vmatrix} \left(\dfrac{p_f}{L_{louv}}\right)^{-0.308} \cdot \left(\dfrac{L_f(L_2)}{L_{louv}}\right)^{-0.308} \cdot \left(\exp\left(-0.1167\dfrac{p_t}{H_t}\right)\right) \\[2em] \quad \cdot \theta^{0.35} \text{ if Re} < 150 \\[2em] \left(\dfrac{p_t}{H_t}\right)^{0.0446} \cdot \left[\ln\left[1.2 + \left(\dfrac{l_p}{p_f}\right)^{1.4}\right]\right]^{-3.553} \cdot \theta^{-0.477} \text{ otherwise} \end{vmatrix}$$

$$\text{(E5.9.53)}$$

$$f_2(L_1, L_2, L_3) := f_a(\text{Re}_2(L_1, L_2, L_3)) \cdot f_b(\text{Re}_2(L_1, L_2, L_3))$$
$$\cdot f_c(\text{Re}_2(L_1, L_2, L_3)) \quad \text{(E5.9.54)}$$

$$f_2(L_1, L_2, L_3) = 0.06104$$

Heat Transfer Coefficient and Fin Efficiency on the Air Side   We compute the heat transfer coefficient from the definition of the j factor given in Equation (5.230) as:

$$h_2(L_1, L_2, L_3) := \frac{j_2(L_1, L_2, L_3) \cdot G_2(L_1, L_3) \cdot c_{p2}}{Pr_2^{\frac{2}{3}}} \quad \text{(E5.9.55)}$$

$$h_2(L_1, L_2, L_3) = 127.217 \cdot \frac{W}{m^2 \cdot K}$$

Now let us calculate the fin efficiency on the air side. Since the louver fins are used on the air side, we will use the concept of overall fin efficiency. From Equation (5.96), m value is given by

$$m_f(L_1, L_2, L_3) := \left(\frac{2 \cdot h_2(L_1, L_2, L_3)}{k_f \cdot \delta}\right)^{0.5} \quad \text{(E5.9.56)}$$

Considering the adiabatic tip at the middle of the fin, the fin length L_s will be half of the fin width s_f.

$$L_s := \frac{s_f}{2} \quad \text{(E5.9.57)}$$

The single fin efficiency is

$$\eta_f(L_1, L_2, L_3) := \frac{\tanh(m_f(L_1, L_2, L_3) \cdot L_s)}{m_f(L_1, L_2, L_3) \cdot L_s} \quad \eta_f(L_1, L_2, L_3) = 0.949 \quad \text{(E5.9.58)}$$

The overall surface (fin) efficiency is then

$$\eta_o\,(L_1, L_2, L_3) := 1 - \frac{A_f\,(L_1, L_2, L_3)}{A_{t2}\,(L_1, L_2, L_3)} \cdot (1 - \eta_f\,(L_1, L_2, L_3)) \tag{E5.9.59}$$

$$\eta_o\,(L_1, L_2, L_3) = 0.961$$

Coolant-side Reynolds Number and Mass Velocity The number of tubes on the coolant side is

$$N_t(L_1) := N_{pg}\,(L_1) + 1 \tag{E5.9.60}$$

The total heat transfer area on the coolant side is given in Equation (5.269) by

$$A_{t1}\,(L_1, L_2, L_3) := [2 \cdot (L_2 - H_t) + \pi \cdot (H_t - 2 \cdot \delta_w)] \cdot L_3 \cdot N_t\,(L_1) \tag{E5.9.61}$$

$$A_{t1}\,(L_1, L_2, L_3) = 1.242\,\text{m}^2$$

The free-flow area on the coolant side is given in Equation (5.270) by

$$A_{c1}\,(L_1, L_2) := \left[(L_2 - H_t) \cdot (H_t - 2 \cdot \delta_w) + \frac{\pi}{4} \cdot (H_t - 2 \cdot \delta_w)^2 \right] \cdot \frac{N_t\,(L_1)}{N_p} \tag{E5.9.62}$$

$$A_{c1}\,(L_1, L_2) = 9.277 \times 10^{-4}\,\text{m}^2$$

Note that the free-flow area on the coolant side is divided by the number of passes N_p. The hydraulic diameter on the coolant side is

$$D_{h1}\,(L_1, L_2, L_3) := \frac{4 \cdot A_{c1}\,(L_1, L_2) \cdot L_3}{A_{t1}\,(L_1, L_2, L_3)} \quad D_{h1}\,(L_1, L_2, L_3) = 0.903\,\text{mm} \tag{E5.9.63}$$

The mass velocity, velocity, and Reynolds number on the coolant side are

$$G_1\,(L_1, L_2) := \frac{mdot_1}{A_{c1}\,(L_1, L_2)} \quad G_1\,(L_1, L_2) = 1.814 \times 10^3\,\frac{\text{kg}}{\text{m}^2 \cdot \text{s}} \tag{E5.9.64}$$

$$v_1\,(L_1, L_2) := \frac{G_1\,(L_1, L_2)}{\rho_1} \quad v_1\,(L_1, L_2) = 1.779\,\frac{\text{m}}{\text{s}} \tag{E5.9.65}$$

$$Re_1\,(L_1, L_2, L_3) := \frac{G_1\,(L_1, L_2) \cdot D_{h1}\,(L_1, L_2, L_3)}{\mu_1} \tag{E5.9.66}$$

$$Re_1\,(L_1, L_2, L_3) = 2047$$

The coolant velocity is increased with increasing the number of passes N_p. The sufficiently high velocity of 1.779 m/s was achieved by $N_p = 3$ in order to reduce the fouling. Low velocities may cause fouling.

Coolant-side Friction Factor and Heat Transfer Coefficient

$$f_1(L_1,L_2,L_3) := \begin{vmatrix} (1.58 \cdot \ln(Re_1(L_1,L_2,L_3)) - 3.28)^{-2} \text{ if} \\ Re_1(L_1,L_2,L_3) \geq 2100 \\ \dfrac{16}{Re_1(L_1,L_2,L_3)} \text{ otherwise} \end{vmatrix} \quad (E5.9.67)$$

$$f_1(L_1,L_2,L_3) = 0.000782$$

$$Nu_1(L_1,L_2,L_3) := \begin{vmatrix} \dfrac{f_1(L_1,L_2,L_3)}{2} \cdot \dfrac{(Re_1(L_1,L_2,L_3) - 1000) \cdot Pr_1}{1 + 12.7 \cdot \left(\dfrac{f_1(L_1,L_2,L_3)}{2}\right)^{0.5} \cdot \left(Pr_1^{\frac{2}{3}} - 1\right)} \\ \text{if } Re_1(L_1,L_2,L_3) \geq 2100 \\ 7.541 \text{ otherwise} \end{vmatrix}$$

$$(E5.9.68)$$

$$Nu_1(L_1,L_2,L_3) = 7.541$$

$$h_1(L_1,L_2,L_3) := \frac{Nu_1(L_1,L_2,L_3) \cdot k_1}{D_{h1}(L_1,L_2,L_3)} \quad h_1(L_1,L_2,L_3) = 3692 \cdot \frac{W}{m^2 \cdot K} \quad (E5.9.69)$$

The total conduction area for the wall thermal resistance is

$$A_w(L_2,L_3) := 2 \cdot L_2 \cdot L_3 \cdot N_t(L_1) \quad A_w(L_2,L_3) = 1.235\,m^2 \quad (E5.9.70)$$

The thermal resistance at the wall is

$$R_w(L_2,L_3) := \frac{\delta_w}{k_w \cdot A_w(L_2,L_3)} \quad R_w(L_2,L_3) = 2.077 \times 10^{-6} \cdot \frac{K}{W} \quad (E5.9.71)$$

The coolant side thermal resistance is

$$R_1(L_1,L_2,L_3) := \frac{1}{h_1(L_1,L_2,L_3) \cdot A_{t1}(L_1,L_2,L_3)}$$
$$R_1(L_1,L_2,L_3) = 2.181 \times 10^{-4} \cdot \frac{K}{W} \quad (E5.9.72)$$

The air-side thermal resistance is

$$R_2(L_1,L_2,L_3) := \frac{1}{\eta_o(L_1,L_2,L_3) \cdot h_2(L_1,L_2,L_3) \cdot A_{t2}(L_1,L_2,L_3)}$$
$$R_2(L_1,L_2,L_3) = 1.56 \times 10^{-3} \cdot \frac{K}{W} \quad (E5.9.73)$$

The overall *UA* value is

$$UA(L_1,L_2,L_3) := \frac{1}{R_1(L_1,L_2,L_3) + R_w(L_2,L_3) + R_2(L_1,L_2,L_3)}$$
$$UA(L_1,L_2,L_3) = 561.6\frac{W}{K} \quad (E5.9.74)$$

Note that the air-side thermal resistance of Equation (5.9.73) dominates the total thermal resistance. This is the reason why the louverfin geometry is adopted on the air side to increase the total heat transfer area.

ε-NTU Method The heat capacity rates are

$$C_1 := mdot_1 \cdot c_{p1} \qquad\qquad C_1 = 6.143 \times 10^3 \cdot \frac{W}{K} \qquad\qquad (E5.9.75)$$

$$C_2 := mdot_2 \cdot c_{p2} \qquad\qquad C_2 = 1.235 \times 10^3 \cdot \frac{W}{K} \qquad\qquad (E5.9.76)$$

$$C_{min} := min\,(C_1, C_2) \qquad\qquad C_{min} = 1.235 \times 10^3 \cdot \frac{W}{K} \qquad\qquad (E5.9.77)$$

$$C_{max} := max\,(C_1, C_2) \qquad\qquad C_{max} = 6.143 \times 10^3 \cdot \frac{W}{K} \qquad\qquad (E5.9.78)$$

The heat capacity ratio is

$$C_r := \frac{C_{min}}{C_{max}} \qquad C_r = 0.201 \qquad\qquad (E5.9.79)$$

The number of transfer unit *NTU* is

$$NTU\,(L_1, L_2, L_3) := \frac{UA\,(L_1, L_2, L_3)}{C_{min}} \qquad NTU\,(L_1, L_2, L_3) = 0.455 \qquad (E5.9.80)$$

The effectiveness is given using Equation (5.77) for both fluids unmixed by

$$\varepsilon_{hx}\,(L_1, L_2, L_3) := 1 - \exp\left[\frac{1}{C_r} \cdot NTU\,(L_1, L_2, L_3)^{0.22}\right.$$

$$\left. \cdot \left(\exp\left(-C_r \cdot NTU\,(L_1, L_2, L_3)^{0.78}\right) - 1\right)\right] \qquad (E5.9.81)$$

$$\varepsilon_{hx}\,(L_1, L_2, L_3) = 0.35$$

The effectiveness appears constant in this problem with the given conditions (T_{1i}, T_{1o}, T_{2i}, and *mdot*$_1$, and *mdot*$_2$). The heat transfer rate is given in Equation (5.35a) by

$$q\,(L_1, L_2, L_3) := \varepsilon_{hx}\,(L_1, L_2, L_3) \cdot C_{min} \cdot (T_{1i} - T_{2i})$$
$$q\,(L_1, L_2, L_3) = 3.027 \times 10^4 \, W \qquad\qquad (E5.9.82)$$

The outlet temperatures are given using Equation (5.38) as

$$T_{1o}\,(L_1, L_2, L_3) := T_{1i} - \varepsilon_{hx}\,(L_1, L_2, L_3) \cdot \frac{C_{min}}{C_1}\,(T_{1i} - T_{2i})$$
$$T_{1o}\,(L_1, L_2, L_3) = 90.073 \cdot {}^{\circ}C \qquad\qquad (E5.9.83)$$

$$T_{2o}\,(L_1, L_2, L_3) := T_{2i} - \varepsilon_{hx}\,(L_1, L_2, L_3) \cdot \frac{C_{min}}{C_2}\,(T_{1i} - T_{2i})$$
$$T_{2o}\,(L_1, L_2, L_3) = 49.497 \cdot {}^{\circ}C \qquad\qquad (E5.9.84)$$

Note if the coolant outlet temperature matches the design requirement ($T_{1o} = 90°C$).

Pressure Drops   The pressure drops for the louver-fin-type heat exchanger can be obtained using the Equation (5.202):

$$\Delta P = \frac{G^2}{2 \cdot \rho_i} \left[\left(1 - \sigma^2 + K_c\right) + 2 \cdot \left(\frac{\rho_i}{\rho_o} - 1\right) \right.$$
$$\left. + \frac{4f \cdot L}{D_h} \cdot \left(\frac{\rho_i}{\rho_m}\right) - \left(1 - \sigma^2 - K_e\right) \cdot \frac{\rho_i}{\rho_o} \right] \qquad (E5.9.85)$$

Contraction and Expansion Coefficients   Jet contraction ratio:

$$C_{c\_tube}(\sigma) := 4.374 \cdot 10^{-4} \cdot e^{6.737 \cdot \sqrt{\sigma}} + 0.621 \qquad (E5.9.86)$$

Friction factors:

$$f_d(Re) := \begin{vmatrix} 0.049\, Re^{-0.2} & \text{if } Re \geq 2300 \\ \dfrac{16}{Re} & \text{otherwise} \end{vmatrix} \qquad (E5.9.87)$$

Velocity-distribution coefficients: For triangular tubes,

$$K_{d\_tube}(Re) := \begin{vmatrix} 1.09068\,(4 \cdot f_d(Re)) + 0.05884\sqrt{4 \cdot f_d(Re)} + 1 & \text{if } Re \geq 2300 \\ 1.33 & \text{otherwise} \end{vmatrix}$$
$$(E5.9.88)$$

$$K_{d\_tri}(Re) := \begin{vmatrix} 1 + 1.29 \cdot \left(K_{d\_tube}(Re) - 1\right) & \text{if } Re \geq 2300 \\ 1.43 & \text{otherwise} \end{vmatrix} \qquad (E5.9.89)$$

Expansion and Contraction Coefficients   For expansion,

$$K_{e\_tri}(\sigma, Re) := 1 - 2 \cdot K_{d\_tri}(Re) \cdot \sigma + \sigma^2 \qquad (E5.9.90)$$

For contraction,

$$K_{c\_tri}(\sigma, Re) := \frac{1 - 2 \cdot C_{c\_tube}(\sigma) + C_{c\_tube}(\sigma)^2 \cdot \left(2 \cdot K_{d\_tri}(Re) - 1\right)}{C_{c\_tube}(\sigma)^2} \qquad (E5.9.91)$$

The expansion and contraction coefficients are obtained assuming that the Reynolds number is turbulent ($Re = 10^7$) due to the louver fins.

$$K_{e2}(L_1, L_3) := K_{e\_tri}\left(\sigma_2(L_1, L_3), 10^7\right) \quad K_{e2}(L_1, L_3) = 0.102 \qquad (E5.9.92)$$

$$K_{c2}(L_1, L_3) := K_{c\_tri}\left(\sigma_2(L_1, L_3), 10^7\right) \quad K_{c2}(L_1, L_3) = 0.187 \qquad (E5.9.93)$$

The air outlet densities are a function of pressure outlet that is unknown:

$$\rho_{2o}(P_{2o}, L_1, L_2, L_3) := \frac{P_{2o}}{R_g \cdot T_{2o}(L_1, L_2, L_3)} \qquad (E5.9.94)$$

Since ρ_m is the mean density and defined using Equation (5.193a) as

$$\frac{1}{\rho_m} = \frac{1}{2} \cdot \left(\frac{1}{\rho_i} + \frac{1}{\rho_o} \right) \tag{E5.9.95}$$

$$\rho_{2m}(P_{2o}, L_1, L_2, L_3) := \left[\frac{1}{2} \cdot \left(\frac{1}{\rho_{2i}} + \frac{1}{\rho_{2o}(P_{2o}, L_1, L_2, L_3)} \right) \right]^{-1} \tag{E5.9.96}$$

The air outlet pressure can be obtained using the MathCAD "root" function. Equation (E5.9.85) is divided into three equations:

$$\Delta P2_{12}(L_1, L_3) := \frac{G_2(L_1, L_3)^2}{2 \cdot \rho_{2i}} \cdot \left(1 - \sigma_2(L_1, L_3)^2 + K_{c2}(L_1, L_3) \right) \tag{E5.9.97}$$

$$K2_{23}(P_{2o}, L_1, L_2, L_3) := 2 \cdot \left(\frac{\rho_{2i}}{\rho_{2o}(P_{2o}, L_1, L_2, L_3)} - 1 \right)$$
$$+ \frac{4f_2(L_1, L_2, L_3) \cdot L_2}{D_{h2}(L_1, L_2, L_3)} \cdot \left(\frac{\rho_{2i}}{\rho_{2m}(P_{2o}, L_1, L_2, L_3)} \right) \tag{E5.9.98}$$

$$\Delta P2_{23}(P_{2o}, L_1, L_2, L_3) := \frac{G_2(L_1, L_3)^2}{2 \cdot \rho_{2i}} \cdot K2_{23}(P_{2o}, L_1, L_2, L_3) \tag{E5.9.99}$$

$$\Delta P2_{34}(P_{2o}, L_1, L_2, L_3) := \frac{G_2(L_1, L_3)^2}{2 \cdot \rho_{2i}} \cdot \left(1 - \sigma_2(L_1, L_3)^2 - K_{e2}(L_1, L_3) \right)$$
$$\cdot \frac{\rho_{2i}}{\rho_{2o}(P_{2o}, L_1, L_2, L_3)} \tag{E5.9.100}$$

The pressure drop on the air side is expressed by

$$\Delta P2(P_{2o}, L_1, L_2, L_3) := \Delta P2_{12}(L_1, L_3) + \Delta P2_{23}(P_{2o}, L_1, L_2, L_3)$$
$$- \Delta P2_{34}(P_{2o}, L_1, L_2, L_3) \tag{E5.9.101}$$

Guess a value that could be a closest value (that you can take best estimate) to the solution.

$$P_{2o} := 100\,\text{kPa} \tag{E5.9.102}$$

$$P_{2o}(L_1, L_2, L_3) := \text{root}\left[\left[(P_{2i} - P_{2o}) - \Delta P2(P_{2o}, L_1, L_2, L_3) \right], P_{2o} \right] \tag{E5.9.103}$$

Solving Equation (E5.9.103), we obtain the air outlet pressure P_{2o} as

$$P_{2o}(L_1, L_2, L_3) = 9.975 \times 10^4 \cdot \text{Pa} \tag{E5.9.104}$$

The pressure drop for the air side is finally obtained as

$$\Delta P2(L_1, L_2, L_3) := P_{2i} - P_{2o}(L_1, L_2, L_3) \quad \Delta P2(L_1, L_2, L_3) = 249.475\,\text{Pa} \tag{E5.9.105}$$

The fan and pump efficiencies are assumed to be 0.8 for each fluid.

$$\eta_f := 0.8 \qquad \eta_\rho := 0.8 \tag{E5.9.106}$$

The fan power is calculated by

$$\text{Power}_{\text{fan}}\,(L_1, L_2, L_3) := \frac{\text{mdot}_2}{\eta_f \cdot \rho_{2i}} \cdot \Delta P_2\,(L_1, L_2, L_3)$$
(E5.9.107)

$$\text{Power}_{\text{fan}}\,(L_1, L_2, L_3) = 0.439\,\text{hp}$$

Note that the fan power of 0.439 hp appears reasonable compared to a typical passenger-car engine power of 150 hp.

Water-side Pressure Drop

$$\Delta P_1\,(L_1, L_2, L_3) := 4 \cdot \left(\frac{f_1\,(L_1, L_2, L_3) \cdot L_3}{D_{h1}\,(L_1, L_2, L_3)} + 1 \right) \cdot N_p \cdot \frac{G_1\,(L_1, L_2,)^2}{2 \cdot \rho_1}$$
(E5.9.108)

$$\Delta P_1\,(L_1, L_2, L_3) = 70.009\,\text{kPa}$$

The pressure drop of 70 kPa is a typical limit for the water side in a shell-and-tube heat exchanger. The pumping power is calculated by

$$\text{Power}_{\text{pump}}\,(L_1, L_2, L_3) := \frac{\text{mdot}_1}{\eta_p \cdot \rho_1} \cdot \Delta P_1\,(L_1, L_2, L_3)$$
(E5.9.109)

$$\text{Power}_{\text{pump}}\,(L_1, L_2, L_3) = 0.194\,\text{hp}$$

Determine the Dimensions of the Plate-Fin Heat Exchanger   The three unknowns of L_1, L_2, and L_3 are now sought using a combination of three requirements. The guess values can be varied with the new results to have a quicker convergence.

   Initial guesses $L_1 := 50\,\text{cm}$ $L_2 := 5\,\text{cm}$ $L_3 := 50\,\text{cm}$ (E5.9.110)

   Given   (E5.9.111)

$$T_{1o}\,(L_1, L_2, L_3) = 90°C$$
(E5.9.112)

$$\Delta P_2\,(L_1, L_2, L_3) = 250\,\text{Pa}$$
(E5.9.113)

$$\Delta P_1\,(L_1, L_2, L_3) = 70\,\text{kPa}$$
(E5.9.114)

$$\begin{pmatrix} L_1 \\ L_2 \\ L_3 \end{pmatrix} := \text{Find}\,(L_1, L_2, L_3)$$
(E5.9.115)

The dimensions are obtained by

$$L_1 = 50.221 \cdot \text{cm}\quad L_2 = 3.381 \cdot \text{cm}\quad L_3 = 30.673\,\text{cm}$$
(E5.9.116)

This set of the dimensions looks reasonable to fit in an automobile. The dimensions are determined to be the ultimate design. Now return to Equation (E5.9.17) and update the values for all the following values in the previous calculations.

Summary of the Results and Geometry

Given information:

$T_{1i} = 95 \cdot^\circ C$	Coolant inlet temperature	(E5.9.117)
$T_{2i} = 25 \cdot^\circ C$	Air inlet temperature	(E5.9.118)
$P_{1i} = 200\,kPa$	Coolant inlet pressure	(E5.9.119)
$P_{2i} = 100\,kPa$	Air inlet pressure	(E5.9.120)
$Q_1 = 1.65 \times 10^{-3}\,\dfrac{m^3}{s}$	Volume flow rate at coolant side	(E5.9.121)
$Q_2 = 1.05\,\dfrac{m^3}{s}$	Volume flow rate at air side	(E5.9.122)

Design requirements:

$T_{1o} = 90^\circ C$	Coolant outlet temperature	(E5.9.123)
$\Delta P_1 \leq 70\,kPa$	Pressure drop at coolant side	(E5.9.124)
$\Delta P_2 \leq 250\,Pa$	Pressure drop at air side	(E5.9.125)

Dimensions of sizing:

$$L_1 = 50.221 \cdot cm \quad L_2 = 3.381 \cdot cm \quad L_3 = 30.673 \cdot cm \qquad \text{(E5.9.126)}$$

Outlet temperatures and pressure drops:

$T_{1o}\,(L_1, L_2, L_3) = 90 \cdot^\circ C$	Coolant outlet temperature	(E5.9.127)
$T_{2o}\,(L_1, L_2, L_3) = 49.86 \cdot^\circ C$	Air outlet temperature	(E5.9.128)
$\Delta P_1\,(L_1, L_2, L_3) = 70 \cdot kPa$	Pressure drop at coolant side	(E5.9.129)
$\Delta P_2\,(L_1, L_2, L_3) = 250 \cdot Pa$	Pressure drop at air side	(E5.9.130)
$\text{Power}_{fan}\,(L_1, L_2, L_3) = 0.44 \cdot hp$	Fan power at air side	(E5.9.131)
$\text{Power}_{pump}\,(L_1, L_2, L_3) = 0.194 \cdot hp$	Pump power at coolant side	(E5.9.132)

Note that the volume flow rate of $Q_2 = 1.05\,m^3/s$ determines the air outlet temperature of $49.86^\circ C$. There might be a practical limit for not overheating the exterior of the engine.

Densities of air at outlet:

$$\rho_{2i} = 1.168\,\frac{kg}{m^3} \quad \rho_{2o}\,(P_{2o}\,(L_1, L_2, L_3)\,, L_1, L_2, L_3) = 1.076\,\frac{kg}{m^3} \qquad \text{(E5.9.133)}$$

Thermal quantities:

$$\text{NTU}(L_1, L_2, L_3) = 0.463 \qquad \text{Number of transfer unit} \qquad \text{(E5.9.134)}$$

$$\varepsilon_{hx}(L_1, L_2, L_3) = 0.3551 \qquad \text{Effectiveness} \qquad \text{(E5.9.135)}$$

$$q(L_1, L_2, L_3) = 3.071 \times 10^4 \text{W} \qquad \text{Heat transfer rate} \qquad \text{(E5.9.136)}$$

$$h_1(L_1, L_2, L_3) = 3.691 \times 10^3 \cdot \frac{\text{W}}{\text{m}^2 \cdot \text{K}} \qquad \text{Heat transfer coefficient} \qquad \text{(E5.9.137)}$$

$$h_2(L_1, L_2, L_3) = 126.715 \frac{\text{W}}{\text{m}^2 \cdot \text{K}} \qquad \text{(E5.9.138)}$$

$$\text{UA}(L_1, L_2, L_3) = 572.519 \frac{\text{W}}{\text{K}} \qquad \text{UA value} \qquad \text{(E5.9.139)}$$

The heat transfer rate would be the cooling loss at the designated load of the engine (usually 15 to 35 percent of the load). For instance, the heat transfer rate of 30,710 Watts in Equation (E5.9.136) is about 27 percent of a designated full load of 150 hp.

Hydraulic quantities:

$$D_{h1}(L_1, L_2, L_3) = 0.903 \cdot \text{mm} \qquad \text{Hydraulic diameter at coolant side} \quad \text{(E5.9.140)}$$

$$D_{h2}(L_1, L_2, L_3) = 2.513 \cdot \text{mm} \qquad \text{Hydraulic diameter at air side} \qquad \text{(E5.9.141)}$$

$$G_1(L_1, L_2) = 1.8 \times 10^3 \frac{\text{kg}}{\text{m}^2 \cdot \text{s}} \qquad v_1(L_1, L_2) = 1.764 \frac{\text{m}}{\text{s}} \qquad \text{(E5.9.142)}$$

$$G_2(L_1, L_3) = 12.307 \frac{\text{kg}}{\text{m}^2 \cdot \text{s}} \qquad v_2(L_1, L_3) = 10.533 \frac{\text{m}}{\text{s}} \qquad \text{(E5.9.143)}$$

$$\text{Re}_1(L_1, L_2, L_3) = 2.032 \times 10^3 \qquad \text{Reynolds number} \qquad \text{(E5.9.144)}$$

$$\text{Re}_2(L_1, L_2, L_3) = 1.676 \times 10^3 \qquad \text{(E5.9.145)}$$

Geometric quantities:

$$b = 6.35 \cdot \text{mm} \qquad \text{Tube spacing} \qquad \text{(E5.9.146)}$$

$$p_f = 2.288 \cdot \text{mm} \qquad \text{Fin pitch} \qquad \text{(E5.9.147)}$$

$$p_t = 8.35 \cdot \text{mm} \qquad \text{Tube pitch} \qquad \text{(E5.9.148)}$$

$$\delta = 0.152 \cdot \text{mm} \qquad \text{Fin thickness} \qquad \text{(E5.9.149)}$$

$$\delta_w = 0.3 \cdot \text{mm} \qquad \text{Wall thickness} \qquad \text{(E5.9.150)}$$

$$l_p = 1 \cdot \text{mm} \qquad \text{Louver length} \qquad \text{(E5.9.151)}$$

$$\theta = 20 \qquad \text{Louver angle} \qquad \text{(E5.9.152)}$$

$$L_{louv} = 5.737 \cdot mm \qquad \text{Louver cut length} \qquad \text{(E5.9.153)}$$

$$H_t = 2 \cdot mm \qquad \text{Tube height} \qquad \text{(E5.9.154)}$$

$$W_t(L_2) = 3.381 \cdot cm \qquad \text{Tube width} \qquad \text{(E5.9.155)}$$

$$N_p = 3 \qquad \text{Number of passes at coolant side} \qquad \text{(E5.9.156)}$$

$$N_{pg}(L_1) = 59.906 \qquad \text{Number of passages at air side} \qquad \text{(E5.9.157)}$$

$$\sigma_2(L_1, L_3) = 0.647 \qquad \text{Porosity} \qquad \text{(E5.9.158)}$$

$$\beta_2(L_1, L_2, L_3) = 1360 \cdot \frac{m^2}{m^3} \qquad \text{Surface area density} \qquad \text{(E5.9.159)}$$

$$A_{t1}(L_1, L_2, L_3) = 1.271 \cdot m^2 \quad \text{Total heat transfer area at coolant side} \quad \text{(E5.9.160)}$$

$$A_{t2}(L_1, L_2, L_3) = 5.364 \cdot m^2 \quad \text{Total heat transfer area at air side} \quad \text{(E5.9.161)}$$

This heat exchanger may be further investigated, mainly by varying the tube spacing b and fin pitch p_f toward a larger surface area density so that the size and weight of the exchanger can be reduced. Louver geometry may be another factor for the compactness.

REFERENCES

1. S. Kakac and H. Liu, *Heat Exchangers*. New York: CRC Press, 1998.

2. W.M. Rohsennow, J.P. Hartnett, and Y.I. Cho, *Handbook of Heat Transfer*, 3rd ed. New York: McGraw-Hill, 1998.

3. E.M. Smith, *Thermal Design of Heat Exchangers*. New York: John Wiley & Sons, 1997.

4. F.P. Incropera, D.P. Dewitt, T.L. Bergman, and A.S. Lavine, *Fundamentals of Heat and Mass Transfer*, 6th ed. Hoboken, NJ: John Wiley & Sons, 2007.

5. W.S. Janna, *Design of Fluid Thermal Systems*, 2nd ed. Boston: PWS Publishing Co., 1998.

6. E.N. Sieder and G.E. Tate, Heat Transfer and Pressure Drop of Liquids in Tubes, *Ind. Eng. Chem.*, 28 (1936): 1429–1453.

7. V. Gnielinski, New Equation for Heat and Mass Transfer in Turbulent Pipe and Channel Flow, *Int. Chem., Eng.*, 16 (1976): 359–368.

8. W.M. Kays and A.L. London, *Compact Heat Exchangers*, 3rd ed. New York: McGraw-Hill, 1984.

9. A.F. Mills, *Heat Transfer* 2nd ed. Englewood Cliffs, NJ: Prentice Hall, 1999.

10. A. Bejan, *Heat Transfer*. New York: John Wiley & Sons, 1993.

11. J.E. Hesselgreaves, *Compact Heat Exchangers* London: Pergamon, 2001.

12. W.M. Kays and A.L. London, *Compact Heat Exchangers*, 2nd ed. New York: McGraw-Hill, 1964.

13. T. Kuppan, *Heat Exchanger Design Handbook*. New York: Marcel Dekker, 2000.

14. J. Mason, Heat Transfer in Cross Flow. Proc. 2nd U.S. National Congress on Applied Mechanics, American Society of Mechanical Engineers, New York, 1955.

15. G.K. Filonenko, Hydraulic Resistance in Pipes (in Russian), *Teplonergetika*, vol. $^1/_4$ (1954): 40–44.

16. L.F. Moody, Friction Factors for Pipe Flow, *Trans. ASME*, 66 (1944): 671–684.

17. R.K. Shah and D.P. Sekulic, *Fundamentals of Heat Exchanger Design*. Hoboken, NJ: John Wiley and Sons, 2003.

18. R.K. Shah and W.W. Focke, Plate Heat Exchangers and Their Design Theory, in *Heat Transfer Equipment Design*, ed. R.K. Shah, E.C. Subbarao, and R.A. Mashelkar, Washington: Hemisphere, 1988, pp. 227–254.

19. H. Martin, A Theoretical Approach to Predict the Performance of Chevron-type Plate Heat Exchanger. *Chemical Engineering and Processing*, 35 (1996): 301–310.

20. K.K. Robinson and D.E. Briggs, Pressure Drop of Air Flowing across Triangular Pitch Banks of Finned Tubes, *Chemical Engineering Progress Symposium Series*, 64 (1966): 177–184.

21. D.E. Briggs and E.H. Young, Convection Heat Transfer and Pressure Drop of Air Flowing across Triangular Pitch Banks of Finned Tubes, *Chem. Eng. Prog. Symp. Ser.* 41, vol. 59 (1963): 1–10.

22. A. Achaichia and T.A. Cowell, Heat Transfer and Pressure Drop Characteristics of Flat Tube Louvered Plate Fin Surfaces. *Experimental Thermal and Fluid Science*, 1 (1988): 147–157.

23. W.M. Kays, Loss Coefficients for Abrupt Changes in Flow Cross Section with Low Reynolds Number Flow in Single and Multiple-Tube Systems, *Transaction of the American Society of Mechanical Engineers*, 72 (1950): 1067–1074.

24. H. Rouse, *Elementary of Fluids*. New York: John Wiley and Sons, 1946.

25. R.L. Webb, *Principles of Enhanced Heat Transfer*, New York: John Wiley and Sons, 1994.

26. L. Wang, B. Sunden, and R.M. Manglik, *Plate Heat Exchangers*. Boston: WIT Press, 2007.

27. R.M. Manglik and A.E. Bergles, Heat Transfer and Pressure Drop Correlations for the Rectangular Offset-Strip-Fin Compact Heat Exchanger. *Exp. Thermal Fluid Sci.*, 10 (1995): 171–180.

28. C.J. Davenport, Correlations for Heat Transfer and Flow Friction Characteristics of Louvered Fin. In *Heat Transfer-Seattle 1983*, AIChE Symposium Series, No.225, vol.79, pp. 19–27, 1983.

29. Y.I. Chang and C.C. Wang, A Generalized Heat Transfer Correlation for Louver Fin Geometry. *Int. J. Heat Mass Transfer*, 40 (1997): 553–544.

30. Y.J. Chang, K.C. Hsu, Y.T. Lin, and C.C. Wang, A Generalized Friction Correlation for Louver Fin Geometry. *Int. J. Heat Mass Transfer*, 43, 2237–2243.

PROBLEMS

Double Pipe Heat Exchanger

5.1 A counterflow double-pipe heat exchanger is to use city water to cool ethylene glycol for a chemical process. Ethylene glycol at a flow rate of 0.63 kg/s is required to be cooled from 80°C to 65°C using water at a flow rate of 1.7 kg/s and 23°C, which is shown in Figure P5.1. The double-pipe heat exchanger is composed of 2-m-long carbon-steel hairpins. The inner and outer pipes are 3/4 and 1 1/2 nominal schedule 40, respectively. The ethylene glycol flows through the inner tube. When the heat exchanger is initially in service (no fouling), calculate the outlet temperatures, the heat transfer rate, and the pressure drops for the exchanger. How many hairpins will be required?

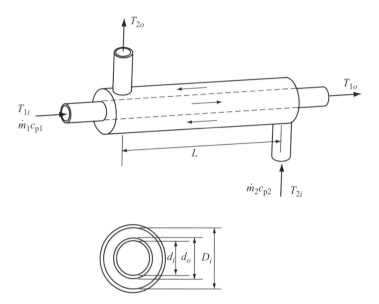

Figure P5.1 and P5.2   Double-pipe heat exchanger

5.2   A counterflow double-pipe heat exchanger is to use city water to cool ethylene glycol for a chemical process. Ethylene glycol at a flow rate of 0.63 kg/s is required to be cooled from 80°C to 65°C using water at a flow rate of 1.7 kg/s and 23°C, which is shown in Figure P5.2. The double-pipe heat exchanger is composed of 2-m-long carbon-steel hairpins. The inner and outer pipes are 3/4 and 1 1/2 nominal schedule 40, respectively. The ethylene glycol flows through the inner tube. Fouling factors of $0.176 \times 10^{-3}\,\text{m}^2\text{K/W}$ for water and $0.325 \times 10^{-3}\,\text{m}^2\text{K/W}$ for ethylene glycol are specified. Calculate the outlet temperatures, the heat transfer rate, and the pressure drops for the exchanger. How many hairpins will be required?

Shell-and-Tube Heat Exchanger

5.3   A miniature shell-and-tube heat exchanger is designed to cool glycerin with cold water. The glycerin at a flow rate of 0.25 kg/s enters the exchanger at 60°C and leaves at 36°C.

The water at a rate of 0.54 kg/s enters at 18°C, as is shown in Figure P5.3. The tube material is carbon steel. Fouling factors of 0.253×10^{-3} m²K/W for water and 0.335×10^{-3} m²K/W for glycerin are specified. Route the glycerin through the tubes. The permissible maximum pressure drop on each side is 30 kPa. The volume of the exchanger is to be minimized. Since the exchanger is custom designed, the tube size may be smaller than NPS 1/8 (DN 6 mm), which that is the smallest size in Table C.1 in Appendix C. The tube pitch ratio of 1.25 and the diameter ratio of 1.3 can be used. Design the shell-and-tube heat exchanger.

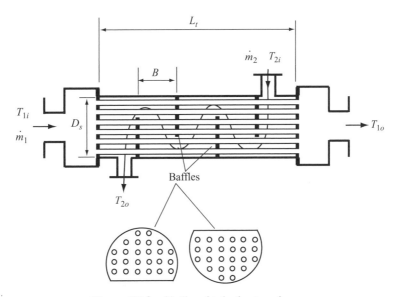

Figure P5.3   Shell-and tube heat exchanger

Plate Heat Exchanger

5.4 Hot water is cooled by cold water using a miniature plate heat exchanger, as shown in Figures P5.4.1 and P5.4.2. Hot water with a flow rate of 1.2 kg/s enters the plate heat exchanger at 75°C and will be cooled to 55°C. Cold water enters at a flow rate of 1.3 kg/s and 24°C. The maximum permissible pressure drop for each stream is 30 kPa (4.5 psi). Using single-pass chevron plates (stainless steel AISI 304) with $\beta = 45°$, determine the rating $(T, q, \varepsilon$ and $\Delta P)$ and dimensions $(W_p, L_p,$ and $H_p)$ of the plate heat exchanger.

Description	value
Number of passes N_p	1
Chevron angle β	$40°$
Total number of plates N_t	40
Plate thickness δ	0.2 mm
Corrugation pitch λ	4 mm
Port diameter D_p	25 mm
Thermal conductivity k_w	14.9 W/mK

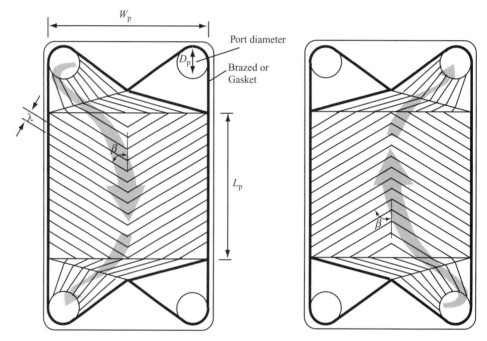

Figure P5.4.1 Plates with chevron-type corrugation pattern for a plate heat exchanger

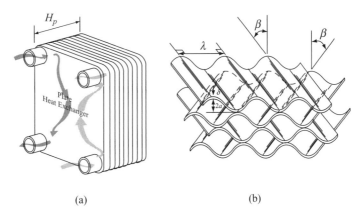

(a)   (b)

Figure P5.4.2   (a) Brazed plate heat exchanger and (b) the inter-plate cross-corrugated flow

5.5   (Time-consuming work) Cold water is heated by wastewater using a plate heat exchanger, as shown in Figures P5.4.1 and P5.4.2. The cold water with a flow rate of 130 kg/s enters the plate heat exchanger at 22°C, and it will be heated to 42°C. The hot wastewater enters at a flow rate of 140 kg/s and 65°C. The maximum permissible pressure drop for each fluid is 70 kPa (10 psi). Using single-pass chevron plates (stainless steel AISI 304) with $\beta = 30°$, determine firstly the corrugation pitch λ and the number of plates N_t by developing a MathCAD model as a function of not only the dimensions (W_p, L_p, and H_p) but also N_t and λ, and then determine the rating (T, q, ε and ΔP) and dimensions of the plate heat exchanger. Show your work in detail for the determination of the number of plates and corrugation pitch. The following data are to be used for the calculations.

Description	value
Number of passes N_p	1
Chevron angle β	30°
Plate thickness δ	0.6 mm
Port diameter D_p	200 mm
Thermal conductivity k_w	14.9 W/mK

Finned-Tube Heat Exchanger

5.6 A circular finned-tube heat exchanger is used to cool hot coolant (50 percent ethylene glycol) with an air stream in an engine. The coolant enters at 96°C with a flow rate of 1.8 kg/s and must leave at 90°C. The air stream enters at a flow rate of 2.5 kg/s and 20°C. The allowable air pressure loss is 250 Pa, and the allowable coolant pressure loss is 3 kPa. The tubes of the exchanger have an equilateral triangular pitch. The thermal conductivities of the aluminum fins and the copper tubes are 200 W/mK and 385 W/mK, respectively. The tube pitch is 2.476 cm and the fin pitch is 3.43 fins/cm. The fin thickness is 0.046 cm and the three diameters are $d_i = 0.765$ cm for the tube inside diameter, $d_o = 0.965$ cm for the tube outside diameter, and $d_e = 2.337$ cm for the fin outside diameter. Design the circular finned-tube heat exchanger by providing the dimensions of the exchanger, outlet temperatures, effectiveness, and pressure drops (See Figure P5.6.1).

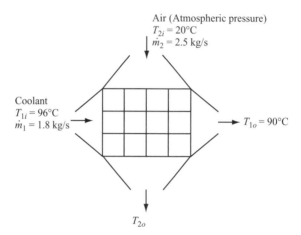

Figure P5.6.1   Flow diagram of a circular finned-tube heat exchanger

Plate-Fin Heat Exchanger

5.7 A gas-to-air single-pass cross-flow heat exchanger is considered for heat recovery from the exhaust gas to preheat incoming air in a portable solid oxide fuel cell (SOFC) system. Offset strip fins of the same geometry are employed on the gas and air sides; the geometrical properties and surface characteristics are provided in Figures P5.7.1 and P5.7.2. Both fins and plates (parting sheets) are made from Inconel 625 with k = 18 W/m.K. The anode gas flows in the heat exchanger at 0.075 m³/s and 750°C. The cathode air on the other fluid side flows at 0.018 m³/s and 90°C. The inlet pressure of the gas is at 130 kPa absolute, whereas that of air is at 170 kPa absolute. Both the gas and air pressure drops are limited to 3 kPa. It is desirable to have an equal length for L_1 and L_2. Design a gas-to-air single-pass cross-flow heat exchanger operating at $\varepsilon = 0.85$. Determine the core dimensions of this exchanger. Then, determine the heat transfer rate, outlet fluid temperatures and pressure drops in each fluid. Use the properties of air for the gas. Use the following geometric information.

Description	Value
Fin thickness δ	0.092 mm
Plate thickness δ_w	0.25 mm
Fin density N_f	980 m^{-1}
Spacing between plates b_1 and b_2	1.3 mm
Offset strip length λ_1 and λ_2	1.7 mm

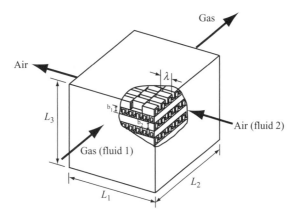

Figure P5.7.1 Plate-fin heat exchanger, employing offset strip fin

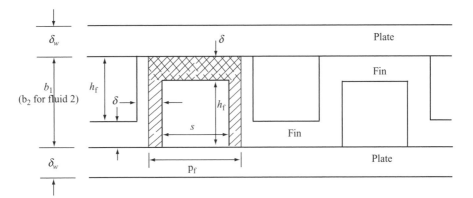

Figure P5.7.2 Schematic of offset strip fin (OSF) geometry

Louver-Fin-Type Plate-Fin Heat Exchanger

5.8 An engine oil-to-air crossflow heat exchanger is designed to cool the oil. Louver fins geometry is employed on the air sides. The geometrical properties and surface characteristics are provided in Figures P5.8.1 and P5.8.2. Both fins and tubes are made from aluminum alloy with k = 117 W/mK. The oil flows in the heat exchanger at $0.5 \times 10^{-3} \, m^3/s$ and 142°C and shall leave at 66°C. Air enters at 0.34 kg/s and 25°C. The inlet pressure of the air is at 827 kPa absolute and the inlet pressure of the oil is at 2,758 kPa absolute. The air pressure drop is limited to 250 Pa. The oil pressure drop is recommended to be less than 70 kPa. Design the oil-to-air crossflow heat exchanger. Determine the core dimensions of this exchanger. Then, determine the heat transfer rate, outlet fluid temperatures, and pressure drops on each fluid. Use the following information in your calculations.

Description	value
Fin thickness δ	0.152 mm
Tube wall thickness δ_w	0.4 mm
Fin density N_f	837 m^{-1}
Tube spacing b	14 mm
Louver pitch l_p	1.0 mm
Louver angle θ	10 degree
Tube outside height H_t	2 mm

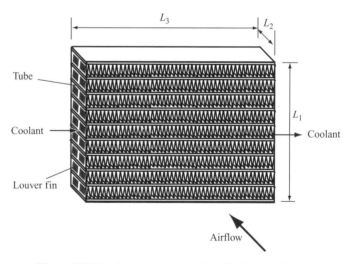

Figure P5.8.1 Louver-fin-type plate-fin heat exchanger

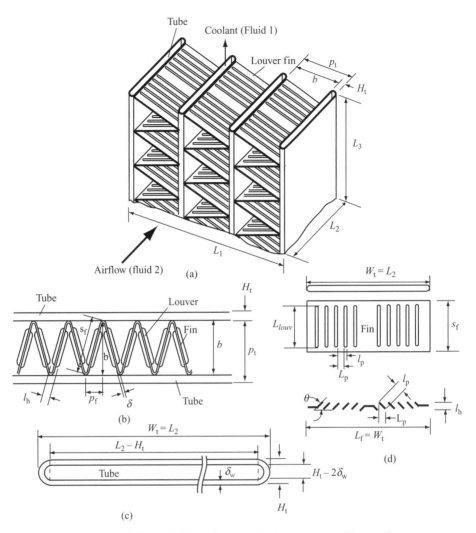

Figure P5.8.2   Definition of geometrical parameters of louver fins

6

Solar Cells

6.1 INTRODUCTION

Solar cells, sometimes referred to as photovoltaic (or PV) cells, are semiconductor devices that convert sunlight directly into electric current to produce useable electric power. As long as light is shining on the solar cell, it generates electrical power. Some have been in continuous outdoor operation on Earth or in space for over 30 years.

The photovoltaic effect in liquid electrolytes was discovered in 1839 by Becquerel, and the first functional PV device was made by Fritts [1] in 1883 using a Selenium film. The modern solar cell was accidently discovered in Bell Labs in 1954 by Chapin et al. [2] with a 6 percent efficient Silicon p–n junction solar cell. The most famous early application of solar cells in space was the *Apollo 11* moon mission in July 1969. In the 1970s, the goal was to break a 10 percent barrier of solar cell efficiency. After continuous development, a world record of 40.7 percent efficiency was reported in 2006 using a concentration of 240 suns by Boeing-Specrolab using triple-junction GaInP/GaAs/Ge solar cells. Commercial solar cells usually appear with a lower efficiency in the years following development in research labs.

The most widely used materials for solar cells are *single crystalline silicon* (c-Si) and *polycrystalline silicon* (poly-Si), which are both referred to as *crystalline silicon*. The single crystalline and polycrystalline materials are also called monocrystalline and multicrystalline materials, respectively. Crystalline silicon is today clearly dominant, with about 91 percent of the PV market in 2001. It will remain dominant at least for the next ten years. The trend is toward the polycrystalline option due to its lower cost, despite the lower efficiency. The atoms of the crystalline silicon are arranged in near-perfect, regular arrays, or lattices. Polycrystalline silicon is composed of a collection of many small groups of crystals or *grains* about few millimeters in size that form grain boundaries, while single crystalline silicon has no such disordered regions (see Figure 6.1(a), (b), (c) and (d)). Thus, polycrystalline silicon can be manufactured at a lower cost than single crystals, but produces less efficient cells, mainly owing to the presence of grain boundaries and other crystal defects.

Silicon is the second-most abundant element on Earth, constituting about 26 percent of the Earth's crust as silicon dioxide. Oxygen is the most abundant. However, the production process from feedstock (crude silica) to crystalline silicon through the process of the metallurgical grade (MG) and the semiconductor grade (SG) is expensive and slow in growth of crystals. The crude silica produces SG silicon through refining processes. Hyperpure silicon is usually required for good efficiency. The semiconductor-grade silicon is then crystallized by the Czochralski (Cz) growth method that is a slow process,

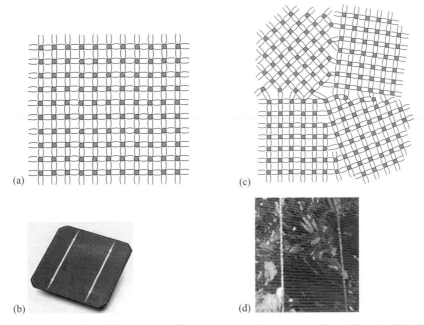

Figure 6.1 Crystalline silicon solar cell structures: a) single crystalline silicon (c-Si), b) c-Si solar cell, c) polycrystalline silicon (poly-Si), and d) poly-Si solar cell

forming an ingot of silicon crystals. The silicon ingot finally produces silicon wafers by a wire-cutting method that wastes as much as 50 percent of the material. The silicon wafers are directly used to form solar cells. For example, the semiconductor-grade silicon through the refining processes was priced at $40 to $60 per kg in 1995, whereas scrap silicon (semiconductor grade) from semiconductor-integrated circuit (IC) industry was between $10 and $15 per kg. As a result, the majority of the silicon supply for solar cells has relied on scrap silicon from the IC industry. Huge governmental and industrial investments were made to improve solar efficiency and manufacturability—or, in other words, to reduce the solar cell cost—since the 1960s, primarily in silicon applications, taking advantage of already established semiconductor IC technology. The cost of energy produced by solar cells is still much higher than that from fossil fuel. The solar cell price had decreased monotonically from about $30/$W_p$ (dollars per Watt peak often used) in early 1980s to about $2/$W_p$ in 2003. The goal was to lower the price below $1/$W_p$. The dilemma is that the supply of scrap silicon is limited, not reaching the large demand of solar cell industry. Hence, the solar cell price began increasing dramatically from 2004, almost doubled in 2008. Furthermore, when foreseeing the rapidly increasing demand of solar cells, the supply of scrap silicon would not reach the ultimate goals required for mass worldwide consumption of solar cells.

Two attempts in parallel with development of crystalline silicon have been made since 1970s by national and international research institutes (NREL, UNSW-Australia, Georgia Tech, Boeing-Spectrolab, and many industrial labs and universities). One attempt is *thin-film solar cells* (TFSC) and the other is *concentrated solar cells*.

The thickness of thin-film solar cells are about $1-50\,\mu m$, while the thickness of typical silicon solar cells is about $300\,\mu m$. Thin-film solar cells would reduce the consumption of silicon. The materials that are considered to be appropriate for the thin-film solar cells are *gallium arsenide* (GaAs), *cadmium telluride* (CdTe), *copper indium gallium di-selenide* (Cu(InGa)Se$_2$), and even other form of silicon, *amorphous silicon* (a-Si). Thin-film solar cells not only use less material but also have shortened travel paths for the free electrons generated. This minimizes recombination that is one of important causes for a low efficiency (see Section 6.1.1 for further explanation of free electrons and recombination).

The other attempt is a *concentrated solar cell* that concentrates sunlight into a small area of the solar cell as much as thousand times (for example, the concentration index can be said 1 sun, 100 sun 1,000 sun, etc.). This concentration reduces the material as much as the concentration index, lowering the concentrated solar cell cost. The materials appropriate for concentrated solar cells are gallium arsenide (GaAs) and silicon (Si). The concentration increases the intensity of sunlight, or the number of photons that work to produce electron and hole pairs. The materials used for concentrated solar cells must have the capability to physically absorb the increased number of photons and to produce the photocurrent.

The parasitic losses (optical losses, ohmic resistances, etc.) due to the concentration are not as high as the concentration index, which improves to some extent the performance of concentrated solar cells. The dilemma is that the concentrated solar cells cannot focus optically scattered or diffused light (reflection from ground or clouds), which causes the efficiency to somewhat be low, so that the concentrated solar cells usually require a sun-tracking system that requires periodic maintenance (which is costly). Actually, concentrators are appropriate for relatively large installations (megaWatts) while PV market has evolved so far in smaller installations such as grid-connected houses, remote homes, or telecommunication applications whose size is seldom bigger than 5 kW.

In summary, it is very likely that photovoltaics will become in the next half century an important source of world electricity. Public support and global environmental concerns will keep photovoltaics viable, visible, and vigorous, both in new technical developments and new user applications. Nations that encourage photovoltaics will be leaders in the new technology, leading the way to a cleaner, more equitable twenty-first century, while those that ignore or suppress photovoltaics will be left behind in the green, economic energy revolution.

6.1.1   Operation of Solar Cells

Figure 6.2 depicts a schematic of a conventional solar cell, which mainly consists of a thin n-type region (order of $1\,\mu m$) and a thick p-type region (order of $300\,\mu m$), which are usually made of *semiconductor* materials. The fingers on the front surface are designated to collect electrons accumulated in the n-type region for the power generation.

Light has a wave-particle *dual nature* and the incident particles of light are called *photons* that have no mass but quantized energy. Light not only propagates effectively through a medium or vacuum as a wave form but also behaves as particles or photons once they reach a surface.

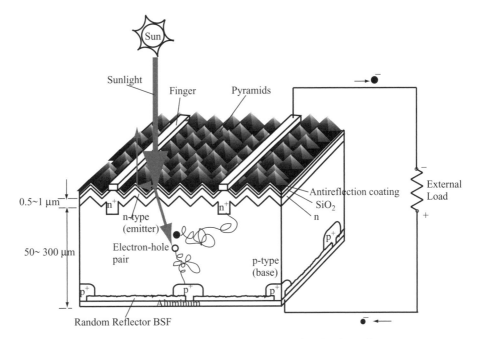

Figure 6.2 A schematic of a conventional solar cell

Consider the photons of sunlight incident on a solar cell surface. A small part of the light would be reflected, while the majority penetrates deeper into the base, as shown in Figure 6.2. The antireflection coating on a textured pyramid surface increases the performance of solar cells. Each photon that has a sufficiently long wavelength mostly penetrates into the deeper p-type region, while each that has a short wavelength is likely absorbed in the thin n-type material. A photon that has greater energy than the *band gap* produces an electron-hole pair, according to the *photoelectric effect*, by giving its energy to an electron, freeing it from its atom and leaving behind a hole that is also free to move. Photons that have less energy than the band gap will turn into *phonon*, or *heat*. The *band gap* is the energy difference between the lower edge of the *conduction band*, where the electrons are free to move, and the upper edge of the *valence band*, where the electrons are bound to an atom. Band gap is an important property of a semiconductor. The free (or ionized) *electrons* and *holes* are called *carriers*, which intrinsically migrate very randomly. Carriers usually have very short *lifetimes*, as they are likely to quickly recombine and not contribute to the useful current. The thickness of the solar cell is an important parameter. The thick solar cells cause the free electrons to travel large distances (more chances for recombination), but there is a limit to making thin solar cells for mechanical reasons (crystalline silicon is brittle). When an electron-hole pair is created or, in other words, *photogenerated*, it can be said that the photon energy is absorbed by the solar cell. The *absorption coefficient* of a semiconductor is an important property also and considerably selected. It is interesting to note that silicon semiconductors have a good value of band gap for efficiency but a poor absorption coefficient that requires a thick solar cell, which decreases efficiency.

One of the primary goals in designing solar cells is for its carriers to live longer—technically, to have the longer *diffusion length*, so that the *carrier lifetime* is extended with minimal *recombination*. The n-type and p-type materials are obtained by a *doping process* that injects impurities (donors in n-type and acceptors in p-type regions, together called *dopants*) into the semiconductor initially during the growth of a crystal ingot or later using the *ion implantation* technique that directly injects dopants to each wafer (a thin slice of silicon crystal). In silicon solar cells, the *donors* are phosphorous and the *acceptors* are boron. The concentrations of donors and acceptors can be controlled by an injection process as required. The areas near the front and rear contacts, which are shown as symbols n^+ and p^+ in Figure 6.2, are heavily doped to reduce the contact resistances.

Most of the donors yield negatively charged electrons, and each donor itself becomes positively charged (because of losing a negative electron), while most of the acceptors yield positively charged holes and each acceptor itself becomes negatively charged. The charged donors and acceptors are immobile due to the covalent bonding with silicon atoms. They create an electrostatic potential, particularly in a *depletion region* near the *p–n junction*, which is called the *built-in voltage*. As a result, the doping directly creates a *built-in voltage* between the n-type region and p-type region. This built-in voltage is to form a region that is called a *space charge region*, or a *depletion region* at the *p–n junction*. The depletion region in conjunction with the built-in voltage restricts the electrons from moving to the p-type region and the holes from moving to the n-type region.

Once the electrons and holes are gathered near the front and rear contacts, the dense carriers form to an electrostatic potential between the contacts. As a result, the electrons start to flow in the wire through the external load, to do useful work for electricity, and eventually to return to recombine or annihilate with the dense holes near the rear contact (this recombination is necessary and desirable). The cycle of an electron is completed.

The performance of a solar cell depends on the balance between photogeneration, or *photocurrent,* and recombination. This process is illustrated in Figure 6.2. The design and the operation of an efficient solar cell have two basic goals: One is the minimization of undesirable recombination rates, and the other is the maximization of the absorption of photons with energy greater than the band gap.

6.1.2   Solar Cells and Technology

Semiconductor solar cells work by using the energy of incoming photons to raise electrons from the semiconductor's valence band to its conduction band. Photons that do not have enough energy turn into heat, or phonon, which limits efficiency. One way in which scientists are trying to overcome this limit is to make cells from billions of tiny pieces of semiconductor known as *quantum dots*, rather than one large piece of semiconductor. Different-sized quantum dots absorb different light frequencies—smaller dots absorb shorter wavelengths and larger dots absorb longer wavelengths. The sizes can be tuned for the incident wavelengths. Cadmium selenide (CdSe) quantum dots (nanoparticles) are sensitized on the porous structure of titanium dioxide (TiO_2), as shown in Figure 6.3. This gives a great potential to create new high-efficiency, low-cost solar cells. There are other approaches such as *nanotubes* or *optical fibers* in

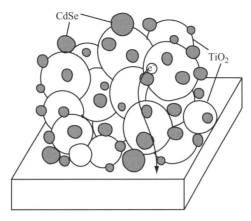

Figure 6.3 Quantum dot nanoparticles (about 25 millionth of an inch) will create electrical current when they are exposed to light

place of nanoparticles. Obviously, these are interesting topics that need a great deal of research.

6.1.3 Solar Irradiance

The sun is the source of our heat and light. Life would not be possible on Earth without the energy provided by the sun. In the interior of the sun, nuclear fusion is taking place, in which hydrogen combines to form helium and the mass reduction resulted from the fusion process converts to energy, providing millions of degree of temperature. However, the sun has an effective blackbody surface temperature of 5,777 K. The blackbody spectral intensity is well known, having first been theoretically determined by Planck [3] in 1901. The sun is continuously releasing an enormous amount of radiant energy into the solar system, an average of 6.316×10^7 W/m^2 based on a simple calculation with Planck's law. The Earth receives a tiny fraction $((r_s/r_0)^2 = 2.164 \times 10^{-5})$ of this energy; only an average of $1,367$ W/m^2 (called *solar constant* G_s) reaches the outer edge of the Earth's atmosphere, illustrated in Figure 6.4. Still, the amount of the sun's energy that reaches the surface of the Earth every hour is greater than the total amount of energy that the world's human population uses in a year.

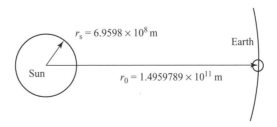

Figure 6.4 Sun radius and Sun–Earth distance

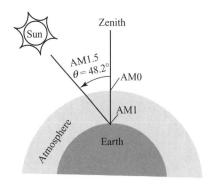

Figure 6.5 Air mass with the zenith angle

6.1.4 Air Mass

An important concept characterizing the effect of atmosphere on clear days is the air mass, defined as the relative length of the direct beam path through the atmosphere, compared with a vertical path directly to sea level, which is shown in Figure 6.5. How much energy does light lose in traveling from the edge of the atmosphere to the surface of the Earth? This energy loss depends on the thickness of the atmosphere that the sun's energy must pass through. As the sun moves lower in the sky, the light passes through a greater thickness (or longer path) of air, losing more energy. The air mass is expressed as AM $= 1/\cos(\theta)$. For example, AM1.5 is based on the calculation of $1/\cos(48.2°) = 1.5$, as shown in Figure 6.5.

The amount of solar radiation incident on a collector on the Earth's surface is affected by a number of factors. The radiation that has not been reflected or scattered in the atmosphere and reaches the surface directly is called *direct radiation*. Part of the incident energy is scattered or absorbed by air molecules, clouds, and other particles in the atmosphere. The scattered radiation that reaches the ground is called *diffuse radiation*. Some of the radiation is reflected from the ground onto the collector. The total radiation consisting of these three components is called *global radiation*. The standard air masses used in the solar cell community are listed in Table 6.1.

The spectral distribution of the radiation emitted from the sun is determined by the temperature of the surface (photosphere) of the sun, which is about 6,000 K. Three standard solar spectra measured on Earth that are commonly used in PV industry are AM0, AM1.5G, and AM1.5D. These are depicted in Figure 6.6 including the blackbody radiation of the sun.

The AM0 spectrum was measured in space outside the Earth's atmosphere. The AM1.5G spectrum was measured on the surface of the Earth, including the scattering

Table 6.1 Standard Air Masses Commonly Used in PV Industry

Air mass[a]	Terrestrial condition	Zenith angle θ, degree	Direct, diffuse, and/or ground reflection (GR)	Measured average irradiance, W/m^2
AM0	Extraterrestrial	N/A	Direct	1,367
AM1.5Gobal	Terrestrial	48.2	Direct, diffuse, and GR	1,000
AM1.5Direct	Terrestrial	48.2	Direct	768

[a]Tables are available from [4], [5], and [6].

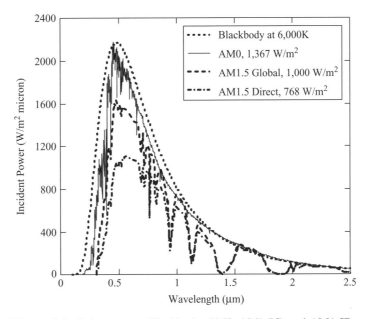

Figure 6.6 Solar spectra: Blackbody, AM0, AM1.5G, and AM1.5D

of air and the reflection of the ground, as well as the direct radiation. The AM1.5D spectrum was measured on the surface of the Earth, which includes only the direct radiation.

The *blackbody radiation* is basically in agreement with the AM0 spectrum, as shown in Figure 6.6. The deviation between the blackbody and the AM0 spectrum is associated with the absorption and dispersion in the outer, cooler layers of the Sun's atmosphere. Earth's atmosphere exhibits the absorption and scattering of atoms and molecules such as oxygen, water vapor, and carbon dioxide, which affect the ground spectra (AM1.5G and D). The fluctuation in the near infrared of a wavelength range $1 \sim 2\,\mu m$ in Figure 6.6 is associated with water vapor and carbon dioxide, while the deviation near the short wavelength $0.2 \sim 0.8\,\mu m$ is due to the oxygen or ozone. The integration of the AM0 spectrum over the entire wavelength provides the average power, or irradiation, of $1,367\,W/m^2$. Likewise, the AM1.5G and D spectra can be obtained to be $1,000\,W/m^2$ and $768\,W/m^2$, respectively.

6.1.5 Nature of Light

Light is known to have a wave-particle *dual nature*. The phenomena of light propagation may best be described by the electromagnetic wave theory, while the interaction of light with matter is best described as a particle phenomenon.

Maxwell (1864) showed theoretically that electrical disturbance should propagate in free space with a speed equal to that of light, which gives that light waves are indeed *electromagnetic waves*. These waves are also known as radio and television transmission, X-ray, and many other examples. Electromagnetic waves are produced whenever electric charges are accelerated. Lights consists of oscillating electric (E)

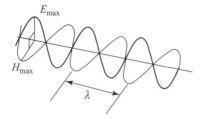

E_{max}

H_{max}

λ

Figure 6.7 Lights consists of oscillating electric (E) and magnetic (H) fields that are perpendicular to each other

and magnetic (H) fields that are perpendicular to each other and to the direction of the light, as shown in Figure 6.7. The wavelength is indicated in the figure.

Hertz (1887) first observed that electrons were emitted when light strokes a metal surface. A modern phototube is shown schematically in Figure 6.8. What was particularly puzzling about this *photoelectric effect* was that no matter how intense the light shines on the metal surface, if the frequency of the light is below a specific minimum value, no electron is emitted. The photoelectric effect could not be explained by classical physics. According to classical theory, the energy associated with electromagnetic radiation depends only on the intensity and not on the frequency.

The correct explanation of the photoelectric effect was given by Einstein in 1905, postulating that a beam of light consisted of small quanta of energy that are now called *photons*, and that an electron in the metal absorbs the incident photon's energy required to escape from the metal, which is an all-or-none process—the electron getting all the photon's energy or none at all. The quantum energy, photon, is the smallest amount of energy that can be transferred in a physical process. The energy E_{ph} of a photon is proportional to the frequency, v, as

$$E_{ph} = hv \tag{6.1}$$

where h is the Planck's constant, whose value is 6.626×10^{-34} Js and v is the frequency.

$$v = \frac{c}{\lambda} \tag{6.2}$$

where c is the speed of light, whose value is 3.0×10^8 m/s and λ is the wavelength.

The photoelectric effect with Equation (6.1) suggests that light behaves like particles (photons). Under Maxwell's wave theory, light should have a wave-particle dual nature. Light propagates like waves but behaves like particles in interaction with matter.

6.2   QUANTUM MECHANICS

Rutherford in 1911 first proposed an almost-modern atomic structure, postulating that each atom contains a massive nucleus whose size is much smaller than the overall size of the atom. The nucleus is surrounded by a swarm of electrons that revolve around the nucleus in orbits, more or less as the planets in the solar system revolve around the sun. The *Rutherford's planetary model* faced a dilemma, however. A charged electron in orbit should be unstable and spiral into the nucleus and come to rest, because a charged particle in an orbit should be accelerated toward the center and emit radiation, owing to the electromagnetic wave theory. Instead, observation showed stable atoms

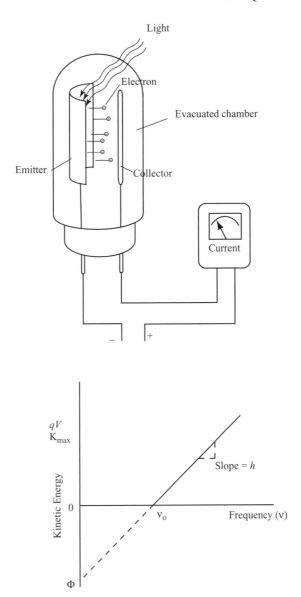

Figure 6.8 Schematic of the experiment for photoelectric effect: (a) a phototube, (b) kinetic energy versus frequency

without emitting radiation from electrons in orbits. There is an obvious defect in the planetary model.

In 1913, Bohr proposed a new model, assuming that the angular momentum is quantized and must be an integer multiple of $h/2\pi$. He postulated that an electron in an atom can revolve in certain stable orbits, each having a definite associated energy, without emitting radiation. *Bohr's model* was successful in accounting for many atomic observations, especially the emission and absorption spectra for hydrogen. Despite these

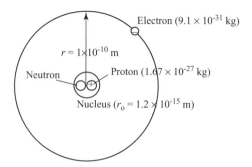

Figure 6.9 Schematic of an atomic structure

successes, it cannot predict the energy levels and spectra of atoms with more than one electron. The postulate of quantized angular momentum had no fundamental basis.

In 1924, de Broglie proposed that not only light but all matter, including electrons, has a *wave-particle dual nature*. De Broglie's work attributed wavelike properties to electrons in atoms, and the *Heisenberg uncertainty principle* (1927) shows that detailed trajectories of electrons cannot be defined. Consequently, it must be dealt with in terms of the *probability* of electrons having certain positions and momenta. These ideas were combined in the fundamental equation of *quantum mechanics*, the *Schrödinger equation*, discovered by the Erwin Schrödinger in 1925. Bohr's theory was replaced by modern quantum mechanics, in which the quantization of energy and angular momentum is a natural consequence of the basic postulates and requires no extra assumptions.

6.2.1 Atomic Structure

Each *atom* consists of a very small *nucleus* which is composed of *protons* and *neutrons*. The nucleus is encircled by moving *electrons*. Both electrons and protons are electrically charged, the charge magnitude being 1.6×10^{-19} Coulomb (C), which is negative in sign for electrons and positive for protons; neutrons are electrically neutral. An atom containing an equal number of protons and electrons is electrically neutral; otherwise, it has a positive or negative charge and is an *ion*. When an electron is bound to an atom, it has a potential energy that is inversely proportional to its distance from the nucleus. This is measured by the amount of energy needed to unbind the electron from the atom, and is usually given in units of electron volt (*eV*).

Masses for these subatomic particles are infinitesimally small; protons and neutrons have approximately the same mass, 1.67×10^{-27} kg, which is significantly larger than that for an electron, 9.11×10^{-31} kg.

Table 6.2 Summary of Information Related to an Atom

Description	Mass	Charge
Electron mass	$m_e = 9.10939 \times 10^{-31}$ kg	$q = 1.6021 \times 10^{-19}$ C = 1 eV/V
Proton mass	1.6726×10^{-27} kg	$+1.6021 \times 10^{-19}$ C
Neutron mass	1.6929×10^{-27} kg	neutral

Table 6.3   Physical Constants and Units

Description	Symbol	Value
Electron volt	eV	$1\,\text{eV} = 1.6021 \times 10^{-19}\,\text{J}$
Planck's constant	h	$h = 6.62608 \times 10^{-34}\,\text{Js} = 4.135 \times 10^{-15}\,\text{eV-s}$
Boltzmann's constant	k_B	$k_B = 1.38066 \times 10^{-23}\,\text{J/K} = 8.617 \times 10^{-5}\,\text{eV/K}$
Permittivity of vacuum (free space)	ε_0	$\varepsilon_0 = 8.854 \times 10^{-12}\,\text{C}^2 \cdot \text{J}^{-1} \cdot \text{m}^{-1} = 8.854 \times 10^{-14}\,\text{CV}^{-1}\,\text{cm}^{-1}$
Dielectric constant of Si	K_s	$K_s = 11.8$
kT	kT	$0.0259\,\text{eV}\ (T = 300\,\text{K})$

The most obvious feature of an atomic nucleus is its size, about 200,000 times smaller than the atom itself. The radius of a nucleus is about 1.2×10^{-15} m, where the electron orbits the nucleus with a radius of 1×10^{-10} m, as shown in Figure 6.9. If the nucleus were the size of a basketball, the diameter of the atom would be about 24 kilometers (15 miles). Tables 6.2 and 6.3 summarize properties of an atom.

6.2.2   Bohr's Model

In 1913, Niels Bohr developed a model for the atom that accounted for the striking regularities seen in the spectrum of the hydrogen atom. Bohr supplemented the Rutherford's planetary model of atom with the assumption that an electron of mass m_e moves in a circular orbit of radius r about a fixed nucleus. Bohr assumed that the angular momentum is quantized and must be an integer multiple of $h/2\pi$.

$$m_e vr = n\frac{h}{2\pi} \qquad (6.3)$$

where m_e is the mass of electron, v the velocity, r the radius, and $n = 1, 2, 3$, etc.

According to Coulomb's law, the attraction force of two charges, Ze and e, separated by a distance r is

$$F = \frac{1}{4\pi\varepsilon_0}\frac{Ze \cdot e}{r^2} \qquad (6.4)$$

From Newton's law,

$$F = \frac{1}{4\pi\varepsilon_0}\frac{Ze \cdot e}{r^2} = m_e a = m_e \frac{v^2}{r} \qquad (6.5)$$

Combining Equations (6.3) and (6.5) and solving for r and v, we obtain

$$r = \frac{\varepsilon_0 n^2 h^2}{\pi Z e^2 m_e} \qquad (6.6)$$

$$v = \frac{Ze^2}{2\varepsilon_0 nh} \qquad (6.7)$$

where Z is the atomic number, e is the electronic charge, m_e is the mass of the electron, and ε_0 is the permittivity of a vacuum. Hence, the radius of the first Bohr's orbit for

$n = 1$ hydrogen is calculated using Tables 6.2 and 6.3 for the physical constants as

$$r = \frac{\left(8.854 \times 10^{-12} \, C^2 \, J^{-1} \, m^{-1}\right) \left(1^2\right) \left(6.626 \times 10^{-34} \, J \cdot s\right)^2}{3.14 \, (1) \left(1.602 \times 10^{-19} \, C\right)^2 \left(9.109 \times 10^{-31} \, kg\right)} = 0.53 \times 10^{-10} \, m \quad (6.8)$$

This is in good agreement with atomic diameters as estimated by other methods, namely about 10^{-10} m (see Figure 6.9). And the velocity of the electron of the first Bohr's orbit for hydrogen is calculated as

$$v = \frac{(1) \left(1.602 \times 10^{-19} \, C\right)^2}{2 \left(8.854 \times 10^{-12} \, C^2 \, J^{-1} \, m^{-1}\right) (1) \left(6.626 \times 10^{-34} \, J \cdot s\right)} = 2.19 \times 10^6 \, m/s \quad (6.9)$$

The velocity of the electron is barely discussed in the literature being much smaller than the speed of light (3×10^8 m/s). According to the Heisenberg uncertainty principle, this velocity is not known precisely (as a cloud-probability). However, the Bohr's velocity gives a good intuition of the atomic structure.

Next, by examining the kinetic energy (K.E) and potential energy (P.E) components of the total electron energy in the various orbits, we find the K.E using Equation (6.7) as

$$K.E = \frac{1}{2} m_e v^2 = \frac{1}{2} m_e \left(\frac{Z e^2}{2 \varepsilon_0 n h}\right)^2 = \frac{Z^2 e^4 m_e}{8 \varepsilon_0^2 n^2 h^2} \quad (6.10)$$

Setting P.E $= 0$ at $r = \infty$ as a boundary condition, we have

$$P.E = \int_r F \, dr = \int_r \frac{1}{4 \pi \varepsilon_0} \frac{Z e \cdot e}{r^2} dr = -\frac{Z e^2}{4 \pi \varepsilon_0} \frac{1}{r} = -\frac{Z^2 e^4 m_e}{4 \varepsilon_0^2 n^2 h^2} \quad (6.11)$$

The total energy that is sometimes referred to as *binding energy* is therefore,

$$E_n = K.E + P.E = -\frac{Z^2 e^4 m_e}{8 \varepsilon_0^2 n^2 h^2} \quad n = 1, 2, 3, \ldots \infty \quad (6.12)$$

6.2.3   Line Spectra

Bohr's model thus predicts a discrete energy level diagram for a one-electron atom, which means that the electron is allowed to stay only in one of the quantized energy levels described in Equation (6.12), which is schematically shown in Figure 6.10. The *ground state* for the system of nucleus plus electron has $n = 1$; *excited states* have higher values of n. When the binding energy turns to zero at $n = \infty$, the electron becomes a *free electron* from the atom. The electron that receives energy from a perturbation such as sunlight, electric field, kinetic energy of other moving electrons, or even thermal energy should jump up to a higher energy level, as shown in the figure. Usually the exited electron is unstable and returns back to a lower level of

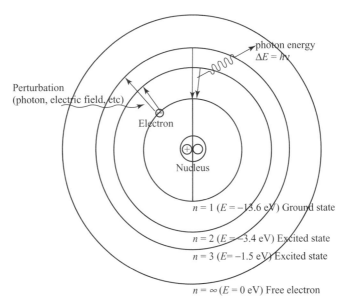

Figure 6.10 Energy level diagram for the hydrogen atom

energy, emitting a photon with energy that is exactly the same as the energy difference of the higher and lower energy levels. The energy difference is easily calculated using Equation (6.12) with Equation (6.1) as

$$\Delta E = E_i - E_f = h\upsilon = h\frac{c}{\lambda} = -\frac{Z^2 e^4 m_e}{8\varepsilon_0^2 h^2}\left(\frac{1}{n_i^2} - \frac{1}{n_f^2}\right) \tag{6.13}$$

Solving for the frequency υ or λ, we have

$$\upsilon = \frac{c}{\lambda} = -\frac{Z^2 e^4 m_e}{8\varepsilon_0^2 h^3}\left(\frac{1}{n_i^2} - \frac{1}{n_f^2}\right) \tag{6.14}$$

Equation (6.14) is in good agreement with the measured line spectra for a hydrogen atom as shown in Figure 6.11. For example, Red 656 nm indicates a spectrum with

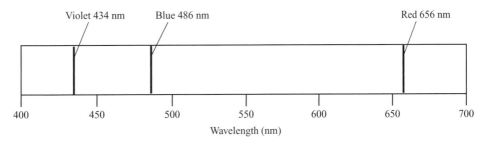

Figure 6.11 Schematic diagrams of the line spectra for a hydrogen atom

the wavelength of 656 nm (nanometer, 10^{-9} m), of which the indicated line may be understandable when considering the visible spectrum of light (rainbow) ranging from 400 nm to 700 nm, where the color ranges from violet to red. The measured spectrum of Red 656 nm in Figure 6.11 can be calculated from Equation (6.14) by varying the quantum number from $n = 3$ to $n = 2$. In the same way, Blue 486 nm from $n = 4$ to $n = 2$ and Violet 434 nm from $n = 5$ to $n = 2$. Equation (6.14) was used to directly identify the line spectra from unknown materials such as a spectrum from stars far away at one-billion-year distance. For example, helium was discovered by studying the line spectra from the sun during the solar eclipse in 1868 even before the helium was found on the Earth in 1895.

6.2.4   De Broglie Wave

Although the Bohr model was immensely successful in explaining the hydrogen spectra, numerous attempts to extend the semiclassical Bohr analysis to more complex atoms such as helium proved to be futile. Moreover, the quantization of angular momentum in the Bohr model had no theoretical basis.

In 1924, de Broglie postulated that if light has the dual properties of waves and particles, electrons should have duality also. He suggested that an electron in an atom might be associated with a circular standing wave, as shown in Figure 6.12. The condition for the circular standing wave is

$$n\lambda = 2\pi r \tag{6.15}$$

Combing Equation (6.15) and the Bohr's assumption of Equation (6.3), we obtain the wavelength as

$$\lambda = \frac{h}{m_e v} = \frac{h}{p} \tag{6.16}$$

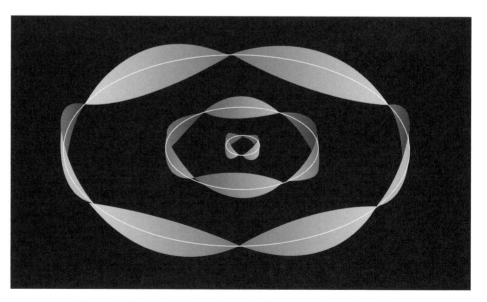

Figure 6.12   The electrons form standing wave about the nucleus

It is very important to note that the wavelength is related to the linear momentum p of the electron or a particle. De Broglie showed with the theory of relativity that exactly the same relationship holds between the wavelength and momentum of a photon. De Broglie therefore proposed as a generalization that any particle moving with linear momentum p has wavelike properties and a wavelength associated with it. Within three years, two different experiments had been performed to demonstrate that electrons do exhibit wave behavior. De Broglie was the first of only two physicists to receive a Nobel Prize for his thesis work.

6.2.5 Heisenberg Uncertainty Principle

In 1927, Werner Heisenberg stated the *uncertainty principle* that the position and the velocity of a particle on an atomic scale cannot be simultaneously determined with precision. Mathematically, it says that the product of the uncertainties of these pairs has a lower limit equal to Planck's constant

$$\Delta x \, \Delta p \geq h \tag{6.17}$$

where Δx is the uncertainty of the position and Δp is the uncertainty of the momentum or the velocity.

Heisenberg said that the experimental inability to determine both the position and the velocity simultaneously is not due to any theoretical flaw in wave mechanics; rather, it is inherent in the dual nature of matter. This principle indicates that the future of matter cannot be predetermined because it cannot be even in the quantum world.

6.2.6 Schrödinger Equation

In 1925, Erwin Schrödinger developed a differential equation, based on de Broglie's notion of dual wave-particle nature of electrons. Although the wave function describes well the motion of a string or electromagnetic waves, the wave function of electron motion was not physically understandable. Nevertheless, the wave function describes the distribution of the particle in space. It is related to the probability of finding the particle in each of various regions—the particle is most likely to be found in regions where the wave function is large, and so on.

The *time-dependent Schrödinger equation* can be formulated as

$$-\frac{\hbar^2}{2m}\nabla^2\Psi + U(x, y, z)\Psi = -\frac{\hbar}{i}\frac{\partial\Psi}{\partial t} \tag{6.18}$$

where $\hbar$ is denoted as $h/2\pi$, m the mass of the particle, Ψ is the wave function, U the potential energy, t the time, i the imaginary unit ($i^2 = -1$).

Equation (6.18) is very difficult to solve due probably to the dual nature of a particle. Instead, the *time-independent Schrödinger equation* developed using the *standing wave* is widely used in quantum mechanics.

$$-\frac{\hbar}{2m}\nabla^2\Psi + U(x, y, z)\Psi = E\Psi \tag{6.19}$$

where E is the total (kinetic) energy of the particle.

According to the Heisenberg uncertainty principle, attempts to determine the position and the velocity of an electron in orbits were futile. We can determine the energy of a particle in a stationary state exactly, but we cannot simultaneously know both the position and the velocity of the particle when it is in the specified state. The Schrodinger equation was applied to the problem of a hydrogen atom. The solutions have quantized values on energy and angular momentum. We recall that angular momentum quantization was put into the Bohr model as an ad hoc assumption with no fundamental justification; with the Schrodinger equation it comes out automatically. The predicted energy levels turned out to be identical to those from the Bohr model and thus to agree with experimental values from spectrum analysis.

The use of wave mechanics has been limited by the difficulty of solving the Schrödinger equation for large atoms. For a system consisting of a single electron and nucleus for a hydrogen atom or a hydrogen-like ion, the Schrodinger equation can be solved exactly.

6.2.7 A Particle in a 1-D Box

Consider a particle in a one-dimensional box the simplest problem for which the Schrödinger equation can be solved exactly. This is schematically shown in Figure 6.13. We begin by considering a particle moving freely in one dimension with classical momentum p. Such a particle is associated with a wave of wavelength $\lambda = h/p$, which is de Broglie's formula of Equation (6.16).

The general wave function is governed by the second-order differential equation.

$$\frac{d^2\psi}{dx^2} + k^2\psi = 0 \tag{6.20}$$

where k is the wavenumber, which is a function of wavelength λ as

$$k = \frac{2\pi}{\lambda} \tag{6.21}$$

The total energy is the sum of the kinetic energy and the potential energy.

$$E = K.E + P.E \tag{6.22}$$

However, we assume that there is no potential energy for a particle in a box. Therefore,

$$E = K.E = \frac{1}{2}mv^2 \tag{6.23}$$

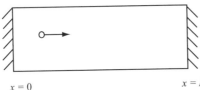

$x = 0$ $x = L$ **Figure 6.13** A particle in a one-dimensional box

Since the momentum is the product of the mass and the velocity, the velocity is expressed as

$$v = \frac{p}{m} \tag{6.24}$$

Substituting Equation (6.24) into Equation (6.23), the kinetic energy of the particle is

$$E = \frac{p^2}{2m} \tag{6.25}$$

We found the relationship between the energy and the linear momentum for a moving particle. Using de Broglie's formula $\lambda = h/p$, Equation (6.16), we have

$$E = \frac{h^2}{2m\lambda^2} \tag{6.26}$$

The kinetic energy of the particle is inversely proportional to the square of the wavelength. Equation (6.26) is rewritten

$$\lambda^2 = \frac{h^2}{2mE} \tag{6.27}$$

Squaring Equation (6.21) and substituting Equation (6.27) into it yields

$$k^2 = \left(\frac{2\pi}{\lambda}\right)^2 = \frac{8\pi^2 mE}{h^2} \tag{6.28}$$

Using the de Broglie's formula $\lambda = h/p$, the wavenumber is proportional to the momentum as

$$k = \frac{2\pi}{\lambda} = \frac{2\pi}{h} p \tag{6.29}$$

Equation (6.20) can be rewritten using Equation (6.28), which is exactly the same as the time-independent Schrödinger equation, Equation (6.19), with $U = 0$. Thus, the time-independent Schrödinger equation is derived for a particle in a box as

$$\frac{d^2\psi}{dx^2} + \frac{8\pi^2 mE}{h^2}\psi = 0 \tag{6.30}$$

The general solution for Equation (6.20) is known to be

$$\psi(x) = A \sin kx + B \cos kx \tag{6.31}$$

The wave function is bounded at the walls ($x = 0$ and $x = L$):

$$\psi(0) = \psi(L) = 0 \tag{6.32}$$

Therefore, for the first boundary condition with $x = 0$,

$$\psi(0) = A \sin k0 + B \cos k0 = B = 0 \tag{6.33}$$

For the second boundary condition with $x = L$,

$$\psi(L) = A \sin kL = 0 \tag{6.34}$$

This can be true only if

$$kL = n\pi \qquad n = 1, 2, 3, \ldots \ldots \infty \tag{6.35}$$

Therefore, the solution of the wave function should be

$$\psi(x) = A \sin \frac{n\pi}{L} x \qquad n = 1, 2, 3, \ldots \ldots \infty \tag{6.36}$$

The wave function gives the distribution of the particle that is the probability where it is likely to be. For the total probability of finding the particle somewhere in the box to be unity,

$$\int_0^L \psi^2(x) dx = 1 \tag{6.37}$$

Substituting Equation (6.36) into (6.37), we have

$$A^2 \int_0^L \sin^2 \left(\frac{n\pi}{L} x\right) dx = 1 \tag{6.38}$$

The integration part is obtained as

$$\int_0^L \sin^2 \left(\frac{n\pi}{L} x\right) dx = \int_0^L \frac{1 - \cos\left(2\frac{n\pi}{L} x\right)}{2} dx = \frac{L}{2} \tag{6.39}$$

Thus, combining the above two equations, we find the coefficient A as

$$A = \sqrt{\frac{2}{L}} \tag{6.40}$$

The wave function is then

$$\psi(x) = \sqrt{\frac{2}{L}} \sin \frac{n\pi}{L} x \qquad n = 1, 2, 3, \ldots \ldots \infty \tag{6.41}$$

To find the energy E_n for a particle with wave function Ψ_n, calculate the second derivative.

$$\frac{d^2\psi}{dx^2} = \frac{d^2}{dx^2} \left[\sqrt{\frac{2}{L}} \sin\left(\frac{n\pi}{L} x\right)\right] = -\left(\frac{n\pi}{L}\right)^2 \sqrt{\frac{2}{L}} \sin\left(\frac{n\pi}{L} x\right) = -\left(\frac{n\pi}{L}\right)^2 \psi(x) \tag{6.42}$$

By comparing this with Equation (6.30), we obtain

$$E_n = \frac{h^2 n^2}{8mL} \qquad n = 1, 2, 3, \ldots \ldots \infty \tag{6.43}$$

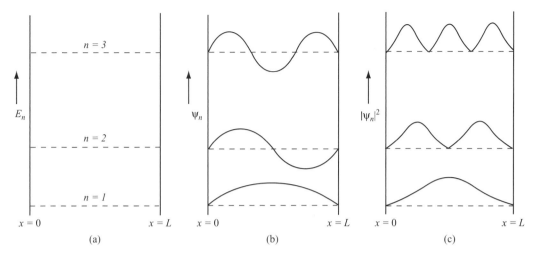

Figure 6.14   (a) The potential energy for a particle in a box of length L, (b) Wave function, and (c) The probability

This solution of the Schrödinger equation demonstrates that a particle in a box is quantized. An energy E_n and a wave function $\Psi_n(x)$ correspond to each quantum number n. These are the only allowed stationary states of the particle in the box.

The solution to the wave equation for a particle in a box of length L provided an important test of quantum mechanics and served to convince scientists of its validity, which is shown in Figure 6.14. The first three quantized energy levels are shown in Figure 6.14(a), the first three wave functions in Figure 6.14(b), and the first three probability of the particle in Figure 6.14(c). Even more importantly, the information derived from studying the wave functions for the hydrogen atom is used to describe and predict the behavior of electrons in many-electron atoms.

6.2.8   Quantum Numbers

A complete theoretical treatment of a hydrogen atom, using a wave equation, yields a set of wave functions, each described by four quantum numbers: n, ℓ, m_ℓ, and m_s. Each set of the first three (n, ℓ, m_ℓ) identifies quantum states of the atom, in which the electron has energy equal to E_n. Each quantum state can hold no more than two electrons, which must have opposite spins. This is called the *Pauli Exclusion Principle*. The fourth quantum number, m_s, accounts for these opposite spins, allowing at most two electrons in each quantum state. The fourth quantum number does not appear when the Schrödinger equation is solved. However, the theory of relativity requires four quantum numbers. The first three quantum numbers are associated with three-dimensional space, while the fourth quantum number is related to time. Orbitals, states, and electrons allowed for a given shell (n) can be obtained in the following steps:

Principal quantum number, n. The principal quantum number determines the sizes of the orbitals and the allowed energy of the electron, having only the values $n = 1, 2, 3, \ldots \ldots \infty$. The principal quantum number specifies shells. Sometimes these shells are designated by the letters K, L, M, N, and so on. The energy obtained using the quantum mechanics for one-electron atoms is exactly the same as Equation (6.12) that Bohr had obtained, namely

$$E_n = -\frac{Z^2 e^4 m_e}{8\varepsilon_0^2 n^2 h^2} \quad n = 1, 2, 3, \ldots \infty \tag{6.44}$$

Angular quantum number, ℓ. The angular quantum number is associated with the shapes of the orbitals (subshell). The allowed values of ℓ are $0, 1, \ldots, (n-1)$ for a given value of n.

If $n = 1$, ℓ can be only be 0.

If $n = 2$, ℓ can be 0 or 1.

If $n = 3$, ℓ can be 0, 1, or 2 and so on.

The letter assigned, respectively, to the ℓ values 0, 1, 2, and 3 are denoted by the *orbitals* as s, p, d, and f.

ℓ	0	1	2	3
Orbital	s	p	d	f

Magnetic quantum number, m_ℓ. The magnetic quantum number involves the orientations of the orbitals. The allowed values of m_ℓ are $0, \pm 1, \pm 2, \ldots, \pm \ell$. Thus,

If $\ell = 0$, $m_\ell = 0$.

If $\ell = 1$, $m_\ell = -1, 0, 1$

If $\ell = 2$, $m_\ell = -2, -1, 0, 1, 2$

Therefore, there are $(2\ell + 1)$ degenerate orbitals for each value of ℓ, which means that the states obtained by the magnetic quantum number have the same energy. Atomic orbitals with the same energy are said to be *degenerate*. The restrictions on ℓ and m give rise to n^2 sets of quantum number for every value of n.

Fourth (Spin) quantum number, m_s. The fourth (spin) quantum number is associated with two spin states, spin up ($+1/2$) and spin down ($-1/2$), which are also described by the *Pauli Exclusion Principle* stating that no more than two electrons can occupy the same quantum state.

For example, we want to determine the orbitals, states, and electrons allowed for a shell of $n = 2$. The number of the allowed states are simply obtained with $n^2 = 4$. The Pauli Exclusion Principle allows only two electrons for each state, leading to $4 \times 2 = 8$. Total eight electrons are allowed to stay on the shell. There are two orbitals ($\ell = 0$ and 1) or (s and p) for $n = 2$ shell. For $\ell = 0$, we have one state ($m_\ell = 0$). For $\ell = 1$, we have three states ($m_\ell = -1, 0, 1$). There are then total four allowed states. All the states can have two electrons that must be in opposite spin. Four times two leads to eight allowed electrons. These are summarized in Table 6.4 for L-shell ($n = 2$). The p-orbital ($\ell = 1$) has three states ($m_\ell = -1, 0, 1$) in one p-orbital, which is said to be *degenerate*.

Table 6.4   Electronic Quantum Numbers and Atomic Orbitals

Shell	n	ℓ	Orbital (subshell)	m_ℓ	Degeneracy	Allowed electron number per subshell	Allowed electron number per shell
K	1	0	1s	0	1	2	2
L	2	0	2s	0	1	2	
	2	1	2p	−1,0,1	3	6	8
M	3	0	3s	0	1	2	
	3	1	3p	−1,0,1	3	6	
	3	2	3d	−2,−1,0,1,2	5	10	18
N	4	0	4s	0	1	2	
	4	1	4p	−1,0,1	3	6	
	4	2	4d	−2,−1,0,1,2	5	10	
	4	3	4f	−3,−2,−1,0,1,2,3	7	14	32

6.2.9   Electron Configurations

Electron configuration is the distribution of electrons in the orbitals of the atom. *Electron configuration* was first conceived of under the Bohr model of the atom, and it is still common to speak of shells and subshells, despite the advances in understanding of the quantum-mechanical nature of electrons. The electron configurations for hydrogen, helium, and sodium are, respectively, $1s^1$, $1s^2$, $1s^2 2s^2 2p^6 3s^1$, where the number in front of orbital letters denotes the principal quantum number n and the superscript of the letters denotes the number of occupied electrons. The electron configuration associated with the lowest energy level of the atom is referred to as *ground state*. Ground state electron configurations for some of the more common elements are listed in Table 6.5. The valence electrons are those that occupy the outermost unfilled shell. These electrons are extremely important—as will be seen, they participate in the bonding between atoms to form atomic and molecular aggregates. Furthermore, many of the physical and chemical properties of solids are based on these valence electrons.

We now consider a somewhat more complicated atom. The nucleus of a silicon atom contains 14 positive charges. Consequently, 14 electrons will locate themselves around this nucleus. If we add the electrons one by one, the first 2 electrons, having opposite spin, will come to rest in the $1s$ shell ($n = 1$). It would be most difficult to tear these away from the nucleus. The Pauli exclusion principle forbids any more electrons in the first shell and dictates that the next 2 electrons will come to rest in the $2s$ shell, while the next 6 electrons occupy the $2p$ shell with magnetic quantum numbers $−1$, 0, 1 and spin quantum numbers of $−1/2$, $1/2$. Because the energy levels specified by the magnetic and spin quantum numbers are very close together, we show it as one energy level with 6 electrons. The next 2 electrons goes into the $3s$ shell, while the last 2 electrons come to rest in the $3p$ shell. The $3s$ and $3p$ shells are valence electrons, which occupy the outermost filled shell. Finally, the electron configuration of the silicon is $1s^2 2s^2 2p^6 3s^2 3p^2$.

Table 6.5 Ground State Electron Configurations for Some of the Common Elements

Element	Symbol	Atomic number	Electron configuration
Hydrogen	H	1	$1s^1$
Helium	He	2	$1s^2$
Lithium	Li	3	$1s^2 2s^1$
Beryllium	Be	4	$1s^2 2s^2$
Boron	B	5	$1s^2 2s^2 2p^1$
Carbon	C	6	$1s^2 2s^2 2p^2$
Nitrogen	N	7	$1s^2\ 2s^2\ 2p^3$
Oxygen	O	8	$1s^2 2s^2 2p^4$
Fluorine	F	9	$1s^2 2s^2 2p^5$
Neon	Ne	10	$1s^2 2s^2 2p^6$
Sodium	Na	11	$1s^2\ 2s^2\ 2p^6\ 3s^1$
Magnesium	Mg	12	$1s^2\ 2s^2\ 2p^6\ 3s^2$
Aluminum	Al	13	$1s^2\ 2s^2\ 2p^6\ 3s^2\ 3p^1$
Silicon	Si	14	$1s^2\ 2s^2\ 2p^6\ 3s^2\ 3p^2$
Phosphorus	P	15	$1s^2\ 2s^2\ 2p^6\ 3s^2\ 3p^3$
Sulfur	S	16	$1s^2\ 2s^2\ 2p^6\ 3s^2\ 3p^4$
Chlorine	Cl	17	$1s^2\ 2s^2\ 2p^6\ 3s^2\ 3p^5$
Argon	Ar	18	$1s^2\ 2s^2\ 2p^6\ 3s^2\ 3p^6$
Potassium	K	19	$1s^2\ 2s^2\ 2p^6\ 3s^2\ 3p^6 4s^1$
Calcium	Ca	20	$1s^2\ 2s^2\ 2p^6\ 3s^2\ 3p^6\ 4s^2$
Scandium	Sc	21	$1s^2\ 2s^2\ 2p^6\ 3s^2\ 3p^6\ 3d^1 4s^2$
Titanium	Ti	22	$1s^2\ 2s^2\ 2p^6\ 3s^2\ 3p^6\ 3d^2 4s^2$
Vanadium	V	23	$1s^2\ 2s^2\ 2p^6\ 3s^2\ 3p^6\ 3d^3 4s^2$
Chromium	Cr	24	$1s^2\ 2s^2\ 2p^6\ 3s^2\ 3p^6\ 3d^5 4s^1$
Manganese	Mn	25	$1s^2\ 2s^2\ 2p^6\ 3s^2\ 3p^6\ 3d^5 4s^2$
Iron	Fe	26	$1s^2\ 2s^2\ 2p^6\ 3s^2\ 3p^6\ 3d^6 4s^2$
Cobalt	Co	27	$1s^2\ 2s^2\ 2p^6\ 3s^2\ 3p^6\ 3d^7 4s^2$
Nickel	Ni	28	$1s^2\ 2s^2\ 2p^6\ 3s^2\ 3p^6\ 3d^8 4s^2$
Copper	Cu	29	$1s^2\ 2s^2\ 2p^6\ 3s^2\ 3p^6\ 3d^{10} 4s^1$
Zinc	Zn	30	$1s^2\ 2s^2\ 2p^6\ 3s^2\ 3p^6\ 3d^{10} 4s^2$
Gallium	Ga	31	$1s^2\ 2s^2\ 2p^6\ 3s^2\ 3p^6\ 3d^{10} 4s^2\ 4p^1$
Germanium	Ge	32	$1s^2\ 2s^2\ 2p^6\ 3s^2\ 3p^6\ 3d^{10} 4s^2\ 4p^2$
Arsenic	As	33	$1s^2\ 2s^2\ 2p^6\ 3s^2\ 3p^6\ 3d^{10} 4s^2\ 4p^3$
Selenium	Se	34	$1s^2\ 2s^2\ 2p^6\ 3s^2\ 3p^6\ 3d^{10} 4s^2\ 4p^4$
Bromine	Br	35	$1s^2\ 2s^2\ 2p^6\ 3s^2\ 3p^6\ 3d^{10} 4s^2\ 4p^5$
Krypton	Kr	36	$1s^2\ 2s^2\ 2p^6\ 3s^2\ 3p^6\ 3d^{10} 4s^2\ 4p^6$

Example 6.2.1 Electronic Configuration of a Silicon Atom   Consider a silicon atom whose atomic number is fourteen ($Z = 14$). Construct the electronic configuration of the silicon atom and sketch the atomic structure indicating the orbitals (subshells) with electrons occupied.

Solution   We want to use Table 6.4 to figure out the electronic configuration for a silicon atom having $Z = 14$. First of all, we know that there are 14 electrons in a silicon atom. We find from Table 6.4 that two electrons are allowed in K-shell ($n = 1$), eight electrons in L-shell ($n = 2$), and eighteen electrons in M-shell ($n = 3$). However, since ten electrons occupy the K- and M-shell, the remaining four electrons out of eighteen allowed electrons can only occupy the M-shell. So we construct the electron configuration for the silicon atom according to Table 6.4 as $1s^2 2s^2 2p^6 3s^2 3p^2$, where the number in front of orbital letters denotes the principal quantum number n and the superscript of the letters denotes the number of occupied electrons. If we sum the superscript numbers as $2 + 2 + 6 + 2 + 2 = 14$, the 14 indicates a total number of electrons occupied in shells. The L-shell ($n = 2$) consists of two subshells and the M-shell ($n = 3$) also consists of two subshells. It is noted that the energy levels specified by the angular quantum numbers (ℓ: s and p subshells) are very close together, we show them as one group in Figure 6.15.

6.2.10   Van der Waals Forces

Weak attraction forces between isolated atoms or molecules are collectively referred to as *van der Waals forces*. These forces arise from the electrostatic attraction of the nuclei of one atom for the electron of a different atom. The repulsions between the electrons of the two atoms and the nuclei of the two atoms counteract the electrostatic attractions, but there is always a small net attractive force. Van der Waals forces are much weaker than the bonding forces that operate in ionic, metallic, and covalent crystals.

6.2.11   Covalent Bonding

Like most semiconductors, silicon crystals display covalent bonding. The isolated Si atom, or a Si atom not interacting with other atoms, contains four valence electrons. Si atoms incorporated in the diamond lattice, by contrast, exhibit a *bonding* that involves

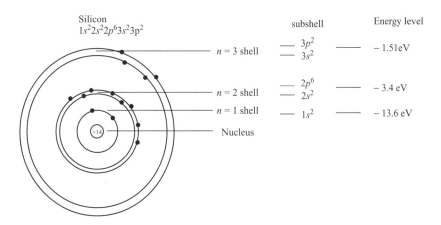

Figure 6.15   A silicon atom with quantum numbers and shells

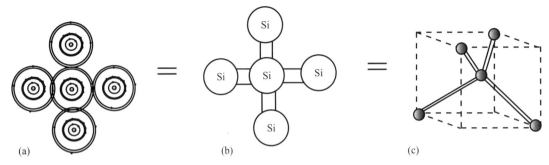

Figure 6.16   Schematic representations of (a) covalent bonding, (b) bonding model, and (c) diamond lattice for a silicon crystal

an attraction between each atom and its four nearest neighbors. The implication here is that, in going from isolated atoms to the collective crystalline state, Si atoms come to share one of their valence electrons with each of the four nearest neighbors, which is shown in Figure 6.16(a). This *covalent bonding* gives rise to the idealized semiconductor representation, the bonding model, shown in Figure 6.16(a), (b), and (c). Each circle in Figure 6.16(b) represents the core of a Si atom, while each line represents a shared valence electron. A diamond lattice of Si crystals is illustrated in Figure 6.16(c).

6.2.12   Energy Band

We bring together several atoms of silicon, creating a crystal of silicon. Once we have formed the crystal we can no longer consider the electrons as being associated with a particular atom; rather, they must be considered as belonging to the crystal at large.

A solid may be thought of as consisting of a large number of N atoms, initially separated from one another, which are subsequently brought together and bonded to form the ordered atomic arrangement found in the crystalline material. At relatively large separation distances, each atom is independent of all the others and will have the atomic energy levels and electron configuration as if isolated. However, as the atoms come within close proximity of one another, electrons are acted upon, or perturbed, by the electrons and nuclei of nearby atoms. This influence is such that each distinct atomic state may split into a series of closely spaced electron states in the solid, to form what is termed an electron *energy band*, which is illustrated in Figure 6.17. The extent of splitting depends on interatomic separation and begins with the outermost electron shells, since they are the first to be disturbed as the atoms coalesce. Bearing in mind that the outermost shells are filled partially (for example, 4 occupied electrons out of 18 allowed electrons in a Si atom), the upper empty state and the lower filled state are, respectively, referred to as a *conduction band* and a *valence band*. The *band gap* (E_g) between the lower edge of the conduction band and the upper edge of the valence band is an important property of the material (see Table 6.6). As discussed in

Table 6.6   Band Gaps of Common Solar Cell Materials at Room Temperature

Material	Band gap (eV)	Material	Band gap (eV)
Si	1.125	CuInSe$_2$	1.02
GaAs	1.422	CdTe	1.5
Ge	0.663		

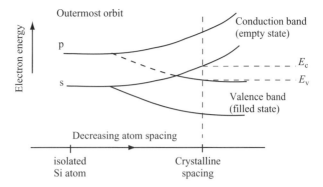

Figure 6.17 Schematic diagram of the energy band vs. interatomic distance

Section 6.1.1 of Operation of Solar Cells, only photons that have greater energy than the band gap produce an electron-hole pair that consists of a free electron and a free hole, which may eventually lead to electricity if they are not recombined earlier.

Within each band, the energy states are discrete, yet the difference between adjacent states is exceedingly small. The number of states within each band will equal the total of all states contributed by the N atoms.

6.2.13 Pseudo-Potential Well

When an electron in a valence band receives sufficient energy by a perturbation such as photons, an electric field, or thermal energy, the electron may jump up to the conduction band. The electron is no longer restricted to the single atom, becoming able to freely migrate in the sea of the conduction band. A free electron in a conduction band may be considered to be in a pool or pseudo-potential well—actually, in a crystal. According to quantum mechanics, the free electron is allowed to stay only in one of the allowed states (quantized energy). Therefore, the first step is to determine how many states per unit volume are allowed in the pool, which is called the *density of states*. The simplified model of energy band in a crystal is illustrated in Figure 6.18, where the lower edge

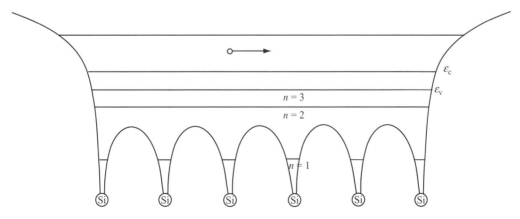

Figure 6.18 Visualization of a conduction band electron moving in a crystal

of the conduction band and the upper edge of the valence band are indicated. The free electron in a crystal has a physical similarity with a particle in a one-dimensional box in Section 6.2.7. Accordingly, the solutions in the section can be applied to this problem.

6.3 DENSITY OF STATES

The *density of states* is the number of allowed quantum states per unit crystal volume between the energy levels, so that electrons or holes may occupy the states. Knowledge of the density of states is required as the first step in determining the carrier (electrons and holes) concentrations and energy distributions of carriers within a semiconductor. As discussed in the previous section, the concepts and the solutions of a particle in a box in Section 6.2.7 can be applied to a free electron in the pseudo-potential well in Section 6.2.13.

6.3.1   Number of States

The density of states can be obtained by evaluating the *number of states* per volume and per energy. Since the number of states is associated with the momentum (see Section 6.2.8), the number of states can be obtained conceptually by dividing a spherical volume of momentum space by the least momentum unit, as much as the Heisenberg uncertainty principle defines. Alternatively, since the momentum is proportional to the wavenumber as shown in Equation (6.29), the wavenumber instead of the momentum can be used to determine the number of states. The smallest wavenumber unit that has the largest wavelength is found from Equation (6.21). Replacing the wavelength λ with the geometric length L of a crystal, the number of states is

$$N_s(k) = \frac{\frac{4}{3}\pi k^3}{\left(\frac{2\pi}{L}\right)^3} \tag{6.45}$$

We express the number of states in terms of energy, also considering that all states can have at least two electrons that have opposite spins, according to the Pauli Exclusion Principle. Using Equation (6.28), the number of states is

$$N_s(E) = \frac{2 \cdot \frac{4}{3}\pi \left(\frac{8\pi^2 m_e E}{h^2}\right)^{3/2}}{\left(\frac{2\pi}{L}\right)^3} = \frac{8\pi (2m_e E)^{3/2} L^3}{3h^3} \tag{6.46}$$

Now we try to formulate the density of states g_c in the conduction band by dividing the number of states by a volume of L^3 and energy E, which is mathematically achieved by taking derivative with respect to E.

$$g_c(E) = \frac{1}{V}\frac{dN_s}{dE} = \frac{8\pi \cdot m_e}{h^3}\sqrt{2m_e E} \tag{6.47}$$

Table 6.7 Density of States Effective Masses at 300 K[a]

Material	m_n^*/m_e	m_p^*/m_e
Si	1.18	0.81
Ge	0.55	0.36
GaAs	0.066	0.52

[a]Reference [7]. $m_e = 9.10939 \times 10^{-31}$ kg

6.3.2 Effective Mass

In reality, the electron experiences complex forces surrounding it, such as other atoms. However, assuming that these forces are small compared to the kinetic energy of the electron, we obtain the electron mass as an effective mass m^* with a subscript for an electron denoted by n and for a hole by p. To obtain the conduction and valence band densities of states near the band edges in real materials, the mass m_e of the particle in the foregoing derivation is replaced by the appropriate carrier *effective mass m^**, such that the effective mass of electron is denoted by m_n^* and the effective mass of hole by m_p^*. This is a very practical assumption that greatly simplifies the analysis of semiconductors. If E_c is taken to be the minimum electron energy in the conduction band and E_v the maximum hole energy in the valence band, the E in Equation (6.47) must be replaced by $E - E_c$ in the conduction band states and $E_v - E$ in the valence band states.

The densities of states are now expressed preferably as a function of energy using the effective mass and the above mentioned minimum and maximum energies E_c and E_v. The density of states in the conduction band is

$$g_c(E) = \frac{8\pi \cdot m_n^*}{h^3} \sqrt{2m_n^*(E - E_c)} \tag{6.48}$$

The density of states in the valence band is

$$g_v(E) = \frac{8\pi \cdot m_p^*}{h^3} \sqrt{2m_p^*(E_v - E)} \tag{6.49}$$

The densities of state effective masses for electrons and holes in Si, Ge, and GaAs at 300 K are listed in Table 6.7.

6.4 EQUILIBRIUM INTRINSIC CARRIER CONCENTRATION

The number of carriers (electrons and holes) per volume in the conduction and valence band with no externally applied electric field or no illumination is called the *equilibrium carrier concentration*. Under equilibrium conditions, the number of carriers occupied in the associated band is obtained by multiplying the density of states with the Fermi function.

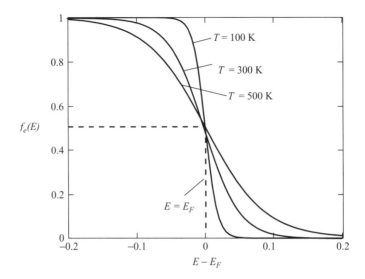

Figure 6.19 The Fermi function at various temperatures

6.4.1 Fermi Function

The *Fermi function* $f_e(E)$ specifies, under equilibrium conditions, the probability that an available state at an energy E will be occupied by an electron.

$$f_e(E) = \frac{1}{1 + \exp\left(\dfrac{E - E_F}{kT}\right)} \qquad (6.50)$$

where E_F is the Fermi energy, k is the Boltzmann's constant (8.617×10^{-5} eV/K) and T is the Kelvin temperature. The Fermi function depends on the temperature, which is shown in Figure 6.19. Nondegenerate semiconductors are defined as semiconductors for which the Fermi energy is at least $3kT$ away from either band edge, as shown in Figure 6.20 (see Section 6.4.2). The reason we restrict ourselves to nondegenerate semiconductors is that this definition allows the Fermi function to be replaced by a

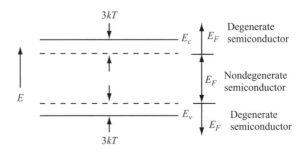

Figure 6.20 Energy band diagram and definition of degenerate/nondegenerate semiconductors

simple exponential function as

$$f_e(E) = \exp\left(\frac{E_F - E}{kT}\right) \tag{6.51a}$$

or equivalently,

$$f_e(E) = \exp\left(\frac{E_F - E_c}{kT}\right)\exp\left(\frac{E_c - E}{kT}\right) \tag{6.51b}$$

6.4.2   Nondegenerate Semiconductor

If the Fermi energy of a semiconductor is confined as Equation (6.52), the semiconductor is *nondegenerate*.

$$E_v + 3kT \leq E_F \leq E_c - 3kT \tag{6.52}$$

Most semiconductors are nondegenerate, while conductors like metals are *degenerate*. Semiconductors doped to very high level act like degenerate conductors. For example, knowing that the band gab E_g of silicon is 1.12 eV, as shown in Table 6.6, and using the information in Table 6.3, under equilibrium conditions (zero electric field and no illumination), we find that

$$E_c - E_F \cong \frac{1}{2}E_g = \frac{1}{2}1.12 \text{ eV} = 0.56 \text{ eV} \tag{6.53}$$

$$kT = (1.38 \times 10^{-23} \text{ J/K} \times (300 \text{ K})/(1.6021 \times 10^{-19} \text{ J/eV}) = 0.026 \text{ eV} \tag{6.54}$$

$$\exp\left(\frac{E_c - E_F}{kT}\right) = \exp\left(\frac{0.56 \text{ eV}}{0.026 \text{ eV}}\right) = \exp(21.5) \gg 1 \tag{6.55}$$

Hence, for Si semiconductors the "1" in the denominator of Equation (6.50) can be neglected compared to the value of exp(21.5) in Equation (6.55), which is referred to nondegenerate semiconductors. Many semiconductors for solar cells exhibit nondegeneracy. In this case, Equation (6.51b) can be used in the place of Equation (6.50). The nondegeneracy is illustrated in Figure 6.20. It is customary to show the lower edge of the conduction band E_c and the upper edge of the valance band E_v in the energy diagram shown in Figure 6.20.

6.4.3   Equilibrium Electron and Hole Concentrations

Under equilibrium conditions, the number of electrons per volume that is occupied in the conduction band is obtained by multiplying the density of states with the Fermi function and integrating from the lower edge of the conduction band E_c to the infinity ∞. The density of electrons and holes under equilibrium conditions is sometimes referred to the *equilibrium carrier concentration*. The equilibrium electron concentration in the conduction band is expressed as

$$n_0 = \int_{E_c}^{\infty} g_c(E)f_e(E)dE \tag{6.56}$$

where the subscript "0" means the equilibrium condition. With Equations (6.48) and (6.51b) under the assumption of nondegenerate semiconductors, the equilibrium electron concentration is derived:

$$n_0 = \int_{E_c}^{\infty} \frac{8\pi \cdot m_n^*}{h^3} \sqrt{2m_n^*(E - E_c)} \exp\left(\frac{E_F - E_c}{kT}\right) \exp\left(\frac{E_c - E}{kT}\right) dE \qquad (6.57)$$

$$n_0 = \frac{8\pi \cdot m_n^*}{h^3} \sqrt{2m_n^*} \exp\left(\frac{E_F - E_c}{kT}\right) \int_{E_c}^{\infty} (E - E_c)^{1/2} \exp\left(\frac{E_c - E}{kT}\right) dE \qquad (6.57a)$$

$$n_0 = \frac{8\pi \cdot m_n^*}{h^3} \sqrt{2m_n^*} \exp\left(\frac{E_F - E_c}{kT}\right) (kT)^{1/2} \int_{E_c}^{\infty} \left(\frac{E - E_c}{kT}\right)^{1/2} \exp\left(\frac{E_c - E}{kT}\right) dE$$

$$(6.57b)$$

Let an arbitrary variable $\xi = \dfrac{E - E_c}{kT}$ and then $d\xi = \dfrac{dE}{kT}$ $\qquad (6.57c)$

$$n_0 = \frac{8\pi \cdot m_n^*}{h^3} \sqrt{2m_n^*} \exp\left(\frac{E_F - E_c}{kT}\right) (kT)^{3/2} \int_{E_c}^{\infty} \left(\frac{E - E_c}{kT}\right)^{1/2} \exp\left(\frac{E_c - E}{kT}\right) \frac{dE}{kT}$$

$$(6.57d)$$

$$n_0 = \frac{8\pi \cdot m_n^*}{h^3} \sqrt{2m_n^*} \exp\left(\frac{E_F - E_c}{kT}\right) (kT)^{3/2} \int_{0}^{\infty} \xi^{1/2} \exp(\xi) d\xi \qquad (6.57e)$$

Equation (6.58) is the gamma function, known to be solvable as

$$\int_{0}^{\infty} \xi^{1/2} \exp(\xi) d\xi = \frac{\sqrt{\pi}}{2} \qquad (6.58)$$

Substituting Equation (6.58) into Equation (6.57e), the equilibrium electron concentration is finally obtained as

$$n_0 = N_c \exp\left(\frac{E_F - E_c}{kT}\right) \qquad (6.59)$$

where N_c is called the effective density of state of the conduction band and defined as

$$N_c = 2\left(\frac{2\pi m_n^* kT}{h^2}\right)^{\frac{3}{2}} \qquad (6.60)$$

As for the equilibrium electron concentration, Equation (6.56), the equilibrium hole concentration can be expressed but with the different form of the Fermi function now counting for the energies lower than the upper edge of the valence band.

$$p_0 = \int_{\infty}^{E_c} g_v(E)(1 - f_e(E)) dE \qquad (6.61)$$

The equilibrium hole concentration in the valence band is obtained in a similar way as the equilibrium electron concentration.

$$p_0 = N_v \exp\left(\frac{E_v - E_F}{kT}\right) \tag{6.62}$$

where N_v is called the effective density of state of the valence band and defined as

$$N_v = 2\left(\frac{2\pi m_p^* kT}{h^2}\right)^{\frac{3}{2}} \tag{6.63}$$

6.4.4   Intrinsic Semiconductors

An *intrinsic semiconductor* is one that contains a negligibly small amount of impurities compared with thermally generated electrons and holes.

We consider a volume of an *intrinsic* semiconductor under an equilibrium condition at room temperature, as shown in Figure 6.21. The intrinsic material indicates that the material is pure—it has no impurities and is *undoped* (will be explained later). And the equilibrium condition implies no external electric field or no sunlight. However, the thermal energy at room temperature is enough to excite some electrons from the upper edge of the valence band to the lower edge of the conduction band, leaving an equal number of holes in the valence band. The electrons in the conduction band, as well as the holes in the valence band, are freely moving but they are allowed to stay on the quantum states, defined previously in Equation (6.59). If the energy band E_g is small enough, substantial numbers of electrons will be excited in this way.

The equilibrium carrier concentration is graphically illustrated in an energy band diagram of Figure 6.21, where the lower edge of the conduction band E_c and the upper edge of the valence band E_v are indicated. The band gap or forbidden gap is simply the energy difference between E_c and E_v. The Fermi energy E_F in this case is very close to the middle of the band gap. It is interesting to note that the equilibrium electron concentration n_0 is mathematically confined to both the density

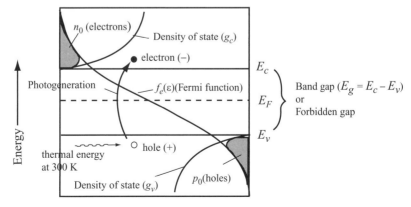

Figure 6.21   Band energy diagram under equilibrium conditions in an intrinsic (undoped) semiconductor at room temperature

of states $g_c(E)$, Equation (6.48), and the Fermi function $f_e(E)$, Equation (6.50), as shown in Figure 6.21. The hole concentration p_0 has the same value as the electron concentration.

6.4.5   Intrinsic Carrier Concentration, n_i

In nondegenerate semiconductors, as discussed in Section 6.4.2, the product of the equilibrium electron and hole concentration, Equation (6.59) and (6.62), is independent of the location of the Fermi energy and is just dependent of the band energy E_g and temperature as

$$n_0 p_0 = N_c N_v \exp\left(\frac{E_F - E_c}{kT}\right) \exp\left(\frac{E_v - E_F}{kT}\right)$$

$$= N_c N_v \exp\left(\frac{E_v - E_c}{kT}\right) = N_c N_v \exp\left(\frac{-E_g}{kT}\right) \tag{6.64}$$

In an intrinsic semiconductor in thermal equilibrium, the number of electrons in the conduction band and the number of holes in the valence band are equal, so that

$$n_0 = p_0 = \sqrt{n_0 p_0} \tag{6.65}$$

The *intrinsic carrier concentration* n_i is defined as

$$n_i = n_0 = p_0 = \sqrt{n_0 p_0} = \sqrt{N_c N_v \exp\left(\frac{-E_g}{kT}\right)} = \sqrt{N_c N_v} \exp\left(\frac{-E_g}{2kT}\right) \tag{6.66}$$

It is noted that n_i is dependent on E_g and T but independent of E_F, E_c and E_v. The intrinsic carrier concentration at thermal equilibrium is an important property, basically dependent on the material, which will be useful in many analyses.

Example 6.4.1 Intrinsic Carrier Concentration   Considering a pure silicon semiconductor in equilibrium at room temperature of 300 K, we want to estimate the number of free electrons and holes per cubic centimeter in the semiconductor for a study of Si solar cells.

Solution   The problem is to determine the intrinsic carrier concentration for Si because of Equation (6.66). The N_c and N_v are evaluated using Equations (6.60) and (6.63) as

$$N_c = 2\left(\frac{2\pi(1.18)\left(9.11 \times 10^{-31}\text{kg}\right)\left(1.38066 \times 10^{-23}\text{ J/K}\right)(300\text{ K})}{\left(6.626 \times 10^{-34}\text{ J}\cdot\text{s}\right)^2}\right)^{\frac{3}{2}}$$

$$= 3.2 \times 10^{19}\text{ cm}^{-3} \tag{6.67}$$

$$N_v = 2\left(\frac{2\pi(0.81)\left(9.11 \times 10^{-31}\text{ kg}\right)\left(1.38066 \times 10^{-23}\text{ J/K}\right)(300\text{ K})}{\left(6.626 \times 10^{-34}\text{ J}\cdot\text{s}\right)^2}\right)^{\frac{3}{2}}$$

$$= 1.8 \times 10^{19}\text{ cm}^{-3} \tag{6.68}$$

The intrinsic carrier concentration for Si semiconductors is calculated using Equation (6.66), Tables 6.6, and 6.7 as

$$n_i = \sqrt{\left(3.2 \times 10^{19} \text{ cm}^{-3}\right)\left(1.8 \times 10^{19} \text{ cm}^{-3}\right)} \exp\left(\frac{-1.12 \text{ eV}}{2\left(8.617 \times 10^{-5} \text{ eV/K}\right)(300 \text{ K})}\right)$$

$$\approx 0.95 \times 10^{10} \text{ cm}^{-3} \tag{6.69}$$

In Si there are 5×10^{22} atoms/cm^3 and four bonds per atom, making a grand total of 2×10^{23} bonds or valence band electrons per cm^3. Since $n_i = 10^{10}$ cm^{-3}, one finds less than one bond out of 10^{13} broken in Si at room temperature.

6.4.6   Intrinsic Fermi Energy

The *intrinsic Fermi energy* is defined as an energy level close to the midgap, being the Fermi energy in intrinsic semiconductors in thermal equilibrium. Hence, the intrinsic Fermi energy in thermal equilibrium is said to be

$$E_i = E_F \tag{6.70}$$

Combining Equations (6.59) and (6.62) with substituting E_i in the place of E_F gives

$$N_c \exp\left(\frac{E_i - E_c}{kT}\right) = N_v \exp\left(\frac{E_v - E_i}{kT}\right) \tag{6.71}$$

This can be reduced to

$$\frac{2E_i - (E_v + E_c)}{kT} = \ln\left(\frac{N_v}{N_c}\right) \tag{6.72}$$

The intrinsic Fermi energy is obtained as

$$E_i = \frac{1}{2}(E_v + E_c) + \frac{1}{2}kT \ln\left(\frac{N_v}{N_c}\right) \tag{6.73}$$

The two terms of Equation (6.73) are estimated for Si semiconductors as

$$\frac{1}{2}(E_c + E_v) \geq \frac{1}{2}(E_c - E_v) = \frac{1}{2}E_g = \frac{1}{2}1.12 \text{ eV} = 0.56 \text{ eV} \tag{6.74}$$

$$\frac{1}{2}kT \ln\left(\frac{N_v}{N_c}\right) = \frac{1}{2}(8.617 \times 10^{-5} \text{ eV/K})(300 \text{ K}) \ln\left(\frac{1.8 \times 10^{19} \text{ cm}^{-3}}{3.2 \times 10^{19} \text{ cm}^{-3}}\right) = -0.0073 \text{ eV} \tag{6.75}$$

We conclude that E_i is very close to the midpoint of band gap E_g, since 0.56 eV in Equation (6.74) is much greater than -0.0073 eV in Equation (6.75). The intrinsic carrier concentration is typically very small compared to the densities of states, and typical doping densities and intrinsic semiconductors behave very much like insulators—that is, they are not very useful as conductors of electricity.

6.4.7 Alternative Expression for n_0 and p_0

Although in closed form, the relationship of Equations (6.59) and (6.62) are not in the simplest form possible, and, more often than not, it is the simpler alternative form of these relationships that one encounters in device analyses. The alternative-form relationships can be obtained by recalling that E_i lies close to the midpoint of the band gap, and hence Equations (6.59) and (6.62) most assuredly apply to an intrinsic semiconductor. If this is the case, then we can specialize Equations (6.59) and (6.62) to an intrinsic semiconductor. Using Equations (6.66) and (6.70), we obtain

$$n_0 = n_i = N_c \exp\left(\frac{E_i - E_c}{kT}\right) \tag{6.76}$$

and

$$p_0 = n_i = N_v \exp\left(\frac{E_v - E_i}{kT}\right) \tag{6.77}$$

Solving these for N_c and N_v, we have

$$N_c = n_i \exp\left(\frac{E_c - E_i}{kT}\right) \tag{6.78}$$

$$N_v = n_i \exp\left(\frac{E_i - E_v}{kT}\right) \tag{6.79}$$

Finally, eliminating N_c and N_v in Equations (6.59) and (6.62) gives

$$n_0 = N_c \exp\left(\frac{E_F - E_c}{kT}\right) = n_i \exp\left(\frac{E_c - E_i}{kT}\right) \exp\left(\frac{E_F - E_c}{kT}\right) \tag{6.80}$$

$$n_0 = n_i \exp\left(\frac{E_F - E_i}{kT}\right) \tag{6.81}$$

Likewise,

$$p_0 = N_v \exp\left(\frac{E_v - E_F}{kT}\right) = n_i \exp\left(\frac{E_i - E_v}{kT}\right) \exp\left(\frac{E_v - E_F}{kT}\right) \tag{6.82}$$

$$p_0 = n_i \exp\left(\frac{E_i - E_F}{kT}\right) \tag{6.83}$$

When the intrinsic material is in thermal equilibrium such as $E_i = E_F$, it is always true that $n_0 = p_0 = n_i$, as discussed in Section 6.4.5.

It is interesting to notice that these equations are valid for any *extrinsic semiconductor* in equilibrium whose doping is such as to give rise to a nondegenerate positioning of the Fermi level (see Section 6.4.2 for nondegenerate semiconductors) (Pierret [10]).

6.5 EXTRINSIC SEMICONDUCTORS IN THERMAL EQUILIBRIUM

An *extrinsic semiconductor* is one that has been doped—that is, into which a doping agent has been introduced, giving it different electrical properties than the intrinsic (pure) semiconductor.

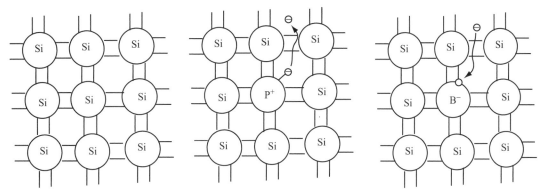

Figure 6.22   Visualization of (a) an intrinsic silicon bonding model, (b) a phosphorus atom (donor) is substituted for a silicon atom, and (c) a boron atom (acceptor) is substituted for a silicon atom, displayed using the bonding model

6.5.1   Doping, Donors, and Acceptors

Doping is a process of intentionally introducing impurities into an extremely pure (also referred to as *intrinsic*) semiconductor to change its electrical properties. Some impurities, or *dopants*, are generally added initially as a crystal ingot is grown, giving each wafer (a thin slice of semiconducting material) an almost uniform initial doping. Further doping can be carried out by such processes as *diffusion* or *ion implantation*, the latter method being more popular in large production runs due to its better controllability. The ion implantation is a process whereby a focused beam of ions (dopants) is directed toward a target wafer. The ions have enough kinetic energy that they can penetrate into the wafer upon impact.

An intrinsic semiconductor is illustrated in Figure 6.22(a) using the bonding model. The extrinsic n-type and p-type semiconductors are illustrated in Figure 6.22(b) and 6.22(c), respectively. The number of electrons and holes in their respective bands can be controlled through the introduction of specific impurities, or dopants, called *donors* and *acceptors*. For example, when a silicon semiconductor is doped with phosphorus, one electron is donated to the conduction band for each atom of phosphorus introduced. From Table 6.8 we can see that phosphorus, in column V of the periodic table of elements, has five valence electrons. Four of these are used to satisfy the four covalent bonds of the silicon lattice, and the fifth is available to fill an empty state in the conduction band, which is illustrated in Figure 6.22(b). If silicon is doped with boron (three valence electrons, since it is in column III), each boron atom

Table 6.8   Abbreviated Periodic Table of the Elements

I	II	III	IV	V	VI
		B	C	N	O
		Al	Si	P	S
Cu	Zn	Ga	Ge	As	Se
Ag	Cd	In	Sn	Sb	Te

accepts an electron from the valence band, leaving behind a hole, which is also shown in Figure 6.22(c).

The unbounded electrons of donors and the unbounded holes of acceptors are very susceptible and are easily excited to produce free (ionized) electrons and holes, even with a thermal energy at room temperature. A donor (phosphorus) in Figure 6.22(b) yields a negatively charged electron and the donor itself becomes a positive charge (because of losing a negative electron from a neutral state). An acceptor (boron) yields a positively charged hole and the acceptor itself becomes a negative charge (called ionized acceptor). The positively charged donors and negatively charged acceptors are immobile due to the covalent bonding with silicon atoms and thus create an electrostatic potential, which is called a *built-in voltage*. As a result, the doping directly creates a *built-in voltage* between the n-type material and p-type material.

This built-in voltage leads to a region that is called a *space charge region* or a *depletion region* at the *p–n junction,* which is shown in Figure 6.23. The p–n junction with the built-in voltage separates the carriers: the electrons in the n-type region and the holes in the p-type region.

6.5.2   Extrinsic Carrier Concentration in Equilibrium

Under an equilibrium condition (no sunlight) at room temperature, all impurities in the n-type and p-type regions introduce localized electric states (built-in voltage) into the band structure within the *space charge region* between E_c and E_v, which is shown in Figure 6.23. Both the electrons n_{N0} in the n-type region and the holes p_{P0} in the p-type region become *majority carriers*, while both the holes p_{N0} in the n-type region and the electrons n_{P0} in the p-type region become *minority carriers*. Often, the donors and acceptors are assumed to be completely ionized so that $n_{N0} \approx N_D$ in

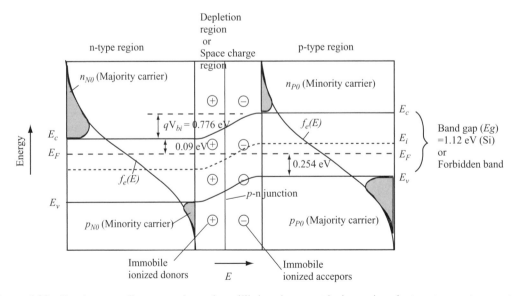

Figure 6.23   Band energy diagram at thermal equilibrium in an extrinsic semiconductor at room temperature

n-type material and $p_{N0} \approx N_A$ in p-type material. The majority *electron concentration in n-type semiconductor* in thermal equilibrium is expressed using Equation (6.59):

$$n_{N0} = N_D = N_c \exp\left(\frac{E_F - E_c}{kT}\right) \tag{6.84}$$

where N_D is the donor impurity concentration (the number of donors per unit volume). After taking logarithms, the Fermi energy in n-type region is obtained:

$$E_F = E_c + kT \ln\left(\frac{N_D}{N_c}\right) \tag{6.85}$$

Likewise, the majority *hole concentration in p-type semiconductor* in thermal equilibrium is expressed using Equation (6.62) by

$$p_{P0} = N_A = N_v \exp\left(\frac{E_v - E_F}{kT}\right) \tag{6.86}$$

where N_A is the acceptor impurity concentration (the number of acceptors per unit volume). After taking logarithms, the Fermi energy in p-type region is obtained:

$$E_F = E_v - kT \ln\left(\frac{N_A}{N_v}\right) \tag{6.89}$$

We want to see the numerical values for the energy diagram in Figure 6.23. Therefore, the realistic donor and acceptor concentration often used in the industry are given as

$$N_D = 1 \times 10^{18} \text{ cm}^{-3} \quad \text{and} \quad N_A = 1 \times 10^{15} \text{ cm}^{-3} \tag{6.90}$$

Then, evaluating the Fermi energy in the n-type region using Equation (6.85) as

$$E_F = E_c + (8.617 \times 10^{-5} \text{ eV/K})(300 \text{ K}) \ln\left(\frac{10^{18} \text{ cm}^{-3}}{3.2 \times 10^{19} \text{ cm}^{-3}}\right) = E_c - 0.09 \text{ eV} \tag{6.91}$$

In the same way, evaluating the Fermi energy in the p-type region using Equation (6.89) as

$$E_F = E_v - \left(8.617 \times 10^{-5} \text{ eV/K}\right)(300 \text{ K}) \ln\left(\frac{10^{15} \text{ cm}^{-3}}{1.8 \times 10^{19} \text{ cm}^{-3}}\right) = E_v + 0.254 \text{ eV} \tag{6.92}$$

These values are used in Figure 6.23, where it is noted that the Fermi energy E_F does not change over the n-type and p-type regions, since the system is in thermal equilibrium. However, according to Equation (6.91), E_c in the n-type region must be located at 0.09 eV above E_F, while E_v in the p-type region must be located at 0.254 eV below E_F. Actually, since we know the band gap of 1.12 eV is a property of silicon, as shown in Table 6.6, we can calculate the rest of the band energy levels. For instance, the difference between E_c in the n-type and E_c in the p-type regions can be obtained graphically, which is called the *built-in voltage* V_{bi}. The energy level (electron volt

eV) of the built-in voltage can be obtained by multiplying the built-in voltage with an electron charge q. From Figure 6.23, the built-in voltage times the charge becomes energy and obtained as follows:

$$qV_{bt} = (E_c - E_v)_P - (E_F - E_v)_P - (E_c - E_F)_N \tag{6.93}$$

The numerical values are obtained as

$$qV_{bi} = E_g - 0.254\,\text{eV} - 0.09\,\text{eV} = 1.12\,\text{eV} - 0.254\,\text{eV} - 0.09\,\text{eV} = 0.776\,\text{eV} \tag{6.93a}$$

The result of Equation (6.93a) is also used in Figure 6.23. It is found that only the change of energy is important in the analyses of solar cells, rather than the absolute values that depends on the reference level. Doping in the n-type and p-type regions creates the built-in voltage at room temperature across the space charge region, even though no sunlight is introduced. The built-in voltage turned out to be the maximum voltage attainable in the operation of solar cells even under sunlight. The built-in voltage is the primary mechanism to separate the electrons in the n-type region from the holes in the p-type region. It is noted that E_i is located close to the midgap along with the band edges, as defined in Equation (6.73).

6.5.3   Built-in Voltage

An explicit expression for the built-in voltage can also be derived from Equation (6.93) using Equations (6.85) and (6.89) with the band gap ($E_c - E_v = E_g$).

$$
\begin{aligned}
qV_{bi} &= (E_c - E_v)_P - (E_F - E_v)_P - (E_c - E_F)_N \\
&= E_g - \left(-kT \ln\left(\frac{N_A}{N_v}\right)\right) - \left(-kT \ln\left(\frac{N_D}{N_c}\right)\right)
\end{aligned}
\tag{6.94}
$$

E_g can be expressed in terms of n_i from Equation (6.66).

$$E_g = 2kT \ln\left(\frac{\sqrt{N_c N_v}}{n_i}\right) \tag{6.95}$$

Substituting this into Equation (6.94) and using logarithms gives

$$qV_{bi} = kT \ln\left(\frac{\sqrt{N_c N_v}}{n_i}\right)^2 + kT \ln\left(\frac{N_A}{N_v}\right) + kT \ln\left(\frac{N_D}{N_c}\right) \tag{6.96}$$

which reduces to the built-in voltage as

$$V_{bi} = \frac{kT}{q} \ln\left(\frac{N_D N_A}{n_i^2}\right) \tag{6.97}$$

where q is the electron charge ($q = 1.6012 \times 10^{-19}$ C or $1\,\text{eV/V}$) (see Table 6.3). It is interesting to note that the built-in voltage is only dependent of the number of dopants, the equilibrium carrier concentration n_i, and the temperature.

Electrons in the n-type region are called the majority carriers, while electrons in the p-type region are called the minority carriers. A majority electron becomes a minority electron when transiting from the n-type region. After some mean-statistical time τ_n (electron lifetime in the p-type semiconductor), the minority electrons have to disappear due to recombination with holes of this region, so new electrons can diffuse from the n-type region. However, at the places where electrons leave the n-type region, the immobile ionized donor atoms are left (similarly, at the places of going away holes, immobile ionized acceptor atoms in the p-type region are left).

Ionized donor and acceptor atoms form the *p-n junction space charge region*, or *depletion region*, which inhibits the directed motion of the charge carriers in the thermodynamic equilibrium. The immobile ionized donor and acceptor atoms produce a built-in voltage V_{bi}, which limits the diffusion of the holes and electrons. In thermal equilibrium, the *diffusion* and *drift* currents exactly balance, so there is no net current flow.

6.5.4   Principle of Detailed Balance

Under equilibrium conditions, each fundamental process and its inverse must self-balance independent of any other process that may be occurring inside the material. This is known as the *principle of detailed balance*.

6.5.5   Majority and Minority Carriers in Equilibrium

As for the intrinsic semiconductor in Section 6.4.5, at thermal equilibrium, the product of the majority (electron) concentration and the minority (hole) concentration in an n-type semiconductor remains constant according to the principle of detailed balance, so it should be equal to the intrinsic carrier concentration n_i^2. In an n-type semiconductor, we have

$$n_{N0}p_{N0} = n_i^2 \qquad (6.98)$$

From Equation (6.84), it is assumed that $n_{N0} = N_D$. The majority concentration in the n-type region is

$$p_{N0} = \frac{n_i^2}{N_D} \qquad (6.99)$$

Likewise, in an p-type semiconductor, we have

$$n_{P0}p_{P0} = n_i^2 \qquad (6.100)$$

For the same reason, we have $p_{P0} = N_A$. The minority concentration in the p-type region is

$$n_{P0} = \frac{n_i^2}{N_A} \qquad (6.101)$$

Using the given information in Equation (6.90), we calculate the numerical values. The majority electron concentration in the n-type region at thermal equilibrium is the same as the number of donors

$$n_{N0} = N_D = 1 \times 10^{18} \text{ cm}^{-3} \qquad (6.102a)$$

The minority electron concentration in the p-type region is obtained by

$$n_{P0} = \frac{n_i^2}{N_A} = \frac{(0.95 \times 10^{10} -\text{cm}^3)^2}{1 \times 10^{15} \text{ cm}^{-3}} = 9.025 \times 10^4 \text{ cm}^{-3} \qquad (6.102b)$$

It is noted that the minority carrier concentration at thermal equilibrium is much less than the majority carrier concentration, due to the dopants. As discussed later, that minority carrier concentration is sensitively affected by generation and recombination, while majority carrier concentration remains unperturbed. This is the main reason why the analysis of a solar cell is based on the minority carriers rather than majority carriers.

6.6   GENERATION AND RECOMBINATION

Generation is a process by which an electron in the valence band absorbs energy and is promoted into the conduction band, leaving behind a hole in the valence band. In other words, generation creates an electron-hole pair in each band. *Recombination* is a process by which an electron and a hole annihilate each other, releasing the same amount of energy as the difference between the initial stage of the electron in the conduction band and the final stage of the electron in the valence band. They both disappear after recombination. The energy released can be given up as a photon, as heat through phonon emission, or as kinetic energy to another free carrier.

For every generation process there is an equivalent recombination process, which is an important principle in understanding the function of solar cells. Generation and recombination are an intrinsic nature of semiconductor materials and are perfectly balanced in thermal equilibrium whether the material is intrinsic or extrinsic. One of the goals is to manage the generated carriers so as to produce as much as electricity before the ultimate (desirable) recombination while minimizing the intermediate (undesirable) recombination.

6.6.1   Direct and Indirect Band Gap Semiconductors

The type of band gap material is important in semiconductors for the selection of material for many electronic devices including solar cells. There are two types of band gap materials in semiconductors, which are *direct* and *indirect* band gap materials. As we discussed in Section 6.2.7, the energy E of a particle is always associated with the momentum p, which implies that, for any transition between bands, both energy and momentum must be conserved. As discussed before, the fundamental absorption takes place when an electron is excited by absorption of a photon. The direct transition involves only a photon, while the indirect transition involves one or more phonons as the photon is absorbed. The *phonon* is associated with vibration or momentum of the crystal lattice.

In *direct band gap semiconductors*, such as GaAs, GaInP, CdTe, and Cu(InGa)Se$_2$, the maximum and minimum of energy versus momentum relationship occur at the same value of the momentum (Figure 6.24(a)), so that a photon energy that has a high energy but a negligible momentum due to zero mass is sufficient for an electron to make a transition from the valence band to the conduction band.

In *indirect band gap semiconductors* like Si and Ge, the maximum and minimum of the energy versus momentum relationship occurs at a different crystal momentum, which is pictured in Figure 6.24(b). An electron transition between the maximum of the valence band and the minimum of the conduction band is not possible with only the absorption of a photon, because the momentum of the photon is too small.

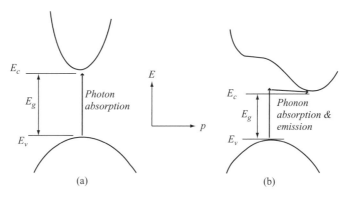

Figure 6.24 Energy versus momentum diagrams in (a) direct and (b) indirect band gap semiconductors

Nevertheless, the photon energy converts to a phonon (crystal lattice vibration) that has momentum enough to make the electron transition from the maximum of the valence band to the minimum of the conduction band. Thus, in an indirect band gap semiconductor, the absorption coefficient $\alpha(\lambda)$ is substantially less than those in the direct band gap semiconductors. This is reflected in Figure 6.25, where Si shows the lowest absorption coefficient, among others.

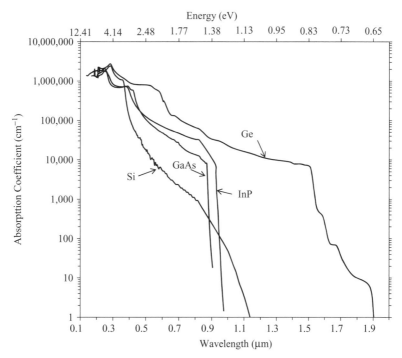

Figure 6.25 Absorption coefficients as a function of wavelength and energy in Si and other selected semiconductors. Modified from the source of Handbook of Optical Constant of Solids, edited by Edward D. Palik (1985), New York: Academic Press [9]

This absorption coefficient directly affects the generation of the electron-hole pairs. However, the direct and indirect transitions occur exactly in an opposite way in recombination, also. Hence, the net gain would depend on the balance of generation and recombination, which may be a good topic for further investigation.

6.6.2   Absorption Coefficient

The absorption coefficient determines how far into a material the light of a particular wavelength can penetrate before it is absorbed. In a material with a low absorption coefficient, light is only poorly absorbed, and if the material is thin enough, it will appear transparent to that wavelength. Semiconductor materials have a sharp edge in their absorption coefficient, since light that has energy below the band gap does not have sufficient energy to raise an electron across the band gap. Consequently, this light is not absorbed. The absorption coefficients for several semiconductors are shown in Figure 6.25. The dependence of absorption coefficient on wavelength causes different wavelengths to penetrate different lengths into a semiconductor before most of the light is absorbed. The absorption length is given by the inverse of the absorption coefficient.

The absorption coefficient $\alpha(\lambda)$ depends on the band structure of the semiconductor. As discussed in the foregoing section, the indirect band gap materials like Si and Ge are compared with the direct band gap materials such as GaAs, InP, and InGaAs, as shown in Figure 6.25. The figure shows the absorption spectra of a number of semiconductors, which are important for solar cells. Notice the absorption edge for the direct band gap materials, silicon, and germanium. Notice how the shape of the curves for GaAs and InP are similar; this is due to their similar crystal structure. It is noted that the band gap of 1.12 eV for Si and the band gap of 1.42 eV for GaAs listed in Table 6.6 are approximately in good agreement in the scale of energy in the figure. Energy lower than about 1.12 eV (or wavelength greater than about 1.1 μm) for Si provides no contribution to the absorptivity. It is our understanding that, when we compare the spectral incident solar radiation in Figure 6.6 with the absorption coefficient of Si in Figure 6.25, the significant portion of the irradiation larger than the wavelength of 1.1 μm is wasted in the case of silicon solar cells.

Half of the solar photons have a wavelength smaller than 1 μm and the other half have a wavelength larger than 1 μm. Hence, $\lambda = 1$ μm is typical solar photon wavelength that corresponds to a photon energy of 1.24 eV. At this energy, Si has an optical absorption length of 156 μm. The optical absorption length is the inverse of the Si absorption coefficient, α.

The absorption length of a solar cell material is a useful quantity. A cell thickness of several hundred μm is thus required for complete optical absorption. That is one reason why conventional Si wafer cells have a thickness of about 300 μm. The other reason is that a thickness of 300 μm permits safe handling of Si wafers. It is also seen in Figure 6.25 that, for instance, at the photon energy of 1.40 eV, the value of α for GaAS is approximately 10^4 cm^{-1}, which comes up with the optical absorption length of 1 μm. This 1 μm for GaAs is very small compared with the 300 μm for Si.

In practice, measured absorption coefficients or empirical expressions for absorption coefficients are used in analysis and modeling.

6.6.3   Photogeneration

Light striking the surface of a solar cell will be partially reflected and partially transmitted into the material. Neglecting the reflection, the amount of light that is absorbed by the material depends on the absorption coefficient (α in cm^{-1}) and the thickness of the absorbing material. The spectral intensity of light $f(x, \lambda)$ at any point in the device can be expressed by

$$f(x, \lambda) = f_0(\lambda)e^{-\alpha(\lambda)x} \tag{6.103}$$

where α is the absorption coefficient, typically in cm^{-1}, x is the distance into the material, and $f_0(\lambda)$ is the spectral incident light intensity at the top surface. Assuming that the loss in light intensity (i.e., the absorption of photons) directly causes the generation of electron-hole pairs, then the spectral photogeneration $G(x, \lambda)$ in a thin slice of material is determined by finding the change in the spectral light intensity across this slice. Consequently, differentiating Equation (6.103) will give the spectral generation at any point in the device. Hence,

$$G(x, \lambda) = -\frac{df(x, \lambda)}{dx} = f_0(\lambda)\alpha(\lambda)e^{-\alpha(\lambda)x} \tag{6.104}$$

If a photon's energy is greater than the band gap energy of the semiconductor, then the light will be absorbed and electron-hole pairs will be created as the light passes through the semiconductor. By integrating up to the wavelength corresponding to the band gap and including the reflectivity and the grid-shading factor, the *photogeneration rate*, the rate of creation of electron-hole pairs (a number/cm^2-sec/cm), is expressed as a function of position within a solar cell.

$$G(x) = (1 - s) \int_{\lambda} (1 - r(\lambda))f_0(\lambda)\alpha(\lambda)e^{-\alpha(\lambda)x} d\lambda \tag{6.105}$$

where s is the grid-shading factor, $r(\lambda)$ is the refection, $\alpha(\lambda)$ is the absorption coefficient, and $f_0(\lambda)$ is the incident photon flux (number of photon incident per unit area per second per wavelength). The photon flux, $f_0(\lambda)$, is obtained by dividing the incident power density (W/cm^2-μm) at each wavelength by the photon energy (J).

6.7 RECOMBINATION

6.7.1 Recombination Mechanisms

Recombination may be grouped into two groups, which are bulk recombination and surface recombination. The bulk recombination mechanisms includes band-to-band (radiative) recombination, trap-assisted (defects) recombination in the forbidden gap, and Auger recombination, which are depicted in Figure 6.26(a), (b), and (c).

Band-to-band recombination, also referred to as radiative recombination, is conceptually the simplest of all recombination processes. As pictured in Figure 6.26(a), it merely involves the direct annihilation of a conduction band electron and a valence band hole, the electron falling from an allowed conduction band state into a vacant valence band state. This process is typically radiative, with the excess energy released during the process going into the production of a photon (light).

Trap-assisted recombination, as shown in Figure 6.26(b), usually occurs due to the impurities or defects in the materials that result in *trap stages* (energy levels) in the band gap. A free electron in the conduction band edge sometime loses energy for a certain reason being captured by a trap in the mid-band gap, which can subsequently be released

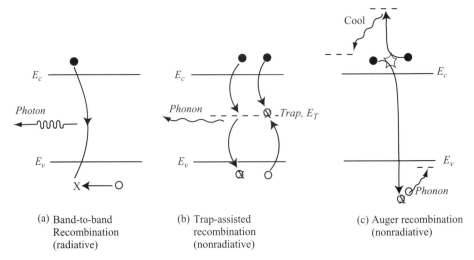

Figure 6.26 Energy-band visualization of bulk recombination processes: (a) Band-to-band recombination, (b) trap-assisted recombination, and (c) Auger recombination

to recombine with a hole in the valence band edge. Alternatively, if the trap captures a hole from the valence band edge before the captured electron is released to the valence band, then the two carriers will, in effect, recombine and the trap will be emptied again, then releasing a phonon (nonradiative) as the result of the recombination. This is pictured in Figure 6.26(b). This is one of the reasons why a pure and nondefective crystal is required for a high-efficiency solar cell.

In *Auger recombination*, a collision between two similar electrons results in the excitation of one electron to a higher kinetic energy and the relaxation of the other electron to a lower kinetic energy. This is pictured in Figure 6.26(c). The electron that has the lower energy directly falls into the valence band and recombines with a hole, releasing extra energy as phonons. Ultimately, that extra energy will be lost as heat. Auger recombination becomes important in heavily doped materials.

Band-to-band recombination and Auger recombination are usually unavoidable while trap-assisted recombination is avoidable, because the trap states are due to impurities in the crystal or defects in the crystal structure. Auger recombination is important in low band gap materials with high carrier densities, where carrier-carrier interactions are stronger. In indirect band gap materials, band-to-band recombination is suppressed (see Section 6.6.1) while auger recombination is important. Trap-assisted recombination is common in all types of materials and plays an important role in the operation of solar cells.

Surface recombination is the annihilation of carriers in the near vicinity of a semiconductor surface via the interaction with the interfacial traps. The front surface of a solar cell has a high concentration of defects due to the abrupt termination of the crystal lattice (dangling silicon bonds). These manifest themselves as a continuum of traps within the forbidden gap at the surface. Electrons and holes can recombine through them as with bulk recombination traps, which are shown in Figure 6.27. Surface recombination can be usually minimized by increasing the doping.

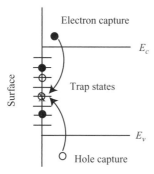

Figure 6.27 Surface recombination

6.7.2 Band Energy Diagram under Nonequilibrium Conditions

We now consider an Si solar cell shown earlier in Figure 6.2 under sunlight, which is now viewed by rotating 90° counterclockwise and is presented in a band energy diagram under nonequilibrium conditions, as shown in Figure 6.28. Sunlight is now illuminating the left-side surface of the solar cell. This is also a continuous discussion with the band energy diagram in *thermal equilibrium* shown in Figure 6.23. Sunlight provides a definitely *nonequilibrium condition*.

Remember from the earlier discussion that doping in each region creates a built-in voltage, as illustrated in Figure 6.28. Photogeneration due to sunlight occurs mainly in the base because the thickness of the base is on the order of $300\,\mu m$, compared to the thickness of the emitter, which is on the order of $1\,\mu m$ (the figure is not to scale). Owing to the *absorption coefficient* of Si, an incident photon whose energy is greater than E_g gives up energy to the bonded electrons in the valence band edge at a somewhat deep point in the base to create an electron-hole pair—a free electron at the conduction band edge and a free hole at the valence band edge.

The conduction band edge E_c in the p-type region was found to have a higher energy level than that in the n-type region as shown in Figure 6.28, due primarily to the *built-in voltage*, allowing electrons in the p-type region to easily flow toward the n-type region as water flows from a high position to a low position. As a result, electrons become very dense in the n-type region while holes become very dense in the p-type region. The dense electrons in the n-type region, either donated or generated, are called majority carriers, while the scarce holes in the n-type region are called minority carriers. The same thing happens in the p-type region. The crowdedness in each region gives rise to an electrostatic potential between the *front contact* and the *rear contact* of the solar cell. This potential, or voltage, results in not only the electrons flowing in the wire through the external load for electricity but also a deviation energy qV associated with the voltage between the Fermi energy in the n-type region and the Fermi energy in the p-type regions. The electrons that lost energy at the external load return to the rear contact to eventually recombine with the dense holes in the p-type region, and the electron-hole pairs come to disappear, so that new electrons are able to be generated for their new cycle, which is schematically pictured in Figure 6.28.

6.7.2.1 *Back Surface Field (BSF)*

Back surface field (BSF) refers to a built-in electric field on the back side that causes a repulsion of the minority electrons approaching the back surface and reduces the

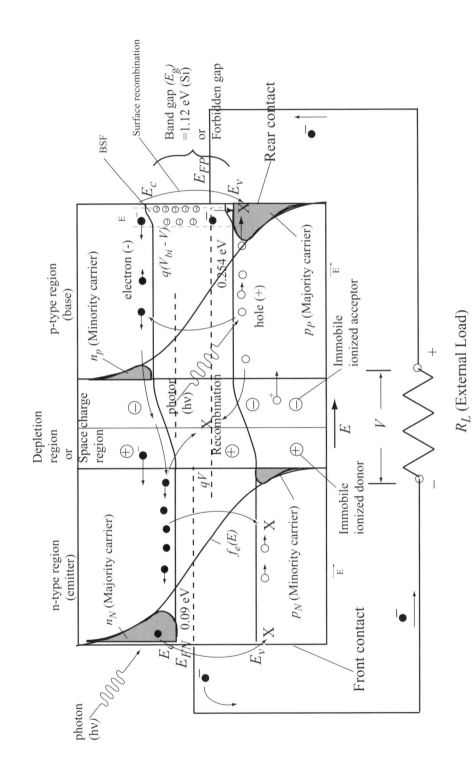

Figure 6.28 Band energy diagram under nonequilibrium conditions (sunlight)

428

recombination rate near the back surface. The BSF can be achieved practically by nonuniformly doping, heavier to the surface, as shown in Figure 6.28, which generates a local electric field and bends slightly the energy band curve, as shown in the figure. This causes the approaching electrons to turn around near the back surface, due to the *Coulomb's forces* (electrons in the opposite direction and holes in the same direction of the electric field). This improves the chance of reaching the emitter through the depletion region for the minority electrons. Most of the effect of the BSF is in the lower value of the dark saturation current I_0 and the significant improved value of the open-circuit voltage V_{OC}. The BSF can be reduced from an initial value of 10^7 cm/s to a passivated value as low as 100 cm/s. The BSF forms a P^+P junction near the back surface diffused by boron or aluminum. Aluminum is a p-type dopant in silicon because it has one less valence electron than silicon.

6.7.3 Low-Level Injection

The level of injection specifies the relative magnitude of changes in the carrier concentrations resulting from a perturbation (sunlight). *Low-level injection* is said to exist if the changes in the carrier concentration are much less than the majority carrier concentration under equilibrium conditions. Conversely, *high-level injection* exists if the changes are much greater than the majority carrier concentration in equilibrium. Low-level injection is viewed as a situation where the majority carrier concentration remains essentially unperturbed. We will use the following notation: n_N is the majority carrier concentration in an n-type region, p_N is the minority carrier concentration. n_P is the minority carrier concentration in p-type region, while p_P is the majority carrier concentration. If "0" is added in the subscript, it means equilibrium.

The *excess minority carrier concentration* in a p-type region is defined by

$$\Delta n_P = n_P - n_{P0} \tag{6.106}$$

The *excess minority carrier concentration* in an n-type region is defined by

$$\Delta p_N = p_N - p_{N0} \tag{6.107}$$

The excess carrier concentration means that an amount of carriers per unit volume generated by illumination is more than the amount of equilibrium carriers.

6.7.3.1 Low-Level Injection

$$\Delta p_N \ll n_{N0}, \quad n_N \cong n_{N0} \quad (\Delta p_N \cong \Delta n_N) \text{ in an n-type material} \tag{6.108}$$

$$\Delta n_P \ll p_{P0}, \quad p_P \cong p_{p0} \quad (\Delta n_P \cong \Delta p_P) \text{ in a p-type material} \tag{6.109}$$

Knowing that

$$n_{N0}p_{N0} = n_i^2 \quad \text{in an n-type material} \tag{6.110}$$

$$n_{P0}p_{P0} = n_i^2 \quad \text{in a p-type material} \tag{6.111}$$

For example, consider a Si semiconductor doped in thermal equilibrium at room temperature with the donor concentration of 10^{18} cm^{-3} subject to a perturbation (sunlight) where the minority carrier concentration in the n-type material is 10^9 cm^{-3}. We want to determine whether the situation is low-level injection. For the given material, we figure out that $n_{N0} = N_D = 10^{18}$ cm^{-3} means equilibrium majority carrier concentration is equal to donor concentration, and $p_N = 10^9$ cm^{-3} means perturbed minority concentration.

From Equation (6.110), the equilibrium minority carrier is obtained by

$$p_{N0} = \frac{n_i^2}{n_{N0}} = \frac{n_i^2}{N_D} = \frac{\left(10^{10}\right)^2}{10^{18}} = 10^2 \text{ cm}^{-3} \tag{6.112a}$$

Using Equation (6.107),

$$\Delta p_N = p_N - p_{N0} = 10^9 \text{ cm}^{-3} - 10^2 \text{ cm}^{-3} \cong 10^9 \text{ cm}^{-3} \tag{6.112b}$$

Therefore, according to Equation (6.108)

$$\Delta p_N = 10^9 \text{ cm}^{-3} \ll n_{N0} = 10^{18} \text{ cm}^{-3} \tag{6.112c}$$

The situation is clearly low-level injection. Under one-sun illumination, solar cells usually operate in low-level injection. However, 10 to 100 suns may be required to reach the high injection condition for concentrated solar cells.

6.7.4 Band-to-Band Recombination

Band-to-band recombination, also referred to as *radiative recombination*, is simply the inverse of the generation process. When band-to-band recombination occurs, an exited electron sometimes loses the energy, drops back to the valence band, and recombines with a hole, and finally emits a photon—this is how semiconductor lasers and light-emitting diodes operate. In an indirect band gap material, some of that energy is shared with a phonon or heat. The net recombination rate due to this radiative process is given by

$$R_{rad} = B_{rad}\left(np - n_i^2\right) \tag{6.113}$$

where B_{rad} is the radiative recombination coefficient that is the property of the material. The expression simplifies further for doped material. If we have a p-type semiconductor in *low-level injection*, the net radiative recombination rate is obtained as

$$R_{rad} = B_{rad}\left(n_p p_p - n_i^2\right) \tag{6.114}$$

Using $\Delta n_P \cong \Delta p_P$ in Equation (6.109) for low-level injection, the parenthesis in Equation (6.114) reduces to

$$\begin{aligned}
\left(n_P p_P - n_i^2\right) &= (n_{P0} + \Delta n_P)(p_{P0} + \Delta p_P) - n_{P0}p_{P0} \\
&= (n_{P0} + \Delta n_P)(p_{P0} + \Delta n_P) - n_{P0}p_{P0} \\
&= n_{P0}p_{P0} + n_{P0}\Delta n_P + \Delta n_P p_{P0} + \Delta n_P \Delta n_P - n_{P0}p_{P0} \\
&= (n_{P0} + p_{P0})\Delta n_P + (\Delta n_P)^2
\end{aligned} \tag{6.114a}$$

Know that $n_{P0} \ll p_{P0} \approx N_A$ (minority carrier is much less than minority carrier in Section 6.5.4) and $(\Delta n_P)^2 \ll p_{P0}\Delta n_P$ using Equation (6.109), Equation (6.114a) further reduces as

$$\left(n_P p_P - n_i^2\right) = (n_{P0} + p_{P0})\Delta n_P + (\Delta n_P)^2 = p_{P0}\Delta n_P = N_A(n_P - n_{P0}) \tag{6.114b}$$

Equation (6.114) can be rewritten as

$$R_{rad} = B_{rad}N_A(n_P - n_{P0}) = \frac{(n_P - n_{P0})}{\dfrac{1}{B_{rad}N_A}} \tag{6.114c}$$

Finally, the radiative recombination rate in a p-type material in low-level injection reduces to a simple expression.

$$R_{rad,P} = \frac{n_P - n_{P0}}{\tau_{rad,n}} = \frac{\Delta n_P}{\tau_{rad,n}} \tag{6.115}$$

where $\tau_{rad,n} = \frac{1}{B_{rad}N_A}$ is the minority carrier lifetime for electrons in p-type material.

In a similar way, the radiative recombination rate in an n-type material in low-level injection is obtained by

$$R_{rad,n} = \frac{p_N - p_{N0}}{\tau_{rad,p}} = \frac{\Delta p_N}{\tau_{rad,p}} \tag{6.116}$$

where $\tau_{rad,p} = \frac{1}{B_{rad}N_D}$ is the minority carrier lifetime for holes in n-type material.

The radiative lifetime can be measured from the time resolved spontaneous emission following instantaneous optical excitation of the semiconductor (called photoluminescence). B_{rad} can be determined experimentally from the variation of τ_{rad} with doping [12].

6.7.5 Trap-Assisted (SRH) Recombination

The *trap-assisted recombination*, also referred to as Shockley-Read-Hall (SRH) recombination, is associated with a trap state that is close to the mid-gap, as shown in Figure 6.29. The trap is also referred to as *recombination and generation* $(R - G)$ *center*. Physically, the traps are lattice defects or special impurities that are present even in semiconductors of the highest available purity. Actually, there could be many traps in the band gap, but a most effective single level trap is usually modeled in the trap-recombination to represent the trap recombination.

As discussed earlier in Section 6.6.4, there are two types of trap recombination, pictured in Figures 6.29(a) and (b). The trap recombination releases a phonon. The thermal energy can result in trap generation, as shown in Figure 6.29(b). Notice that

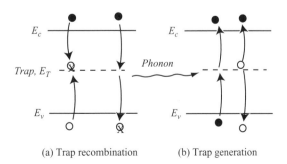

(a) Trap recombination   (b) Trap generation

Figure 6.29   Trap-assisted recombination: (a) trap recombination and (b) trap generation

trap generation occurs independently for electrons and holes. An electron in the valence band edge happens to move into the trap state (or level) in the mid-gap. The electron then receives the thermal energy in the band gap released during the previous trap recombination and jumps into the conduction band edge. In contrast, a hole can be generated when a hole in the trap state drops into the valence band edge, thus generating a hole. The *SRH net electron and hole recombination rates* can be obtained as

$$r_n = c_n p_T n - c_n n_T n_1 \tag{6.117}$$

$$r_p = c_p n_T p - c_p p_T p_1 \tag{6.118}$$

where c_n and c_p are the electron capture coefficient and the hole capture coefficient (cm^3/sec), respectively. n_T and p_T are the number of traps per cm^3 that are filled with electrons and the number of traps per cm^3 that are empty, respectively. n_1 and p_1 are defined as

$$n_1 = n_i e^{(E_T - E_i)/kT} \tag{6.119}$$

$$p_1 = n_i e^{(E_i - E_T)/kT} \tag{6.120}$$

For an approximate evaluation of these parameters, we find that $n_1 = n_0$, $p_1 = p_0$, and eventually $n_1 = p_1 = n_i$ if we assume that $E_F \approx E_T$ (the trap state is close to the Fermi level) using Equations (6.82), (6.83) and (6.66).

For steady-state conditions according to the principle of detailed balance in Section 6.5.4, we can set the equations to be equal as

$$r_n = r_p \tag{6.121}$$

And we find the total number of trap states N_T by summing n_T and p_T.

$$N_T = n_T + p_T \tag{6.122}$$

Combining Equations (6.119) to (6.122) and after some algebra, we obtain

$$n_T = \frac{c_n N_T n + c_p N_T p_1}{c_n (n + n_1) + c_p (p + p_1)} \tag{6.123}$$

Substituting this into Equation (6.117) using Equation (6.122), after some algebra, yields

$$R = r_n = r_p = \frac{np - n_1 p_1}{\dfrac{1}{N_T c_p} (n + n_1) + \dfrac{1}{N_T c_n} (p + p_1)} \tag{6.124}$$

The product of n_1 and p_1 in Equations (6.119) and (6.120) can be obtained as

$$n_1 p_1 = n_i^2 \tag{6.125}$$

The SRH trap recombination rate under steady state conditions is finally expressed from Equation (6.124) by

$$R_{SRH} = \frac{np - n_i^2}{\tau_{SRH,p} (n + n_1) + \tau_{SRH,n} (p + p_1)} \tag{6.126}$$

where

$$\tau_{SRH,p} = \frac{1}{c_p N_T} : \text{the hole minority carrier lifetime (sec).}$$

and

$$\tau_{SRH,n} = \frac{1}{c_n N_T} : \text{the electron minority carrier lifetime (sec).}$$

Equation (6.126) is an important result that is encountered often in the device literature. It should be emphasized that the expression applies to any steady-state situation and gives the net recombination rate for both electrons and holes. The minority carrier lifetimes are important material parameters and are interpreted as the average time an excess minority carrier will live in a sea of majority carriers. The minority carrier lifetimes are obtained experimentally, τ_n for a p-type semiconductor and τ_p for an n-type semiconductor [10].

6.7.6 Simplified Expression of the SRH Recombination Rate

The expression for the SRH net steady-state recombination rate can be drastically simplified under low-level injection conditions. We herein consider a p-type semiconductor under low-level injection conditions. It is also assume that the trap energy level E_T is fairly close to the mid-gap low-level injection, so that $n_1 = p_1 = n_i$. We derive a simplified expression of Equation (6.126) for a specific p-type material, first eliminating a denominator term having $\tau_p(n_p + n_i)$ that is irrelevant with p-type material and replacing p_1 with n_i. Omitting the SRH in the subscript of the lifetime, rewrite Equation (6.126) as

$$R_{SRH} = \frac{np - n_i^2}{\tau_p(n + n_1) + \tau_n(p + p_1)} = \frac{n_p p_p - n_i^2}{\tau_n(p_p + n_i)} \tag{6.127}$$

The numerator of Equation (6.127) for p-type material has already been obtained in Equation (6.114b), which is

$$(n_P p_P - n_i^2) \cong p_{P0}\Delta n_P \tag{6.127a}$$

Now the denominator of Equation (6.127) is expanded using the definition of excess majority carrier concentration, $\Delta p_p = p_P - p_{P0}$ and assuming $\Delta n_p = \Delta p_p$ in Equation (6.109) as

$$\tau_n(p_P + n_i) = \tau_n(\Delta p_P + p_{P0} + n_i) = \tau_n(\Delta n_P + p_{P0} + n_i) \tag{6.127b}$$

From Equation (6.109), we have $\Delta n_P \ll p_{P0}$ and know that majority carrier concentration is much greater than intrinsic carrier concentration, as $p_{P0} \gg n_i$. Therefore,

$$\tau_n(p_P + n_i) = \tau_n(\Delta n_P + p_{P0} + n_i) \cong \tau_n p_{P0} \tag{6.127c}$$

Inserting Equations (6.127a) and (6.127c) into Equation (6.127), we therefore arrive at the drastically simplified *trap recombination rate in p-type material*, including the SRH in the subscript of the lifetime.

$$R_{SRH,P} = \frac{\Delta n_P}{\tau_{SRH,n}} \tag{6.128}$$

where $\tau_{SRH,n} = \dfrac{1}{c_n N_T}$ is the electron minority carrier lifetime (sec) $\qquad$ (6.128a)

In a similar way, the simplified *trap recombination rate in n-type material* is obtained as

$$R_{SRH,N} = \frac{\Delta p_N}{\tau_{SRH,p}} \qquad (6.129)$$

where $\tau_{SRH,p} = \dfrac{1}{c_p N_T}$ is the hole minority carrier lifetime (sec) $\qquad$ (6.129a)

The recombination rate expressions obtained here are used extensively in device analyses.

6.7.7 Auger Recombination

Auger recombination was discussed in Section 6.7.1. Here we derive an expression for use in modeling. For Auger recombination, an electron and two holes or a hole and two electrons are involved. By a similar argument to that, the rate is proportional to the concentrations of all three carriers [12].

The Auger recombination rate for two-hole collisions is given by

$$R_{Aug} = A_n(np^2 - n_0 p_0^2) \qquad (6.130)$$

The Auger recombination rate for two-electron collisions is given by

$$R_{Aug} = A_p(n^2 p - n_0^2 p_0) \qquad (6.131)$$

Auger processes are most important where carrier concentrations are high—for instance, in low band gap and doped materials, or at high temperature. The dependence on doping concentration is strong. In p-type doped material, the Auger recombination rate for two-hole collisions is expressed as

$$R_{Aug,P} = A_n \left(n_P p_P^2 - n_{P0} p_{P0}^2 \right) \qquad (6.132)$$

Under low-level injection conditions, knowing that $p_P \cong p_{P0} \cong N_A$ from Equations (6.109) and (6.86) and also knowing that $n_{P0} p_{P0} = n_i^2$ from Equation (6.111), we have

$$R_{Aug,P} = A_n(n_P p_P^2 - n_{P0} p_{P0}^2) = A_n p_{P0}(n_P p_P - n_{P0} p_{P0}) = A_n N_A(n_P p_P - n_i^2)$$
$$(6.132a)$$

The expression in the parenthesis for p-type material is found in Equation (6.114b), which is

$$(n_P p_P - n_i^2) \cong N_A \Delta n_P \qquad (6.132b)$$

Substituting this into Equation (6.132a) gives

$$R_{Aug,P} = A_n N_A^2 \Delta n_P = \frac{\Delta n_P}{\dfrac{1}{A_n N_A^2}} \qquad (6.132c)$$

Finally, the Auger recombination rate in p-type material is expressed by

$$R_{Aug,P} = \frac{\Delta n_P}{\tau_{Aug,n}} \tag{6.133}$$

where $\tau_{Aug,n} = \frac{1}{A_n N_A^2}$ is the electron lifetime for Auger recombination in p-type material.

In a similar way, the Auger recombination rate in n-type material is obtained by

$$R_{Aug,N} = \frac{\Delta p_N}{\tau_{Aug,p}} \tag{6.134}$$

where $\tau_{Aug,p} = \frac{1}{A_p N_D^2}$ is the electron lifetime for Auger recombination in p-type material.

6.7.8 Total Recombination Rate

Each of these recombination processes occurs in parallel, and there can be multiple and/or distributed traps in the forbidden gap; thus the total recombination rate in p-type material is the sum of rates due to each process:

$$R_P = R_{rad,P} + R_{SRH,P} + R_{Aug,P} = \frac{\Delta n_P}{\tau_{rad,n}} + \frac{\Delta n_P}{\tau_{SRH,n}} + \frac{\Delta n_P}{\tau_{Aug,n}} \tag{6.135}$$

An effective minority carrier lifetime in p-type material is given as

$$\frac{1}{\tau_n} = \frac{1}{\tau_{rad,n}} + \frac{1}{\tau_{SRH,n}} + \frac{1}{\tau_{Aug,n}} \tag{6.136}$$

Finally, the *total recombination rate* in p-type material in low-level injection is given by

$$R_P = \frac{\Delta n_P}{\tau_n} \tag{6.137}$$

where τ_n is the *effective minority carrier lifetime*.

In a similar way, the total recombination rate in n-type material in low-level is obtained by

$$R_N = \frac{\Delta p_N}{\tau_p} \tag{6.138}$$

where τ_p is the effective minority carrier lifetime.

For clarity, the capitalized subscripts, P and N, are used to indicate quantities in p-type and n-type regions, respectively, when it may not be otherwise apparent. Lower-case subscripts, p and n, refers to quantities associated with holes and electrons, respectively.

6.8   CARRIER TRANSPORT

In this section, we try to formulate the basic equations required in the operational *modeling* of a solar cell. The carrier motion and the current are related with an electric field and carrier gradient, which are known as drift and diffusion. The basic equations combine the *drift*, *diffusion*, and *generation-recombination* satisfying the *conservation of mass* and the *Poisson's equation*. The basic equations must be solved subject to imposed boundary conditions under non-equilibrium conditions like sunlight.

6.8.1   Drift

Drift, by definition, is charged-particle motion in response to an *applied electric field*. In the absence of an electric field, the photogenerated free electrons and holes in each band move at a velocity in a random direction, colliding with and bouncing off objects in a crystal. When an electric field is applied across the crystal of a doped semiconductor, the electrons in the conduction band, negatively charged, still move in a complex way, colliding and bouncing off but resultantly at a constant *drift velocity* in the opposite direction of the applied electric field. Holes in the valence band, positively charged, move in the same way with the electrons at a constant drift velocity in the same direction of the applied field. The hole current density, the current per unit area, is simply the product of an electron charge q, the hole density (concentration) p, and the hole drift velocity $v_{d,p}$. The *drift current density* for holes can be thus expressed as

$$J_p^{drift} = qpv_{d,p} \qquad (6.139)$$

Likewise, the drift current density for electrons can be written as

$$J_n^{drift} = qnv_{d,n} \qquad (6.140)$$

The drift velocities for holes and electrons is proportional to the electric field using the *carrier mobility* μ_p (see Section 6.8.2 for the details) and are written as

$$v_{d,p} = \mu_p E \qquad (6.141)$$

$$v_{d,n} = \mu_n E \qquad (6.142)$$

Finally, the drift current densities for holes and electrons are rewritten in terms of an electric field as

$$J_p^{drift} = q\mu_p p E \qquad (6.143)$$

$$J_n^{drift} = q\mu_n n E \qquad (6.144)$$

6.8.2   Carrier Mobility

Carrier mobility depends on both doping and temperature. The carrier mobilities in silicon are well expressed with measurements.

For electrons,

$$\mu_n (N_D, T) = \mu_{n1}(T) + \frac{\mu_{n2}(T)}{1 + \left(\dfrac{N_D}{N_{D1}(T)}\right)^{\alpha_n(T)}} \frac{\text{cm}^2}{\text{V} \cdot \text{s}} \qquad (6.145)$$

where

$$\mu_{n1}(T) = 92 \left(\frac{T}{300 \text{ K}}\right)^{-0.57} \qquad (6.145a)$$

$$\mu_{n2}(T) = 1{,}268 \left(\frac{T}{300 \text{ K}}\right)^{-2.33} \qquad (6.145b)$$

$$N_{D1}(T) = 1.3 \times 10^{17} \left(\frac{T}{300 \text{ K}}\right)^{2.4} \qquad (6.145c)$$

$$\alpha_n(T) = 0.91 \left(\frac{T}{300 \text{ K}}\right)^{-0.146} \qquad (6.145d)$$

For holes,

$$\mu_p(N_A, T) = \mu_{p1}(T) + \frac{\mu_{p2}(T)}{1 + \left(\dfrac{N_A}{N_{A1}(T)}\right)^{\alpha_p(T)}} \frac{\text{cm}^2}{\text{V} \cdot \text{s}} \qquad (6.146)$$

where

$$\mu_{p1}(T) = 54.3 \left(\frac{T}{300 \text{ K}}\right)^{-0.57} \qquad (6.146a)$$

$$\mu_{p2}(T) = 406.9 \left(\frac{T}{300 \text{ K}}\right)^{-2.23} \qquad (6.146b)$$

$$N_{A1}(T) = 2.35 \times 10^{17} \left(\frac{T}{300 \text{ K}}\right)^{2.4} \qquad (6.146c)$$

$$\alpha_p(T) = 0.88 \left(\frac{T}{300 \text{ K}}\right)^{-0.146} \qquad (6.146d)$$

The carrier mobilities for electrons and holes are a function of dopants and temperature, which are plotted in Figure 6.30. At low impurity levels, the mobility is governed by intrinsic lattice scattering, while at high levels the mobility is governed by ionized impurity scattering.

6.8.3   Diffusion

Diffusion is a process whereby particles tend to spread out or redistribute as a result of their random thermal motion, migrating on a microscopic scale from regions of high

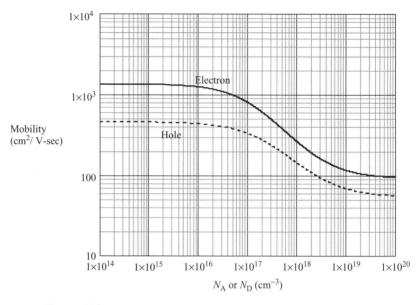

Figure 6.30   Electron and hole mobilities in silicon for $T = 300\,\text{K}$

particle concentration into regions of low particle concentration. The *diffusion current densities* are given

$$J_p^{diff} = -q D_p \nabla p \qquad (6.147)$$

$$J_n^{diff} = q D_n \nabla n \qquad (6.148)$$

where D_p and D_n are the hole and electron diffusion coefficients (cm²/s), respectively. ∇ is the gradient derivative.

6.8.4   Total Current Densities

The total hole and electron current densities in a semiconductor arise as the combined result of drift and diffusion.

$$J_p = J_p^{drift} + J_p^{diff} = q \mu_p p E - q D_p \nabla p \qquad (6.149)$$

$$J_n = J_n^{drift} + J_n^{diff} = q \mu_n n E + q D_n \nabla n \qquad (6.150)$$

The total current density is expressed by

$$J = J_p + J_n \qquad (6.151)$$

This expression gives the power of a solar cell with the voltage formed across the solar cell. The total current density consists of the hole and electron current densities that depend mainly on the number of the carriers and the gradient of the carriers.

6.8.5   Einstein Relationship

Under equilibrium conditions, the total current density is equal to zero. Because the electron and hole current densities are totally decoupled under equilibrium conditions, they must independently vanish.

$$J_p = J_n = 0 \tag{6.152}$$

For the hole current density,

$$J_p = J_p^{drift} + J_p^{diff} = q\mu_p p E - q D_p \nabla p = q\mu_p p E - q D_p \frac{dp}{dx} = 0 \tag{6.153}$$

Since it is known $p = p_0$ in equilibrium from Equation (6.83) and $\frac{dE_F}{dx} = 0$, we have

$$p = n_i \exp\left(\frac{E_i - E_F}{kT}\right) \tag{6.154}$$

$$\frac{dp}{dx} = \frac{n_i}{kT}\exp\left(\frac{E_i - E_F}{kT}\right)\frac{dE_i}{dx} = \frac{p}{kT}\frac{dE_i}{dx} \tag{6.155}$$

Substituting this into Equation (6.153) yields

$$q\mu_p p E - q D_p \frac{p}{kT}\frac{dE_i}{dx} = 0 \tag{6.156}$$

It is known that the electric field is generated due to the gradient of the intrinsic Fermi energy, which can be seen in Figure 6.23. Therefore, we must have

$$E = \frac{1}{q}\frac{dE_i}{dx} \tag{6.157}$$

Substituting this into Equation (6.156) gives

$$q\mu_p p E - q^2 D_p \frac{p}{kT}E = 0 \tag{6.158}$$

Finally, we arrive at an expression that is called the *Einstein relationship* for holes as

$$\frac{D_p}{\mu_p} = \frac{kT}{q} \tag{6.159}$$

Similarly, the Einstein relationship for electrons is

$$\frac{D_n}{\mu_n} = \frac{kT}{q} \tag{6.160}$$

Although it was established while assuming equilibrium conditions, we can present more elaborate arguments that show the Einstein relationship to be valid even under nonequilibrium conditions (Pierret [17]).

6.8.6 Semiconductor Equations

The basic equations of device physics, the semiconductor equations, are based on two simple principles: that the electrostatic potential due to the carrier charges obeys *Poisson's equation* and that the number of carriers of each type must be conserved. Poisson's equation in differential form is

$$\nabla E = \frac{q}{\varepsilon}(p - n + N) \tag{6.161}$$

where E is the electric field, N is the net charge due to dopants and other trapped charges, and ε is the electric permittivity depending on the material.

$$\varepsilon = K_s \varepsilon_0 \tag{6.162}$$

where ε_0 is the *permittivity of vacuum* (free space) and K_s is the *semiconductor dielectric constant* (relative permittivity). These values are found in Table 6.3. For a semiconductor containing electrons and holes, the conservation equations, or also called the *continuity equations*, are given

$$\frac{\partial p}{\partial t} + \frac{1}{q}\nabla J_p = G - R \tag{6.163}$$

$$-\frac{\partial p}{\partial t} + \frac{1}{q}\nabla J_n = R - G \tag{6.164}$$

where G is the generation rate and R the recombination rate.

6.8.7 Minority-Carrier Diffusion Equations

We consider a uniformly doped thin semiconductor. The carrier mobilities and diffusion coefficients are independent of position. For the net charge, N, remember that the donors N_D and acceptors N_A have the positive and negative charges, respectively, as discussed earlier in Section 6.5.1. And actually they form the net charge. As we are herein mainly interested in the *one-dimensional steady-state* operation of the thin solar cell, the semiconductor equations reduce to the following.

From Equation (6.161), we have

$$\frac{dE}{dx} = \frac{q}{\varepsilon}(p - n + N_D - N_A) \tag{6.165}$$

From Equation (6.163) using Equation (6.149), we have

$$q\mu_p \frac{d}{dx}(pE) - qD_p \frac{d^2 p}{dx^2} = q(G - R_p) \tag{6.166}$$

From Equation (6.164) using Equation (6.150), we have

$$q\mu_n \frac{d}{dx}(nE) - qD_n \frac{d^2 n}{dx^2} = q(R_n - G) \tag{6.167}$$

We also consider the region sufficiently far from the p-n junction of the solar cell, so-called *quasi-neutral region* (QNR), where the electric field is very small. The semiconductor equations are applied to the quasi-neutral region, so the drift current that involves the electric field is neglected with respect to the diffusion current. Basically, we assume that the doped material is in *low-level injection* (this means that the excess carriers are much fewer than the majority carriers of which the number is close to that of dopants), thus the minority carriers are more important in presenting the carrier transform than the majority carriers, although the majority carriers play an important role for useful work.

The definitions of Δn_P and Δp_N are found in Equation (6.106) and (6.107), respectively. Mathematically, there is no difference between the second derivative of the excess minority carrier and the second derivative of the minority carrier, so

$$\frac{d^2 \Delta p_N}{dx^2} = \frac{d^2 p_N}{dx^2} \tag{6.168}$$

Under low-level injection, the total recombination simplifies to Equation (6.137) and (6.138) for p-type and n-type materials, respectively.

For n-type material, Equation (6.166) finally reduces to

$$D_p \frac{d^2 \Delta p_N}{dx^2} - \frac{\Delta p_N}{\tau_p} = -G(x) \tag{6.169}$$

where Δp_N is the excess minority carrier (hole) concentration in n-type material.

For p-type material, Equation (6.167) finally reduces to

$$D_n \frac{d^2 \Delta n_P}{dx^2} - \frac{\Delta n_P}{\tau_n} = -G(x) \tag{6.170}$$

where Δn_P is the excess minority carrier (electron) concentration in p-type material.

The above two equations with the appropriate boundary conditions can be solved to provide the excess carrier distributions in the quasi-neutral regions, which allow us to determine the minority carrier concentrations. In order to do that, we need to first determine the size of the deplete region or the space charge region.

6.8.8 P−n Junction

Immobile ionized donors and acceptors are uncovered initially by the carrier diffusion, creating an electric field at the p-n junction, as indicated in Figure 6.31. In other words, doping creates a built-in voltage in the depletion region. The electric field causes more to deplete carriers in the depletion region due to the drifting mechanism (Section 6.8.1). Notice that the electric field E is strong in the depletion region but very small in the quasi-neutron regions.

The relationship between an electric field and an electrostatic potential (voltage) is given by

$$E = -\nabla\phi \tag{6.171}$$

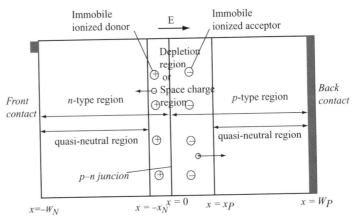

Figure 6.31 A depletion region near the p–n junction for a solar cell

Applying this to Poisson's equation, Equation (6.161), in thermal equilibrium, we have

$$\nabla^2\phi = \frac{q}{\varepsilon}(n_0 - p_0 + N_A - N_D) \tag{6.172}$$

Within the depletion region defined by $-x_N < x < x_P$, it can be assumed that p_0 and n_0 are both negligible compared to $|N_A - N_D|$ so that Equation (6.172) can be simplified to

$$\nabla^2\phi = -\frac{q}{\varepsilon}N_D \quad \text{for } -x_N < x < 0 \tag{6.173}$$

$$\nabla^2\phi = \frac{q}{\varepsilon}N_A \quad \text{for } 0 < x < x_P \tag{6.174}$$

Outside the depletion region, charge neutrality is assumed and

$$\nabla^2\phi = 0 \quad \text{for } x \leq -x_N \text{ and } x \geq x_P \tag{6.175}$$

These regions are called the *quasi-neutral regions*. The electrostatic potential difference across the junction is the built-in voltage as obtained in Equation (6.97). Therefore,

$$\phi(x_P) = 0 \tag{6.176}$$

$$\phi(x) = 0 \quad \text{for } x \geq x_P \tag{6.176a}$$

$$\phi(-x_N) = V_{bi} \tag{6.177}$$

$$\phi(x) = V_{bi} \quad \text{for } x \leq -x_N \tag{6.177a}$$

By taking derivative of Equations (6.176) and (6.177), we have

$$\frac{d\phi(x_P)}{dx} = 0 \tag{6.178}$$

$$\frac{d\phi(-x_N)}{dx} = 0 \tag{6.179}$$

From Equation (6.173), separating the variable and integrating from $-x_N$ to an arbitrary x gives

$$\int_0^{\frac{d\phi}{dx}} d\left(\frac{d\phi}{dx}\right) = -\frac{q}{\varepsilon}N_D \int_{-x_N}^x dx \tag{6.180}$$

$$\frac{d\phi}{dx} = -\frac{q}{\varepsilon}N_D(x + x_N) \tag{6.180a}$$

Integrating this again, we have

$$\int_{V_{bi}}^{\phi} d\phi = -\frac{q}{\varepsilon}N_D \frac{1}{2} \int_{-x_N}^x (x_N + x)dx \tag{6.180b}$$

Therefore, the electrostatic potential for $-x_N < x < 0$ is expressed with respect to the arbitrary point x.

$$\phi(x) = V_{bi} - \frac{qN_D}{2\varepsilon}(x_N + x)^2 \quad \text{for } -x_N < x < 0 \tag{6.181}$$

Now we do the similar procedure with Equation (6.174) as

$$\int_{\frac{d\phi}{dx}}^{0} d\left(\frac{d\phi}{dx}\right) = \frac{q}{\varepsilon}N_A \int_x^{x_P} dx \tag{6.182}$$

$$-\frac{d\phi}{dx} = \frac{q}{\varepsilon}N_A(x_P - x) \tag{6.182a}$$

Integrating this again, we have

$$-\int_{\phi}^{0} d\phi = \frac{q}{\varepsilon}N_A \frac{1}{2} \int_x^{x_P} (x_P - x)dx \tag{6.182b}$$

Thus, the electrostatic potential for $0 < x < x_P$ is expressed with respect to the arbitrary point x.

$$\phi(x) = \frac{qN_A}{2\varepsilon}(x_P - x)^2 \quad \text{for } 0 < x < x_P \tag{6.183}$$

We summarize this derivation for the electrostatic potential as

$$\phi(x) = \begin{cases} V_{bi}, \ x \leq -x_N & (6.177a) \\ V_{bi} - \dfrac{qN_D}{2\varepsilon}(x_N + x)^2, \ -x_N \leq x \leq 0 & (6.181) \\ \dfrac{qN_A}{2\varepsilon}(x_P - x)^2, \ 0 \leq x \leq x_P & (6.183) \\ 0, \ x \geq x_P & (6.176a) \end{cases} \tag{6.184}$$

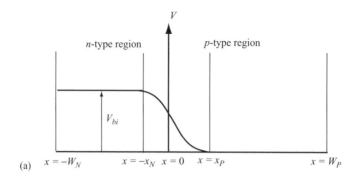

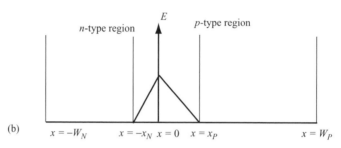

Figure 6.32   (a) Electrostatic potential and (b) electric field

Equation (6.184) is plotted in Figure 6.32(a), where the electrostatic potential is illustrated with respect to the point x. This reflects the concept of the built-in voltage. Reminding that $E = -\nabla\phi$, Equation (180a) and (182a) are plotted in Figure 6.32(b), where the electric field is illustrated with respect to the point x. This reflects well the concept of the very small electric fields in the quasi-neutral regions.

6.8.9   Calculation of Depletion Width

We will derive the expressions for x_N and x_P in the depletion region. The two equations (6.181) and (6.183) must be continuous at $x = 0$. Therefore, we have

$$V_{bi} - \frac{qN_D}{2\varepsilon}x_N^2 = \frac{qN_A}{2\varepsilon}x_P^2 \tag{6.185}$$

The two equations (6.180a) and (6.182a) must also be continuous at $x = 0$. Therefore,

$$N_D x_N = N_A x_P \tag{6.186}$$

Combining Equations (6.185) and (6.186) and solving for x_N and x_P gives

$$x_N = \left(\frac{2\varepsilon}{q}\frac{N_A V_{bi}}{N_D(N_A + N_D)}\right)^{1/2} \tag{6.187}$$

$$x_P = \left(\frac{2\varepsilon}{q}\frac{N_D V_{bi}}{N_A(N_A + N_D)}\right)^{1/2} \tag{6.188}$$

The depletion width can be obtained

$$W_D = x_N + x_P = \left(\frac{2\varepsilon}{q} \frac{N_A + N_D}{N_A N_D} V_{bi} \right)^{1/2} \tag{6.189}$$

Notice that the dimensions of the depletion region are dependent of the built-in voltage, donors N_A, and acceptors N_D. Under nonequilibrium conditions, the electrostatic potential difference across the p–n junction is modified by the applied voltage, V, which is zero in thermal equilibrium. As a consequence, x_N, x_P, and W_D are dependent on the applied voltage.

$$x_N = \left(\frac{2\varepsilon}{q} \frac{N_A(V_{bi} - V)}{N_D(N_A + N_D)} \right)^{1/2} \tag{6.187a}$$

$$x_P = \left(\frac{2\varepsilon}{q} \frac{N_D(V_{bi} - V)}{N_A(N_A + N_D)} \right)^{1/2} \tag{6.188a}$$

$$W_D = x_N + x_P = \left(\frac{2\varepsilon}{q} \frac{N_A + N_D}{N_A N_D} (V_{bi} - V) \right)^{1/2} \tag{6.189a}$$

For example, we want to calculate the depletion width W_D including x_N and x_P. Assume a Si operated at $300\,\text{K}$ with the given data in Equation (6.90): $N_D = 1 \times 10^{18}\,\text{cm}^{-3}$ and $N_A = 1 \times 10^{15}\,\text{cm}^{-3}$.

From Equation (6.162),

$$\varepsilon = K_s \varepsilon_0 = 11.8 \times 8.85 \times 10^{-14}\,\text{C V}^{-1}\,\text{cm}^{-1} = 1.044 \times 10^{-12}\,\text{C V}^{-1}\,\text{cm}^{-1}$$

From Equation (6.187a),

$$x_N = \left(\frac{2 \times \left(1.044 \times 10^{-12}\,\text{C V}^{-1}\,\text{cm}^{-1}\right) \left(10^{15}\,\text{cm}^{-3}\right) (0.776\,\text{V} - 0.35\,\text{V})}{\left(1.6021 \times 10^{-19}\,\text{C}\right) \left(10^{18}\,\text{cm}^{-3}\right) \left(10^{15} + 10^{18}\,\text{cm}^{-3}\right)} \right)^{1/2}$$

$$= 7.45 \times 10^{-8}\,\text{cm} = 7.45 \times 10^{-4}\,\mu\text{m}$$

From Equations (6.186) and (6.189)

$$x_P = \frac{N_D}{N_A} x_N = \frac{10^{18}\,\text{cm}^{-3}}{10^{15}\,\text{cm}^{-3}} \left(7.45 \times 10^{-4}\,\mu\text{m}\right) = 0.745\,\mu\text{m}$$

$$W_D = x_N + x_P \cong x_P \cong 0.745\,\mu\text{m}$$

The depletion region lies almost exclusively on the lightly doped side of the metallurgical boundary.

6.8.10 Energy Band Diagram with a Reference Point

Although the energy band has been discussed many times in the foregoing sections, the construction of the energy band distribution with respect to the point x is not trivial.

The relationship between the bottom of the conduction band E_C and the electrostatic potential $\phi(x)$ with the aid of the reference point E_0 is given by

$$E_C = E_0 - q\phi(x) - X \qquad (6.190)$$

where X is the *electron affinity* and E_0 the *vacuum energy*. The electron affinity is the minimum energy needed to free an electron from the bottom of the conduction band and take it to the vacuum level. The electron affinity X is an invariant fundamental property of the specified semiconductor, $X = 4.0\,\text{eV}$, $4.03\,\text{EV}$, and $4.07\,\text{eV}$ for Ge, Si, and GaAs, respectively. E_0 is defined as the vacancy energy, serves as a convenient reference point, and is universally constant with position. An electron at the vacuum energy is completely free of influence from all external forces.

The equilibrium energy band diagram with the previously given data for a silicon semiconductor in the vicinity of the depletion region is pictured in Figure 6.33. The electrostatic potential $\phi(x)$ determines the shape of the energy band edge using the values obtained previously, particularly in the depletion region. E_i is approximately at the midgap between E_C and E_V.

6.8.11 Quasi-Fermi Energy Levels

Quasi-energy levels are energy used to specify the carrier concentrations inside a semiconductor under nonequilibrium conditions. You may find the energy levels in Figure 6.28. In fact, the Fermi level is defined only for a system under equilibrium conditions and cannot be used in nonequilibrium conditions. Nevertheless, the convenience of being able to deduce the carrier concentrations from the energy band diagram is extended to nonequilibrium conditions through the use of quasi-Fermi energy levels. This is accomplished by introducing two energies, E_{FN} (the quasi-Fermi level for electrons) and E_{FP} (the quasi-Fermi energy level for holes). Therefore, from Equation (6.81) and (6.83), assuming the semiconductor to be nondegenerate, we have

$$n = n_i \exp\left(\frac{E_{FN} - E_i}{kT}\right) \qquad (6.191)$$

$$p = n_i \exp\left(\frac{E_i - E_{FP}}{kT}\right) \qquad (6.192)$$

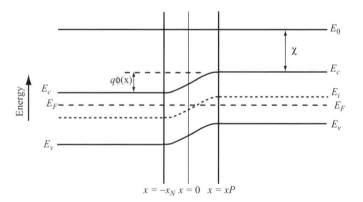

Figure 6.33 Energy band under equilibrium conditions in a solar cell

If the left- and right-hand sides of Equations (6.191) and (6.192) are multiplied together, we obtain

$$np = n_i^2 \exp\left(\frac{E_{FN} - E_i}{kT}\right) \exp\left(\frac{E_i - E_{FP}}{kT}\right) = n_i^2 \exp\left(\frac{E_{FN} - E_{FP}}{kT}\right) \quad (6.193)$$

Assuming that both the quasi-Fermi energies remain constant inside the depletion region as $E_{FN} - E_{FP} = qV$, we have

$$np = n_i^2 \exp\left(\frac{qV}{kT}\right) \quad \text{for } -x_N \le x \le x_P \quad (6.194)$$

This has been referred to as the *law of the junction*. The product of majority and minority carrier concentrations in either n-type or p-type region is constant with position and depends on the *bias*, or also called the external voltage.

6.9   MINORITY CARRIER TRANSPORT

Minority carrier concentration gradients are important for determining the current densities, depending on many quantities such as donors, acceptors, minority carrier diffusion length, and material. Despite the built-in voltage limits, the diffusion across the depletion region is always forced to take place from the high carrier density to the low carrier concentration, which is pictured in Figure 6.34. For example, the majority electrons in the n-type region diffuse into the p-type region over the depletion region and become minority electrons that decay with respect to position, depending on the recombination or equivalently the diffusion length. The steeper the gradient of the minority carriers, the more the recombination. Therefore, it is essential to first understand the mechanisms of the *minority excess carrier concentration gradients* and to optimize the design of a solar cell. This is related to *minority carrier transport*.

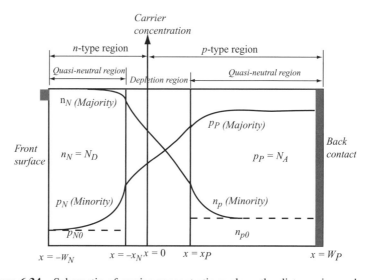

Figure 6.34   Schematic of carrier concentrations along the distance in a solar cell

6.9.1 Boundary Conditions

The basic current-voltage characteristic of the solar cell can be derived by solving the minority-carrier diffusion equations with appropriate boundary conditions.

At $x = -W_N$, the usual assumption is that the front contact can be treated as an ideal ohmic contact. Hence,

$$\Delta p(-W_N) = 0 \qquad (6.195)$$

However, since the front contact is usually a grid with metal contacting the semiconductor on only a small percentage of the front surface, modeling the front surface with an effective surface recombination velocity takes account the combined effects of the ohmic contact and the antireflective passivation layer (SiO_2). In this case, the boundary condition at $x = -W_N$ is

$$\left. \frac{d\Delta p}{dx} \right|_{x=-W_N} = \frac{S_{F,eff}}{D_p} \Delta p(-W_N) \qquad (6.196)$$

where $S_{F,eff}$ is the effective front surface recombination velocity. As $S_{F,eff} \to \infty$, $\Delta p \to 0$ and the boundary condition given by Equation (6.196) reduces to that of an ideal ohmic contact, Equation (1.195). In reality, $S_{F,eff}$ depends on a number of parameters and is bias-dependent.

The back contact could also be treated as an ideal ohmic contact, so that

$$\Delta p(W_P) = 0 \qquad (6.197)$$

However, solar cells are often fabricated with a back-surface field (BSF), a thin, more heavily doped region at the back of the base region. The BSF keeps minority carriers away from the back ohmic contact and increases their chances of being collected. It can be modeled by an effective surface recombination velocity. The boundary condition is then

$$\left. \frac{d\Delta n}{dx} \right|_{x=W_P} = -\frac{S_{BSF}}{D_n} \Delta n(W_P) \qquad (6.198)$$

where S_{BSF} is the effective back surface recombination velocity at the back-surface field.

Under equilibrium conditions, zero applied voltage, and no illumination, the Fermi energy E_F is constant with position. When a bias voltage is applied, it is convenient to introduce the concept of quasi-Fermi energies. Using Equation (6.194), we set the boundary conditions used at the edges of the depletion region.

$$n_N(-x_N)p_N(-x_N) = n_i^2 \exp\left(\frac{qV}{kT}\right) \qquad (6.199)$$

Under the assumption of low-level injection, the majority electron concentration in n-type region is close to the donor concentration (all donors are ionized). Therefore,

$$n_N(-x_N) \cong n_{N0}(-x_N) = N_D \qquad (6.200)$$

Substituting this into Equation (6.199) provides another boundary condition at $x = -x_N$.

$$p_N(-x_N) = \frac{n_i^2}{N_D} \exp\left(\frac{qV}{kT}\right) \tag{6.201}$$

The product of the equilibrium carrier concentrations is

$$n_{N0}(-x_N)p_{N0}(-x_N) = n_i^2 \tag{6.202}$$

$$p_{N0}(-x_N) = \frac{n_i^2}{N_D} \tag{6.203}$$

The definition of the excess minority carrier concentration for holes is

$$\Delta p_N(-x_N) = p_N(-x_N) - p_{N0}(-x_N) \tag{6.204}$$

Substituting Equations (6.201) and (6.203) into Equation (6.204) yields

$$\Delta p_N(-x_N) = \frac{n_i^2}{N_D}\left[\exp\left(\frac{qV}{kT}\right) - 1\right] \tag{6.205}$$

Similarly, the excess minority carrier concentration for electrons is derived as

$$\Delta n_P(x_P) = \frac{n_i^2}{N_A}\left[\exp\left(\frac{qV}{kT}\right) - 1\right] \tag{6.206}$$

The summary of the boundary conditions obtained is rewritten as follows:

$$\left.\frac{d\Delta p}{dx}\right|_{x=-W_N} = \frac{S_{F,eff}}{D_p}\Delta p(-W_N) \quad \text{at } x = -W_N \tag{6.196}$$

$$\Delta p_N(-x_N) = \frac{n_i^2}{N_D}\left[\exp\left(\frac{qV}{kT}\right) - 1\right] \quad \text{at } x = -x_N \tag{6.205}$$

$$\Delta n_P(x_P) = \frac{n_i^2}{N_A}\left[\exp\left(\frac{qV}{kT}\right) - 1\right] \quad \text{at } x = x_P \tag{6.206}$$

$$\left.\frac{d\Delta n}{dx}\right|_{x=W_P} = -\frac{S_{BSF}}{D_n}\Delta n(W_P) \quad \text{at } x = W_P \tag{6.198}$$

6.9.2 Minority Carrier Lifetimes

The *minority carrier lifetimes* (τ_p and τ_n) are the average time that the excess minority carriers will live in a sea of majority carriers. They are important material parameters invariably required in the modeling of solar cells. In fact, experimental measurements must be performed to determine the minority carrier lifetimes in a given semiconductor sample. The reason for the lack of cataloged information can be traced to the extreme variability of the lifetime parameters. The minority carrier lifetimes were defined in

Equations (6.128a) and (6,129a). We see that the minority carrier lifetimes depend on the often poorly controlled *R-G center concentration*, not on the concentrations of dopants. A fabrication procedure called *gettering* (removing metal impurities by heat treatment) can reduce the R-G center concentration and give rise to τ_p and $\tau_n \sim 1$ ms in Si [7].

6.9.3 Minority Carrier Diffusion Lengths

Diffusion lengths physically represent the average distance that the minority carrier can diffuse into a sea of majority carriers before being recombined. The *minority carrier diffusion lengths* for holes and electrons are defined, respectively, as

$$L_p = \sqrt{D_p \tau_p} \tag{6.207}$$

$$L_n = \sqrt{D_n \tau_n} \tag{6.208}$$

where D_p and D_n are the *minority carrier diffusion coefficients* for holes and electrons, respectively. τ_p and τ_n are the *minority carrier lifetimes* for holes and electrons, respectively.

For example, considering a semiconductor with $N_A = 10^{15}\ \text{cm}^{-3}$ and $\tau_p = 10^{-6}$ sec, we want to estimate the minority carrier diffusion length L_p for holes at room temperature (300 K).

If we determine the minority carrier diffusivity D_p in Equation (6.207), we can compute the diffusion length. The Einstein relationship provides Equation (6.159) and the hole mobility is found to be about $500\ \text{cm}^2/\text{V-sec}$ at $N_A = 10^{15}\ \text{cm}^{-3}$ from Figure 6.30. Thus,

$$D_p = \frac{kT}{q}\mu_p \tag{6.209}$$

$$L_p = \sqrt{D_p \tau_p} = \sqrt{\frac{kT}{q}\mu_p \tau_p}$$

$$= \sqrt{\frac{\left(8.617 \times 10^{-5}\ \text{eV/K}\right)(300\ \text{K})}{1\,\text{eV/V}}\left(500\ \text{cm}^2/\text{V} \cdot \text{sec}\right)\left(10^{-6}\ \text{sec}\right)} = 35.9\ \mu\text{m} \tag{6.209a}$$

Although the computed value is fairly representative, it should be understood that diffusion lengths can vary over several orders of magnitude because of wide variations in the carrier lifetime.

6.9.4 Minority Carrier Diffusion Equation for Holes

To start, we rewrite Equation (6.169):

$$D_p \frac{d^2 \Delta p_N}{dx^2} - \frac{\Delta p_N}{\tau_p} = -G(x) \tag{6.169}$$

Dividing this by D_p, we have

$$\frac{d^2 \Delta p_N}{dx^2} - \frac{\Delta p_N}{D_p \tau_p} = -\frac{G(x)}{D_p} \tag{6.210}$$

Substituting Equation (6.207) into this gives an expression that can be solved.

$$\frac{d^2 \Delta p_N}{dx^2} - \frac{\Delta p_N}{L_p^2} = -\frac{G(x)}{D_p} \tag{6.211}$$

The general solution of Equation (6.211) consists of the homogeneous and particular solutions.

$$\Delta p_N(x) = \Delta p_N|_{\text{hom}} + \Delta p_N|_{par} \tag{6.212}$$

The homogeneous equation is

$$\frac{d^2 \Delta p_N}{dx^2} - \frac{\Delta p_N}{L_p^2} = 0 \tag{6.213}$$

The general solution for the homogeneous equation is easily found.

$$\Delta p_N|_{\text{hom}o} = A_N \sinh\left(\frac{x + x_N}{L_p}\right) + B_N \cosh\left(\frac{x + x_N}{L_p}\right) \tag{6.214}$$

It is noted that the minority hole carrier diffusion occurs at $x = -x_N$, so the coordinate encounters the shift. And the photogeneration (incident irradiation) starts at the front surface where $x = -W_N$, so the coordinate encounter the shift. These can be pictured in Figure 6.34. Thus, the photogeneration rate obtained previously in Equation (6.105) is rewritten for the coordinate.

$$G(x) = (1 - s) \int_\lambda (1 - r(\lambda)) f_0(\lambda) \alpha(\lambda) e^{-\alpha(\lambda)(x + W_N)} d\lambda \tag{6.215}$$

We look for a particular solution that satisfies Equation (6.211). Since $G(x)$ is a form of integration having the exponential term, the following particular solution is assumed to be pertinent.

$$\Delta p_N|_{par} = \int_\lambda C_N (1 - r) f_0 \alpha e^{-\alpha(x + W_N)} d\lambda \tag{6.216}$$

For convenience, the wavelength notation (λ) is eliminated from Equation (6.216) and herein. Taking the derivative twice, we have

$$\frac{d(\Delta p_N|_{par})}{dx} = -\int_\lambda C_N (1 - r) f_0 \alpha^2 e^{-\alpha(x + W_N)} d\lambda \tag{6.216a}$$

$$\frac{d^2(\Delta p_N|_{par})}{dx^2} = -\int_\lambda C_N (1 - r) f_0 \alpha^3 e^{-\alpha(x + W_N)} d\lambda \tag{6.216b}$$

Substituting Equation (6.216) and (6.216b) into Equation (6.211) gives

$$-\int_\lambda C_N(1-r)f_0\alpha^3 e^{-\alpha(x+W_N)}d\lambda + \int_\lambda \frac{C_N}{L_p^2}(1-r)f_0\alpha e^{-\alpha(x+W_N)}d\lambda$$

$$= \int_\lambda \frac{(1-s)}{D_p}(1-r)f_0\alpha e^{-\alpha(x+W_N)}d\lambda \tag{6.217}$$

Equating integrands and rearranging Equation (6.217) reduces to

$$C_N = -\frac{L_p^2(1-s)}{D_p(\alpha^2 L_p^2 - 1)} = -\frac{\tau_p(1-s)}{\alpha^2 L_p^2 - 1} \tag{6.218}$$

Thus, Equation (6.216) can be written as

$$\Delta p_N|_{par} = -(1-s)\int_\lambda \frac{\tau_p}{\alpha^2 L_p^2 - 1}(1-r)f_0\alpha e^{-\alpha(x+W_N)}d\lambda \tag{6.219}$$

Now the general solution of Equation (6.212) that describes the distribution of the excess minority carrier concentration in the n-type region is expressed as

$$\Delta p_N(x) = A_N \sinh\left(\frac{x+x_N}{L_p}\right) + B_N \cosh\left(\frac{x+x_N}{L_p}\right)$$

$$- (1-s)\int_\lambda \frac{\tau_p}{\alpha^2 L_p^2 - 1}(1-r)f_0\alpha\ e^{-a(x+W_N)}d\lambda \tag{6.220}$$

where A_N and B_N are the to be determined by the boundary conditions, Equation (6.196) at $x = -W_N$ and Equation (6.205) at $x = -x_N$. The boundary condition at $x = -W_N$ is

$$\left.\frac{d\Delta P_N}{dx}\right|_{x=-W_N} = \frac{S_{F,eff}}{D_p}\Delta p(-W_N) \tag{6.221}$$

We need to take derivative of Equation (6.220).

$$\frac{d(\Delta p_N)}{dx} = \frac{A_N}{L_p}\cosh\left(\frac{x+x_N}{L_p}\right) + \frac{B_N}{L_p}\cosh\left(\frac{x+x_N}{L_p}\right)$$

$$+ (1-s)\int_\lambda \frac{\tau_p}{\alpha^2 L_p^2 - 1}(1-r)f_0\alpha^2 e^{-\alpha(x+W_N)}d\lambda \tag{6.222}$$

$$\left.\frac{d(\Delta p_N)}{dx}\right|_{x=-W_N} = \frac{A_N}{L_p}\cosh\left(\frac{-W_N+x_N}{L_p}\right) + \frac{B_N}{L_p}\sinh\left(\frac{-W_N+x_N}{L_p}\right)$$

$$+ (1-s)\int_\lambda \frac{\tau_p}{\alpha^2 L_p^2 - 1}(1-r)f_0\alpha^2 e^{-\alpha(-W_N+W_N)}d\lambda \tag{6.223}$$

Using Equation (6.220)

$$\Delta p_N(-W_N) = A_N \sinh\left(\frac{-W_N + x_N}{L_p}\right) + B_N \cosh\left(\frac{-W_N + x_N}{L_p}\right)$$
$$- (1-s) \int_\lambda \frac{\tau_p}{\alpha^2 L_p^2 - 1}(1-r) f_0 \alpha \, e^{-\alpha(-W_N + W_N)} d\lambda \qquad (6.224)$$

Plugging Equation (6.223) and (6.224) into Equation (6.221) and reducing gives

$$\frac{A_N}{L_p}\cosh\left(\frac{-W_N + x_N}{L_p}\right) + \frac{B_N}{L_p}\sinh\left(\frac{-W_N + x_N}{L_p}\right) + (1-s)\int_\lambda \frac{\tau_p}{\alpha^2 L_p^2 - 1}(1-r) f_0 \alpha^2 d\lambda$$
$$= \frac{S_{F,eff}}{D_p}\left(A_N \sinh\left(\frac{-W_N + x_N}{L_p}\right) + B_N \cosh\left(\frac{-W_N + x_N}{L_p}\right)\right.$$
$$\left. -(1-s)\int_\lambda \frac{\tau_p}{\alpha^2 L_p^2 - 1}(1-r) f_0 \alpha \, d\lambda\right) \qquad (6.225)$$

Now we consider the other boundary condition at $x = -x_N$,

$$\Delta p_N(-x_N) = \frac{n_i^2}{N_D}\left[\exp\left(\frac{qV}{kT}\right) - 1\right] \qquad (6.226)$$

Using Equation (6.220) at $x = -x_N$,

$$\Delta p_N(-x_N) = A_N \sinh\left(\frac{-x_N + x_N}{L_p}\right) + B_N \cosh\left(\frac{-x_N + x_N}{L_p}\right)$$
$$- (1-s) \int_\lambda \frac{\tau_p}{\alpha^2 L_p^2 - 1}(1-r) f_0 \alpha \, e^{-\alpha(-x_N + W_N)} d\lambda \qquad (6.227)$$
$$= B_N - (1-s) \int_\lambda \frac{\tau_p}{\alpha^2 L_p^2 - 1}(1-r) f_0 \alpha \, e^{-\alpha(-x_N + W_N)} d\lambda$$

Plugging this into Equation (6.226) gives

$$B_N = \frac{n_i^2}{N_D}\left[\exp\left(\frac{qV}{kT}\right) - 1\right] + (1-s)\int_\lambda \frac{\tau_p}{\alpha^2 L_p^2 - 1}(1-r) f_0 \alpha \, e^{-\alpha(-x_N + W_N)} d\lambda$$

$$(6.228)$$

Now we are back to Equation (6.225). For the physical interpretation, we make the sinh and cosh terms positive, knowing that $W_N > x_N$ (since $\sinh(-x) = -\sinh(x)$ and

$\cosh(-x) = \cosh(x)$). Solving for A_N, we have

$$
A_N = \left[\frac{S_{s,eff}}{D_p} \sinh\left(\frac{W_N - x_N}{L_p} \right) + \frac{1}{L_p} \cosh\left(\frac{W_N - x_N}{L_p} \right) \right]^{-1}
$$

$$
\times \left\{ B_N \left[\frac{1}{L_p} \sinh\left(\frac{W_N - x_N}{L_p} \right) + \frac{S_{s,eff}}{D_p} \cosh\left(\frac{W_N - x_N}{L_p} \right) \right] - (1 - s) \right.
$$

$$
\left. \times \int_\lambda \frac{\tau_p}{\alpha^2 L_p^2 - 1} (1 - r) f_0 \alpha^2 d\lambda - \frac{S_{s,eff}}{D_p} (1 - s) \int_\lambda \frac{\tau_p}{\alpha^2 L_p^2 - 1} (1 - r) f_0 \alpha \, d\lambda \right\}
$$

$$
\tag{6.229}
$$

A_N can now be obtained by plugging Equation (6.228) into Equation (6.229). Finally, the excess minority carrier hole concentration has been solved using Equation (6.220) with Equations (6.228) and (6.229).

6.9.5 Minority Carrier Diffusion Equation for Electrons

We copy Equation (6.170) here:

$$
D_n \frac{d^2 \Delta n_p}{dx^2} - \frac{\Delta n_p}{\tau_n} = -G(x) \tag{6.170}
$$

Dividing this by D_n, we have

$$
\frac{d^2 \Delta n_p}{dx^2} - \frac{\Delta n_p}{D_n \tau_n} = -\frac{G(x)}{D_n} \tag{6.230}
$$

Substituting Equation (6.208) into this gives an expression to be solved.

$$
\frac{d^2 \Delta n_p}{dx^2} - \frac{\Delta n_p}{L_n^2} = -\frac{G(x)}{D_n} \tag{6.231}
$$

The general solution of Equation (6.231) consists of the homogeneous and particular solutions as

$$
\Delta n_p = \Delta n_p \big|_{hom} + \Delta n_p \big|_{par} \tag{6.232}
$$

The homogeneous equation is

$$
\frac{d^2 \Delta n_p}{dx^2} - \frac{\Delta n_p}{L_n^2} = 0 \tag{6.233}
$$

The general solution for the homogeneous equation is easily found as

$$
\Delta n_p \big|_{hom} = A_P \sinh\left(\frac{x - x_p}{L_n} \right) + B_P \cosh\left(\frac{x - x_P}{L_n} \right) \tag{6.234}
$$

The photogeneration rate obtained previously in Equation (6.105) is written for the shifted coordinate.

$$G(x) = (1-s) \int_\lambda (1-r(\lambda)) f_0(\lambda) \alpha(\lambda) e^{-\alpha(\lambda)(x+W_N)} d\lambda \qquad (6.235)$$

Since $G(x)$ is a form of integration having the exponential term, the following particular solution is found to be pertinent.

$$\Delta n_p\big|_{par} = \int_\lambda C_P (1-r) f_0 \alpha e^{-\alpha(x+W_N)} d\lambda \qquad (6.236)$$

Taking the derivative twice, we have

$$\frac{d(\Delta n_P|_{par})}{dx} = - \int_\lambda C_P (1-r) f_0 \alpha^2 e^{-\alpha(x+W_N)} d\lambda \qquad (6.236a)$$

$$\frac{d^2(\Delta n_P|_{par})}{dx^2} = - \int_\lambda C_P (1-r) f_0 \alpha^3 e^{-\alpha(x+W_N)} d\lambda \qquad (6.236b)$$

Substituting Equation (6.236) and (6.236b) into Equation (6.231) gives

$$
-\int_\lambda C_P (1-r) f_0 \alpha^3 e^{-\alpha(x+W_N)} d\lambda + \int_\lambda \frac{C_P}{L_n^2} (1-r) f_0 \alpha e^{-\alpha(x+W_N)} d\lambda
$$
$$
= \int_\lambda \frac{(1-s)}{D_n} (1-r) f_0 \alpha e^{-\alpha(x+W_N)} d\lambda \qquad (6.237)
$$

Canceling the integral parts and rearranging Equation (6.237) reduces to

$$C_P = -\frac{L_n^2(1-s)}{D_n(\alpha^2 L_n^2 - 1)} = -\frac{\tau_n(1-s)}{\alpha^2 L_n^2 - 1} \qquad (6.238)$$

Thus, Equation (6.236) can be written as

$$\Delta n_P|_{par} = -(1-s) \int_\lambda \frac{\tau_n}{\alpha^2 L_n^2 - 1} (1-r) f_0 \alpha e^{-\alpha(x+W_N)} d\lambda \qquad (6.239)$$

Now the general solution of Equation (6.231) that describes the distribution of the excess minority carrier concentration in the n-type region is expressed as

$$
\Delta n_P(x) = A_P \sinh\left(\frac{x-x_P}{L_n}\right) + B_P \cosh\left(\frac{x-x_P}{L_n}\right)
$$
$$
-(1-s) \int_\lambda \frac{\tau_n}{\alpha^2 L_n^2 - 1} (1-r) f_0 \alpha e^{-\alpha(x+W_N)} d\lambda \qquad (6.240)
$$

where A_P and B_P are to be determined by the boundary conditions, Equation (6.198) at $x = W_P$ and Equation (6.206) at $x = x_P$. The boundary condition at $x = W_P$ is

$$\left.\frac{d\Delta n_P}{dx}\right|_{x=W_P} = -\frac{S_{BSF}}{D_n}\Delta p(W_P) \tag{6.241}$$

We need to take derivative of Equation (6.240).

$$\frac{d(\Delta n_P)}{dx} = \frac{A_P}{L_n}\cosh\left(\frac{x-x_P}{L_n}\right) + \frac{B_P}{L_n}\cosh\left(\frac{x-x_P}{L_n}\right)$$
$$+ (1-s)\int_\lambda \frac{\tau_n}{\alpha^2 L_n^2 - 1}(1-r)f_0\alpha^2 e^{-\alpha(x+W_N)}d\lambda \tag{6.242}$$

$$\left.\frac{d(\Delta n_P)}{dx}\right|_{x=W_p} = \frac{A_P}{L_n}\cosh\left(\frac{W_P-x_P}{L_n}\right) + \frac{B_N}{L_P}\sinh\left(\frac{W_P-x_P}{L_n}\right)$$
$$+ (1-s)\int_\lambda \frac{\tau_n}{\alpha^2 L_n^2 - 1}(1-r)f_0\alpha^2 e^{-\alpha(W_P+W_N)}d\lambda \tag{6.243}$$

And using Equation (6.240)

$$\Delta n_P(W_P) = A_P\sinh\left(\frac{W_P-x_P}{L_n}\right) + B_P\cosh\left(\frac{W_P-x_P}{L_n}\right)$$
$$- (1-s)\int_\lambda \frac{\tau_n}{\alpha^2 L_n^2 - 1}(1-r)f_0\alpha e^{-\alpha(W_P+W_N)}d\lambda \tag{6.244}$$

Plugging Equation (6.243) and (6.244) into Equation (6.241) and reducing it gives

$$\frac{A_P}{L_n}\cosh\left(\frac{W_P-x_P}{L_n}\right) + \frac{B_P}{L_n}\sinh\left(\frac{W_P-x_P}{L_n}\right) + (1-s)\int_\lambda \frac{\tau_n}{\alpha^2 L_n^2 - 1}(1-r)f_0\alpha^2 d\lambda$$
$$= \frac{S_{BSF}}{D_n}\left(A_P\sinh\left(\frac{W_P-x_P}{L_n}\right) + B_P\cosh\left(\frac{W_P-x_P}{L_n}\right)\right.$$
$$\left. -(1-s)\int_\lambda \frac{\tau_n}{\alpha^2 L_n^2 - 1}(1-r)f_0\alpha\, d\lambda\right)$$

$$\tag{6.245}$$

Now we consider the other boundary condition at $x = x_P$

$$\Delta n_P(x_P) = \frac{n_i^2}{N_A}\left[\exp\left(\frac{qV}{kT}\right) - 1\right] \tag{6.246}$$

Using Equation (6.240) at $x = x_P$,

$$\Delta n_P(x_P) = A_P \sinh\left(\frac{W_P - x_P}{L_n}\right) + B_P \cosh\left(\frac{W_P - x_P}{L_n}\right)$$

$$- (1-s) \int_\lambda \frac{\tau_n}{\alpha^2 L_n^2 - 1}(1-r) f_0 \alpha \, e^{-\alpha(x_P + W_N)} d\lambda \qquad (6.247)$$

$$= B_P - (1-s) \int_\lambda \frac{\tau_n}{\alpha^2 L_n^2 - 1}(1-r) f_0 \alpha \, e^{-\alpha(x_P + W_N)} d\lambda$$

Plugging this into Equation (6.246) gives

$$B_P = \frac{n_i^2}{N_A}\left[\exp\left(\frac{qV}{kT}\right) - 1\right] + (1-s)\int_\lambda \frac{\tau_n}{\alpha^2 L_n^2 - 1}(1-r) f_0 \alpha \, e^{-\alpha(x_P + W_N)} d\lambda$$

$$(6.248)$$

Now we return to Equation (6.245). Solving Equation (6.245) for A_P, we have

$$A_P = \left[\frac{S_{BSF}}{D_n}\sinh\left(\frac{W_P - x_P}{L_n}\right) + \frac{1}{L_n}\cosh\left(\frac{W_P - x_P}{L_n}\right)\right]^{-1}$$

$$\times \left\{-B_P\left[\frac{1}{L_n}\sinh\left(\frac{W_P - x_P}{L_n}\right) + \frac{S_{BSF}}{D_n}\cosh\left(\frac{W_P - x_P}{L_n}\right)\right]\right.$$

$$- (1-s)\int_\lambda \frac{\tau_n}{\alpha^2 L_n^2 - 1}(1-r) f_0 \alpha^2 \, e^{-\alpha(W_P + W_n)} d\lambda$$

$$\left. + \frac{S_{BSF}}{D_n}(1-s)\int_\lambda \frac{\tau_n}{\alpha^2 L_n^2 - 1}(1-r) f_0 \alpha \, e^{-\alpha(W_P + W_n)} d\lambda\right\} \qquad (6.249)$$

A_P is found explicitly by plugging Equation (6.248) into Equation (6.249). Finally, the excess minority carrier hole concentration has been solved using Equation (6.240) with Equations (6.248) and (6.249).

6.10   CHARACTERISTICS OF SOLAR CELLS

6.10.1   Current Density

The total current density (Ampere/cm^2) is the sum of the hole and electron current densities, as discussed in Equation (6.151). Now we express it as a function of position. Notice that the total current density is constant in the x direction (only one current exists across the cell at a steady-state condition). This is depicted in Figure 6.35. The total current is then expressed by

$$J = J_P(x) + J_N(x) \qquad (6.250)$$

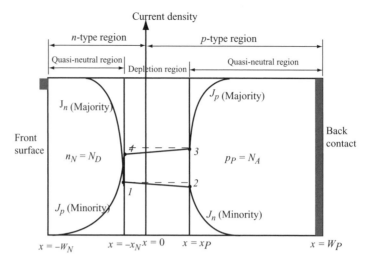

Figure 6.35   Schematic of current densities in a solar cell

As the minority carrier diffusions in the quasi-neutral regions are considered in the previous section, the minority current densities at the edges of points 2 and 4 are readily obtained using the carrier diffusion equations developed so far. However, we need to compute the total current densities at either $x = -x_N$ or $x = x_P$. We just pick the current density at $x = -x_N$. The other choice would require the same degree of efforts. In some other studies they assumed that there is no generation and recombination in the depletion region, for simplicity. Then the current densities in the region would be flat, as shown as the dotted lines in Figure 6.35. Then, the analysis would be much easier. Nevertheless, we want to develop a more general model encountering the generation and recombination in the depletion region. Therefore, we need the condition such as

$$J = J_P(-x_N) + J_n(-x_N) = \text{point } 4 + \text{point } 1 \tag{6.251}$$

where $J_P(-x_N)$ at point 4 is readily obtained from the minority carrier diffusion equation, as shown in Figure 6.35. However, $J_n(-x_N)$ at point 1 requires us to compute the net generation- recombination rate in the depletion region. Considering the continuity equation for electrons in the depletion region from Equation (6.164) with 1-D steady state conditions, we have

$$\frac{dJ_n}{dx} = q(R_n - G) \tag{6.252}$$

Integrating this over the depletion region gives

$$\int_{-x_N}^{x_P} \frac{dJ_n}{dx}\,dx = q \int_{-x_N}^{x_P} [R_n(x) - G(x)]\,dx \tag{6.253}$$

The left-hand side is expressed as

$$\int_{-x_N}^{x_P} \frac{dJ_n}{dx} dx = \int_{J_n(-x_N)}^{J_n(x_P)} dJ_n = J_n(x_P) - J_n(-x_N) = \text{point 2} - \text{point 1} \qquad (6.254)$$

Since $J_n(x_P)$ at point 2 is readily obtained from the minority carrier diffusion equation, $J_n(-x_N)$ at point 1 can be computed if the quantities in the right-hand side of Equation (6.253) are obtained. R_n is the recombination rate within the depletion region which can be approximated by assuming that the recombination rate is constant and maximum in the region, where $p_D = n_D$ and $E_i = E_T$ (Gray [18]). The general trap recombination of Equation (6.126) with Equations (6.119) and (6.120) can be used.

$$R_n = \frac{np - n_i^2}{\tau_p(n + n_1) + \tau_n(p + p_1)} \qquad (6.255)$$

$$n_1 = n_i e^{(E_T - E_i)/kT} \qquad (6.255a)$$

$$p_1 = n_i e^{(E_T - E_i)/kT} \qquad (6.255b)$$

Rewrite Equation (6.255) using the previous assumptions:

$$R_n = \frac{n_D p_D - n_i^2}{\tau_p(n_D + n_i) + \tau_n(p_D + p_i)} = \frac{n_D^2 - n_i^2}{\tau_p(n_D + n_i) + \tau_n(p_D + p_i)}$$
$$= \frac{n_D^2 - n_i^2}{(\tau_n + \tau_p)(n_D + n_i)} = \frac{n_D - n_i}{\tau_n + \tau_p} \qquad (6.256)$$

From Equation (6.194) under non-equilibrium conditions,

$$n_D p_D = n_D^2 = n_i^2 \exp\left(\frac{qV}{kT}\right) \qquad (6.257)$$

$$n_D = n_i \exp\left(\frac{qV}{kT}\right) \qquad (6.258)$$

Substituting this into Equation (6.256) gives the recombination rate in the depletion region.

$$R_n = \frac{n_i \left(\exp\left(\frac{qV}{2kT}\right) - 1\right)}{\tau_D} \qquad (6.259)$$

where $\tau_D = \tau_n + \tau_p$.

The first term of the right-hand side of Equation (6.253) can now be rewritten as

$$q \int_{-x_N}^{x_P} R_n(x)dx = q \int_{-x_N}^{x_P} \frac{n_i \left(\exp\left(\frac{qV}{2kT}\right) - 1 \right)}{\tau_D} dx$$

$$= q \frac{(x_p + x_N)n_i}{\tau_D} \left(\exp\left(\frac{qV}{2kT}\right) - 1 \right)$$

$$= q \frac{W_D W_i}{\tau_D} \left(\exp\left(\frac{qV}{2kT}\right) - 1 \right)$$

(6.260)

The second term of the right-hand side of Equation (6.253) can be rewritten using Equation (6.215).

$$q \int_{-x_N}^{x_P} G(x)dx = q \int_{-x_N}^{x_P} (1-s) \int_\lambda (1-r(\lambda)) f_0(\lambda)\alpha(\lambda)e^{-\alpha(\lambda)(x+W_N)} d\lambda dx$$

$$= q(1-s) \int_\lambda (1-r(\lambda)) f_0(\lambda)\alpha(\lambda) \int_{-x_N}^{x_P} e^{-\alpha(\lambda)(x+W_N)} dx d\lambda$$

(6.261)

Computing the inside integral,

$$\int_{-x_N}^{x_P} e^{-\alpha(\lambda)(x+W_N)} dx = \left[-\frac{1}{\alpha(\lambda)} e^{-\alpha(\lambda)(x+W_N)} \right]_{-x_N}^{x_P}$$

$$= -\frac{1}{\alpha(\lambda)} \left[e^{-\alpha(\lambda)(x_P+W_N)} - e^{-\alpha(\lambda)(-x_N+W_N)} \right]$$

(6.262)

Then, inserting this into Equation (6.261) gives

$$q \int_{-x_N}^{x_P} G(x)dx = -q(1-s) \int_\lambda (1-r(\lambda)) f_0(\lambda) \left[e^{-\alpha(\lambda)(x_P+W_N)} - e^{-\alpha(\lambda)(-x_N+W_N)} \right] d\lambda$$

(6.263)

Substituting Equations (6.254), (6.260), and (6.263) into Equation (6.253) yields

$$J_n(x_P) - J_n(-x_N) = q \frac{W_D n_i}{\tau_D} \left(\exp\left(\frac{qV}{2kT}\right) - 1 \right)$$

$$+ q(1-s) \int_\lambda (1-r(\lambda)) f_0(\lambda) \left[e^{-\alpha(\lambda)(x_P+W_N)} - e^{-\alpha(\lambda)(-x_N+W_N)} \right] d\lambda$$

(6.264)

Solving for $J_n(-x_N)$,

$$J_n(-x_N) = J_n(x_P) - q\frac{W_D n_i}{\tau_D}\left(\exp\left(\frac{qV}{2kT}\right) - 1\right)$$

$$- q(1-s)\int_\lambda (1-r(\lambda))f_0(\lambda)\left[e^{-\alpha(\lambda)(x_P+W_N)} - e^{-\alpha(\lambda)(-x_N+W_N)}\right]d\lambda$$

$$(6.265)$$

Inserting this into Equation (6.251), the total current density is now expressed as

$$J = J_p(-x_N) + J_n(x_P) - q\frac{W_D n_i}{\tau_D}\left(\exp\left(\frac{qV}{2kT}\right) - 1\right)$$

$$- q(1-s)\int_\lambda (1-r(\lambda))f_0(\lambda)\left[e^{-\alpha(\lambda)(x_P+W_N)} - e^{-\alpha(\lambda)(-x_N+W_N)}\right]d\lambda$$

$$(6.266)$$

This is solvable because both $J_p(-x_N)$ at point 4 and $J_n(x_P)$ at point 1 can be obtained from the solvable minority carrier diffusion equation for the depletion region, as shown in Figure 6.35. The former is obtained from the minority current density given by Equation (6.149), which is just the diffusion current, since the electric field is negligible. Therefore,

$$J_p(-x_N) = -qD_p\frac{dp}{dx}\bigg|_{x=-x_N} = -qD_p\frac{d\Delta p_N}{dx}\bigg|_{x=-x_N} \tag{6.267}$$

Taking derivative of Equation (6.220), we have

$$\frac{d\Delta p_N}{dx}\bigg|_{x=-x_N} = \frac{A_N}{L_p}\cosh\left(\frac{-x_N+x_N}{L_p}\right) + \frac{B_N}{L_p}\sinh\left(\frac{-x_N+x_N}{L_p}\right)$$

$$+ (1-s)\int_\lambda \frac{\tau_p}{\alpha^2 L_p^2 - 1}(1-r)f_0\alpha^2 e^{-\alpha(-x_N+W_N)}d\lambda \tag{6.268}$$

$$J_p(-x_N) = -q\frac{D_p}{L_p}A_N - qD_p(1-s)\int_\lambda \frac{\tau_p}{\alpha^2 L_p^2 - 1}(1-r)f_0\alpha^2 e^{-\alpha(-x_N+W_N)}d\lambda$$

$$(6.269)$$

The second term of Equation (6.266) is now obtained using Equation (6.150).

$$J_n(x_P) = qD_n\frac{dn}{dx}\bigg|_{x=x_P} = qD_n\frac{d\Delta n_P}{dx}\bigg|_{x=x_P} \tag{6.270}$$

Taking derivative of Equation (6.240), we have

$$\frac{d\Delta n_P}{dx}\bigg|_{x=x_P} = \frac{A_P}{L_n}\cosh\left(\frac{x_P-x_P}{L_n}\right) + \frac{B_P}{L_n}\sinh\left(\frac{x_P-x_P}{L_n}\right)$$

$$+ (1-s)\int_\lambda \frac{\tau_n}{\alpha^2 L_n^2 - 1}(1-r)f_0\alpha^2 e^{-\alpha(x_P+W_N)}d\lambda \tag{6.271}$$

From Equation (270),

$$J_n(x_P) = q\frac{D_n}{L_n}A_P + qD_n(1-s)\int_\lambda \frac{\tau_n}{\alpha^2 L_n^2 - 1}(1-r)f_0\alpha^2 e^{-\alpha(-x_P+W_N)}d\lambda \quad (6.272)$$

Finally, the total current density from Equation (6.266) is obtained as

$$
\begin{aligned}
J = &-q\frac{D_p}{L_p}A_N - qD_p(1-s)\int_\lambda \frac{\tau_p}{\alpha^2 L_p^2 - 1}(1-r)f_0\alpha^2 e^{-\alpha(-x_N+W_N)}d\lambda \\
&+ q\frac{D_n}{L_n}A_P + qD_n(1-s)\int_\lambda \frac{\tau_n}{\alpha^2 L_n^2 - 1}(1-r)f_0\alpha^2 e^{-\alpha(x_P+W_N)}d\lambda \\
&- q\frac{W_D n_i}{\tau_D}\left(\exp\left(\frac{qV}{2kT}\right) - 1\right) \\
&- q(1-s)\int_\lambda (1-r)f_0\left[e^{-\alpha(x_P+W_N)} - e^{-\alpha(-x_N+W_N)}\right]d\lambda
\end{aligned}
\quad (6.273)
$$

where A_N and A_P were derived and presented in Equations (6.249) and (6.248). The total current density in Equation (6.273) provides the current-voltage characteristic for a given solar cell. This single formula literally takes into account all the phenomena and mechanisms that we discussed up to now from quantum physics to minority carrier diffusion. We obviously want to take a look at the meaning of each term and its effect to see if there are the critical points that might significantly change the characteristics and the figures of merit and so influence the performance. In order to do that, we need to rearrange all the terms into meaningful groups that have similar behaviors. This rearrangement will be carried out as follows.

For convenience, we define two quantities (positive) associated with the front surface recombination as

$$X_P = \frac{S_{F,eff}}{D_p}\sinh\left(\frac{W_N - x_N}{L_p}\right) + \frac{1}{L_p}\cosh\left(\frac{W_N - x_N}{L_p}\right) \quad (6.274)$$

$$Y_P = \frac{S_{F,ef}}{D_p}\cosh\left(\frac{W_N - x_N}{L_p}\right) + \frac{1}{L_p}\sinh\left(\frac{W_N - x_N}{L_p}\right) \quad (6.275)$$

And two quantities (positive) associated with the back surface recombination,

$$X_N = \frac{S_{BSF}}{D_n}\sinh\left(\frac{W_P - x_P}{L_n}\right) + \frac{1}{L_n}\cosh\left(\frac{W_P - x_P}{L_n}\right) \quad (6.276)$$

$$Y_N = \frac{S_{BSF}}{D_n}\cosh\left(\frac{W_P - x_P}{L_n}\right) + \frac{1}{L_n}\sinh\left(\frac{W_P - x_P}{L_n}\right) \quad (6.277)$$

And also define four quantities (positive or negative) related to the photogeneration as shown in Equations (6.278) to (6.281). We note that the quantities have a singularity at $\alpha L = 1$, where α is the absorption coefficient and L is the minority diffusion length. When $(\alpha^2 L^2 - 1) \geq 0$, the quantity positively explodes as the value approach the critical point. When $(\alpha^2 L^2 - 1) \leq 0$, the quantity changes its polarity and negatively explodes, completely reversing the contribution of the photogeneration mathematically. Obviously, this is not desirable. It is speculated from the formula that although the critical point occurs at a specific wavelength of the integration in the equations, the resultant effect of the critical point depends on the range of the wavelength at the vicinity of the critical point. The product of the absorption coefficient and the diffusion length indicates that the distribution of the absorption coefficient along the wavelength becomes very important when the diffusion length remains constant, or vice versa.

$$G_{p1}(x) = (1-s)\int_\lambda \frac{\tau_p}{\alpha^2 L_p^2 - 1}(1-r)f_0\alpha\, e^{-\alpha(x+W_N)}d\lambda \tag{6.278}$$

$$G_{p2}(x) = (1-s)\int_\lambda \frac{\tau_p}{\alpha^2 L_p^2 - 1}(1-r)f_0\alpha^2\, e^{-\alpha(x+W_N)}d\lambda \tag{6.279}$$

$$G_{n1}(x) = (1-s)\int_\lambda \frac{\tau_n}{\alpha^2 L_n^2 - 1}(1-r)f_0\alpha\, e^{-\alpha(x+W_N)}d\lambda \tag{6.280}$$

$$G_{n2}(x) = (1-s)\int_\lambda \frac{\tau_n}{\alpha^2 L_n^2 - 1}(1-r)f_0\alpha^2\, e^{-\alpha(x+W_N)}d\lambda \tag{6.281}$$

We are then able to rewrite A_N in Equation (6.229) to a simpler form using the quantities already defined:

$$A_N = \frac{Y_P}{X_P}\frac{n_i^2}{N_D}\left[\exp\left(\frac{qV}{kT}\right)-1\right] + \frac{Y_P}{X_P}G_{p1}(-x_N) - \frac{G_{p2}(-W_N)}{X_P}$$
$$- \frac{S_{F,eff}}{D_p X_P}G_{p1}(-W_N) \tag{6.282}$$

We also rewrite A_P in Equation (6.249) to a simpler form using the quantities defined:

$$A_P = -\frac{Y_N}{X_N}\frac{n_i^2}{N_A}\left[\exp\left(\frac{qV}{kT}\right)-1\right] - \frac{Y_N}{X_N}G_{n1}(x_P) - \frac{G_{n2}(W_P)}{X_N} + \frac{S_{BSF}}{D_n X_n}G_{n1}(W_P) \tag{6.283}$$

6.10.2   Current-Voltage Characteristics

Finally, substituting Equations (6.282) and (6.283) into Equation (6.273) and rearranging, the total current density in Equation (6.273) can be written in a more compact form as

$$J = J_{SC} - J_{01}\left(\exp\left(\frac{qV}{kT}\right)-1\right) - J_{02}\left(\exp\left(\frac{qV}{2kT}\right)-1\right) \tag{6.284}$$

where J_{SC} is called the *short-circuit current density* and is the sum of contributions from each of the three regions: The n-type region J_{SCN}, the p-type region J_{SCP}, and the depletion region J_{SCD}. J_{01} is the *dark saturation current density* due to recombination in the quasi-neutral regions. J_{02} is the *dark saturation current density* due to recombination in the depletion region. Therefore, the short-circuit current density is expressed by

$$J_{SC} = J_{SCN} + J_{SCP} + J_{SCD} \tag{6.285}$$

where

$$J_{SCN} = \frac{q S_{F,eff}}{L_p X_P} G_{p1}(-W_N) + \frac{q D_p}{L_p X_P} G_{p2}(-W_N) - \frac{q D_p}{L_p} \frac{Y_P}{X_P} G_{p1}(-x_N)$$
$$- q D_p G_{p2}(-x_N) \tag{6.286}$$

$$J_{SCP} = \frac{q S_{BSF}}{L_n X_N} G_{n1}(W_P) - \frac{q D_n}{L_n X_N} G_{n2}(W_P) - \frac{q D_n}{L_n} \frac{Y_N}{X_N} G_{n1}(x_P) + q D_n G_{n2}(x_P) \tag{6.287}$$

$$J_{SCD} = q(1-s) \int_\lambda (1-r) f_0 \left[e^{-\alpha(-x_N+W_N)} - e^{-\alpha(x_P+W_N)} \right] d\lambda \tag{6.288}$$

The dark saturation current densities are expressed, respectively.

$$J_{01} = \left[\frac{q D_p}{L_p} \frac{Y_P}{X_P} \frac{n_i^2}{N_D} + \frac{q D_n}{L_n} \frac{Y_N}{X_N} \frac{n_i^2}{N_A} \right] \tag{6.289}$$

$$J_{02} = q \frac{W_D n_i}{\tau_D} \tag{6.290}$$

The current is obtained by multiplying J by the area so that

$$I = AJ \tag{6.291}$$

The total current obtained from Equation (6.284) is an expression for the current produced by a solar cell.

$$I = I_{SC} - I_{01}\left(\exp\left(\frac{qV}{kT}\right) - 1 \right) - I_{02}\left(\exp\left(\frac{qV}{2kT}\right) - 1 \right) \tag{6.292}$$

where I_{SC} is called the *short-circuit current*. I_{01} is the *dark saturation current* due to recombination in the quasi-neutral regions. I_{02} is the *dark saturation current density* due to recombination in the depletion region. The dark current due to the depletion region is usually very small and negligible (a reasonable and common assumption for a good silicon solar cell). The simple solar cell circuit is

$$I = I_{SC} - I_0\left(\exp\left(\frac{qV}{kT}\right) - 1 \right) \tag{6.292a}$$

The equivalent circuit diagram models the solar cell and its external load as a lumped parameter circuit, as shown in Figure 6.36(a). The photogeneration mechanism is presented by the current generator of the left-hand side of the circuit. The dark recombination current mechanisms are represented by two diodes. The magnitudes of their current are I_{SC}, I_{01}, and I_{02}, respectively. Note that the current generator and

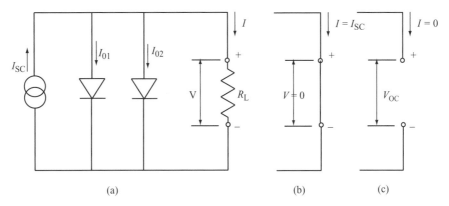

Figure 6.36   Equivalent circuit diagram for a solar cell, (a) external load, (b) short circuit, and (c) open circuit

diodes are oriented in opposite directions since the diode currents detract from the photogenerated current. The *photocurrent* is the photogenerated current that is equal to the short-circuit current I_{SC}, where the voltage is zero, which is seen in Figure 6.36(b).

When a load is present, a potential develops between the terminals of the cell. This potential difference generates a current that acts in the opposite direction to the photocurrent, and the net current is reduced from its short-circuit value. This reverse current is usually called the dark current.

At open circuit ($I = 0$) in Figure 6.36(c), all the short circuit current or photocurrent is equal to the dark current. From Equation (6.292a), letting $V = V_{OC}$, we have,

$$0 = I_{SC} - I_0 \left(\exp \left(\frac{q V_{OC}}{kT} \right) - 1 \right) \tag{6.293}$$

Solving for V_{OC}, the open-circuit voltage is

$$V_{OC} = \frac{kT}{q} \ln \left(\frac{I_{SC}}{I_0} + 1 \right) \tag{6.294}$$

Usually, the short-circuit current I_{SC} is much greater than the dark current I_0. Therefore, the open-circuit current may be expressed in such case as

$$V_{OC} = \frac{kT}{q} \ln \left(\frac{I_{SC}}{I_0} \right) \tag{6.294a}$$

The current-voltage curve for a solar cell is obtained by plotting the output current as a function of the output voltage. This is done experimentally by varying the external load from zero ohms to several kilo-ohms while the cell is illuminated. Analytically, the curve may be estimated with Equation (6.292) for the input parameters shown in Table 6.9. These parameters are used in the following figures unless otherwise indicated. The analytic current-voltage curve is shown in Figure 6.37. Of particular interest is the point on the current-voltage curve where the power produced is at a maximum. This is referred to as the maximum power point with $V = V_{mp}$ and $I = I_{mp}$. The short-circuit current I_{SC} and the open-circuit voltage V_{OC} are indicated in the figure. The effect of

Table 6.9   Silicon Solar Cell Model Parameters Used at 300 K
for AM1.5 Global Spectrum

Parameter	Value	Parameter	Value
A(Area)	$100\ \text{cm}^2$	W_N	$0.35\ \mu\text{m}$
N_D	$1 \times 10^{18}\ \text{cm}^{-3}$	W_P	$300\ \mu\text{m}$
N_A	$1 \times 10^{15}\ \text{cm}^{-3}$	τ_n	$350\ \mu\text{s}$
$S_{F,\text{eff}}$	$3 \times 10^4\ \text{cm/s}$	τ_p	$1\ \mu\text{s}$
S_{BSF}	$100\ \text{cm/s}$	s (shade factor)	0.05

temperature is also shown in Figure 6.37. On the one hand, the open-circuit voltage
considerably reduces as the temperature increases from 300 K to 350 K. On the other
hand, the sort-circuit current increases slightly.

6.10.3   Figures of Merit

The maximum power point is found by solving

$$\left.\frac{dP}{dV}\right|_{V=V_{mp}} = \frac{d(IV)}{dV} = 0 \tag{6.295}$$

Using Equation (6.292a), the power is expressed by

$$P = IV = I_{SC}V - I_0 V \left(\exp\left(\frac{qV}{kT}\right) - 1\right) \tag{6.296}$$

Taking derivative of this equation gives

$$\left.\frac{dP}{dV}\right|_{V=V_{mp}} = I_{SC} - \left[I_0 V \frac{q}{kT} e^{\frac{qV_{mp}}{kT}} + I_0 \left(e^{\frac{qV_{mp}}{kT}} - 1\right)\right] = 0 \tag{6.297}$$

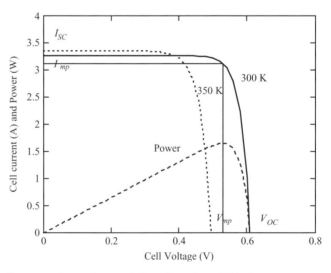

Figure 6.37   Current-voltage characteristics of a solar cell for input parameters in Table 6.9

Dividing this by I_0 and taking logarithms yields

$$V_{mp} = V_{OC} - \frac{kT}{q} \ln\left(\frac{qV_{mp}}{kT} + 1\right) \tag{6.298}$$

Since the logarithm depends only weakly on its argument, we can substitute V_{OC} for V_{mp} in the logarithm. We arrive at the voltage at the maximum power point. The maximum power voltage is

$$V_{mp} = V_{OC} - \frac{kT}{q} \ln\left(\frac{qV_{OC}}{kT} + 1\right) \tag{6.299}$$

Substituting this into Equation (6.292a), we find the current at the maximum power point as

$$I_{mp} = I_{SC} - I_0 \left\{ \exp\left[\frac{q}{kT}\left(V_{OC} - \frac{kT}{q}\ln\left(\frac{qV_{OC}}{kT}+1\right)\right)\right] - 1 \right\} \tag{6.300}$$

$$I_{mp} = I_{SC} - I_0 \left\{ \exp\left[\frac{q}{kT}V_{OC} - \ln\left(\frac{qV_{OC}}{kT}+1\right)\right] - 1 \right\} \tag{6.300a}$$

Assuming, as is commonly done in the literature, that

$$\exp\left[\frac{q}{kT}V_{OC} - \ln\left(\frac{qV_{OC}}{kT}+1\right)\right] \gg 1 \tag{6.300b}$$

The maximum power current is obtained by

$$I_{mp} \cong I_{SC} - I_0 \exp\left[\frac{q}{kT}V_{OC} - \ln\left(\frac{qV_{OC}}{kT}+1\right)\right] \tag{6.301}$$

The *fill factor*, FF, of a current-voltage curve is a parameter evaluating the quality of the current-voltage curve.

$$FF = \frac{P_{mp}}{I_{SC}V_{OC}} = \frac{I_{mp}V_{mp}}{I_{SC}V_{OC}} \tag{6.302}$$

The fill factor is the ratio of the maximum power rectangle ($I_{mp} \times V_{mp}$) to the maximum rectangle ($I_{SC} \times V_{OC}$). In other words, the fill factor is a measure of *squareness* of the I-V curve. Typically, it has a value of 0.7 to 0.9 for solar cells with a reasonable efficiency. Fill factor reflects the effects of series and shunt resistances on the I-V characteristic (see Section 6.10.8). The series and shunt resistances degrade the fill factor, depending on the shape of the knee of the curve in Figure 6.37. As the knee of the curve becomes more rounded, the output power decreases.

$$FF = \frac{I_{mp}V_{mp}}{I_{SC}V_{OC}}$$

$$= \left\{I_{SC} - I_0 \exp\left[\frac{q}{kT}V_{OC} - \ln\left(\frac{qV_{OC}}{kT}+1\right)\right]\right\}\frac{1}{I_{SC}}\left[V_{OC} - \frac{kT}{q}\ln\left(\frac{qV_{OC}}{kT}+1\right)\right]\frac{1}{V_{OC}}$$

$$= \left\{1 - \frac{I_0}{I_{SC}} \exp\left[\frac{q}{kT}V_{OC} - \ln\left(\frac{qV_{OC}}{kT}+1\right)\right]\right\}\left[1 - \frac{kT}{qV_{OC}}\ln\left(\frac{qV_{OC}}{kT}+1\right)\right]$$

$$\tag{6.302a}$$

Using the assumption of Equation (6.294a) and solving for I_{SC}/I_0, we have

$$\frac{I_{SC}}{I_0} = \exp\left(\frac{kT}{qV_{OC}}\right) \qquad (6.303)$$

Substituting this into Equation (6.302a) gives

$$FF \cong \left\{1 - \exp\left(\frac{qV_{OC}}{kT}\right)\exp\left[\frac{q}{kT}V_{OC} - \ln\left(\frac{qV_{OC}}{kT}+1\right)\right]\right\}\left[1 - \frac{kT}{qV_{OC}}\ln\left(\frac{qV_{OC}}{kT}+1\right)\right] \qquad (6.304)$$

After some algebra, the *fill factor* is expressed by

$$FF \cong \frac{\dfrac{qV_{OC}}{kT} - \ln\left(\dfrac{qV_{OC}}{kT}+1\right)}{\dfrac{qV_{OC}}{kT}+1} \qquad (6.305)$$

Notice that the fill factor is a function of V_{OC}.

The most important *figure of merit* for a solar cell is its *power conversion efficiency*, η.

$$\eta = \frac{P_{mp}}{P_{in}} = \frac{I_{mp}V_{mp}}{P_{in}} = \frac{FF \cdot I_{SC}V_{OC}}{P_{in}} \qquad (6.306)$$

The incident power P_{in} is determined by the properties of the light spectrum incident upon the solar cell.

$$P_{in} = A \int_0^\infty h\nu(\lambda)f_0(\lambda)d\lambda \qquad (6.307)$$

where $f_0(\lambda)$ is the incident photon flux (number of photon incident per unit area per second per wavelength, also defined in Section 6.6.3.

6.10.4 Effect of Minority Electron Lifetime on Efficiency

The minority electron concentration was previously discussed in Section 6.9.2. The minority electron concentration gradient in the p-type region is of great importance for the efficiency of a solar cell. The minority electron concentration gradient depends primarily on the recombination taking place in the quasi-neutron region or, in other words, on the minority electron lifetime.

The effect of the *minority electron lifetime* τ_n on the performance of a silicon solar cell is illustrated in Figure 6.38 in terms of the *figures of merit* such as FF, I_{SC}, and V_{OC} from the results of the computations in the previous section. The efficiency of the solar cell is shown in Figure 6.39. These indicate that the short-circuit current I_{SC} and the open-circuit voltage V_{OC} increase with increasing minority electron lifetime τ_n. Since FF is a function of V_{OC}, the FF increases as well with increasing τ_n.

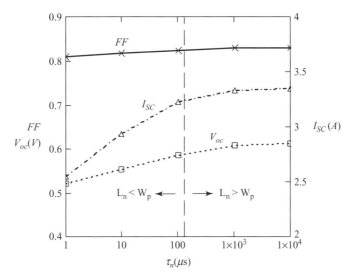

Figure 6.38   Effect of minority electron lifetime on FF, I_{SC}, and V_{OC} for a solar cell at $W_P = 300\,\mu m$ and $S_{BSF} = 100\,cm/s$

Since the minority electron diffusion length is defined as $L_n = \sqrt{D_n \tau_n}$ in Equation (6.208), the base width W_P is used to calculate a specific lifetime ($\tau_n = W_P^2/D_n$) that lies at 132 μs in the figure. There are two aspects. When $L_n < W_p$, the diffusion length in the base is much less than the base thickness and the electrons created deeper than the diffusion length are unlikely to be collected. This consequently reduces the short-circuit current. In this case, the back surface recombination velocity S_{BSF} has no

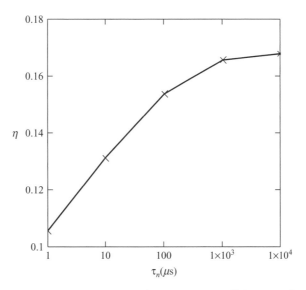

Figure 6.39   Effect of minority electron lifetime on the efficiency η for a solar cell at $W_P = 300\,\mu m$ and $S_{BSF} = 100\,cm/s$

effect on the dark saturation current I_0. When $L_n > W_p$, the minority electron diffusion length is much larger than the base thickness and the electrons created anywhere in the base can either transfer to the emitter region through the depletion region or move to the back surface and annihilate by *surface recombination* (Section 6.7.1) unless a special treatment of keeping away from the back surface is applied. This treatment can be achieved by controlling the effective *back surface recombination velocity*, S_{BSF} (see Section 6.9.1).

The impact of the minority electron lifetime on the power conversion efficiency is depicted in Figure 6.39. The efficiency drastically increases with increasing the minority electron lifetime. The lifetime of 350 μs used in other figures is a large value and requires a high-quality semiconductor material.

6.10.5 Effect of Minority Hole Lifetime on Efficiency

The minority hole lifetime τ_p is usually small due to the short emitter width W_N. Figure 6.40 shows that the effect of τ_p on the efficiency is minimal and the effect of τ_p does not change with varying the emitter width from 0.1 μm to 1.0 μm. We notice that the efficiency is greatly affected by the emitter width.

6.10.6 Effect of Back Surface Recombination Velocity on Efficiency

The trend in the solar cell industry is to move to thin solar cells (below 50 μm), reducing the material volume, and hence lowering price. In general, a thin solar cell faces three problems: decrease of photon absorptivity, increase of back surface recombination, and brittleness of the material (mechanical) due to the thinness. When a solar cell is thin, some minority electrons can reach the back surface, which usually annihilate

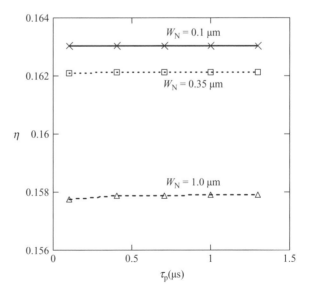

Figure 6.40 Effect of minority hole lifetime, τ_p, on the efficiency

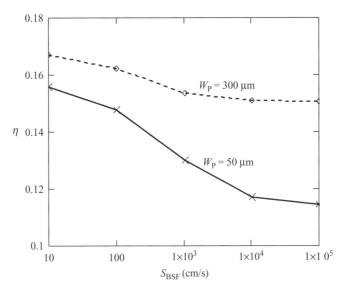

Figure 6.41   Effect of S_{BSF} on the efficiency of a solar cell for $W_P = 50\,\mu$m and $W_P = 300\,\mu$m

due to the *surface recombination*. However, the electrons that reach the back surface can be turned away from the back surface by means of a *back surface field* (BSF), which is achieved by nonuniform heavy doping near the back surface. The effect of the back-surface recombination velocity S_{BSF} on the efficiency is illustrated with two plots for base widths of $W_P = 50\,\mu$m and $W_P = 300\,\mu$m, as shown in Figure 6.41. We see that both the efficiencies increase with decreasing S_{BSF} and the gap between the two base widths gradually becomes close together with decreasing S_{BSF}. This implies that, when S_{BSF} has a low value, the thin solar cell at $W_P = 50\,\mu$m would have the similar efficiency with the typical solar cell at $W_P = 300\,\mu$m. This is an advantage for the thin solar cells with a low S_{BSF}. The *figures of merit* such as FF, I_{SC}, and V_{OC} are plotted in Figure 6.42 at $W_P = 50\,\mu$m with respect to S_{BSF}, showing that both I_{SC} and V_{OC} increase with decreasing S_{BSF}. It is good to know whether I_{SC} and V_{OC} are in the same direction of increasing.

6.10.7   Effect of Base Width on Efficiency

As said before, the back surface recombination velocity S_{BSF} is an important parameter on a solar cell thickness that is mostly the base width, since the emitter width is very thin. The effect of base width on the efficiency as a function of S_{BSF} is illustrated in Figure 6.43. We see that the base width significantly affects the performance of a solar cell, due primarily to the silicon material limits of the absorption coefficient. Notice that there is nearly no effect of S_{BSF} on the efficiency at $W_p = 1,000\,\mu$m in Figure 6.43. Conclusively, the effect of the back surface recombination velocity becomes significant as the base width decreases.

The effect of base width W_p on the efficiency of a solar cell as a function of the minority electron diffusion length L_n or lifetime τ_n is illustrated in Figure 6.44. Obviously, the longer lifetimes show an improved efficiency. It is interesting to note

472 Solar Cells

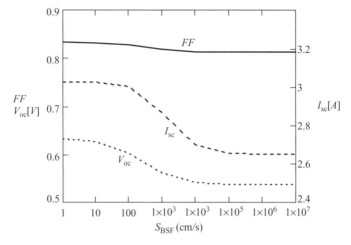

Figure 6.42 Effect of S_{BSF} on the performance of a solar cell at $W_P = 50\,\mu m$

that the efficiency has a maximum value for a certain base width that depends on the minority electron lifetime. Figure 6.45 shows the effect of W_p on FF, I_{SC}, and V_{OC} at $\tau_n = 15\,\mu s$. As W_p increases, I_{SC} increases but V_{OC} decreases. This produces a maximum point that can be an important design concept of a solar cell. Conclusively, as the base width increases, the photogenerated current increases but the open-circuit voltage decreases.

6.10.8 Effect of Emitter Width W_N on Efficiency

As we discussed previously, the emitter width, W_N, is very small compared to the base width, W_P. However, its effect on the performance should not be belittled. The

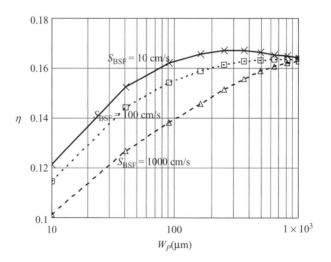

Figure 6.43 Effect of base width on the efficiency as a function of S_{BSF}

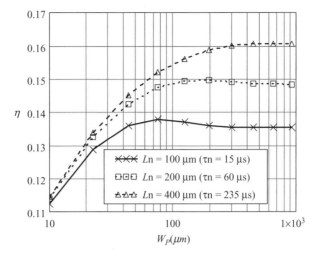

Figure 6.44   Effect of base width on the efficiency as a function of the minority electron diffusion length or lifetime at $S_{BSF} = 100$ cm/s

effect of the emitter width on the short-circuit current I_{SC} and the open-circuit voltage V_{OC} is shown in Figure 6.46. I_{SC} decreases as W_N increases, while V_{OC} remains constant as W_N increases. When we look at Equation (6.294), we deduce that the dark saturation current I_0 must decrease at the same rate of I_{SC} in order for V_{OC} to be constant. The dark saturation current occurs due to the recombination in the quasi-neutral region. Therefore, we conclude that, *decreasing the emitter width not only increases the photogenerated current but also decreases the recombination in the*

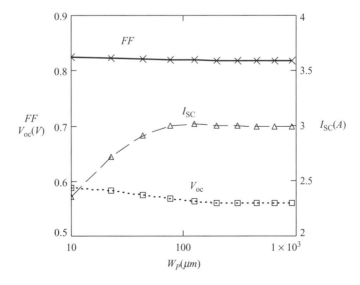

Figure 6.45   Effect of base width on FF, I_{SC}, and V_{OC} for a solar cell at $\tau_n = 15\,\mu$s

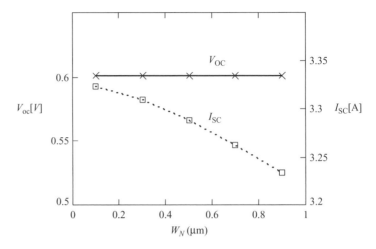

Figure 6.46 Effect of emitter width on I_{SC} and V_{OC}

n-type quasi-neutral region. This should be an important feature and implemented into the design concept of a solar cell.

The effect of the emitter width, W_N, on the efficiency is a function of the front surface recombination velocity, S_{Feff}, which is shown in Figure 6.47. When S_{Feff} decreases from $S_{\text{Feff}} = 3 \times 10^4$ cm/s to $S_{\text{Feff}} = 1 \times 10^3$ cm/s, its effect becomes negligible. The effect of S_{Feff} on the efficiency is shown in Figure 6.48. We find that S_{Feff} of about 3×10^3 cm/s appears to be a good number for a minimum passivation process with various emitter widths.

6.10.9 Effect of Acceptor Concentration on Efficiency

The accepters are dopants producing holes in p-type region. The effect of the acceptor concentration N_A on FF, I_{SC}, and V_{OC} is presented in Figure 6.49. The *figure of merit* of Equation (6.306) tells us that the power conversion efficiency η is strongly related to only both I_{SC} and V_{OC}. We see in Figure 6.49 that, as the acceptor concentration N_A increases, the open-circuit voltage V_{OC} increases while the short-circuit current I_{SC} decreases. The net gives a maximum point for the efficiency that is shown in Figure 6.50. For the data given in Table 6.9 used for the present computations, a maximum efficiency is found at about $N_A = 1 \times 10^{17}$ cm^{-3} in the figure.

The decrease in the short-circuit current in Figure 6.49 must be mainly associated with the *minority electron diffusion coefficient* D_n in the p-type region, which is shown in Equation (6.287). From Equation (6.160), D_n is proportional to the *electron mobility* μ_n. From Figure 6.30, the mobility μ_n decreases with increasing N_A. Therefore, D_n must decrease with increasing N_A. The decrease of D_n results in lowering the short-circuit current. Consequently, the short-circuit current I_{SC} decreases with increasing the acceptors concentration N_A, which is illustrated in Figure 6.49.

The open-circuit voltage interestingly increases with increasing N_A, as shown in Figure 6.49. The open-circuit voltage was expressed in Equation (6.294). V_{OC} increases logarithmically with increasing the ratio I_{SC}/I_0. Since we know that I_{SC} decreases

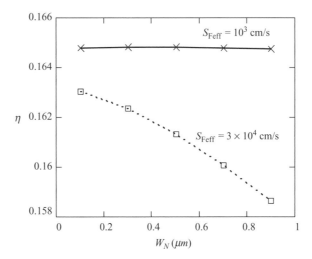

Figure 6.47   Effect of S_{Feff} on the efficiency as a function of W_N

with increasing N_A in the foregoing paragraph, I_0 must decrease at a rate faster than I_{SC} in order to have the increase of the ratio. I_0 is the *dark-saturation current* due to the recombination in the quasi-neutral region, which was presented in Equation (6.289), where I_0 decreases directly with increasing N_A. Conclusively, *as acceptor concentration increases, the photo generated current decreases while the open-circuit voltage increases*.

6.10.10   Effect of Donor Concentration on Efficiency

It is interesting how the computational results show the effect of donor concentration on the short-circuit current and the open-circuit voltage. As for the acceptor concentration

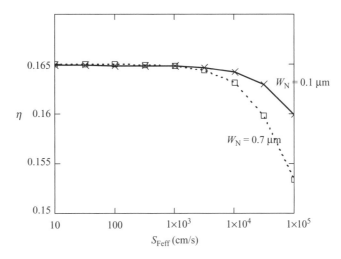

Figure 6.48   Effect of emitter width on the efficiency as a function of S_{Feff}

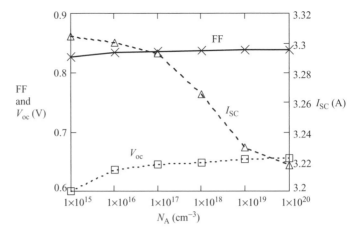

Figure 6.49 *FF*, I_{SC}, and V_{OC} versus N_A for a solar cell

in the p-type region, the donor concentration in the n-type region behaves in the similar way. As N_D increases, V_{OC} increases but I_{SC} decreases. This produces a maximum point in the efficiency. This is shown in Figures 6.51 and 6.52. We find an optimum donor concentration N_D at about 1×10^{18} cm^{-3} toward a little lower part of the maximum from the figure because the high donor concentration causes other problems, such as the increase of the lateral sheet resistance in the emitter (see Section 6.13.4) and the increase of Auger recombination (see Section 6.7.1).

6.10.11 Band Gap Energy with Temperature

The band gap energy E_g is perhaps the most important parameter in semiconductor physics. With increasing temperature a contraction of the crystal lattice usually leads to a strengthening of the interatomic bonds and an associated increase in the bans gap

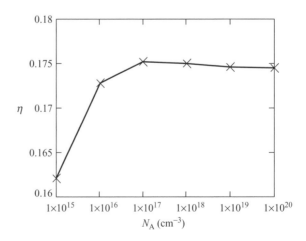

Figure 6.50 Efficiency versus acceptor concentration for a solar cell

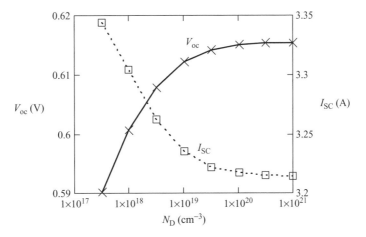

Figure 6.51   The effect of N_D on I_{SC} and V_{OC}

energy. To a very good approximation, the cited variation of band gap energy with temperature can be modeled by the universal empirical relationship

$$E_g(T) = E_g(0) - \frac{\alpha T^2}{T + \beta} \tag{6.308}$$

where α and β are constants chosen to obtain the best fit to experimental data and $E_g(0\,\mathrm{K})$ is the limiting value of the band gap at zero Kelvin. The constants and the limiting values are found in Table 6.10.

6.10.12   Effect of Temperature on Efficiency

Solar cells deliver only a small part of the absorbed energy as electrical energy to a load. The remainder is dissipated as heat and the solar cell must therefore have a higher

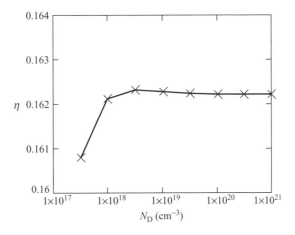

Figure 6.52   The effect of N_D on the efficiency

Table 6.10   Band Gap Energy Temperature Coefficients for Ge, Si, and GaAs
Materials (Pierret [10])

Material	$E_g(300\,K)$	$E_g(0\,K)$	α	β
Ge	0.663	0.7437	4.774×10^{-4}	235
Si	1.125	1.170	4.730×10^{-4}	636
GaAs	1.422	1.519	5.405×10^{-4}	204

temperature than the environment. Heating reduces the size of the band gap energy. The
absorbed photocurrent increases, leading to a slight increase in the short-circuit current.
The heating has a detrimental effect on the open-circuit voltage, which is shown in
Figure 6.37.

Many parameters in Equation (6.292) actually depend on temperature such as band
gap energy, diffusion coefficients, diffusion lengths, built-in voltage, depletion region
width, and intrinsic carrier concentration. It is difficult to explicitly elucidate the effect
of temperature term by term. However, the computational results give us a solid solution
for the effect of temperature on FF, I_{SC}, and V_{OC}, as shown in Figure 6.53. With
increasing temperature, I_{SC} increases almost linearly while V_{OC} decreases as expected.
The net result is shown in Figure 6.54 for the effect of temperature on the efficiency.
Cell temperature is of great importance for the energy conversion efficiency.

6.11   ADDITIONAL TOPICS

6.11.1   Parasitic Resistance Effects (Ohmic Losses)

In real solar cells, power is dissipated through the resistance of the contacts and through
leakage currents around the sides of the device. These effects are equivalent electrically
to two parasitic resistances in series (R_s) and in parallel (R_{sh}) with the solar cell.

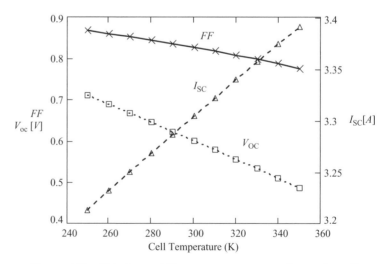

Figure 6.53 FF, I_{SC}, and V_{OC} versus temperature for a solar cell

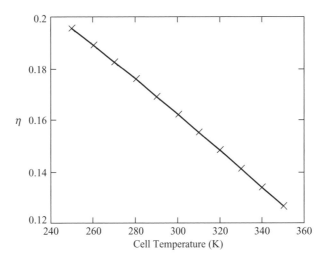

Figure 6.54 Efficiency versus temperature for a solar cell

The solar cell two-diode circuit model, now including the two parasitic resistances, is presented in Figure 6.55.

The *series resistance* arises from the bulk resistance of the semiconductor and the resistance of contacts and interconnection. Series resistance is a particular problem at high current, for instance, under concentrated light. The *parallel or shunt resistance* arises from leakage current around the edges of the solar cell and extended lattice defects in the depletion region. Series and parallel resistances reduce the fill factor. For an efficient solar cell we want R_s to be as small and R_{sh} to be as large as possible. When parasitic resistances are included, the total current is modified from Equation (6.292) as

$$I = I_{SC} - I_{01}\left(\exp\left(\frac{q(V + IR_s)}{kT}\right) - 1\right) - I_{02}\left(\exp\left(\frac{q(V + IR_s)}{2kT}\right) - 1\right) - \frac{V + IR_s}{R_{sh}}$$

$$(6.309)$$

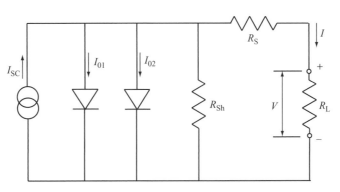

Figure 6.55 Solar cell two-diode circuit model including the parasitic series and shunt resistances

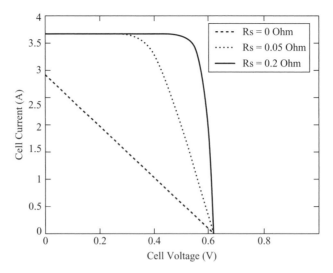

Figure 6.56   Effect of series resistances on the I-V characteristic of a solar cell ($R_{sh} \rightarrow \infty$)

The effect of these parasitic resistances on the I-V characteristic is shown in Figures 6.56 and 6.57. As can also be seen in Equation (6.309), the series resistance R_s has no effect on the open-circuit voltage, but reduces the short-circuit current, which is seen in Figure 6.56. Conversely, the shunt resistance R_{sh} has no effect on the short-circuit current, but reduces the open-circuit voltage, which is seen in Figure 6.48. Sources of series resistance include the metal contacts, particularly the front grid, and the transverse flow of current in solar cell emitter to the front grid.

In particular, as shown in Figure 6.56, the series resistance influences the gradient of the I-V characteristic near the open-circuit voltage and can be graphically determined

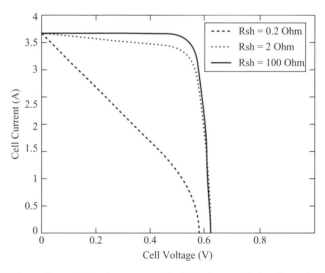

Figure 6.57   Effect of parallel resistances on the I-V characteristic of a solar cell ($R_s \rightarrow 0$)

in a simplified model by

$$R_s \cong \left. \frac{dV}{dI} \right|_{V=V_{OC}} \tag{6.310}$$

As shown in Figure 6.57, the shunt resistance influences the gradient of the I-V characteristic within the area of the short-circuit current and can be determined graphically approximately by

$$R_{sh} \cong \left. \frac{dV}{dI} \right|_{V=0} \tag{6.311}$$

The short-circuit current will be degraded by series resistance. Any series resistance will bias the dark diode in Figure 6.55, and cause the current through the load to be less than the short-circuit current (photocurrent) in Equation (6.309). The open-circuit voltage will be degraded by shunt resistance. When the circuit is opened at the load, the current path through the shunt resistance lowers the bias across the dark diode in Figure 6.55, thereby degrading the open-circuit voltage.

6.11.2   Quantum Efficiency

Quantum efficiency (*QE*), often called *spectral response* (*SR*), is the probability that an incident photon of energy E_{ph} will deliver one electron to the external circuit. The *external quantum efficiency* (*EQE*) of a solar cell includes the effect of optical losses such as grid shading and reflection. However, it is often useful to look at the quantum efficiency of the light left after the reflected and shaded light has been lost. *Internal quantum efficiency* (*IQE*) refers to the efficiency with which photons that are not reflected or shaded out of the cell can generate collectable carriers. By measuring the reflection and grid-shading factor of a device, the external quantum efficiency curve can be corrected to obtain the internal quantum efficiency. Hence, we can write the short-circuit current I_{SC} in terms of EQE, defined previously as

$$I_{SC} = qA \int_\lambda f_0(\lambda) EQE(\lambda) d\lambda \tag{6.312}$$

Since

$$I_{SC} = \int_\lambda I_{SC}(\lambda) d\lambda, \tag{6.313}$$

The external quantum efficiency is expressed by

$$EQE(\lambda) = \frac{I_{SC}(\lambda)}{qAf_0(\lambda)} \tag{6.314}$$

where $f_0(\lambda)$ is the incident spectral photon flux. And the internal quantum efficiency (*IQE*) can be expressed by

$$IQE(\lambda) = \frac{EQE(\lambda)}{(1-s)(1-r(\lambda))} \tag{6.315}$$

where s is the grid-shading factor and $r(\lambda)$ is the spectral reflection.

The quantum efficiency (QE) depends on the absorption coefficient of the solar cell material, the efficiency of charge separation, and the efficiency of charge collection in the device. It does *not* depend on the incident spectrum. It is therefore a key quantity in describing solar cell performance in comparison with the spectrum of solar photons. The external quantum efficiency is plotted along with the incident spectral power using Equation (6.314) with the aid of Equation (6.285) through Equation (6.288), as shown in Figure 6.58. Notice that the quantum efficiency shows zero after the wavelength of 1.1 μs because no light is absorbed below the band gap (1.12 eV ∼ 1.1 μs by $E_{ph} = hc/\lambda$).

The effects of both effective front surface recombination velocity S_{Feff} and back surface recombination velocity S_{BSF} on external quantum efficiency are illustrated in Figure 6.59. The short wavelength (violet spectrum) response improves dramatically by decreasing S_{Feff} from 3×10^5 cm/s to 100 cm/s since the high energy (short wavelength) photon are absorbed at the vicinity of the front surface in the emitter region.

This decrease can be practically achieved by *passivation* (coating) at the front surface. The long wavelength (red spectrum) response improves by decreasing S_{BSF} from 10^7 cm/s to 100 cm/s, since low energy (long wavelength) photons are absorbed deep in the base region. This is shown in Figure 6.59. This also can be achieved by passivation at the back surface.

The space between unity and the external quantum efficiency curve indicates losses such as the reflection/transmission and the shading factor at the front surface and the recombination in the vicinities of the front and back surfaces.

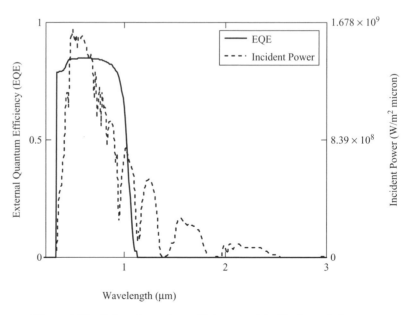

Figure 6.58 External quantum efficiency and incident spectral power

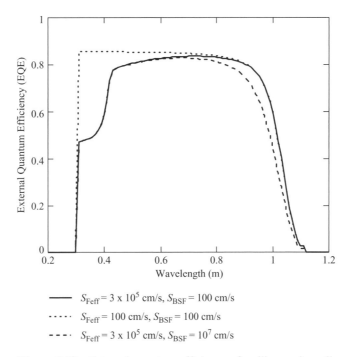

Figure 6.59   External quantum efficiency of a silicon solar cell

6.11.3   Ideal Solar Cell Efficiency

The *ideal solar cell efficiency* or the limit of solar cell efficiency is a fundamental question that has been driven since the invention of solar cells. There are many approaches that include thermodynamic limits with a blackbody source of sun. However, we would rather try to obtain more practical limits of a solar cell for the AM1.5G spectral irradiance (Figure 6.6) for 1 sun and 1,000 sun (concentrated solar cell).

We have seen in the previous sections that the photocurrent is largely determined by the band gap energy E_g. As a result, the photocurrent reduces with increasing band gap energy (see Figure 6.25). Assuming small ohmic losses, the photocurrent is equal to the short-circuit current I_{SC}. Conversely, a high band gap energy causes a high output voltage, which results in a high open-circuit voltage. As a result, the open-circuit voltage V_{OC} increases with increasing band gap energy. When we consider the definition of the energy conversion efficiency in Equation (6.306), the efficiency is dependent on both the short-circuit current and the open-circuit voltage, since the fill factor FF is a function of the open-circuit voltage, as shown in Equation (6.305). Therefore, the two counteracting features determine the maximum efficiency. We are going to develop these two parameters in terms of the band gap energy for an ideal solar cell using already-developed equations.

The ideal solar cell would be achieved by assuming that the *external quantum efficiency* is unity. This means that every incident photon that has greater energy than the band gap energy creates an electron-hole pair. Therefore, from Equation (6.314)

we have the spectral short-circuit current as

$$I_{SC}(\lambda) = q A f_0(\lambda) \tag{6.316}$$

Then, the ideal short-circuit current becomes

$$I_{SC} = \int_\lambda q A f_0(\lambda) d\lambda \tag{6.317}$$

By converting the band gap energy into the wavelength using Equations (6.1) and (6.2), the band gap wavelength is expressed by

$$\lambda_{BG}(E_g) = \frac{hc}{E_g} \tag{6.318}$$

We rewrite Equation (6.137) in terms of this as

$$I_{SC}(E_g) = \int_0^{\lambda_{BG}(E_g)} q A f_0(\lambda) d\lambda \tag{6.319}$$

This indicates that I_{SC} is a function of E_g. Now we take a look at the open-circuit voltage in Equation (6.294a):

$$V_{OC} = \frac{kT}{q} \ln\left(\frac{I_{SC}}{I_0}\right) \tag{6.294a}$$

From Equation (6.292a),

$$I = I_{SC} - I_0 \left(\exp\left(\frac{qV}{kT}\right) - 1\right) \tag{6.292a}$$

J_0 is found in Equation (6.289). Multiplying J_0 by the area gives the dark saturation current I_0 as

$$I_0 = A\left[\frac{q D_p}{L_p} \frac{Y_P}{X_P} \frac{n_i^2}{N_D} + \frac{q D_n}{L_n} \frac{Y_N}{X_N} \frac{n_i^2}{N_A}\right] \tag{6.320}$$

Using Equation (6.66) for the intrinsic carrier concentration n_i, we rewrite Equation (6.320).

$$I_0 = q A N_c N_v \left[\frac{D_p}{L_p} \frac{Y_P}{X_P} \frac{1}{N_D} + \frac{D_n}{L_n} \frac{Y_N}{X_N} \frac{1}{N_A}\right] \left(\exp\left(\frac{-E_g}{kT}\right)\right) \tag{6.321}$$

For the ideal solar cell, some assumptions deduced from the primary assumption $EQE = 1$ follow. First, we assume that the base width W_P is infinite and the depletion width W_D is zero so that the absorption is perfect and no recombination exists in the depletion region.

$$W_P = \infty \tag{6.322a}$$

$$W_D = 0, \quad x_N = 0 \quad \text{and} \quad x_P = 0 \tag{6.322b}$$

The ideal solar cell has no front and back surface recombination, so we have

$$S_{F,eff} = S_{BSF} = 0 \tag{6.322c}$$

It is assumed for the ideal solar cell that the diffusion length is sufficiently large and uniform in both the emitter and base regions.

$$L_n = L_p = L \tag{6.322d}$$

Likewise, the carrier lifetimes are long and the diffusion lengths are large, so that we have

$$\tau_n = \tau_p = \tau \quad \text{and} \quad D_n = D_p = D \tag{6.322e}$$

We can reasonably assume that the emitter width W_N is much smaller than the diffusion length, L.

$$W_N \ll L \tag{6.322f}$$

Also the product of the high absorption coefficient and the large diffusion length lead to

$$\alpha L \gg 1 \tag{6.322g}$$

where α is the absorption coefficient. Actually, if we conduct the derivation for the short-circuit current I_{SC} using Equations (6.286) to (6.288) with the above assumptions, we come up with the external quantum efficiency as we already assumed in the earlier part of this section. Equations (6.274) to (6.277), after some algebra, lead to

$$\frac{Y_N}{X_N} = \tanh\left(\frac{W_P}{L}\right) \cong 1 \tag{6.323}$$

$$\frac{Y_P}{X_P} = \tanh\left(\frac{W_N}{L}\right) \cong 0 \tag{6.324}$$

Equation (6.321) becomes

$$I_0 = qAN_cN_v\frac{D}{LN_A}\left(\exp\left(\frac{-E_g}{kT}\right)\right) = C_{S0}\left(\exp\left(\frac{-E_g}{kT}\right)\right) \tag{6.325}$$

where C_{S0} is the coefficient as

$$C_{S0} = qAN_cN_v\frac{D}{LN_A} \tag{6.325a}$$

The fraction term can be expressed using Equations (6.207) and (6.159) as

$$\frac{D}{LN_A} = \frac{D}{N_A\sqrt{D\tau}} = \frac{1}{N_A}\sqrt{\frac{D}{\tau}} = \frac{1}{N_A}\sqrt{\frac{\mu\dfrac{kT}{q}}{\tau}} = \sqrt{\frac{kT}{q}}\sqrt{\frac{\mu}{\tau \cdot N_A^2}} \tag{6.325b}$$

The lifetimes of electrons and holes are intrinsically limited by radiative combination in the semiconductor. One can assign favorable values to the mobility and lifetime and calculate the efficiency as a function of the band gap energy.

Combining this with Equation (6.294a), the open-circuit voltage is derived by

$$V_{OC} = \frac{kT}{q} \ln\left(\frac{I_{SC}}{I_0}\right) = \frac{kT}{q} \ln\left(\frac{I_{SC}}{C_{S0}\left(\exp\left(\frac{-E_g}{kT}\right)\right)}\right)$$

$$= \frac{kT}{q} \ln\left(\frac{I_{SC}}{C_{S0}} \exp\left(\frac{E_g}{kT}\right)\right) = \frac{kT}{q}\left[\ln\left(\frac{I_{SC}}{C_{S0}}\right) + \ln\left(\exp\left(\frac{E_g}{kT}\right)\right)\right] \qquad (6.326)$$

Finally, we have an expression for the open-circuit voltage as a function of E_g.

$$V_{OC}(E_g) = \frac{1}{q}\left[E_g + kT \ln\left(\frac{I_{SC}(E_g)}{C_{S0}}\right)\right] \qquad (6.327)$$

V_{OC} is usually less than E_g so that I_{SC} should be less than C_{S0} (it is true). Hence, the logarithm is negative. Now we recall the definition of the fill factor FF in Equation (6.305) and rewrite it as

$$FF(V_{OC}) \cong \frac{\dfrac{qV_{OC}(E_g)}{kT} - \ln\left(\dfrac{qV_{OC}(E_g)}{kT} + 1\right)}{\dfrac{qV_{OC}(E_g)}{kT} + 1} \qquad (6.328)$$

The power conversion efficiency defined in Equation (6.306) is expressed as

$$\eta(E_g) = \frac{I_{mp}V_{mp}}{P_{in}} = \frac{FF(V_{OC}) \cdot I_{SC}(E_g)V_{OC}(E_g)}{p_{in}} \qquad (6.329)$$

where P_{in} is the incident power that is obtained from Equation (6.307) as

$$P_{in} = A \int_0^\infty h\nu(\lambda) f_0(\lambda) d\lambda \qquad (6.307)$$

The summary for the power conversion efficiency for the AM1.5G spectral irradiance with 1 sun at 300 K is presented as follows

$$\eta(E_g) = \frac{I_{mp}V_{mp}}{P_{in}} = \frac{FF(V_{OC}) \cdot I_{SC}(E_g)V_{OC}(E_g)}{P_{in}} \qquad (6.330a)$$

$$I_{SC}(E_g) = qA \int_0^{\lambda_{BG}(E_g)} f_0(\lambda) d\lambda \qquad (6.330b)$$

$$P_{in} = A \int_0^\infty h\nu(\lambda) f_0(\lambda) d\lambda \qquad (6.330c)$$

$$V_{OC}(E_g) = \frac{1}{q}\left[E_g + kT\ln\left(\frac{I_{SC}(E_g)}{C_{S0}}\right)\right] \tag{6.330d}$$

$$C_{S0} = qAN_cN_v\frac{D}{LN_A} \tag{6.330e}$$

$$FF(V_{OC}) \cong \frac{\dfrac{qV_{OC}(E_g)}{kT} - \ln\left(\dfrac{qV_{OC}(E_g)}{kT} + 1\right)}{\dfrac{qV_{OC}(E_g)}{kT} + 1} \tag{6.330f}$$

We now consider the concentration of 1,000 suns on the ideal solar cell. The power conversion efficiency can be easily obtained by noting that the number of incident photons $f_0(\lambda)$ is simply multiplied by 1,000 due to the concentration of the solar irradiation, which is shown in Equations (6.331b) and (6.331c), respectively. The summary for the power conversion efficiency for the AM1.5G spectral irradiance with 1,000 sun at 300 K is presented here.

$$\eta(E_g) = \frac{I_{mp}V_{mp}}{P_{in}} = \frac{FF(V_{OC})\cdot I_{SC}(E_g)V_{OC}(E_g)}{P_{in}} \tag{6.331a}$$

$$I_{SC}(E_g) = qA\int_0^{\lambda_{BG}(E_g)}[1{,}000\times f_0(\lambda)]d\lambda \tag{6.331b}$$

$$P_{in} = A\int_0^{\infty}h\nu(\lambda)[1{,}000\times f_0(\lambda)]d\lambda \tag{6.331c}$$

$$V_{OC}(E_g) = \frac{1}{q}\left[E_g + kT\ln\left(\frac{I_{SC}(E_g)}{C_{S0}}\right)\right] \tag{6.331d}$$

$$C_{S0} = qAN_cN_v\frac{D}{LN_A} \tag{6.331e}$$

$$FF(V_{OC}) \cong \frac{\dfrac{qV_{OC}(E_g)}{kT} - \ln\left(\dfrac{qV_{OC}(E_g)}{kT} + 1\right)}{\dfrac{qV_{OC}(E_g)}{kT} + 1} \tag{6.331f}$$

The ideal solar cell efficiencies for the AM1.5G spectral irradiance with 1 sun and 1,000 sun at 300 K are presented in Figure 6.60. In considering the AM1.5G 1 sun, the maximum possible efficiency of about 31 percent occurs between 1.2 eV and 1.5 eV in band gap energy for a non-concentrated ideal solar cell. More interestingly, the GaAs band gap of 1.42 eV closely matches the band gap of the ideal maximum possible efficiency. The Si band gap of 1.12 eV shows about 30 percent efficiency that is slightly less than the maximum. However, these two materials show a good selection as solar cells. Actually, the most efficient single-junction solar cell, made of

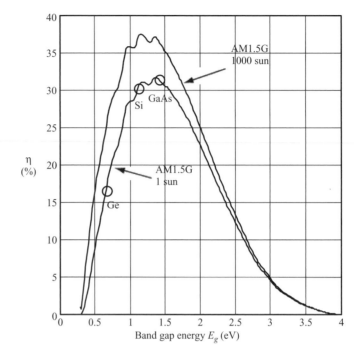

Figure 6.60 Ideal solar cell efficiency as a function of band gap energy for spectral distribution AM1.5 with 1 sun and 1,000 sun at 300 K

GaAs, had achieved an efficiency of 25.1 percent of AM1.5G spectrum by Green [19] in 2001. The world record of a single-junction crystalline Si solar cell was reported to be 24.7 percent by UNSW (Sydney, Australia) in 1999.

Since the ratio I_{SC}/P_{in} in Equation (6.331a) is independent of the concentration of the incident power (1,000 is canceled in the fraction), only V_{OC} in Equation (6.331a) contributes to the increase of η. Equation (6.331) suggests that the power conversion efficiency can be increased by enhancing the intensity of the incident sunlight. The increase is most effective near the maximum. Figure 6.60 shows that the maximum increases to about 37 percent.

In 1961, Shockley and Queisser [20] studied the efficiency upper limit of an ideal solar cell. Their results for both AM1.5G and AM1.5D spectrum with 1 sun are in good agreement with the present work, although both used different methods in analysis.

6.12   MODELING

6.12.1   Modeling for a Silicon Solar Cell

We want to develop a MathCAD model to solve the solar cell equations derived in the preceding sections, so that we can study the current-voltage characteristic of a solar cell and the figures of merit, and develop the design of a solar cell. Typically, nine parameters are linked each other and govern the performance of a solar cell, which are cell temperature, minority carrier lifetimes, emitter and base widths, front and back surface recombination velocities, and donor and acceptor concentrations. In this model, we want to express the cell current as a function of only cell temperature among the nine parameters. We consider a silicon solar cell with typical operating conditions.

MathCAD Format Solution We first define the *physical constants* for a silicon solar cell. The equation numbers newly begin with model number as M.1.

$$e_c := 1.60217733 \cdot 10^{-19} C \qquad \text{Electron charge}$$

$$k_B := 8.617386 \cdot 10^{-5} \frac{e_c \cdot V}{K} \qquad \text{Boltzmann constant}$$

$$q := e_c \qquad \text{Electron charge}$$

$$m_e := 9.10939 \cdot 10^{-31} \text{ kg} \qquad \text{Electron mass}$$

$$h_P := 6.62608 \cdot 10^{-34} \text{ J} \cdot \text{s} \qquad \text{Plancks constant}$$

$$m_n := 1.18 \cdot m_e \qquad \text{Effective mass of electron}$$

$$m_p := 0.81 \cdot m_e \qquad \text{Effective mass of hole}$$

$$\varepsilon_0 = 8.854 \times 10^{-14} \cdot C \cdot V^{-1} \cdot cm^{-1} \qquad \text{Permittivity of vacuum (default)}$$

$$K_S := 11.8 \qquad \text{Semiconductor dielectric constant of Si}$$

$$\varepsilon_p := K_S \cdot \varepsilon_0 \qquad \text{Electric permittivity for Si (definition)}$$

(M.1)

The following *silicon solar cell model parameters* are defined as

$$A_{cell} := 100 \text{ cm}^2 \qquad \text{Solar cell illumination area}$$

$$N_D := 1 \cdot 10^{18} \text{ cm}^{-3} \qquad \text{Number of donors per unit volume}$$

$$N_A := 1 \cdot 10^{15} \text{ cm}^{-3} \qquad \text{Number of acceptors per unit volume}$$

$$S_{Feff} := 3 \cdot 10^4 \text{ cm/s} \qquad \text{Effective front surface recombination velocity}$$

$$S_{BSF} := 100 \text{ cm/s} \qquad \text{Effective back surface recombination velocity}$$

(M.2)

$$W_N := 0.35 \, \mu m \qquad \text{Emitter width of a solar cell}$$

$$W_P := 300 \, \mu m \qquad \text{Base width of a solar cell}$$

$$\tau_n := 350 \, \mu s \qquad \text{Lifetimes of electrons}$$

$$\tau_p := 1 \, \mu s \qquad \text{Lifetime of holes}$$

$$s_f := 0.05 \qquad \text{Grid shading factor}$$

Blackbody Solar Irradiance The sun has an effective blackbody surface temperature of 5,777 K.

$$T_{sol} := 5777 \text{ K} \qquad (M.3)$$

The sun's blackbody emittance (W/m²μm) is obtained from Planck's law.

$$f_b(\lambda) := \frac{2 \cdot \pi \cdot h_P \cdot c^2}{\lambda^5 \left(\exp\left(\frac{h_P \cdot c}{\lambda \cdot k_B \cdot T_{sol}} \right) - 1 \right)} \qquad (M.4)$$

The Sun's radius and Earth's orbit radius are known as

$$r_s := 6.9598 \cdot 10^8 \text{ m} \qquad \text{Sun's radius} \qquad (M.5)$$

$$r_o := 1.4959789 \cdot 10^{11} \text{ m} \qquad \text{Earth's orbit radius} \qquad (M.6)$$

Earth receives a small fraction of the Sun's energy. The fraction is

$$F_{SE} := \left(\frac{r_s}{r_o}\right)^2 \qquad F_{SE} = 2.164 \times 10^{-5} \qquad (M.7)$$

Extraterrestrial blackbody solar irradiation between the sun and Earth is called the *solar constant* Gsc. We calculated the solar constant to be 1,367 W/m² at space outside the Earth's atmosphere.

$$G_{SC\lambda}(\lambda) := F_{SE} \cdot f_b(\lambda) \qquad (M.8)$$

$$G_{SC} := \int_{0.01\mu m}^{100\mu m} G_{SC\lambda}(\lambda)d\lambda \qquad G_{SC} = 1{,}367 \cdot \frac{W}{m^2} \qquad (M.9)$$

Incident Photon Flux $f_0(\lambda)$ Measured incident photon fluxes are imported from a structured ASCII data file, where f_1 and f_2 are the incident photon fluxes for AM1.5 Global and AM1.5 Direct, respectively. READPRN (MathCAD commend) reads the file and returns a matrix M. The data in the file are tabulated in Table A.1 in Appendix A. The readers need to create an appropriate text file from the table to conduct the following instruction.

By default, ORIGIN is 0, but we want to change its value to 1. The corrugated underline under the ORIGIN indicates when we redefine an already defined quantity in MathCAD as

$$\text{ORIGIN:} = 1 \qquad (M.10)$$

$$M := \text{READPRN(“solar radiation AM1-5.txt”)} \qquad (M.11)$$

A portion of the data matrix is displayed while the full data can be seen by clicking and scrolling the matrix in MathCAD.

tx reads column 1 and has the unit of nanometer (nm). The superscript symbol $\langle\rangle$ can be obtained in the Programming Toolbar icon and the number in the symbol indicates the column number in the matrix M (see figure M.1).

$$tx := M^{\langle 1 \rangle}nm \qquad (M.12)$$

The *cubic spline interpolation* is applied for a smooth interpolation between two data points. And the unit of spectral incident power is added as:

$$hy := M^{\langle 2 \rangle} \cdot W \cdot m^{-2} \cdot \mu m^{-1} \qquad (M.13)$$

$$hs := \text{lspline(tx,hy)} \qquad (M.14)$$

$f_1(\lambda)$ interpolates the AM1.5G spectrum data for a specific wavelength.

$$f_1(\lambda) := \text{interp(hs,tx,hy,}\lambda) \qquad (M.15)$$

	λ	AM1.5G	AM1.5D
	1	2	3
1	305	9.5	3.4
2	310	42.3	15.6
3	315	107.8	41.1
4	320	181	71.2
5	325	246.8	100.2
6	330	395.3	152.4
7	335	390.1	155.6
M 8	340	435.3	179.4
9	345	438.9	186.7
10	350	483.7	212
11	360	520.3	240.5
12	370	666.2	324
13	380	712.5	362.4
14	390	720.7	381.7
15	400	$1.013 \cdot 10^3$	556
16	410	$1.158 \cdot 10^3$	...

Figure M.1 Imported data for AM1.5G spectrum

The ranges of the wavelengths for the AM1.5G spectrum and the Sun's blackbody irradiances are defined for plotting (see Figure M.2)

$$\lambda 1 := M^{(1)} \cdot nm \quad \lambda 2 := 0.1 \mu m, 0.102 \mu m .. 3 \mu m \tag{M.16}$$

The AM1.5G spectrum measured on Earth shows good agreement with the theoretical blackbody irradiation. The more detailed description for the figure may be found in Section 6.1.4.

The accidental incident photon flux less than zero from either the Matrix data or the cubic spline interpolation has no physical meaning (actually deteriorates the results) and should be forced to be zero for correct numerical computations. You should select "Add Line," "if," and "otherwise" from the programming tool rather than typing in MathCAD.

$$\underset{\sim}{f_1}(\lambda) := \left| \begin{array}{l} f_1(\lambda) \text{ if } f_1(\lambda) > 0 \cdot W \cdot m^{-2} \cdot \mu m^{-1} \\ 0 \cdot W \cdot m^{-2} \cdot \mu m^{-1} \quad \text{otherwise} \end{array} \right. \tag{M.17}$$

We can obtain the incident photon flux (number of photon per unit area per second per unit wavelength) by dividing the AM1.5G data $f_1(\lambda)$ by the photon energy shown in Equation (6.1). The incident photon flux is written by

$$f_0(\lambda) := \frac{f_1(\lambda)}{h_P \cdot \dfrac{c}{\lambda}} \tag{M.18}$$

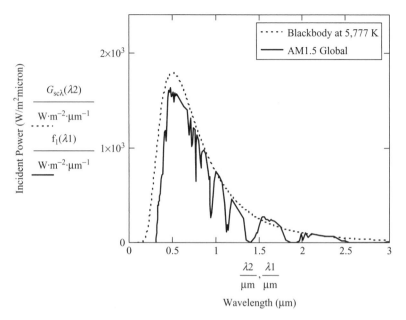

Figure M.2   Solar spectra, blackbody irradiance at 5,777 K and AM1.5G spectrum

Assume that the front surface of the solar cell is passivated with anti-reflective material, which gives approximately the low reflectance. The reflectance $r(\lambda)$ is assumed by

$$r(\lambda) := 0.1 \qquad\qquad\qquad\qquad\text{(M.19)}$$

Band gap energy dependence on temperature   The band gap energy is a function of temperature as shown in Equation (6.308). The band gap coefficients for a silicon semiconductor are found in Table 6.10.

$$Eg_0 := 1.170\, e_c \cdot V \qquad \alpha_1 := 4.730 \times 10^{-4} \frac{e_c \cdot V}{K} \qquad \beta_1 := 636\,K \qquad \text{(M.20)}$$

From Equation (6.308),

$$E_g(T_o) := Eg_0 - \frac{\alpha_1 \cdot T_o^2}{T_o + \beta_1} \qquad E_g(300\,K) = 1.125 \cdot e_c \cdot V \qquad \text{(M.21)}$$

Absorption Coefficient α (cm^{-1})   We herein develop a semiempirical correlation based on the theory and compare it with the measured absorption coefficients. The following semiempirical correlation is obtained by curve fitting.

$$A_d := 2.3 \times 10^6\, cm^{-1} \cdot (e_c \cdot V)^{\frac{-1}{2}} \qquad E_{gd} := 2.9 \cdot e_c \cdot V \qquad\qquad \text{(M.22)}$$

$$A_i := 1391\, cm^{-1} \cdot (e_c \cdot V)^{-2} \qquad h\nu_1 := 1.827 \times 10^2 \cdot e_c \cdot V \qquad\qquad \text{(M.23)}$$

$$\alpha_d(\lambda) := A_d \cdot \frac{\left(h_P \cdot \dfrac{c}{\lambda} - E_{gd} \right)^{\frac{3}{2}}}{h_P \cdot \dfrac{c}{\lambda}} \qquad\qquad\qquad \text{(M.24)}$$

$$\alpha_{in}(\lambda, T_o) := \frac{\left(h_P \cdot \dfrac{c}{\lambda} - E_g(T_o) + h\nu_1\right)^2}{\exp\left(\dfrac{h\nu_1}{k_B \cdot T_o}\right) - 1} + \frac{\left(h_P \cdot \dfrac{c}{\lambda} - E_g(T_o) - h\nu_1\right)^2}{1 - \exp\left(-\dfrac{h\nu_1}{k_B \cdot T_o}\right)} \qquad (M.25)$$

The spectral absorption coefficient as a function of both wavelength and temperature is expressed by

$$\alpha(\lambda, T_o) := \mathrm{Re}\left(\alpha_d(\lambda)\right) + A_i \cdot \alpha_{in}(\lambda, T_o) \qquad (M.26)$$

where Re is the real part of the value. The first term in Equation (M.26) takes only the real part of the term for the curve fitting. The absorption coefficient that its photon energy is less than the band gap energy is vanished.

$$\underset{\sim}{\alpha}(\lambda, T_o) := \left| \begin{array}{l} \alpha(\lambda, T_o) \ \text{if} \ h_P \cdot \dfrac{c}{\lambda} > E_g(T_o) \\[2ex] 0 \, \mathrm{cm}^{-1} \quad \text{otherwise} \end{array} \right. \qquad (M.27)$$

Experimental data of absorption coefficients are imported by reading an external ASCII file for Silicon and compared with the semiempirical correlation. The data in the file are tabulated in Table A.14 in Appendix A.

$$Z := \mathrm{READPRN}(\text{``absorption\_coeff\_c\_Si.txt''}) \qquad (M.28)$$

$$\lambda data := Z^{\langle 1 \rangle} \cdot \mu m \qquad (M.29)$$

$$\alpha data := Z^{\langle 2 \rangle} \, cm^{-1} \qquad (M.30)$$

$$\underset{\sim}{\lambda_1} := 0.1\mu m, 0.102\mu m \cdot 1.3\mu m \qquad (M.31)$$

The semiempirical correlation developed is compared with the measurement with good agreement as shown in Figure M.3, so we have a realistic absorption coefficient for our model. There was actually an available correlation in the literature [13], but somehow

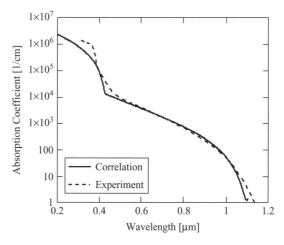

Figure M.3 Absorption coefficient: semiempirical correlation and measurements

no agreement between their proposed correlation and the measurements was found in the present work. Therefore, a new correlation had to be developed for this model.

Intrinsic Carrier Concentration   The *intrinsic carrier concentration* was presented in Equation (6.66). Also the effective density of states of the conduction band is found in Equation (6.60).

$$N_c(T_o) := 2 \cdot \left(\frac{2 \cdot \pi \cdot m_n \cdot k_B \cdot T_o}{h_P^2} \right)^{\frac{3}{2}} \qquad N_c(300\,K) = 3.217 \times 10^{19} \cdot cm^{-3} \quad (M.32)$$

The effective density of states of the valence band is found in Equation (6.63)

$$N_v(T_o) := 2 \cdot \left(\frac{2 \cdot \pi \cdot m_p \cdot k_B \cdot T_o}{h_P^2} \right)^{\frac{3}{2}} \qquad N_v(300\,K) = 1.829 \times 10^{19} \cdot cm^{-3}$$
(M.33)

The intrinsic carrier concentration from Equation (6.66) is written here.

$$n_i(T_o) := \sqrt{N_c(T_o) \cdot N_v(T_o)} \cdot \exp\left(\frac{-E_g(T_o)}{2 \cdot k_B \cdot T_o} \right) \quad n_i(300\,K) = 8.697 \times 10^9 \, cm^{-3}$$
(M.34)

The numerical value of the intrinsic carrier concentration for silicon semiconductor is in agreement with the measurements.

Built-in Voltage Vbi   The built-in voltage from Equation (6.97) is written here and calculated but cannot be measured.

$$V_{bi}(T_o) := \frac{k_B \cdot T_o}{q} \cdot \ln\left(\frac{N_D \cdot N_A}{n_i(T_o)^2} \right) \quad V_{bi}(300\,K) = 0.781\,V \qquad (M.35)$$

Dimensions of the Depletion Region or the Space Charge Region   The schematic of the depletion region can be found in Figure 6.31. x_N, x_P, and W_D in Equations (6.187a), (6.188a), and (6.189a) are rewritten here

$$x_N(T_o, V_L) := \left[\frac{2 \cdot \varepsilon_p}{q} \cdot \frac{N_A}{N_D \cdot (N_A + N_D)} \cdot (V_{bi}(T_o) - V_L) \right]^{\frac{1}{2}} \qquad (M.36)$$

$$x_P(T_o, V_L) := x_N(T_o, V_L) \cdot \frac{N_D}{N_A} \qquad (M.37)$$

$$W_D(T_o, V_L) := x_N(T_o, V_L) + x_P(T_o, V_L) \qquad (M.38)$$

Assuming that $V_L = 0.527\,V$, the numerical values are

$$x_N(300\,K, 0.527V) = 5.754 \times 10^{-4} \mu m \qquad (M.39)$$

$$x_N(300\,K, 0.527V) = 0.575 \mu m \qquad (M.40)$$

$$W_D(300\,K, 0.527V) = 0.576 \mu m \qquad (M.41)$$

Carrier Mobility, μ The carrier mobility μ depends on both doping and temperature. The carrier mobility in silicon is well known from measurements. The empirical correlation was developed as follows. You may find more information in Section 6.8.2.

$$\mu_{n1}(T_o) := 92 \cdot \left(\frac{T_o}{300\,\text{K}}\right)^{-0.57} \cdot \frac{\text{cm}^2}{\text{V} \cdot \text{s}} \qquad \mu_{n2}(T_o) := 1268 \cdot \left(\frac{T_o}{300\,\text{K}}\right)^{-2.33} \cdot \frac{\text{cm}^2}{\text{V} \cdot \text{s}} \tag{M.42}$$

$$N_{D1}(T_o) := 1.3 \cdot 10^{17} \cdot \left(\frac{T_o}{300\,\text{K}}\right)^{2.4} \cdot \text{cm}^{-3} \qquad \alpha_n(T_o) := 0.91 \cdot \left(\frac{T_o}{300\,\text{K}}\right)^{-0.146} \tag{M.43}$$

$$\mu_{p1}(T_o) := 54.3 \cdot \left(\frac{T_o}{300\,\text{K}}\right)^{-0.57} \cdot \frac{\text{cm}^2}{\text{V} \cdot \text{s}} \qquad \mu_{p2}(T_o) := 406.9 \cdot \left(\frac{T_o}{300\,\text{K}}\right)^{-2.23} \cdot \frac{\text{cm}^2}{\text{V} \cdot \text{s}} \tag{M.44}$$

$$N_{A1}(T_o) := 2.35 \cdot 10^{17} \cdot \left(\frac{T_o}{300\,\text{K}}\right)^{2.4} \cdot \text{cm}^{-3} \qquad \alpha_p(T_o) := 0.88 \cdot \left(\frac{T_o}{300\,\text{K}}\right)^{-0.146} \tag{M.45}$$

For electron mobility,

$$\mu_n(N_D, T_o) := \mu_{n1}(T_o) + \frac{\mu_{n2}(T_o)}{1 + \left(\dfrac{N_D}{N_{D1}(T_o)}\right)^{\alpha_n(T_o)}} \tag{M.46}$$

For hole mobility (see Figure M.4),

$$\mu_p(N_A, T_o) := \mu_{p1}(T_o) + \frac{\mu_{p2}(T_o)}{1 + \left(\dfrac{N_A}{N_{A1}(T_o)}\right)^{\alpha_p(T_o)}} \tag{M.47}$$

$$N1_D := 10^{14}\,\text{cm}^{-3}, 10^{15}\,\text{cm}^{-3}..10^{20}\,\text{cm}^{-3} \qquad N1_A := 10^{14}\,\text{cm}^{-3}, 10^{15}\,\text{cm}^{-3}..10^{20}\,\text{cm}^{-3} \tag{M.48}$$

We will express the carrier mobilities as a function of temperature only for convenience.

$$\underset{\wwbar}{\mu_n}(T_o) := \mu_n(N_D, T_o) \tag{M.49}$$

$$\underset{\wwbar}{\mu_p}(T_o) := \mu_p(N_A, T_o) \tag{M.50}$$

Minority Carrier Diffusion Coefficients D_n and D_p   The diffusion coefficients for electrons and holes are defined by the Einstein relationship in Equations (6.160) and (6.159).

$$D_n(T_o) := \mu_n(T_o) \cdot \frac{k_B \cdot T_o}{q} \qquad D_n(300\,\text{K}) = 6.807 \cdot \frac{\text{cm}^2}{\text{s}} \tag{M.51}$$

$$D_p(T_o) := \mu_p(T_o) \cdot \frac{k_B \cdot T_o}{q} \qquad D_p(300\,\text{K}) = 11.838 \cdot \frac{\text{cm}^2}{\text{s}} \tag{M.52}$$

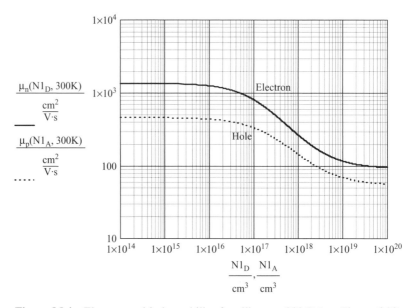

Figure M.4   Electron and hole mobility for silicon at 300 K (see Figure 6.30)

Minority Carrier Diffusion Lengths L_n and L_p   The minority carrier diffusion lengths for electrons and holes are defined as in Equations (6.208) and (6.207).

$$L_n (T_o) := \sqrt{D_n (T_o) \cdot \tau_n} \qquad L_n (300\ K) = 488.105 \cdot \mu m \qquad (M.53)$$

$$L_p (T_o) := \sqrt{D_p (T_o) \cdot \tau_p} \qquad L_p (300\ K) = 34.406 \cdot \mu m \qquad (M.54)$$

Some groups of quantities related to the recombination as shown in Equations (6.274) through (6.277) are defined herein.

$$X_P (T_o, V_L) := \frac{S_{Feff}}{D_p (T_o)} \cdot \sinh\left(\frac{W_N - x_N (T_o, V_L)}{L_p (T_o)} \right) + \frac{1}{L_p (T_o)} \cdot \cosh\left(\frac{W_N - x_N (T_o, V_L)}{L_p (T_o)} \right)$$
$$(M.55)$$

$$Y_P (T_o, V_L) := \frac{S_{Feff}}{D_p (T_o)} \cdot \cosh\left(\frac{W_N - x_N (T_o, V_L)}{L_p (T_o)} \right) + \frac{1}{L_p (T_o)} \cdot \sinh\left(\frac{W_N - x_N (T_o, V_L)}{L_p (T_o)} \right)$$
$$(M.56)$$

$$X_N (T_o, V_L) := \frac{S_{BSF}}{D_n (T_o)} \cdot \sinh\left(\frac{W_P - x_P (T_o, V_L)}{L_n (T_o)} \right) + \frac{1}{L_n (T_o)} \cdot \cosh\left(\frac{W_P - x_P (T_o, V_L)}{L_n (T_o)} \right)$$
$$(M.57)$$

$$Y_N (T_o, V_L) := \frac{S_{BSF}}{D_n (T_o)} \cdot \cosh\left(\frac{W_P - x_P (T_o, V_L)}{L_n (T_o)} \right) + \frac{1}{L_n (T_o)} \cdot \sinh\left(\frac{W_P - x_P (T_o, V_L)}{L_n (T_o)} \right)$$
$$(M.58)$$

By considering photon energy in Equation (6.1), determine the band gap wavelength that is an absorbable limit and contributes to the photocurrent.

$$\lambda_{BG}\,(T_o) := \frac{h_P \cdot c}{E_g\,(T_o)} \tag{M.59}$$

Define the ranges of wavelength for definite integrals, treating the singularity in the denominator by dividing it into two integrals to avoid the singularity.

$$\lambda_i := 0.01 \mu m \qquad \lambda_{BG}\,(300\ K) = 1.103 \cdot \mu m \tag{M.60}$$

The groups of quantities related to photogeneration were presented in Equations (6.278) through (6.281), where a singularity appears in each denominator of the equations. This causes a problem in smoothing numerical computations. Therefore, the singularity point must be found by using a MathCAD built-in function called "root function." And the point is eliminated by separating the integrals at the point.

An initial guess (any reasonable number) is needed for the MathCAD root function. You may always try a most reasonable guess for the best result.

$$\lambda := 0.6\,\mu m \tag{M.61}$$

The root functions and the singularity points $\lambda_P(T_o)$ and $\lambda_n(T_o)$ for holes and electrons are written by

$$\lambda_p\,(T_o) := \mathrm{root}\left(\alpha\,(\lambda,\,T_o)^2 \cdot L_p\,(T_o)^2 - 1,\,\lambda\right) \qquad \lambda_p\,(300\ K) = 0.888 \cdot \mu m \quad (M.62)$$

$$\lambda_n\,(T_o) := \mathrm{root}\left(\alpha\,(\lambda,\,T_o)^2 \cdot L_n\,(T_o)^2 - 1,\,\lambda\right) \qquad \lambda_n\,(300\ K) = 1.034 \cdot \mu m \quad (M.63)$$

With the parameters given in the beginning of this section, the computed singularity points appear less than the band gap wavelength of $1.103\,\mu m$ calculated in Equation (M.60). Therefore, the singularity points are in effective during the computations. However, note that the singularity points will be changing with changing temperature, minority carrier lifetimes, or dopants concentrations. This must be accounted for in the computations. In this section, since we have the fixed dopants concentrations (N_A and N_D) and the fixed lifetimes (τ_n and τ_p), the singularity points are dependent only on temperature. The groups of quantities related to photogeneration shown in Equations (6.278) through (6.281) are defined here:

$$G_{p1\lambda}\,(T_o,\,x,\,\lambda) := \frac{\tau_p}{\left(\alpha\,(\lambda,\,T_o)^2 \cdot L_p\,(T_o)^2 - 1\right)} \cdot (1 - r\,(\lambda)) \cdot f_0\,(\lambda) \cdot \alpha\,(\lambda,\,T_o)$$

$$\times \exp\left[-\alpha\,(\lambda,\,T_o) \cdot (x + W_N)\right] \tag{M.64}$$

$$G_{p1}\,(T_o,\,x) := \left| \begin{array}{l} (1 - s_f) \cdot \displaystyle\int_{\lambda_i}^{\lambda_{BG}(T_o)} G_{p1\lambda}\,(T_o,\,x,\,\lambda)\,d\lambda \quad \text{if} \quad \lambda_{BG}\,(T_o) < \lambda_p\,(T_o) \\[2ex] (1 - s_f) \cdot \left(\displaystyle\int_{\lambda_i}^{\lambda_p(T_o)\cdot 0.999} G_{p1\lambda}\,(T_o,\,x,\,\lambda)\,d\lambda \right. \\[2ex] \qquad\qquad \left. + \displaystyle\int_{\lambda_p(T_o)\cdot 1.001}^{\lambda_{BG}(T_o)} G_{p1\lambda}\,(T_o,\,x,\,\lambda)\,d\lambda \right) \quad \text{otherwise} \end{array} \right.$$

$$\tag{M.65}$$

The vertical line (*Add Line*) in Equation (M.65) allows the MathCAD built-in program where we can execute if-do-otherwise-do task:

$$
G_{p2\lambda}(T_o, x, \lambda) := \frac{\tau_p}{\left(\alpha(\lambda, T_o)^2 \cdot L_p(T_o)^2 - 1\right)} \cdot (1 - r(\lambda)) \cdot f_0(\lambda) \cdot \alpha(\lambda, T_o)^2
$$
$$
\times \exp\left[-\alpha(\lambda, T_o) \cdot (x + W_N)\right] \tag{M.66}
$$

$$
G_{p2}(T_o, x) := \begin{vmatrix} (1 - s_f) \cdot \displaystyle\int_{\lambda_i}^{\lambda_{BG}(T_o)} G_{p2\lambda}(T_o, x, \lambda)\, d\lambda & \text{if} \quad \lambda_{BG}(T_o) < \lambda_p(T_o) \\[2ex] (1 - s_f) \cdot \left(\displaystyle\int_{\lambda_i}^{\lambda_p(T_o)\cdot 0.999} G_{p2\lambda}(T_o, x, \lambda)\, d\lambda \right. \\[2ex] \qquad\qquad \left. + \displaystyle\int_{\lambda_p(T_o)\cdot 1.001}^{\lambda_{BG}(T_o)} G_{p2\lambda}(T_o, x, \lambda)\, d\lambda \right) & \text{otherwise} \end{vmatrix}
$$
$$
\tag{M.67}
$$

$$
G_{n1\lambda}(T_o, x, \lambda) := \frac{\tau_n}{\left(\alpha(\lambda, T_o)^2 \cdot L_n(T_o)^2 - 1\right)} \cdot (1 - r(\lambda)) \cdot f_0(\lambda) \cdot \alpha(\lambda, T_o)
$$
$$
\times \exp\left[-\alpha(\lambda, T_o) \cdot (x + W_N)\right] \tag{M.68}
$$

$$
G_{n1}(T_o, x) := \begin{vmatrix} (1 - s_f) \cdot \displaystyle\int_{\lambda_i}^{\lambda_{BG}(T_o)} G_{n1\lambda}(T_o, x, \lambda)\, d\lambda & \text{if} \quad \lambda_{BG}(T_o) < \lambda_n(T_o) \\[2ex] (1 - s_f) \cdot \left(\displaystyle\int_{\lambda_i}^{\lambda_n(T_o)\cdot 0.999} G_{n1\lambda}(T_o, x, \lambda)\, d\lambda \right. \\[2ex] \qquad\qquad \left. + \displaystyle\int_{\lambda_n(T_o)\cdot 1.001}^{\lambda_{BG}(T_o)} G_{n1\lambda}(T_o, x, \lambda)\, d\lambda \right) & \text{otherwise} \end{vmatrix}
$$
$$
\tag{M.69}
$$

$$
G_{n2\lambda}(T_o, x, \lambda) := \frac{\tau_n}{\left(\alpha(\lambda, T_o)^2 \cdot L_n(T_o)^2 - 1\right)} \cdot (1 - r(\lambda)) \cdot f_0(\lambda) \cdot \alpha(\lambda, T_o)^2
$$
$$
\times \exp\left[-\alpha(\lambda, T_o) \cdot (x + W_N)\right] \tag{M.70}
$$

$$
G_{n2}(T_o, x) := \begin{vmatrix} (1 - s_f) \cdot \displaystyle\int_{\lambda_i}^{\lambda_{BG}(T_o)} G_{n2\lambda}(T_o, x, \lambda)\, d\lambda & \text{if} \quad \lambda_{BG}(T_o) < \lambda_n(T_o) \\[2ex] (1 - s_f) \cdot \left(\displaystyle\int_{\lambda_i}^{\lambda_n(T_o)\cdot 0.999} G_{n2\lambda}(T_o, x, \lambda)\, d\lambda \right. \\[2ex] \qquad\qquad \left. + \displaystyle\int_{\lambda_n(T_o)\cdot 1.001}^{\lambda_{BG}(T_o)} G_{n2\lambda}(T_o, x, \lambda)\, d\lambda \right) & \text{otherwise} \end{vmatrix}
$$
$$
\tag{M.71}
$$

Current-Voltage Characteristic   From Equation (6.286), we define

$$
J_{SCN1}(T_o, V_L) := \frac{q \cdot S_{Feff}}{L_p(T_o) \cdot X_P(T_o, V_L)} \cdot G_{p1}(T_o, -W_N)
$$
$$
+ \frac{qD_p(T_o)}{L_p(T_o) \cdot X_P(T_o, V_L)} \cdot G_{p2}(T_o, -W_N) \tag{M.72}
$$

$$J_{SCN2}(T_o, V_L) := -\frac{q \cdot D_p(T_o) \cdot Y_P(T_o, V_L)}{L_p(T_o) \cdot X_P(T_o, V_L)} \cdot G_{p1}(T_o, -x_N(T_o, V_L))$$

$$- q \cdot D_p(T_o) \cdot G_{p2}(T_o, -x_N(T_o, V_L)) \quad \text{(M.73)}$$

$$J_{SCN}(T_o, V_L) := J_{SCN1}(T_o, V_L) + J_{SCN2}(T_o, V_L) \quad \text{(M.74)}$$

From Equation (6.287),

$$J_{SCP1}(T_o, V_L) := \frac{q \cdot S_{BSF}}{L_n(T_o) \cdot X_N(T_o, V_L)} \cdot G_{n1}(T_o, W_P)$$

$$- \frac{qD_n(T_o)}{L_n(T_o) \cdot X_N(T_o, V_L)} \cdot G_{n2}(T_o, W_P) \quad \text{(M.75)}$$

$$J_{SCP2}(T_o, V_L) := -\frac{q \cdot D_n(T_o) \cdot Y_N(T_o, V_L)}{L_n(T_o) \cdot X_N(T_o, V_L)} \cdot G_{n1}(T_o, x_P(T_o, V_L))$$

$$+ q \cdot D_n(T_o) \cdot G_{n2}(T_o, x_P(T_o, V_L)) \quad \text{(M.76)}$$

$$J_{SCP}(T_o, V_L) := J_{SCP1}(T_o, V_L) + J_{SCP2}(T_o, V_L) \quad \text{(M.77)}$$

From Equation (6.288),

$$\text{DEXP}(T_o, V_L, \lambda) := \exp[-\alpha(\lambda, T_o) \cdot (-x_N(T_o, V_L) + W_N)] \quad \text{(M.78)}$$

$$- \exp[-\alpha(\lambda, T_o) \cdot (x_P(T_o, V_L) + W_N)]$$

$$J_{SCD}(T_o, V_L) := q \cdot (1 - s_f) \cdot \int_{\lambda_i}^{\lambda_{BG}(T_o)} (1 - r(\lambda)) \cdot f_0(\lambda) \cdot \text{DEXP}(T_o, V_L, \lambda)d\lambda$$

$$\text{(M.79)}$$

The short-circuit current density that equals the photocurrent is the sum of Equations (M.74), (M.77), and (M.79), as shown in Equation (6.285).

$$J_{SC}(T_o, V_L) := J_{SCN}(T_o, V_L) + J_{SCP}(T_o, V_L) + J_{SCD}(T_o, V_L) \quad \text{(M.80)}$$

We want to take a look at the numerical values for the photocurrents in the n-type and p-type regions. It is interesting to note from the following values in Equations (M.80b) and (M.80c) that the photocurrent of the p-type region is about six times that of the n-type region, while the width of the p-type region is about a thousand times that of the n-type region. This is associated with the solar cell parameters such as the surface recombination velocities, minority carrier lifetimes, or diffusion lengths.

$$J_{SC}(300\,\text{K}, 0.527\,\text{V}) = 0.0327\,\text{A} \cdot \text{cm}^{-2} \quad \text{(M.80a)}$$

$$A_{cell} \cdot J_{SCN}(300\,\text{K}, 0.527\,\text{V}) = 0.437\,\text{A} \quad \text{(M.80b)}$$

$$A_{cell} \cdot J_{SCP}(300\,\text{K}, 0.527\,\text{V}) = 2.448\,\text{A} \quad \text{(M.80c)}$$

The dark saturation current density due to the recombination in the quasi-neutral region shown in Equation (6.289) is

$$J_{01}(T_o, V_L) := \left(\frac{q \cdot D_p(T_o)}{L_p(T_o)} \cdot \frac{Y_P(T_o, V_L)}{X_P(T_o, V_L)} \cdot \frac{n_i(T_o)^2}{N_D} + \frac{q \cdot D_n(T_o)}{L_n(T_o)} \cdot \frac{Y_N(T_o, V_L)}{X_N(T_o, V_L)} \cdot \frac{n_i(T_o)^2}{N_A} \right)$$

(M.81)

$$J_{01}(300K, 0.527\,V) = 1.869 \times 10^{-12}\,A \cdot cm^{-2}$$

(M.81a)

The dark saturation current density due to the recombination in the depletion region shown in Equation (6.290) is

$$J_{02}(T_o, V_L) := \frac{q \cdot W_D(T_o, V_L)}{\tau_p + \tau_n} \cdot n_i(T_o)$$

(M.82)

The cell current density which is the current produced by solar cell shown in Equation (6.284) is

$$J_{cell}(T_o, V_L) := J_{SC}(T_o, V_L) - J_{01}(T_o, V_L) \cdot \left(\exp\left(\frac{q \cdot V_L}{k_B \cdot T_o} \right) - 1 \right)$$

$$- J_{02}(T_o, V_L) \cdot \left(\exp\left(\frac{q \cdot V_L}{2 \cdot k_B \cdot T_o} \right) - 1 \right)$$

(M.83)

Currents are now obtained by multiplying the solar cell area by density current:

$$I_{SC}(T_o, V_L) := A_{cell} \cdot J_{SC}(T_o, V_L)$$

(M.84)

$$I_{01}(T_o, V_L) := A_{cell} \cdot J_{01}(T_o, V_L)$$

(M.85)

$$I_{02}(T_o, V_L) := A_{cell} \cdot J_{02}(T_o, V_L)$$

(M.86)

$$I_{cell}(T_o, V_L) := A_{cell} \cdot J_{cell}(T_o, V_L)$$

(M.87)

$$Power(T_o, V_L) := I_{cell}(T_o, V_L) \cdot V_L$$

(M.88)

The numerical values are shown here:

$$I_{SC}(300\,K, 0.527\,V) = 3.266\,A$$

(M.89)

$$I_{01}(300\,K, 0.527\,V) \cdot \left(\exp\left(\frac{q \cdot 0.527\,V}{k_B \cdot 300\,K} \right) - 1 \right) = 0.133\,A$$

(M.90)

$$I_{02}(300\,K, 0.527\,V) \cdot \left(\exp\left(\frac{q \cdot 0.527\,V}{2 \cdot k_B \cdot 300\,K} \right) - 1 \right) = 6.105 \times 10^{-4}\,A$$

(M.91)

$$I_{cell}(300\,K, 0.527\,V) = 3.132\,A$$

(M.92)

$$Power(300\,K, 0.527\,V) = 1.651\,W$$

(M.93)

Notice that the dark current density in the depletion region I_{02} is very small and usually neglected. The range of voltages for the I-V curve is given. We better estimate

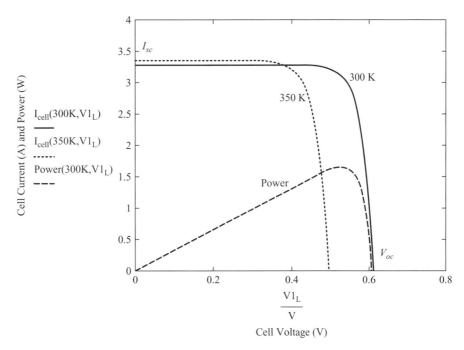

Figure M.5 Current-voltage curve for a silicon solar cell

the number of iteration as dividing the full scale of 1 V by the interval of 0.02 V, which gives 50 iterations. Assuming that each iteration takes 5 seconds, we have total 250 seconds (4 minutes 10 seconds) to complete the computations.

$$V1_L := 0V, 0.02V..1V \tag{M.94}$$

Note that the I-V characteristic varies with temperature and the maximum power point on the power curve. (See Figure 6.37 and Figure M.5).

Figures of Merit We found numerically that the open-circuit voltage is independent of the cell voltage. So zero voltage is used to compute the open-circuit voltage. From Equation (6.294), we have

$$V_{OC}(T_o) := \frac{k_B \cdot T_o}{q} \cdot \ln\left(\frac{I_{SC}(T_o, 0V)}{I_{01}(T_o, 0V)} + 1\right) \tag{M.95}$$

The short-circuit current in Equation (6.292) is equal to the cell current when the voltage is set to zero, so that the short-circuit current is no longer a function of voltage. Therefore, we can write

$$I_{SC}(T_o) := I_{cell}(T_o, 0V) \qquad I_{SC}(300\,K) = 3.268\,A \tag{M.96}$$

The maximum power voltage and current in Equations (6.299) and (6.301) are rewritten here

$$V_{mp}(T_o) := V_{OC}(T_o) - \frac{k_B \cdot T_o}{q} \cdot \ln\left(\frac{q \cdot V_{OC}(T_o)}{k_B \cdot T_o} + 1\right) \tag{M.97}$$

$$I_{mp}(T_o) := I_{SC}(T_o) - I_{01}(T_o, V_{mp}(T_o)) \cdot \exp\left(\frac{q \cdot V_{OC}(T_o)}{k_B \cdot T_o} - \ln\left(\frac{q \cdot V_{OC}(T_o)}{k_B \cdot T_o} + 1\right)\right) \tag{M.98}$$

The numerical values are

$$V_{OC}(300\,K) = 0.61\,V \tag{M.99}$$

$$V_{mp}(300\,K) = 0.527\,V \tag{M.100}$$

$$I_{mp}(300\,K) = 3.135\,A \tag{M.101}$$

The explicit approximate expressions such as V_{OC}, I_{SC}, V_{MP}, and I_{MP} significantly reduce the computational time with tolerable uncertainties (within about 1 percent error). There is usually good agreement between the exact numerical calculations and the approximate calculations. The maximum power will be

$$P_{mp}(T_o) := I_{mp}(T_o) \cdot V_{mp}(T_o) \qquad\qquad P_{mp}(300\,K) = 1.652\,W \tag{M.102}$$

The total incident solar power is obtained using Equation (6.307) with Equation (6.2).

$$P_{in} := A_{cell} \cdot \int_{0.1\mu m}^{4\mu m} h_P \cdot \frac{c}{\lambda} \cdot f_0(\lambda)d\lambda \qquad\qquad P_{in} = 10.125\,W \tag{M.103}$$

The power conversion efficiency of the solar cell in Equation (6.306) is

$$\eta(T_o) := \frac{P_{mp}(T_o)}{P_{in}} \qquad\qquad \eta(300\,K) = 0.163 \tag{M.104}$$

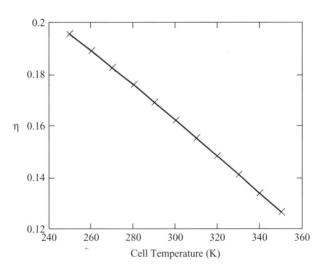

Figure M.6   Efficiency versus temperature for a Si solar cell (see Figure 6.45)

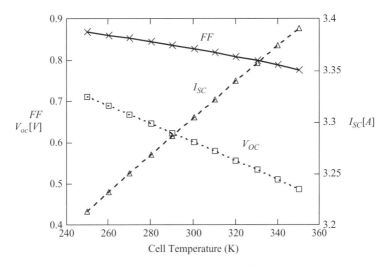

Figure M.7   *FF*, I_{SC}, and V_{OC} versus temperature for a Si solar cell (see Figure 6.44)

The fill factor *FF* in Equation (6.302) is an important parameter for evaluating the quality of the I-V characteristic. See Section 6.10.3 for more detailed discussion for the fill factor.

$$\text{FF}(T_o) := \frac{P_{mp}(T_o)}{I_{SC}(T_o) \cdot V_{OC}(T_o)} \qquad\qquad \text{FF}(300\,\text{K}) = 0.829 \qquad (\text{M}.105)$$

Effect of Temperature on the Performance of a Solar Cell   Since the cell current is a function of both voltage and temperature, we can plot the power conversion efficiency versus temperature as

$$T1_o := 250\,\text{K}, 270\,\text{K} .. 350\,\text{K} \qquad\qquad (\text{M}.106)$$

The detailed discussion for the effect of temperature on the performance of a solar cell was presented in Section 6.10.12 (See Figures M.6 and M.7).

6.12.2   Comparison of the Solar Cell Model with a Commercial Product

The numerical results for a solar cell obtained in Section 6.12.1 were simply compared with the parameters of a commercial product, Siemens Solar Module SM55, that consists of 36 single-crystalline solar cells, which is shown in Table 6.11. In typical modules such as Siemens SM55, the cells are connected in series to produce a voltage sufficient to charge a 12 V battery. In this case, the voltages of a cell are simply multiplied by 36 to calculate the module voltages. Since the power is the product of current and voltage, the cell power is also multiplied by 36 to calculate the module power. The module current is the same as the individual cell current, depending not on the number of solar cells but on the size of a solar cell.

We note that the input data such as doping concentrations, surface recombination velocities, and minority lifetimes for the commercial Siemens module SM55 were not known (the typical values were just used in the model) and no attempt was made to

Table 6.11   Comparison between the Present Solar Model and a Commercial Module

Parameters	Symbol	Present model (based on Section 6.12.1)	Siemens solar module SM55	Discrepancy (%)
Solar cell Material		Single-crystalline Si	Single-crystalline Si	—
Solar irradiance		AM1.5G (1000 W/m^2)	AM1.5G (1000 W/m^2)	—
Cell Area [cm^2]	A_{cell}	100	103	3%
Cell number		36	36	—
Temperature [K]		300 K	298 K	—
Maximum peak power [Watt]	P_{mp}	$1.652 \times 36 = 59.5$	55	8.1%
Rated current [A]	I_{mp}	3.135	3.15	0.5%
Rated voltage [V]	V_{mp}	$0.527 \times 36 = 18.9$	17.4	8.6%
Short-circuit current [A]	I_{SC}	3.266	3.45	5.3%
Open-circuit voltage [V]	V_{OC}	$0.61 \times 36 = 21.96$	21.7	1.2%
Efficiency (cell and module)	η	16.32%	14.83%	10.0%

fit the parameters between the present model and the Siemens module. Nevertheless, the comparison indicates actually good agreement when considering that the present model in Section 6.12.1 does not include any ohmic losses taking place at the contacts and connections of the solar module. In general, the cell efficiency is 2 to 3 percent higher than the module efficiency, which is fortuitously in agreement with the efficiencies shown in Table 6.11. In conclusion, the present one-dimensional analytical model simulates the real solar cell with approximate 10 percent uncertainty.

Example 6.12.1 Solar Cell Design   A 160-watt source is required in a remote area where the incident solar power is 1000 W/m^2. Using the information in Section 6.12.1, Modeling for a Silicon Solar Cell, we can find the active cell surface area that is required when operating under conditions of maximum power at 300 K. What is the overall efficiency of the converter?

Solution   Since the incident solar power is 1,000 W/m^2, we find from Equations (M.80a) and (M.81a) that $J_{SC} = 0.0327$ Acm^{-2} and $J_{01} = J_0 = 1.869 \times 10^{-12}$ Acm^{-2}, respectively.

From Equation (6.294), we calculate the open-circuit voltage as

$$V_{OC} = \frac{kT}{q} \ln\left(\frac{J_{SC}}{J_0} + 1\right) = \frac{\left(8.617 \times 10^{-5}\, \text{eV/K}\right)(300\,\text{K})}{1\text{eV/V}}$$

$$\times \ln\left(\frac{0.0327\, \text{Acm}^{-2}}{1.869 \times 10^{-12}\text{Acm}^{-2}} + 1\right) = 0.61\,\text{V}$$

From Equation (6.299), the maximum power voltage is obtained

$$V_{mp} = V_{OC} - \frac{kT}{q} \ln\left(\frac{qV_{OC}}{kT} + 1\right)$$

$$= 0.61\,\text{V} - \frac{\left(8.617 \times 10^{-5}\text{eV/K}\right)(300\,\text{K})}{1\,\text{eV/V}} \ln\left(\frac{(1\,\text{eV/V})(0.61\,\text{V})}{\left(8.617 \times 10^{-5}\, \text{eV/K}\right)(300\,\text{K})} + 1\right)$$

$$= 0.527\,\text{V}$$

From Equation (6.301), the maximum power current is obtained as

$$I_{mp} = A_{SC} J_{mp}$$

where A_{SC} is the required cell surface area. Since

$$\frac{q V_{OC}}{kT} = \frac{(1 \, eV/V)(0.61 \, V)}{(8.617 \times 10^{-5} \, eV/K)(300 \, K)} = 23.56$$

The maximum power current density is

$$J_{mp} \cong J_{SC} - J_0 \exp\left[\frac{q}{kT} V_{OC} - \ln\left(\frac{q V_{OC}}{kT} + 1\right)\right]$$

$$= \left(0.0327 \, Acm^{-2}\right) - \left(1.869 \times 10^{-12} \, Acm^{-2}\right) \exp\left(23.56 - \ln\left(23.56 + 1\right)\right)$$

$$= 0.0314 \, Acm^{-2}$$

The maximum power current is

$$I_{mp} = A_{sc} J_{mp} = A_{sc}\left(0.0314 \, Acm^{-2}\right)$$

The maximum power is obtained as

$$P_{mp} = I_{mp} V_{mp} = A_{SC}\left(0.0314 \, Acm^{-2}\right)(0.527 \, V) = A_{SC} 0.016547 \, W/cm^2$$

Now, in order to calculate the required cell surface area, the power requirement is equal to the maximum power. The required solar cell surface area is calculated by

$$A_{sc} = \frac{160 \, Watts}{0.016547 \, W/\, cm^2} = 9{,}669.4 \, cm^2 \quad \Longleftarrow$$

The incident power was defined in Equation (6.307) as

$$P_{in} = A \int_0^\infty h\upsilon\left(\lambda\right) f_0\left(\lambda\right) d\lambda$$

where it is given that

$$\int_0^\infty h\upsilon\left(\lambda\right) f_0\left(\lambda\right) d\lambda = \text{ incident power per unit area} = 1000 \, W/m^2$$

Therefore, the power conversion efficiency from Equation (6.306) is

$$\eta = \frac{P_{mp}}{P_{in}} = \frac{A_{sc} 0.016547 \, W/cm^2}{A_{sc} 1000 \, W/m^2} = 0.16547 \quad \Longleftarrow$$

Recall that the power conversion efficiency obtained herein does not include the ohmic losses of the cells and the modules if they are used. Therefore, the real efficiency would be approximately 10 percent less than the efficiency of 16.5 percent, which is 14.85 percent. The corresponding required cell surface area would increase to 10,743 cm^2.

6.13   DESIGN OF A SOLAR CELL

Solar cell design involves specifying the parameters of the solar cell in order to maximize efficiency, given a certain set of constraints. These constraints will be defined by the working environment in which solar cells are produced. For example, in a commercial environment where the objective is to produce a competitively priced solar cell, the cost of fabricating a particular solar cell must be taken into consideration. However, in a research environment where the objective is to produce a highly efficient laboratory-type cell, maximizing efficiency rather than minimizing cost is the main consideration. It is desired to design a silicon solar cell using the present solar cell model in Section 6.12.1, but the conclusions are valid for other semiconductors as well.

We want to design a typical silicon solar cell for one sun and single junction. In practice, the solar cell design requires the optimum values of the design parameters such as the cell geometry (emitter and base widths), donor and acceptor concentrations, the front- and back-surface recombination velocities, the minority carrier diffusion lengths (or lifetimes), as well as lastly the contact metal grid spacing on the front surface.

6.13.1   Solar Cell Geometry with Surface Recombination Velocities

The geometry of a solar cell consists of the emitter and base widths. Since the emitter width is very small (less than $1\,\mu m$), the base width (usually more than $300\,\mu m$) is considered first. The determination of the base width W_P is basically dependent on the absorption coefficient of the silicon semiconductor. We want to first estimate the optic diffusion lengths from the absorption coefficient. From Figure 6.25, the Si curve is considered among others. The critical points for photogeneration are wavelengths (or energy levels) near the edge of the curve (the band gap energy) which must be absorbed in the solar cell. These wavelengths are found from the figure to be at an absorption coefficient of about 10 to 100 cm^{-1}. Since the optic diffusion length is the inverse of the absorption coefficient, we obtain correspondingly 100 to $1000\,\mu m$ for the optic diffusion lengths, which we take as the base width.

The effect of base width on the performance of a solar cell was discussed in Section 6.10.7. For the reasonable back-surface recombination velocity of $S_{BSF} = 100$ cm/s, we find an optimum base width of about $300\,\mu m$ with an efficiency of about 16 percent (Figure 6.43). Nevertheless, it requires a great amount of effort and technique to achieve the reasonable back-surface recombination velocity, which also depends on the manufacturability. However, if we are obliged to go to a very thin solar cell with lower efficiency, it is theoretically possible to have a solar cell thickness of $50\,\mu m$ with an efficiency of about 14 percent as shown in Figure 6.44. Then, it is questionable whether the technique could support the required design parameters such as a low value ($\tau_n = 15\mu s$) of the minority electron lifetime.

Now consider the emitter width W_N. The emitter width is very small but has a great effect on the performance of the solar cell, as discussed in Section 6.10.8. We found that, by decreasing the emitter width, not only photogenerated current increases but also the recombination in the n-type quasi-neutral region decreases. Therefore, we design the emitter width as small as possible. However, the emitter width depends on the front surface recombination velocity. Then, in Figure 6.48 we find almost no effect of the emitter width between 0.1 μm and 0.7 μm on the efficiency when the front surface recombination velocity is lower than 3×10^3 cm/s. Therefore, we conclude that any emitter width between 0.1 μm and 0.7 μm with an efficiency of 16.5 percent at front-surface recombination velocity less than 3×10^3 cm/s can be considered as an optimum point. In practice, production techniques allow us to have an emitter width of 0.35 μm.

6.13.2 Donor and Acceptor Concentrations

Dopants comprise donors in an n-type region and acceptor in a p-type region, as discussed in Sections 6.10.10 and 6.10.9, respectively. We want to first determine the donor concentration N_D for a solar cell design as discussed in Section 6.10.10. In general, an increase in donor concentration reduces electron mobility, and reduction of electron mobility decreases the photocurrent. Hence, an increase in donor concentration decreases the photocurrent, or equivalently, the short-circuit current, without ohmic losses. However, an increase in donor concentration increases the open-circuit voltage. This gives a maximum point in the efficiency. We find the optimum point ($N_D = 1 \times 10^{18}$ cm^{-3}) in donor concentration in Figure 6.52 with the efficiency of about 16 percent.

Now we want to determine the acceptor concentration. The acceptor concentration is considered in Figures 6.49 and 6.50. As for the donor concentration, we find the optimum point ($N_A = 1 \times 10^{17}$ cm^{-3}) of the acceptor concentration with the efficiency of about 17.5 percent for the given input conditions.

Since high doping causes other problems such as an increase in lateral sheet resistance in the emitter layer (see Section 6.13.4) and an increase in Auger recombination (see Section 6.7.1), the optimum donor and acceptor concentrations must be determined in parallel.

6.13.3 Minority Carrier Diffusion Lifetimes

Minority carrier diffusion lifetimes comprise the minority electron and hole diffusion lifetimes, as discussed in Sections 6.10.4 and 6.10.5, respectively. It was shown previously in Figures 6.39 and 6.40 that the minority electron diffusion lifetime is more important than the minority hole diffusion lifetime. Hence, we now focus on the minority electron diffusion lifetime in the base.

From Figure 6.38, we see that the short-circuit current, the open-circuit voltage, and the efficiency increase with increasing the electron lifetime. The longer the lifetime, the better the performance. The minority electron diffusion length L_n can be computed using Equation (6.208). A minority electron diffusion lifetime produces the corresponding diffusion length. The dashed line in Figure 6.38 indicates the base width ($W_P = 300$ μs). When the diffusion length L_n is less than the base width W_P, the minority electrons created deeper than the diffusion length are unlikely to reach either the depletion region or the back surface of the base, thus recombining in the base. This reduces the photocurrent. However, when the diffusion length L_n is much larger than the base width W_P, the electrons either transfer to the n-type region through the depletion region or move to the back surface, annihilating at the surface if a back-surface field

(BSF) is not applied. Therefore, the BSF is an important treatment when $L_n \gg W_P$, which is a typical condition, particularly in a thin solar cell. The BSF is controlled by the back-surface recombination velocity S_{BSF}. Since a large lifetime or large diffusion length, for instance, $L_n > 300\,\mu m$, requires a high-quality semiconductor material, minority electron diffusion lifetime of about $350\,\mu s$ is chosen for the solar cell design.

Since the minority hole diffusion lifetime τ_p has no effect on the efficiency (Figure 6.40), we randomly chose a minority diffusion lifetime of $1\,\mu s$ at this time until we have a reason for some other value.

6.13.4 Grid Spacing

Currents generally flow perpendicular to the surface, whereas they flow laterally in the emitter, as shown in Figure 6.61. Since the series resistance of a solar cell is a critical parameter for the performance, the resistance of the emitter must be kept small. The sheet resistivity in the emitter has an inverse dependency on the doping concentration.

The standard geometry for a contact metal grid on a solar cell is shown in Figure 6.61, where the fingers collect the electrons and deliver them to the busbars, which are heavier areas of metallization and are directly connected to the external loads. The vertical current in a unit cell of the size $D \times S$ flows upward in the base. The lateral current I is zero at the midpoint ($x = 0$) and increases linearly with the distance x. Ignoring losses due to the busbars, the total fractional power loss in the single unit cell consists of the power loss due to the sheet resistance, the power losses due to the metal fingers, the power losses due to the contact resistances, and the power loss due to the shading of the fingers [13].

In general, the resistance is defined as

$$R = \frac{\rho \cdot \text{Length}}{\text{Cross-sectional area}} \tag{6.332}$$

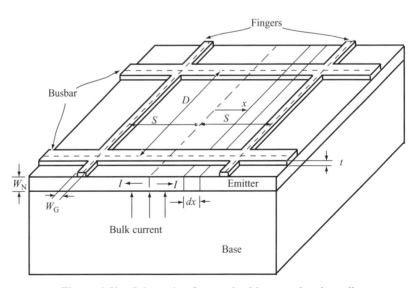

Figure 6.61   Schematic of a metal grid on a unit solar cell

The *lateral sheet resistance* R_S in the emitter shown in Figure 6.61 is a function of the distance x and defined as

$$R_S(x) = \frac{\rho \cdot x}{W_N D} = \frac{\rho_S x}{D} \tag{6.333}$$

where ρ_S is the *sheet resistivity* that equals to $\rho_S = \rho/W_N$ and has a unit of Ω/Sqaure (the square has no unit, but it is used to distinguish with the resistivity ρ). The lateral current I can be written as a function of the distance x and the current density J.

$$I(x) = AJ = Dx \cdot J \tag{6.334}$$

Using Equations (6.333) and (6.334) and knowing that $dR_s = (\rho_s/D)dx$, the resistive power loss due to the lateral sheet resistance for a current is obtained as

$$P_{loss,S} = \int I(x)^2 dR_s = \int_0^S D^2 x^2 J^2 \frac{\rho_s}{D} dx = \frac{1}{3} \rho_s J^2 D S^3 \tag{6.335}$$

The maximum power is the product of the maximum power current and voltage and can be written as

$$P_{mp} = I_{mp} V_{mp} = D S J_{mp} V_{mp} \tag{6.336}$$

The ratio of the *resistive power loss* to the maximum power is called the *fractional power loss* due to the lateral sheet resistance, P_S. By letting $J = J_{mp}$ at the maximum point, the fractional power loss due to the sheet resistance is

$$P_S = \frac{P_{loss,s}}{P_{mp}} = \frac{\frac{1}{3}\rho_S J_{mp}{}^2 D S^3}{D S J_{mp} V_{mp}} = \frac{\rho_S J_{mp} S^2}{3 V_{mp}} \tag{6.337}$$

The resistive power losses due to the fingers P_G can be derived as follows. The resistance in the fingers is

$$R_G = \frac{\rho \cdot D}{t W_G} = \frac{\rho_G D}{W_G} \tag{6.338}$$

where ρ_G is the *finger resistivity*, $\rho_G = \rho/t$ with units Ω/sqaure. Using the current collected in the unit cell ($I = DSJ$), the power loss due to a metal finger is obtained as

$$P_{loss,G} = I^2 R_G = J^2 D^2 S^2 \frac{\rho_G D}{W_G} \tag{6.339}$$

The fractional power loss due to the metal finger P_G is obtained as

$$P_G = \frac{P_{loss,G}}{P_{mp}} = \frac{J_{mp}{}^2 D^2 S^2 \frac{\rho_G D}{W_G}}{D S J_{mp} V_{mp}} = \frac{\rho_G J_{mp} D^2 S}{W_G V_{mp}} \tag{6.340}$$

The contact resistance is given by

$$R_C = \frac{\rho_C}{D W_G} \tag{6.341}$$

where ρ_C is the contact resistivity that has a unit of $\Omega \, cm^2$. The power loss due to the *contact resistance* $P_{loss,C}$ is written as

$$P_{loss,C} = I^2 R_C = D^2 S^2 J^2 \frac{\rho_C}{D W_G} \tag{6.342}$$

The fractional power loss due to the contact resistance P_C is

$$P_C = \frac{P_{loss,C}}{P_{mp}} = \frac{D^2 S^2 J_{mp}^2 \frac{\rho_C}{D W_G}}{D S J_{mp} V_{mp}} = \frac{\rho_C J_{mp} S}{W_G V_{mp}} \tag{6.343}$$

Finally, the power loss due to *the shading of the fingers* is

$$P_{SH} = \frac{D W_G J_{mp} V_{mp}}{D S J_{mp} V_{mp}} = \frac{W_G}{S} \tag{6.344}$$

The total fractional power loss in a single solar cell is obtained by summing those individual fractional losses as

$$P = P_S + P_G + P_C + P_{SH} \tag{6.345}$$

$$P = \frac{\rho_S J_{mp} S^2}{3 V_{mp}} + \frac{\rho_G J_{mp} D^2 S}{W_G V_{mp}} + \frac{\rho_C J_{mp} S}{W_G V_{mp}} + \frac{W_G}{S} \tag{6.345a}$$

The minimum value of the fractional power loss is obtained by differentiating with respect to S from Equation (6.345) and setting it zero.

$$\frac{dP}{dS} = 0 \tag{6.346}$$

$$\frac{2 \rho_S J_{mp}}{3 V_{mp}} S + \frac{\rho_G J_{mp} D^2}{W_G V_{mp}} + \frac{\rho_C J_{mp}}{W_G V_{mp}} - \frac{W_G}{S^2} = 0 \tag{6.347}$$

Dividing by $\frac{2 \rho_S J_{mp}}{3 V_{mp} S^2}$ and rearranging it gives

$$S^3 + \frac{3}{2} \frac{(\rho_G D^2 + \rho_C)}{\rho_S W_G} S^2 - \frac{3}{2} \frac{W_G V_{mp}}{\rho_S J_{mp}} = 0 \tag{6.348}$$

This equation depends on the distance D and the finger width W_G. In practice, the technology used for the top contacts will determine the gross features of the metal grid and place limits on the dimensions of these parameters. For instance, the application of the screen printing technique for the metallization does not allow finer structures than about $50 \, \mu m$ [13]. If one assumes that the metal finger and contact resistivities are negligible, approximately from Equation (6.348) the minimum finger width S is found to be

$$S_{opt} = \left(\frac{3 W_G V_{mp}}{2 \rho_s J_{mp}} \right)^{1/3} \tag{6.349}$$

where W_G is the finger width, ρ_S the sheet resistivity, V_{mp} the maximum power voltage, and J_{mp} the maximum power current density. The *minimum total fractional power loss* is obtained from Equation (6.345a) with Equation (6.349).

$$P_{loss,min} = \left(\frac{\rho_S J_{mp}}{V_{mp}}\right)^{1/3} \left(\frac{3W_G}{2}\right)^{2/3} \tag{6.350}$$

The optimum value for the finger distance $2S$ is thus primarily determined by the sheet resistivity of the emitter layer and the finger width W_G. Lower sheet resistivities reduce the power loss and allow large finger distances. Smaller finger widths reduce the power loss. For instance, the application of the screen-printing technique for the metallization does not allow finer structures than about $50\,\mu m$.

It should be noted that in practice more complex contact schemes are used, which require a more sophisticated analysis of the best parameters for the dimensions of the unit cell and the width of fingers and busbars.

Example 6.13.1 Grid Spacing   Consider the optimum grid spacing for a silicon solar cell for use in an area where the incident solar power is $1000\,W/m^2$. It is given the cell that the sheet resistivity is $40\,\Omega$/square and the finger width W_G is $300\,\mu m$. Find the optimum finger width and the minimum fractional power loss.

Solution   Since we consider a silicon solar cell under the incident solar power of $1{,}000\,W/m^2$, we can use the information in Section 6.12.1, Modeling for a Silicon Solar Cell.

From Equation (M.100), we find the maximum power voltage is

$$V_{mp} = 0.527\,V$$

And from Equation (M.92), we calculate the maximum power current is

$$I_{mp} = I_{cell}(300\,K, 0.527\,V) = 3.132A$$

The maximum power current density is obtained using the cell area used in the model is

$$J_{mp} = \frac{3.132A}{100\,cm^2} = 0.03132\,Acm^{-2}$$

Now from Equation (6.349),

$$S_{opt} = \left(\frac{3W_G V_{mp}}{2\rho_S J_{mp}}\right)^{1/3} = \left(\frac{3\left(300\,\mu m \times \dfrac{10^{-4}\,cm}{1\,\mu m}\right)0.527\,V}{2(40\,\Omega)(0.03132\,Acm^{-2})}\right)^{1/3} = 2.67\,mm$$

From Equation (6.350), we find the minimum total fractional power loss is

$$P_{loss,min} = \left(\frac{\rho_S J_{mp}}{V_{mp}}\right)^{1/3}\left(\frac{3W_G}{2}\right)^{2/3} = \left(\frac{40\,\Omega \cdot 0.03132\,Acm^{-2}}{0.527\,V}\right)^{1/3}\left(\frac{3 \cdot 300\,\mu m \cdot \dfrac{10^{-4}\,cm}{1\,\mu m}}{2}\right)$$

$$= 0.168$$

From this calculation, we find that the optimum finger spacing is $2S = 2 \times 2.67$ mm $=$ 5.34 mm and the minimum fractional power loss is 16.8 percent. This seems high compared to the typical values for commercial silicon solar cells of an average spacing of the fingers of about 5 mm and minimum fractional power loss below 5 percent.

6.13.5   Anti-Reflection, Light Trapping and Passivation

It is evident that the incident light cannot be fully utilized because of finite reflectivity. A bare silicon surface reflects almost 30 percent of the incident solar light. Therefore, the reflection should be reduced in order to improve the solar cell efficiency. This usually can be achieved by texturing the surface and depositing *anti-reflection (AR) coatings* on the surface. One of the most commonly used textures is a *pyramidal structure* (see Figure 6.2) of height 4–6 μm, which is achieved by etching the surface with sodium hydroxide (NaOH)-water-isopropyl alcohol (IPA) solution even before the p–n junction formation. The commonly used anti-reflection coatings with 0.06-μm TiO_2 are applied after completing the doping process (p–n junction formation). 0.11-μm SiO_2 is usually sandwiched between the AR coating and the silicon substrate to improve the interfacial contacts. The pyramidal structure allows multiple reflections on the textured surface, reducing the reflection coefficient. The two combined techniques can reduce the reflectivity down to 3 percent.

Optical losses also occur because of the finite thickness of the solar cell. The absorption coefficient of the material property requires a minimal optical thickness (about 1,000 μm), which is usually much thicker than those of commercial solar cells (100–300 μm). However, these losses can be reduced by a technique called *light trapping*. The light inside the crystal is reflected several times between the front and back surfaces before it is finally absorbed. Light trapping requires both a passivated surface at the back side and a textured (random) surface that reflects the light at oblique angles (the top surface should already have the AR coating and textured surface anyway). The *passivation* at the back surface can be done in a similar way to the AR coating at the top surface.

REFERENCES

1. C.E. Fritts, *Proc. Am. Assoc. Adv. Sci.* vol. 33, pp. 97, 1883.

2. D.M. Chapin, C.S. Fuller, and G.L. Pearson, *J. Appl. Phys.*, 25(1954): 676.

3. M. Planck, On the Law of Distribution of Energy in the Normal Spectrum. *Annalen der Physik*, 4(1901): 553ff.

4. Standard ASTM E490-2000. Standard for Solar Constant and Air Mass Zero Solar Spectral Irradiance Tables. American Society for Testing and Materials, West Conshocken, PA.

5. Standard ASTM E891-92. *Standard for Terrestrial Solar Direct Normal Solar Spectral Irradiance Tables for Air Mass 1.5*. American Society for Testing and Materials, West Conshocken, PA.

6. Standard ASTM E892-92. *Standard for Terrestrial Solar Spectral Irradiance Tables at Air Mass 1.5 for a 37° Tilted Surface*. American Society for Testing and Materials, West Conshocken, PA.

7. R.F. Pierret, *Semiconductor Device Fundamentals*. Reading, MA: Addison-Wesley Publishing Company, 1996.

8. A. Luque and S. Hegedus, *Handbook of Photovoltaic Science and Engineering*, West Sussex, England: John Wiley & Sons, 2003.

9. Courtesy of Handbook of Optical Constant of Solids, edited by Edward D. Palik. New York: Academic Press, 1985.

10. R. Pierret and G. Neudeck (eds.) *Modular Series on Solid State Devices. Volume VI: Advanced Semiconductor Fundamentals*, Chapter 5. Upper Saddle River, NJ: Prentice Hall, 2003.

11. A. Luque, and S. Hegedus, *Handbook of Photovoltaic Science and Engineering*. West Sussex, England: John Wiley & Sons, 2003.

12. J. Nelson, *The Physics of Solar Cells*. London: Imperial College Press, 2003.

13. H.J. Moller, *Semiconductors for Solar Cells*. Boston: Artech House, 1993.

14. J. Mazer, *Solar Cells: An Introduction to Crystalline Photovoltaic Technology*. Boston: Kluwer Academic Publishers, 1997.

15. V.M. Andreev, V.A. Grilikhes, and V.D. Rumyantsev, *Photovolyaic Conversion of Concentrated Sunlight*. New York: John Wiley & Sons, 1997.

16. P. Wurfel, *Physics of Solar Cells*. Weinheim: Wiley-VCH, 2005.

17. R.F. Pierret, *Semiconductor Device Fundamentals*. Reading, MA: Addison-Wesley Publishing Company, 1996.

18. J.L. Gray, The Physics of the Solar Cell, Chapter 3 of *Handbook of Photovoltaic Science and Engineering*. West Sussex, England: John Wiley & Sons, 2003.

19. M. Green, *Prog. Photovolt*, 9 (2001): 137–144.

20. W. Shockley and H. Queisser, *J. Appl. Phys.*, 32 (1961): 510–519.

PROBLEMS

Introduction

6.1 Read Section 6.1 and provide two-page summary.

6.2 Read Section 6.2 and provide two-page summary.

6.3 Describe the reason why silicon solar cells are dominant in the PV market.

6.4 Describe a couple of solutions for the reduction of silicon material for a solar cell.

6.5 Describe the operation of a typical silicon solar cell.

6.6 Explain the Sun's blackbody spectrum, AM0, AM1.5G, and AM1.5D spectral irradiances.

6.7 Describe the nature of light including photon energy.

6.8 Describe the significance of the Bohr's model in the light of line spectra.

Quantum Mechanics

6.9 Calculate the Bohr's radius and the total energy for a hydrogen atom with the quantum number $n = 3$.

6.10   Consider that we receive some line spectra from an unknown object in the sky, some of which are $0.434\,\mu\text{m}$, $0.486\,\mu\text{m}$, and $656\,\mu\text{m}$. Determine the element that composes the object and show your detailed calculation indicating the initial and final quantum numbers for each line spectrum.

6.11   Consider a copper atom whose atomic number is twenty-nine ($Z = 29$). Construct the electronic configuration of the atom and sketch the atomic structure indicating the orbitals (subshells) with electrons occupied.

6.12   Consider a selenium atom whose atomic number is thirty-four ($Z = 34$). Construct the electronic configuration of the atom and sketch the atomic structure indicating the orbitals (subshells) with electrons occupied.

Density of State

6.13   Show the detailed derivation of Equations (6.48) and (6.49).

Equilibrium Intrinsic Carrier Concentration

6.14   Considering a pure gallium arsenide semiconductor under an equilibrium condition at room temperature of $300\,\text{K}$, we want to know the number of free electrons and holes per cubic centimeter for a study of GaAs solar cells.

6.15   Considering a pure germanium semiconductor under an equilibrium condition at room temperature of $300\,\text{K}$, we want to estimate the number of electrons and holes per cubic centimeter for a study of Ge solar cells.

6.16   Show the detailed derivation of Equation (6.66).

Extrinsic Semiconductor in Thermal Equilibrium

6.17   Describe the doping for a Si solar cell, including donors and acceptors.

6.18   Describe how the built-in voltage is created and why it is created in the depletion region for a Si solar cell.

6.19   Calculate the built-in voltage for a semiconductor doped with $N_D = 1 \times 10^{18}\,\text{cm}^{-3}$ and $N_A = 1 \times 10^{15}\,\text{cm}^{-3}$.

6.20   Calculate the majority and minority electron concentrations if a silicon semiconductor was doped with $N_D = 1 \times 10^{18}\,\text{cm}^{-3}$ and $N_A = 1 \times 10^{15}\,\text{cm}^{-3}$.

Generation and Recombination

6.21   Describe the absorption coefficients in terms of both the wavelength and the band gap, including the optical absorption length for solar cells.

6.22   Calculate the maximum (band gap) wavelength absorbed by the following solar cell materials: crystalline silicon with a band gap of $1.125\,\text{eV}$, CdTe with a band gap of $1.5\,\text{eV}$, and amorphous silicon with a band gap of $1.7\,\text{eV}$.

Recombination

6.23 Describe all recombination mechanisms for solar cells.

6.24 Describe the primary mechanism that separates the electrons and holes generated by light.

6.25 Consider a Si semiconductor doped with donor concentration of 10^{17} cm^{-3} subject to sunlight where the excess minority hole is 10^8 cm^{-3}. We want to determine the situation is low-level injection.

6.26 Explain why the Auger recombination increases with increasing the dopants and how it affects the photocurrent.

Carrier Transport

6.27 Calculate a depletion width W_D including x_N and x_P. Assume that a Si semiconductor receives sunlight at 300 K with $N_D = 1 \times 10^{16}$ cm^{-3} and $N_A = 1 \times 10^{14}$ cm^{-3}. And also assume that the load voltage is 0.564 V.

6.28 Derive $\Delta n_P(x_P) = \frac{n_i^2}{N_A}\left[\exp\left(\frac{qV}{kT}\right) - 1\right]$ of Equation (6.206) with the appropriate explanation and proofs.

6.29 Considering a semiconductor with $N_D = 10^{16}$ cm^{-3} and $\tau_n = 100$ µs, estimate the minority carrier diffusion length L_n for electrons at room temperature (300 K).

6.30 Considering a semiconductor with $N_A = 10^{18}$ cm^{-3} and $\tau_p = 10$ µs, estimate the minority carrier diffusion length L_p for holes at room temperature (300 K).

Minority Carrier Lifetime

6.31 For a silicon solar cell, first describe what the minority electron lifetime is and why the electron lifetime is more important than hole lifetime. Second, explain the effect of the minority electron lifetime with respect to the diffusion length in the light of the back-surface recombination velocity (with at least 100 words).

Characteristic of Solar Cell

6.32 Show the detailed derivation of Equations (6.299) and (6.301).

6.33 Show the detailed derivation of Equations (6.305) and (6.307).

6.34 Show the detailed derivation of Equations (6.331a–f).

Modeling

6.35 Provide a complete MathCAD modeling for a silicon solar cell including Figures 6.37, 6.53, and 6.54 using the physical constants and solar cell parameters shown in Section 6.12.1, Modeling for a Silicon Solar Cell. You must provide your own work of MathCAD program for the complete modeling. Discuss the results with respect to solar cell design.

6.36 Compute the external quantum efficiency (EQE) for the given information in Section 6.12.1, Modeling for a Silicon Solar Cell and plot the EQE along with incident spectral power, as shown in Figure 6.58. Also plot the effect of surface recombination velocities on the EQE, as shown in Figure 6.59. You must provide your own work of MathCAD program with the above two plots. You should modify properly the MathCAD modeling shown in Section 6.12.1 for your work. Discuss the results with respect to solar cell design.

6.37 Consider the effect of acceptor concentration on the performance for a silicon solar cell and provided Figures 6.49 and 6.50 by modifying properly the MathCAD modeling in Section 6.12.1 for a silicon solar cell. Discuss the results with respect to solar cell design.

6.38 A 54-watt solar power is required in a remote area where the incident solar power is $1,000\,\text{W/m}^2$. Using the modeling in Section 6.12.1, find the active cell surface area that is required when operating under conditions of maximum power at 300 K. What is the overall efficiency of the converter?

6.39 (This problem requires that students have a working MathCAD program of Section 6.12.1.) A 120-W solar power is required in a remote area where the incident solar power is $768\,\text{W/m}^2$. Modifying the modeling in Section 6.12.1, find the active cell surface area required when operating under conditions of maximum power at 300 K. What is the overall efficiency of the converter?

6.40 (This problem requires that students have a working MathCAD program of Section 6.12.1.) A 90-W solar power is required in a hot remote area where the incident solar power is $1,000\,\text{W/m}^2$. Using the modeling in Section 6.12.1, find the active cell surface area that is required when operating under conditions of maximum power at 320 K. What is the overall efficiency of the converter?

Solar Cell Design

6.41 Consider the optimum grid spacing for a silicon solar cell at an area where the incident solar power is $1,000\,\text{W/m}^2$. It is given for the solar cell that the finger resistivity is 35 Ω/square and the finger width W_G is $50\,\mu\text{m}$. Find the optimum finger width and the minimum fractional power loss.

6.42 (This problem requires that students have a working MathCAD program of Section 6.12.1.) Consider the optimum grid spacing for a silicon solar cell at an area where the incident solar power is $768\,\text{W/m}^2$. It is given for the solar cell that the finger resistivity is 30 Ω/square and the finger width W_G is $100\,\mu\text{m}$. Find the optimum finger width and the minimum fractional power loss.

6.43 Show the detailed derivation of Equation (6.350).

$$P_{loss,\min} = \left(\frac{\rho_S J_{mp}}{V_{mp}}\right)^{1/3} \left(\frac{3W_G}{2}\right)^{2/3}$$

6.44 Design a typical *solar cell* with the AM1.5G solar spectrum using a silicon material at room temperature. Provide the design parameters such as the cell geometry (emitter and base widths), donor and acceptor concentrations, the front-1 and back-surface recombination velocities, the minority carrier diffusion lengths (or lifetimes), and lastly, the contact metal grid spacing on the front surface, with the appropriate explanations and equations. Provide the power conversion efficiency. Sketch the 3-D solar cell for 10×10 cm surface area using a CAD program such as SolidWorks.

6.45 Design a *thin solar cell* with the AM1.5G solar spectrum using a silicon material at room temperature. Provide the design parameters such as the cell geometry (emitter and base widths), donor and acceptor concentrations, the front-1 and back-surface recombination velocities, the minority carrier diffusion lengths (or lifetimes), and the contact metal grid spacing on the front surface, with the appropriate explanations and equations. Provide the power conversion efficiency. Sketch the 3-D solar cell for 4×7 cm surface area using a CAD program such as SolidWorks.

6.46 Design a solar cell running at a high temperature of 330 K under the AM1.5D solar spectrum using a silicon material. Provide the design parameters such as the cell geometry (emitter and base widths), donor and acceptor concentrations, the front-1 and back-surface recombination velocities, the minority carrier diffusion lengths (or lifetimes), and the contact metal grid spacing on the front surface, with the appropriate explanations and equations. Provide the power conversion efficiency. Sketch the 3-D solar cell for a 10×10 cm surface area using a CAD program such as SolidWorks.

Appendix A

Thermophysical Properties

Table A.1 Molar, Gas, and Critical Properties

Gas	Chemical formula	Molecular mass	R kJ/kg-K	ρ kg/m^3	C_{p*} kJ/kg-K	C_{p*} kJ/kg-K	k
		Properties of various ideal gases at 25°C, 100 kPa (SI Units)*					
Steam	H_2O	18.015	0.4615	0.0231	1.872	1.410	1.327
Acetylene	C_2H_2	26.038	0.3193	1.05	1.699	1.380	1.231
Air	–	28.97	0.287	1.169	1.004	0.717	1.400
Ammonia	NH_3	17.031	0.4882	0.694	2.130	1.642	1.297
Argon	Ar	39.948	0.2081	1.613	0.520	0.312	1.667
Butane	C_4H_{10}	58.124	0.1430	2.407	1.716	1.573	1.091
Carbon monoxide	CO	28.01	0.2968	1.13	1.041	0.744	1.399
Carbon dioxide	CO_2	44.01	0.1889	1.775	0.842	0.653	1.289
Ethane	C_2H_4	30.07	0.2765	1.222	1.766	1.490	1.186
Ethanol	C_2H_5OH	46.069	0.1805	1.883	1.427	1.246	1.145
Ethylene	C_2H_4	28.054	0.2964	1.138	1.548	1.252	1.237
Helium	He	4.003	2.0771	0.1615	5.193	3.116	1.667
Hydrogen	H_2	2.016	4.1243	0.0813	14.209	1.008	1.409
Methane	CH_4	16.043	0.5183	0.648	2.254	1.736	1.299
Methanol	CH_2OH	32.042	0.2595	1.31	1.405	1.146	1.227
Neon	Ne	20.183	0.4120	0.814	1.03	0.618	1.667
Nitric oxide	NO	30.006	0.2771	1.21	0.993	0.716	1.387
Nitrogen	N_2	28.013	0.2968	1.13	1.042	0.745	1.400
Nitrous oxide	N_2O	44.013	0.1889	1.775	0.879	0.690	1.274
n-octane	C_8H_{18}	114.23	0.07279	0.092	1.711	1.638	1.044
Oxygen	O_2	31.999	0.2598	1.292	0.922	0.662	1.393
Propane	C_3H_2	44.094	0.1886	1.808	1.679	1.490	1.126
R-12	CCl_2F_2	120.914	0.06876	4.98	0.616	0.547	1.126
R-22	$CHCLF_2$	86.469	0.09616	3.54	0.658	0.562	1.171
R-134a	CF_3CH_2F	102.03	0.08149	4.20	0.852	0.771	1.106
Sulfur dioxide	SO_2	64.059	0.1298	2.618	0.624	0.494	1.263
Sulfur trioxide	SO_3	80.053	0.10386	3.272	0.635	0.531	1.196

Adapted from Reference [1].

Table A.2 Thermophysical Properties of Gases at atmospheric Pressure

T (K)	ρ (kg/m^3)	c_p (kJ/kg · K)	$\mu \cdot 10^7$ (N · s/m^2)	$\nu \cdot 10^6$ (m^2/s)	$k \cdot 10^3$ (W/m · K)	$\alpha \cdot 10^6$ (m^2/s)	Pr
Air							
100	3.5562	1.032	71.1	2.00	9.34	2.54	0.786
150	2.3364	1.012	103.4	4.426	13.8	5.84	0.758
200	1.7458	1.007	132.5	7.590	18.1	10.3	0.737
250	1.3947	1.006	159.6	11.44	22.3	15.9	0.720
300	1.1614	1.007	184.6	15.89	26.3	22.5	0.707
350	0.9950	1.009	208.2	20.92	30.0	29.9	0.700
400	0.8711	1.014	230.1	26.41	33.8	38.3	0.690
450	0.7740	1.021	250.7	32.39	37.3	47.2	0.686
500	0.6964	1.030	270.1	38.79	40.7	56.7	0.684
550	0.6329	1.040	288.4	45.57	43.9	66.7	0.683
600	0.5804	1.051	305.8	52.69	46.9	76.9	0.685
650	0.5356	1.063	322.5	60.21	49.7	87.3	0.690
700	0.4975	1.075	338.8	68.10	52.4	98.0	0.695
750	0.4643	1.087	354.6	76.37	54.9	109	0.702
800	0.4354	1.099	369.8	84.93	57.3	120	0.709
850	0.4097	1.110	384.3	93.80	59.6	131	0.716
900	0.3868	1.121	398.1	102.9	62.0	143	0.720
950	0.3666	1.131	411.3	112.2	64.3	155	0.723
1000	0.3482	1.141	424.4	121.9	66.7	168	0.726
1100	0.3166	1.159	449.0	141.8	71.5	195	0.728
1200	0.2902	1.175	473.0	162.9	76.3	224	0.728
1300	0.2679	1.189	496.0	185.1	82	238	0.719
1400	0.2488	1.207	530	213	91	303	0.703
1500	0.2322	1.230	557	240	100	350	0.685
1600	0.2177	1.248	584	268	106	390	0.688
1700	0.2049	1.267	611	298	113	435	0.685
1800	0.1935	1.286	637	329	120	482	0.683
1900	0.1833	1.307	663	362	128	534	0.677
2000	0.1741	1.337	689	396	137	589	0.672
2100	0.1658	1.372	715	431	147	646	0.667
2200	0.1582	1.417	740	468	160	714	0.655
2300	0.1513	1.478	766	506	175	783	0.647
2400	0.1448	1.558	792	547	196	869	0.630
2500	0.1389	1.665	818	589	222	960	0.613
3000	0.1135	2.726	955	841	486	1570	0.536
Ammonia (NH$_3$)							
300	0.6894	2.158	101.5	14.7	24.7	16.6	0.887
320	0.6448	2.170	109	16.9	27.2	19.4	0.870
340	0.6059	2.192	116.5	19.2	29.3	22.1	0.872
360	0.5716	2.221	124	21.7	31.6	24.9	0.872
380	0.5410	2.254	131	24.2	34.0	27.9	0.869

(continued)

Table A.2 (*continued*)

T (K)	ρ (kg/m^3)	c_p (kJ/kg · K)	$\mu \cdot 10^7$ (N · s/m^2)	$\nu \cdot 10^6$ (m^2/s)	$k \cdot 10^3$ (W/m · K)	$\alpha \cdot 10^6$ (m^2/s)	Pr
Ammonia (NH$_3$) (*continued*)							
400	0.5136	2.287	138	26.9	37.0	31.5	0.853
420	0.4888	2.322	145	29.7	40.4	35.6	0.833
440	0.4664	2.357	152.5	32.7	43.5	39.6	0.826
460	0.4460	2.393	159	35.7	46.3	43.4	0.822
480	0.4273	2.430	166.5	39.0	49.2	47.4	0.822
500	0.4101	2.467	173	42.2	52.5	51.9	0.813
520	0.3942	2.504	180	45.7	54.5	55.2	0.827
540	0.3795	2.540	186.5	49.1	57.5	59.7	0.824
560	0.3708	2.577	193	52.0	60.6	63.4	0.827
580	0.3533	2.613	199.5	56.5	63.8	69.1	0.817
Carbon Dioxide (CO$_2$)							
280	1.9022	0.830	140	7.36	15.20	9.63	0.765
300	1.7730	0.851	149	8.40	16.55	11.0	0.766
320	1.6609	0.872	156	9.39	18.05	12.5	0.754
340	1.5618	0.891	165	10.6	19.70	14.2	0.746
360	1.4743	0.908	173	11.7	21.2	15.8	0.741
380	1.3961	0.926	181	13.0	22.75	17.6	0.737
400	1.3257	0.942	190	14.3	24.3	19.5	0.737
450	1.1782	0.981	210	17.8	28.3	24.5	0.728
500	1.0594	1.02	231	21.8	32.5	30.1	0.725
550	0.9625	1.05	251	26.1	36.6	36.2	0.721
600	0.8826	1.08	270	30.6	40.7	42.7	0.717
650	0.8143	1.10	288	35.4	44.5	49.7	0.712
700	0.7564	1.13	305	40.3	48.1	56.3	0.717
750	0.7057	1.15	321	45.5	51.7	63.7	0.714
800	0.6614	1.17	337	51.0	55.1	71.2	0.716
Carbon Monoxide (CO)							
200	1.6888	1.045	127	7.52	17.0	9.63	0.781
220	1.5341	1.044	137	8.93	19.0	11.9	0.753
240	1.4055	1.043	147	10.5	20.6	14.1	0.744
260	1.2967	1.043	157	12.1	22.1	16.3	0.741
280	1.2038	1.042	166	13.8	23.6	18.8	0.733
300	1.1233	1.043	175	15.6	25.0	21.3	0.730
320	1.0529	1.043	184	17.5	26.3	23.9	0.730
340	0.9909	1.044	193	19.5	27.8	26.9	0.725
360	0.9357	1.045	202	21.6	29.1	29.8	0.725
380	0.8864	1.047	210	23.7	30.5	32.9	0.729
400	0.8421	1.049	218	25.9	31.8	36.0	0.719
450	0.7483	1.055	237	31.7	35.0	44.3	0.714
500	0.67352	1.065	254	37.7	38.1	53.1	0.710
550	0.61226	1.076	271	44.3	41.1	62.4	0.710
600	0.56126	1.088	286	51.0	44.0	72.1	0.707

Table A.2   *(continued)*

T (K)	ρ (kg/m^3)	c_p (kJ/kg · K)	$\mu \cdot 10^7$ (N · s/m^2)	$\nu \cdot 10^6$ (m^2/s)	$k \cdot 10^3$ (W/m · K)	$\alpha \cdot 10^6$ (m^2/s)	Pr
Carbon Monoxide (CO) *(continued)*							
650	0.51806	1.101	301	58.1	47.0	82.4	0.705
700	0.48102	1.114	315	65.5	50.0	93.3	0.702
750	0.44899	1.127	329	73.3	52.8	104	0.702
800	0.42095	1.140	343	81.5	55.5	116	0.705
Helium (He)							
100	0.4871	5.193	96.3	19.8	73.0	28.9	0.686
120	0.4060	5.193	107	26.4	81.9	38.8	0.679
140	0.3481	5.193	118	33.9	90.7	50.2	0.676
160	—	5.193	129	—	99.2	—	—
180	0.2708	5.193	139	51.3	107.2	76.2	0.673
200	—	5.193	150	—	115.1	—	—
220	0.2216	5.193	160	72.2	123.1	107	0.675
240	—	5.193	170	—	130	—	—
260	0.1875	5.193	180	96.0	137	141	0.682
280	—	5.193	190	—	145	—	—
300	0.1625	5.193	199	122	152	180	0.680
350	—	5.193	221	—	170	—	—
400	0.1219	5.193	243	199	187	295	0.675
450	—	5.193	263	—	204	—	—
500	0.09754	5.193	283	290	220	434	0.668
550	—	5.193	—	—	—	—	—
600	—	5.193	320	—	252	—	—
650	—	5.193	332	—	264	—	—
700	0.06969	5.193	350	502	278	768	0.654
750	—	5.193	364	—	291	—	—
800	—	5.193	382	—	304	—	—
900	—	5.193	414	—	330	—	—
1000	0.04879	5.193	446	914	354	1400	0.654
Hydrogen (H$_2$)							
100	0.24255	11.23	42.1	17.4	67.0	24.6	0.707
150	0.16156	12.60	56.0	34.7	101	49.6	0.699
200	0.12115	13.54	68.1	56.2	131	79.9	0.704
250	0.09693	14.06	78.9	81.4	157	115	0.707
300	0.08078	14.31	89.6	111	183	158	0.701
350	0.06924	14.43	98.8	143	204	204	0.700
400	0.06059	14.48	108.2	179	226	258	0.695
450	0.05386	14.50	117.2	218	247	316	0.689
500	0.04848	14.52	126.4	261	266	378	0.691
550	0.04407	14.53	134.3	305	285	445	0.685

(continued)

Table A.2 (*continued*)

T (K)	ρ (kg/m³)	c_p (kJ/kg · K)	$\mu \cdot 10^7$ (N · s/m²)	$\nu \cdot 10^6$ (m²/s)	$k \cdot 10^3$ (W/m · K)	$\alpha \cdot 10^6$ (m²/s)	Pr
Hydrogen (H₂) (*continued*)							
600	0.04040	14.55	142.4	352	305	519	0.678
700	0.03463	14.61	157.8	456	342	676	0.675
800	0.03030	14.70	172.4	569	378	849	0.670
900	0.02694	14.83	186.5	692	412	1030	0.671
1000	0.02424	14.99	201.3	830	448	1230	0.673
1100	0.02204	15.17	213.0	966	488	1460	0.662
1200	0.02020	15.37	226.2	1120	528	1700	0.659
1300	0.01865	15.59	238.5	1279	568	1955	0.655
1400	0.01732	15.81	250.7	1447	610	2230	0.650
1500	0.01616	16.02	262.7	1626	655	2530	0.643
1600	0.0152	16.28	273.7	1801	697	2815	0.639
1700	0.0143	16.58	284.9	1992	742	3130	0.637
1800	0.0135	16.96	296.1	2193	786	3435	0.639
1900	0.0128	17.49	307.2	2400	835	3730	0.643
2000	0.0121	18.25	318.2	2630	878	3975	0.661
Nitrogen (N₂)							
100	3.4388	1.070	68.8	2.00	9.58	2.60	0.768
150	2.2594	1.050	100.6	4.45	13.9	5.86	0.759
200	1.6883	1.043	129.2	7.65	18.3	10.4	0.736
250	1.3488	1.042	154.9	11.48	22.2	15.8	0.727
300	1.1233	1.041	178.2	15.86	25.9	22.1	0.716
350	0.9625	1.042	200.0	20.78	29.3	29.2	0.711
400	0.8425	1.045	220.4	26.16	32.7	37.1	0.704
450	0.7485	1.050	239.6	32.01	35.8	45.6	0.703
500	0.6739	1.056	257.7	38.24	38.9	54.7	0.700
550	0.6124	1.065	274.7	44.86	41.7	63.9	0.702
600	0.5615	1.075	290.8	51.79	44.6	73.9	0.701
700	0.4812	1.098	321.0	66.71	49.9	94.4	0.706
800	0.4211	1.22	349.1	82.90	54.8	116	0.715
900	0.3743	1.146	375.3	100.3	59.7	139	0.721
1000	0.3368	1.167	399.9	118.7	64.7	165	0.721
1100	0.3062	1.187	423.2	138.2	70.0	193	0.718
1200	0.2807	1.204	445.3	158.6	75.8	224	0.707
1300	0.2591	1.219	466.2	179.9	81.0	256	0.701
Oxygen (O₂)							
100	3.945	0.962	76.4	1.94	9.25	2.44	0.796
150	2.585	0.921	114.8	4.44	13.8	5.80	0.766
200	1.930	0.915	147.5	7.64	18.3	10.4	0.737
250	1.542	0.915	178.6	11.58	22.6	16.0	0.723
300	1.284	0.920	207.2	16.14	26.8	22.7	0.711

Table A.2 (*continued*)

T (K)	ρ (kg/m^3)	c_p (kJ/kg · K)	$\mu \cdot 10^7$ (N · s/m^2)	$\nu \cdot 10^6$ (m^2/s)	$k \cdot 10^3$ (W/m · K)	$\alpha \cdot 10^6$ (m^2/s)	Pr
Oxygen (O$_2$) (*continued*)							
350	1.100	0.929	233.5	21.23	29.6	29.0	0.733
400	0.9620	0.942	258.2	26.84	33.0	36.4	0.737
450	0.8554	0.956	281.4	32.90	36.3	44.4	0.741
500	0.7698	0.972	303.3	39.40	41.2	55.1	0.716
550	0.6998	0.988	324.0	46.30	44.1	63.8	0.726
600	0.6414	1.003	343.7	53.59	47.3	73.5	0.729
700	0.5498	1.031	380.8	69.26	52.8	93.1	0.744
800	0.4810	1.054	415.2	86.32	58.9	116	0.743
900	0.4275	1.074	447.2	104.6	64.9	141	0.740
1000	0.3848	1.090	477.0	124.0	71.0	169	0.733
1100	0.3498	1.103	505.5	144.5	75.8	196	0.736
1200	0.3206	1.115	532.5	166.1	81.9	229	0.725
1300	0.2960	1.125	588.4	188.6	87.1	262	0.721
Water Vapor (Steam)							
380	0.5863	2.060	127.1	21.68	24.6	20.4	1.06
400	0.5542	2.014	134.4	24.25	26.1	23.4	1.04
450	0.4902	1.980	152.5	31.11	29.9	30.8	1.01
500	0.4405	1.985	170.4	38.68	33.9	38.8	0.998
550	0.4005	1.997	188.4	47.04	37.9	47.4	0.993
600	0.3652	2.026	206.7	56.60	42.2	57.0	0.993
650	0.3380	2.056	224.7	66.48	46.4	66.8	0.996
700	0.3140	2.085	242.6	77.26	50.5	77.1	1.00
750	0.2931	2.119	260.4	88.84	54.9	88.4	1.00
800	0.2739	2.152	278.6	101.7	59.2	100	1.01
850	0.2579	2.186	296.9	115.1	63.7	113	1.02

Adapted from Reference [2].

Table A.3 Properties of Solid Materials

Composition	Melting point (K)	Properties at 300 K				Properties at various temperatures (K) k (W/m · K)/c_p (J/kg · K)									
		ρ (kg/m³)	c_p (J/kg · K)	k (W/m · K)	$\alpha \cdot 10^6$ (m²/s)	100	200	400	600	800	1000	1200	1500	2000	2500
Aluminum															
Pure	933	2702	903	237	97.1	302	237	240	231	218					
						482	798	949	1033	1146					
Alloy 2024-T6 (4.5% Cu, 1.5% Mg, 0.6% Mn)	775	2770	875	177	73.0	65	163	186	186						
						473	787	925	1042						
Alloy 195, Cast (4.5% Cu)		2790	883	168	68.2			174	185						
								—	—						
Beryllium	1550	1850	1825	200	59.2	990	301	161	126	106	90.8	78.7			
						203	1114	2191	2604	2823	3018	3227	3519		
Bismuth	545	9780	122	7.86	6.59	16.5	9.69	7.04							
						112	120	127							
Boron	2573	2500	1107	27.0	9.76	190	55.5	16.8	10.6	9.60	9.85				
						128	600	1463	1892	2160	2338				
Cadmium	594	8650	231	96.8	48.4	203	99.3	94.7							
						198	222	242							
Chromium	2118	7160	449	93.7	29.1	159	111	90.9	80.7	71.3	65.4	61.9	57.2	49.4	
						192	384	484	542	581	616	682	779	937	
Cobalt	1769	8862	421	99.2	26.6	167	122	85.4	67.4	58.2	52.1	49.3	42.5		
						236	379	450	503	550	628	733	674		
Copper															
Pure	1358	8933	385	401	117	482	413	393	379	366	352	339			
						252	356	397	417	433	451	480			
Commercial bronze (90% Cu, 10% Al)	1293	8800	420	52	14		42	52	59						
							785	460	545						

Composition	Melting Point (K)	ρ (kg/m³)	cp (J/kg·K)	k (W/m·K)	α·10⁶ (m²/s)	100 K	200 K	400 K	600 K	800 K	1000 K	1200 K	1500 K
Phosphor gear bronze (89% Cu, 11% Sn)	1104	8780	355	54	17		41 / —	65 / —	74 / —				
Cartridge brass (70% Cu, 30% Zn)	1188	8530	380	110	33.9	75 / —	95 / 360	137 / 395	149 / 425				
Constantan (55% Cu, 45% Ni)	1493	8920	384	23	6.71	17 / 237	19 / 362						
Germanium	1211	5360	322	59.9	34.7	232 / 190	96.8 / 290	43.2 / 337	27.3 / 348	19.8 / 357	17.4 / 375	17.4 / 395	
Gold	1336	19300	129	317	127	327 / 109	323 / 124	311 / 131	298 / 135	284 / 140	270 / 145	255 / 155	
Iridium	2720	22500	130	147	50.3	172 / 90	153 / 122	144 / 133	138 / 138	132 / 144	126 / 153	120 / 161	111 / 172
Iron Pure	1810	7870	447	80.2	23.1	134 / 216	94.0 / 384	69.5 / 490	54.7 / 574	43.3 / 680	32.8 / 975	28.3 / 609	32.1 / 654
Armco (99.75% pure)		7870	447	72.7	20.7	95.6 / 215	80.6 / 384	65.7 / 490	53.1 / 574	42.2 / 680	32.3 / 975	28.7 / 609	31.4 / 654
Carbon steels Plain carbon (Mn ≤ 1%, Si ≤ 0.1%)		7854	434	60.5	17.7			56.7 / 487	48.0 / 559	39.2 / 685	30.0 / 1169		
AISI 1010		7832	434	63.9	18.8			58.7 / 487	48.8 / 559	39.2 / 685	31.3 / 1168		
Carbon–silicon (Mn ≤ 1%, 0.1% < Si ≤ 0.6%)		7817	446	51.9	14.9			49.8 / 501	44.0 / 582	37.4 / 699	29.3 / 971		
Carbon–manganese–silicon (1% < Mn ≤ 1.65%, 0.1% < Si ≤ 0.6%)		8131	434	41.0	11.6			42.2 / 487	39.7 / 559	35.0 / 685	27.6 / 1090		

(continued)

Table A.3 (continued)

Composition	Melting point (K)	Properties at 300 K ρ (kg/m³)	c_p (J/kg·K)	k (W/m·K)	$\alpha \cdot 10^6$ (m²/s)	Properties at various temperatures (K) k (W/m·K)/c_p (J/kg·K) 100	200	400	600	800	1000	1200	1500	2000	2500
Chromium (low) steels															
Cr–Mo–Si (0.18% C, 0.65% Cr, 0.23% Mo, 0.6% Si)		7822	444	37.7	10.9			38.2 / 492	36.7 / 575	33.3 / 688	26.9 / 969				
1 Cr–Mo (0.16% C, 1% Cr, 0.54% Mo, 0.39% Si)		7858	442	42.3	12.2			42.0 / 492	39.1 / 575	34.5 / 688	27.4 / 969				
1 Cr–V (0.2% C, 1.02% Cr, 0.15% V)		7836	443	48.9	14.1			46.8 / 492	42.1 / 575	36.3 / 688	28.2 / 969				
Stainless steels															
AISI 302		8055	480	15.1	3.91			17.3 / 512	20.0 / 559	22.8 / 585	25.4 / 606				
AISI 304	1670	7900	477	14.9	3.95	9.2 / 272	12.6 / 402	16.6 / 515	19.8 / 557	22.6 / 582	25.4 / 611	28.0 / 640	31.7 / 682		
AISI 316		8238	468	13.4	3.48			15.2 / 504	18.3 / 550	21.3 / 576	24.2 / 602				
AISI 347		7978	480	14.2	3.71			15.8 / 513	18.9 / 559	21.9 / 585	24.7 / 606				
Lead	601	11340	129	35.3	24.1	39.7 / 118	36.7 / 125	34.0 / 132	31.4 / 142						
Magnesium	923	1740	1024	156	87.6	169 / 649	159 / 934	153 / 1074	149 / 1170	146 / 1267					
Molybdenum	2894	10240	251	138	53.7	179 / 141	143 / 224	134 / 261	126 / 275	118 / 285	112 / 295	105 / 308	98 / 330	90 / 380	86 / 459

(continued)

Composition	Melting Point (K)	ρ (kg/m³)	c_p (J/kg·K)	k (W/m·K)	$\alpha \cdot 10^6$ (m²/s)	100 K (k/c_p)	200 K	400 K	600 K	800 K	1000 K	1200 K	1500 K	2000 K	2500 K
Nickel Pure	1728	8900	444	90.7	23.0	164/232	107/383	80.2/485	65.6/592	67.6/530	71.8/562	76.2/594	82.6/616		
Nichrome (80% Ni, 20% Cr)	1672	8400	420	12	3.4			14/480	16/525	21/545					
Inconel X-750 (73% Ni, 15% Cr, 6.7% Fe)	1665	8510	439	11.7	3.1	8.7/—	10.3/372	13.5/473	17.0/510	20.5/546	24.0/626	27.6/—	33.0/—		
Niobium	2741	8570	265	53.7	23.6	55.2/188	52.6/249	55.2/274	58.2/283	61.3/292	64.4/301	67.5/310	72.1/324	79.1/347	
Palladium	1827	12020	244	71.8	24.5	76.5/168	71.6/227	73.6/251	79.7/261	86.9/271	94.2/281	102/291	110/307		
Platinum Pure	2045	21450	133	71.6	25.1	77.5/100	72.6/125	71.8/136	73.2/141	75.6/146	78.7/152	82.6/157	89.5/165	99.4/179	
Alloy 60Pt–40Rh (60% Pt, 40% Rh)	1800	16630	162	47	17.4			52/—	59/—	65/—	69/—	73/—	76/—		
Rhenium	3453	21100	136	47.9	16.7	58.9/97	51.0/127	46.1/139	44.2/145	44.1/151	44.6/156	45.7/162	47.8/171	51.9/186	
Rhodium	2236	12450	243	150	49.6	186/147	154/220	146/253	136/274	127/293	121/311	116/327	110/349	112/376	
Silicon	1685	2330	712	148	89.2	884/259	264/556	98.9/790	61.9/867	42.2/913	31.2/946	25.7/967	22.7/992		
Silver	1235	10500	235	429	174	444/187	430/225	425/239	412/250	396/262	379/277	361/292			
Tantalum	3269	16600	140	57.5	24.7	59.2/110	57.5/133	57.8/144	58.6/146	59.4/149	60.2/152	61.0/155	62.2/160	64.1/172	65.6/189
Thorium	2023	11700	118	54.0	39.1	59.8/99	54.6/112	54.5/124	55.8/134	56.9/145	56.9/156	58.7/167			
Tin	505	7310	227	66.6	40.1	85.2/188	73.3/215	62.2/243							

Table A.3 (*continued*)

Composition	Melting point (K)	Properties at 300 K				Properties at various temperatures (K) k (W/m · K)/c_p (J/kg · K)									
		ρ (kg/m³)	c_p (J/kg · K)	k (W/m · K)	$\alpha \cdot 10^6$ (m²/s)	100	200	400	600	800	1000	1200	1500	2000	2500
Titanium	1953	4500	522	21.9	9.32	30.5	24.5	20.4	19.4	19.7	20.7	22.0	24.5		
						300	465	551	591	633	675	620	686		
Tungsten	3660	19300	132	174	68.3	208	186	159	137	125	118	113	107	100	95
						87	122	137	142	145	148	152	157	167	176
Uranium	1406	19070	116	27.6	12.5	21.7	25.1	29.6	34.0	38.8	43.9	49.0			
						94	108	125	146	176	180	161			
Vanadium	2192	6100	489	30.7	10.3	35.8	31.3	31.3	33.3	35.7	38.2	40.8	44.6	50.9	
						258	430	515	540	563	597	645	714	867	
Zinc	693	7140	389	116	41.8	117	118	111	103						
						297	367	402	436						
Zirconium	2125	6570	278	22.7	12.4	33.2	25.2	21.6	20.7	21.6	23.7	26.0	28.8	33.0	
						205	264	300	322	342	362	344	344	344	

Adapted from Reference [2].

Table A.4 Properties of Working Fluids

Helium

Temp °C	Latent heat kJ/kg	Liquid density kg/m³	Vapour density kg/m³	Liquid thermal conductivity W/m °C × 10⁻²	Liquid viscos. cP × 10²	Vapour viscos. cP × 10³	Vapour press. bar	Vapour specific heat kJ/kg °C	Liquid surface tension N/m × 10³
−271	22.8	148.3	26.0	1.81	3.90	0.20	0.06	2.045	0.26
−270	23.6	140.7	17.0	2.24	3.70	0.30	0.32	2.699	0.19
−269	20.9	128.0	10.0	2.77	2.90	0.60	1.00	4.619	0.09
−268	4.0	113.8	8.5	3.50	1.34	0.90	2.29	6.642	0.01

Nitrogen

Temp °C	Latent heat kJ/kg	Liquid density kg/m³	Vapour density kg/m³	Liquid thermal conductivity W/m °C	Liquid viscos. cP × 10¹	Vapour viscos. cP × 10²	Vapour press. bar	Vapour specific heat kJ/kg °C	Liquid surface tension N/m × 10²
−203	210.0	830.0	1.84	0.150	2.48	0.48	0.48	1.083	1.054
−200	205.5	818.0	3.81	0.146	1.94	0.51	0.74	1.082	0.985
−195	198.0	798.0	7.10	0.139	1.51	0.56	1.62	1.079	0.870
−190	190.5	778.0	10.39	0.132	1.26	0.60	3.31	1.077	0.766
−185	183.0	758.0	13.68	0.125	1.08	0.65	4.99	1.074	0.662
−180	173.7	732.0	22.05	0.117	0.95	0.71	6.69	1.072	0.561
−175	163.2	702.0	33.80	0.110	0.86	0.77	8.37	1.070	0.464
−170	152.7	672.0	45.55	0.103	0.80	0.83	1.07	1.068	0.367
−160	124.2	603.0	80.90	0.089	0.72	1.00	19.37	1.063	0.185
−150	66.8	474.0	194.00	0.075	0.65	1.50	28.80	1.059	0.110

(*continued*)

Table A.4 *(continued)*

Ammonia

Temp °C	Latent heat kJ/kg	Liquid density kg/m³	Vapour density kg/m³	Liquid thermal conductivity W/m °C	Liquid viscos. cP	Vapour viscos. cP × 10²	Vapour press. bar	Vapour specific heat kJ/kg °C	Liquid surface tension N/m × 10²
−60	1343	714.4	0.03	0.294	0.36	0.72	0.27	2.050	4.062
−40	1384	690.4	0.05	0.303	0.29	0.79	0.76	2.075	3.574
−20	1338	665.5	1.62	0.304	0.26	0.85	1.93	2.100	3.090
0	1263	638.6	3.48	0.298	0.25	0.92	4.24	2.125	2.480
20	1187	610.3	6.69	0.286	0.22	1.01	8.46	2.150	2.133
40	1101	579.5	12.00	0.272	0.20	1.16	15.34	2.160	1.833
60	1026	545.2	20.49	0.255	0.17	1.27	29.80	2.180	1.367
80	891	505.7	34.13	0.235	0.15	1.40	40.90	2.210	0.767
100	699	455.1	54.92	0.212	0.11	1.60	63.12	2.260	0.500
120	428	374.4	113.16	0.184	0.07	1.89	90.44	2.292	0.150

Pentane

Temp °C	Latent heat kJ/kg	Liquid density kg/m³	Vapour density kg/m³	Liquid thermal conductivity W/m °C	Liquid viscos. cP	Vapour viscos. cP × 10²	Vapour press. bar	Vapour specific heat kJ/kg °C	Liquid surface tension N/m × 10²
−20	390.0	663.0	0.01	0.149	0.344	0.51	0.10	0.825	2.01
0	378.3	644.0	0.75	0.143	0.283	0.53	0.24	0.874	1.79
20	366.9	625.5	2.20	0.138	0.242	0.58	0.76	0.922	1.58
40	355.5	607.0	4.35	0.133	0.200	0.63	1.52	0.971	1.37
60	342.3	585.0	6.51	0.128	0.174	0.69	2.28	1.021	1.17
80	329.1	563.0	10.61	0.127	0.147	0.74	3.89	1.050	0.97
100	295.7	537.6	16.54	0.124	0.123	0.81	7.19	1.088	0.83
120	269.7	509.4	25.20	0.122	0.120	0.90	13.81	1.164	0.68

Acetone

Temp °C	Latent heat kJ/kg	Liquid density kg/m³	Vapour density kg/m³	Liquid thermal conductivity W/m°C	Liquid viscos. cP	Vapour viscos. cP × 10²	Vapour press. bar	Vapour specific heat kJ/kg°C	Liquid surface tension N/m × 10²
−40	660.0	860.0	0.03	0.200	0.800	0.68	0.01	2.00	3.10
−20	615.6	845.0	0.10	0.189	0.500	0.73	0.03	2.06	2.76
0	564.0	812.0	0.26	0.183	0.395	0.78	0.10	2.11	2.62
20	552.0	790.0	0.64	0.181	0.323	0.82	0.27	2.16	2.37
40	536.0	768.0	1.05	0.175	0.269	0.86	0.60	2.22	2.12
60	517.0	744.0	2.37	0.168	0.226	0.90	1.15	2.28	1.86
80	495.0	719.0	4.30	0.160	0.192	0.95	2.15	2.34	1.62
100	472.0	689.6	6.94	0.148	0.170	0.98	4.43	2.39	1.34
120	426.1	660.3	11.02	0.135	0.148	0.99	6.70	2.45	1.07
140	394.4	631.8	18.61	0.126	0.132	1.03	10.49	2.50	0.81

Methanol

Temp °C	Latent heat kJ/kg	Liquid density kg/m³	Vapour density kg/m³	Liquid thermal conductivity W/m°C	Liquid viscos. cP	Vapour viscos. cP × 10²	Vapour press. bar	Vapour specific heat kJ/kg°C	Liquid surface tension N/m × 10²
−50	1194	843.5	0.01	0.210	1.700	0.72	0.01	1.20	3.26
−30	1187	833.5	0.01	0.208	1.300	0.78	0.02	1.27	2.95
−10	1182	818.7	0.04	0.206	0.945	0.85	0.04	1.34	2.63
10	1175	800.5	0.12	0.204	0.701	0.91	0.10	1.40	2.36
30	1155	782.0	0.31	0.203	0.521	0.98	0.25	1.47	2.18
50	1125	764.1	0.77	0.202	0.399	1.04	0.55	1.54	2.01
70	1085	746.2	1.47	0.201	0.314	1.11	1.31	1.61	1.85
90	1035	724.4	3.01	0.199	0.259	1.19	2.69	1.79	1.66
110	980	703.6	5.64	0.197	0.211	1.26	4.98	1.92	1.46
130	920	685.2	9.81	0.195	0.166	1.31	7.86	1.92	1.25
150	850	653.2	15.90	0.193	0.138	1.38	8.94	1.92	1.04

(continued)

Table A.4 (*continued*)

Flutec PP2

Temp °C	Latent heat kJ/kg	Liquid density kg/m³	Vapour density kg/m³	Liquid thermal conductivity W/m °C	Liquid viscos. cP	Vapour viscos. cP × 10²	Vapour press. bar	Vapour specific heat kJ/kg °C	Liquid surface tension N/m × 10²
−30	106.2	1942	0.13	0.637	5.200	0.98	0.01	0.72	1.90
−10	103.1	1886	0.44	0.626	3.500	1.03	0.02	0.81	1.71
10	99.8	1829	1.39	0.613	2.140	1.07	0.09	0.92	1.52
30	96.3	1773	2.96	0.601	1.435	1.12	0.22	1.01	1.32
50	91.8	1716	6.43	0.588	1.005	1.17	0.39	1.07	1.13
70	87.0	1660	11.79	0.575	0.720	1.22	0.62	1.11	0.93
90	82.1	1599	21.99	0.563	0.543	1.26	1.43	1.17	0.73
110	76.5	1558	34.92	0.550	0.429	1.31	2.82	1.25	0.52
130	70.3	1515	57.21	0.537	0.314	1.36	4.83	1.33	0.32
160	59.1	1440	103.63	0.518	0.167	1.43	8.76	1.45	0.01

Ethanol

Temp °C	Latent heat kJ/kg	Liquid density kg/m³	Vapour density kg/m³	Liquid thermal conductivity W/m °C	Liquid viscos. cP	Vapour viscos. cP × 10²	Vapour press. bar	Vapour specific heat kJ/kg °C	Liquid surface tension N/m × 10²
−30	939.4	825.0	0.02	0.177	3.40	0.75	0.01	1.25	2.76
−10	928.7	813.0	0.03	0.173	2.20	0.80	0.02	1.31	2.66
10	904.8	798.0	0.05	0.170	1.50	0.85	0.03	1.37	2.57
30	888.6	781.0	0.38	0.168	1.02	0.91	0.10	1.44	2.44
50	872.3	762.2	0.72	0.166	0.72	0.97	0.29	1.51	2.31
70	858.3	743.1	1.32	0.165	0.51	1.02	0.76	1.58	2.17
90	832.1	725.3	2.59	0.163	0.37	1.07	1.43	1.65	2.04
110	786.6	704.1	5.17	0.160	0.28	1.13	2.66	1.72	1.89
130	734.4	678.7	9.25	0.159	0.21	1.18	4.30	1.78	1.75

Heptane

Temp °C	Latent heat kJ/kg	Liquid density kg/m³	Vapour density kg/m³	Liquid thermal conductivity W/m°C	Liquid viscos. cP	Vapour viscos. cP × 10²	Vapour press. bar	Vapour specific heat kJ/kg°C	Liquid surface tension N/m × 10²
−20	384.0	715.5	0.01	0.143	0.69	0.57	0.01	0.83	2.42
0	372.6	699.0	0.17	0.141	0.53	0.60	0.02	0.87	2.21
20	362.2	683.0	0.49	0.140	0.43	0.63	0.08	0.92	2.01
40	351.8	667.0	0.97	0.139	0.34	0.66	0.20	0.97	1.81
60	341.5	649.0	1.45	0.137	0.29	0.70	0.32	1.02	1.62
80	331.2	631.0	2.31	0.135	0.24	0.74	0.62	1.05	1.43
100	319.6	612.0	3.71	0.133	0.21	0.77	1.10	1.09	1.28
120	305.0	592.0	6.08	0.132	0.18	0.82	1.85	1.16	1.10

Water

Temp °C	Latent heat kJ/kg	Liquid density kg/m³	Vapour density kg/m³	Liquid thermal conductivity W/m°C	Liquid viscos. cP	Vapour viscos. cP × 10²	Vapour press. bar	Vapour specific heat kJ/kg°C	Liquid surface tension N/m × 10²
20	2448	998.2	0.02	0.603	1.00	0.96	0.02	1.81	7.28
40	2402	992.3	0.05	0.630	0.65	1.04	0.07	1.89	6.96
60	2359	983.0	0.13	0.649	0.47	1.12	0.20	1.91	6.62
80	2309	972.0	0.29	0.668	0.36	1.19	0.47	1.95	6.26
100	2258	958.0	0.60	0.680	0.28	1.27	1.01	2.01	5.89
120	2200	945.0	1.12	0.682	0.23	1.34	2.02	2.09	5.50
140	2139	928.0	1.99	0.683	0.20	1.41	3.90	2.21	5.06
160	2074	909.0	3.27	0.679	0.17	1.49	6.44	2.38	4.66
180	2003	888.0	5.16	0.669	0.15	1.57	10.04	2.62	4.29
200	1967	865.0	7.87	0.659	0.14	1.65	16.19	2.91	3.89

(continued)

Table A.4 (continued)

Flutec PP9

Temp °C	Latent heat kJ/kg	Liquid density kg/m³	Vapour density kg/m³	Liquid thermal conductivity W/m°C	Liquid viscos. cP	Vapour viscos. cP × 10²	Vapour press. bar	Vapour specific heat kJ/kg°C	Liquid surface tension N/m × 10²
−30	103.0	2098	0.01	0.060	5.77	0.82	0.00	0.80	2.36
0	98.4	2029	0.01	0.059	3.31	0.90	0.00	0.87	2.08
30	94.5	1960	0.12	0.057	1.48	1.06	0.01	0.94	1.80
60	90.2	1891	0.61	0.056	0.94	1.18	0.03	1.02	1.52
90	86.1	1822	1.93	0.054	0.65	1.21	0.12	1.09	1.24
120	83.0	1753	4.52	0.053	0.49	1.23	0.28	1.15	0.95
150	77.4	1685	11.81	0.052	0.38	1.26	0.61	1.23	0.67
180	70.8	1604	25.13	0.051	0.30	1.33	1.58	1.30	0.40
225	59.4	1455	63.27	0.049	0.21	1.44	4.21	1.41	0.01

High Temperature Organic (Diphenyl–Diphenyl Oxide Eutectic)

Temp °C	Latent heat kJ/kg	Liquid density kg/m³	Vapour density kg/m³	Liquid thermal conductivity W/m°C	Liquid viscos. cP	Vapour viscos. cP × 10	Vapour press. bar	Vapour specific heat kJ/kg°C	Liquid surface tension N/m × 10²
100	354.0	992.0	0.03	0.131	0.97	0.67	0.01	1.34	3.50
150	338.0	951.0	0.22	0.125	0.57	0.78	0.05	1.51	3.00
200	321.0	905.0	0.94	0.119	0.39	0.89	0.25	1.67	2.50
250	301.0	858.0	3.60	0.113	0.27	1.00	0.88	1.81	2.00
300	278.0	809.0	8.74	0.106	0.20	1.12	2.43	1.95	1.50
350	251.0	755.0	19.37	0.099	0.15	1.23	5.55	2.03	1.00
400	219.0	691.0	41.89	0.093	0.12	1.34	10.90	2.11	0.50
450	185.0	625.0	81.00	0.086	0.10	1.45	19.00	2.19	0.03

Mercury

Temp °C	Latent heat kJ/kg	Liquid density kg/m³	Vapour density kg/m³ × 10²	Liquid thermal conductivity W/m°C	Liquid viscos. cP	Vapour viscos. cP × 10²	Vapour press. bar	Vapour specific heat kJ/kg°C	Liquid surface tension N/m × 10²
150	308.8	13230	0.01	9.99	1.09	0.39	0.01	1.04	4.45
250	303.8	12995	0.60	11.23	0.96	0.48	0.18	1.04	4.15
300	301.8	12880	1.73	11.73	0.93	0.53	0.44	1.04	4.00
350	298.9	12763	4.45	12.18	0.89	0.61	1.16	1.04	3.82
400	296.3	12656	8.75	12.58	0.86	0.66	2.42	1.04	3.74
450	293.8	12508	16.80	12.96	0.83	0.70	4.92	1.04	3.61
500	291.3	12308	28.60	13.31	0.80	0.75	8.86	1.04	3.41
550	288.8	12154	44.92	13.62	0.79	0.81	15.03	1.04	3.25
600	286.3	12054	65.75	13.87	0.78	0.87	23.77	1.04	3.15
650	283.5	11962	94.39	14.15	0.78	0.95	34.95	1.04	3.03
750	277.0	11800	170.00	14.80	0.77	1.10	63.00	1.04	2.75

Caesium

Temp °C	Latent heat kJ/kg	Liquid density kg/m³	Vapour density kg/m³ × 10²	Liquid thermal conductivity W/m°C	Liquid viscos. cP	Vapour viscos. cP × 10²	Vapour press. bar	Vapour specific heat kJ/kg°C	Liquid surface tension N/m × 10²
375	530.4	1740	0.01	20.76	0.25	2.20	0.02	1.56	5.81
425	520.4	1730	0.01	20.51	0.23	2.30	0.04	1.56	5.61
475	515.2	1720	0.02	20.02	0.22	2.40	0.09	1.56	5.36
525	510.2	1710	0.03	19.52	0.20	2.50	0.16	1.56	5.11
575	502.8	1700	0.07	18.83	0.19	2.55	0.36	1.56	4.81
625	495.3	1690	0.10	18.13	0.18	2.60	0.57	1.56	4.51
675	490.2	1680	0.18	17.48	0.17	2.67	1.04	1.56	4.21
725	485.2	1670	0.26	16.83	0.17	2.75	1.52	1.56	3.91
775	477.8	1655	0.40	16.18	0.16	2.28	2.46	1.56	3.66
825	470.3	1640	0.55	15.53	0.16	2.90	3.41	1.56	3.41

(continued)

Table A.4 (*continued*)

Potassium

Temp °C	Latent heat kJ/kg	Liquid density kg/m^3	Vapour density kg/m^3	Liquid thermal conductivity W/m °C	Liquid viscos. cP	Vapour viscos. cP × 10^2	Vapour press. bar	Vapour specific heat kJ/kg °C	Liquid surface tension N/m × 10^2
350	2093	763.1	0.002	51.08	0.21	0.15	0.01	5.32	9.50
400	2078	748.1	0.006	49.08	0.19	0.16	0.01	5.32	9.04
450	2060	735.4	0.015	47.08	0.18	0.16	0.02	5.32	8.69
500	2040	725.4	0.031	45.08	0.17	0.17	0.05	5.32	8.44
550	2020	715.4	0.062	43.31	0.15	0.17	0.10	5.32	8.16
600	2000	705.4	0.111	41.81	0.14	0.18	0.19	5.32	7.86
650	1980	695.4	0.193	40.08	0.13	0.19	0.35	5.32	7.51
700	1969	685.4	0.314	38.08	0.12	0.19	0.61	5.32	7.12
750	1938	675.4	0.486	36.31	0.12	0.20	0.99	5.32	6.72
800	1913	665.4	0.716	34.81	0.11	0.20	1.55	5.32	6.32
850	1883	653.1	1.054	33.31	0.10	0.21	2.34	5.32	5.92

Sodium

Temp °C	Latent heat kJ/kg	Liquid density kg/m^3	Vapour density kg/m^3	Liquid thermal conductivity W/m °C	Liquid viscos. cP	Vapour viscos. cP × 10	Vapour press. bar	Vapour specific heat kJ/kg °C	Liquid surface tension N/m × 10^2
500	4370	828.1	0.003	70.08	0.24	0.18	0.01	9.04	1.51
600	4243	805.4	0.013	64.62	0.21	0.19	0.04	9.04	1.42
700	4090	763.5	0.050	60.81	0.19	0.20	0.15	9.04	1.33
800	3977	757.3	0.134	57.81	0.18	0.22	0.47	9.04	1.23
900	3913	745.4	0.306	53.35	0.17	0.23	1.25	9.04	1.13
1000	3827	725.4	0.667	49.08	0.16	0.24	2.81	9.04	1.04
1100	3690	690.8	1.306	45.08	0.16	0.25	5.49	9.04	0.95
1200	3577	669.0	2.303	41.08	0.15	0.26	9.59	9.04	0.86
1300	3477	654.0	3.622	37.08	0.15	0.27	15.91	9.04	0.77

Lithium

Temp °C	Latent heat kJ/kg	Liquid density kg/m^3	Vapour density kg/m^3	Liquid thermal conductivity W/m °C	Liquid viscos. cP	Vapour viscos. cP × 10^2	Vapour press. bar	Vapour specific heat kJ/kg °C	Liquid surface tension N/m × 10^2
1030	20500	450	0.005	67	0.24	1.67	0.07	0.532	2.90
1130	20100	440	0.013	69	0.24	1.74	0.17	0.532	2.85
1230	20000	430	0.028	70	0.23	1.83	0.45	0.532	2.75
1330	19700	420	0.057	69	0.23	1.91	0.96	0.532	2.60
1430	19200	410	0.108	68	0.23	2.00	1.85	0.532	2.40
1530	18900	405	0.193	65	0.23	2.10	3.30	0.532	2.25
1630	18500	400	0.340	62	0.23	2.17	5.30	0.532	2.10
1730	18200	398	0.490	59	0.23	2.26	8.90	0.532	2.05

Adapted from Reference [3].

Table A.5 Thermophysical Properties

Composition	Melting point (K)	Properties at 300 K				Properties at various temperatures (K) k (W/m·K)/c_p (J/kg·K)									
		ρ (kg/m³)	c_p (J/kg·K)	k (W/m·K)	$\alpha \cdot 10^6$ (m²/s)	100	200	400	600	800	1000	1200	1500	2000	2500
Aluminum oxide, sapphire	2323	3970	765	46	15.1	450	82	32.4	18.9	13.0	10.5				
						—	—	940	1110	1180	1225				
Aluminum oxide, polycrystalline	2323	3970	765	36.0	11.9	133	55	26.4	15.8	10.4	7.85	6.55	5.66	6.00	
						—	—	940	1110	1180	1225				
Beryllium oxide	2725	3000	1030	272	88.0	196	111	70	47	33	21.5	15			
						1350	1690	1865	1975	2055	2145	2750			
Boron	2573	2500	1105	27.6	9.99	190	52.5	18.7	11.3	8.1	6.3	5.2			
						—	—	1490	1880	2135	2350	2555			
Boron fiber epoxy (30% vol) composite	590	2080													
k, ∥ to fibers				2.29		2.10	2.23	2.28							
k, ⊥ to fibers				0.59		0.37	0.49	0.60							
c_p			1122			364	757	1431							
Carbon Amorphous	1500	1950	—	1.60	—	0.67	1.18	1.89	2.19	2.37	2.53	2.84	3.48		
						—	—	—	—	—	—	—	—		
Diamond, type IIa insulator	—	3500	509	2300	—	10,000	4000	1540							
						21	194	853							
Graphite, pyrolytic	2273	2210													
k, ∥ to layers				1950		4970	3230	1390	892	667	534	448	357	262	
k, ⊥ to layers				5.70		16.8	9.23	4.09	2.68	2.01	1.60	1.34	1.08	0.81	
c_p			709			136	411	992	1406	1650	1793	1890	1974	2043	

Table continued (Thermophysical Properties of Selected Nonmetallic Solids). The "at 300 K" property columns are ρ (kg/m³), c_p (J/kg·K), k (W/m·K) and $\alpha \cdot 10^6$ (m²/s); the remaining columns give k (W/m·K) with c_p (J/kg·K) listed beneath, at the indicated temperatures (K).

Composition	Melting Point (K)	ρ	c_p	k	$\alpha \cdot 10^6$	100	200	300	400	600	800	1000	1200	1500
Graphite fiber epoxy (25% vol) composite	450	1400	935											
k, heat flow ∥ to fibers				11.1			5.7	8.7	13.0					
k, heat flow ⊥ to fibers				0.87			0.46	0.68	1.1					
c_p							337	642	1216					
Pyroceram, Corning 9606	1623	2600	808	3.98	1.89	5.25	4.78		3.64	3.28	3.08	2.96	2.87	2.79
c_p									908	1038	1122	1197	1264	1498
Silicon carbide	3100	3160	675	490	230							87	58	30
c_p									880	1050	1135	1195	1243	1310
Silicon dioxide, crystalline (quartz)	1883	2650	745											
k, ∥ to c axis				10.4		39	16.4		7.6	5.0	4.2			
k, ⊥ to c axis				6.21		20.8	9.5		4.70	3.4	3.1			
c_p									885	1075	1250			
Silicon dioxide, polycrystalline (fused silica)	1883	2220	745	1.38	0.834	0.69	1.14		1.51	1.75	2.17	2.87	4.00	
c_p									905	1040	1105	1155	1195	
Silicon nitride	2173	2400	691	16.0	9.65			13.9	11.3	9.88	8.76	8.00	7.16	6.20
c_p							578	778	937	1063	1155	1226	1306	1377
Sulfur	392	2070	708	0.206	0.141	0.165	0.185							
c_p						403	606							
Thorium dioxide	3573	9110	235	13	6.1				10.2	6.6	4.7	3.68	3.12	2.73
c_p									255	274	285	295	303	315
Titanium dioxide, polycrystalline	2133	4157	710	8.4	2.8				7.01	5.02	3.94	3.46	3.28	2.5
c_p									805	880	910	930	945	330

Adapted from Reference [2].

Table A.6   Thermophysical Properties of Common Materials

Structural building materials

	Typical properties at 300 K		
Description/composition	Density, ρ (kg/m^3)	Thermal conductivity, k (W/m · K)	Specific heat, c_p (J/kg · K)
Building Boards			
Asbestos–cement board	1920	0.58	—
Gypsum or plaster board	800	0.17	—
Plywood	545	0.12	1215
Sheathing, regular density	290	0.055	1300
Acoustic tile	290	0.058	1340
Hardboard, siding	640	0.094	1170
Hardboard, high density	1010	0.15	1380
Particle board, low density	590	0.078	1300
Particle board, high density	1000	0.170	1300
Woods			
Hardwoods (oak, maple)	720	0.16	1255
Softwoods (fir, pine)	510	0.12	1380
Masonry Materials			
Cement mortar	1860	0.72	780
Brick, common	1920	0.72	835
Brick, face	2083	1.3	—
Clay tile, hollow			
1 cell deep, 10 cm thick	—	0.52	—
3 cells deep, 30 cm thick	—	0.69	—
Concrete block, 3 oval cores			
Sand/gravel, 20 cm thick	—	1.0	—
Cinder aggregate, 20 cm thick	—	0.67	—
Concrete block, rectangular core			
2 cores, 20 cm thick, 16 kg	—	1.1	—
Same with filled cores	—	0.60	—
Plastering Materials			
Cement plaster, sand aggregate	1860	0.72	—
Gypsum plaster, sand aggregate	1680	0.22	1085
Gypsum plaster, vermiculite aggregate	720	0.25	—

Table A.6 (*continued*)

Insulating Materials and systems

	Typical properties at 300 K		
Description/composition	Density, ρ (kg/m^3)	Thermal conductivity, k (W/m·K)	Specific heat, c_p (J/kg·K)
Blanket and Batt			
Glass fiber, paper faced	16	0.046	—
	28	0.038	—
	40	0.035	—
Glass fiber, coated; duct liner	32	0.038	835
Board and Slab			
Cellular glass	145	0.058	1000
Glass fiber, organic bonded	105	0.036	795
Polystyrene, expanded			
Extruded (R-12)	55	0.027	1210
Molded beads	16	0.040	1210
Mineral fiberboard; roofing material	265	0.049	—
Wood, shredded/cemented	350	0.087	1590
Cork	120	0.039	1800
Loose Fill			
Cork, granulated	160	0.045	—
Diatomaceous silica, coarse	350	0.069	—
Powder	400	0.091	—
Diatomaceous silica, fine powder	200	0.052	—
	275	0.061	—
Glass fiber, poured or blown	16	0.043	835
Vermiculite, flakes	80	0.068	835
	160	0.063	1000
Formed/Foamed-in-Place			
Mineral wool granules with asbestos/inorganic binders, sprayed	190	0.046	—
Polyvinyl acetate cork mastic; sprayed or troweled	—	0.100	—
Urethane, two-part mixture; rigid foam	70	0.026	1045
Reflective			
Aluminum foil separating fluffy glass mats; 10–12 layers, evacuated; for cryogenic applications (150 K)	40	0.00016	—
Aluminum foil and glass paper laminate; 75–150 layers; evacuated; for cryogenic application (150 K)	120	0.000017	—
Typical silica powder, evacuated	160	0.0017	—

(*continued*)

Table A.6 (*continued*)

Industrial insulation

Description/composition	Maximum service temperature (K)	Typical density (kg/m³)	Typical thermal conductivity, k (W/m · K), at various temperatures (K)													
			200	215	230	240	255	270	285	300	310	365	420	530	645	750
Blankets																
Blanket, mineral fiber, metal reinforced	920	96–192								0.038	0.046	0.056	0.078			
	815	40–96								0.035	0.045	0.058	0.088			
Blanket, mineral fiber, glass; fine fiber, organic bonded	450	10				0.036	0.038	0.040	0.043	0.048	0.052	0.076				
		12				0.035	0.036	0.039	0.042	0.046	0.049	0.069				
		16				0.033	0.035	0.036	0.039	0.042	0.046	0.062				
		24				0.030	0.032	0.033	0.036	0.039	0.040	0.053				
		32				0.029	0.030	0.032	0.033	0.036	0.038	0.048				
		48				0.027	0.029	0.030	0.032	0.033	0.035	0.045				
Blanket, alumina–silica fiber	1530	48												0.071	0.105	0.150
		64												0.059	0.087	0.125
		96												0.052	0.076	0.100
		128												0.049	0.068	0.091
Felt, semirigid; organic bonded	480	50–125	0.023	0.025	0.026	0.027	0.029	0.035	0.036	0.038	0.039	0.051	0.063			
Felt, laminated; no binder	730	50						0.030	0.032	0.033	0.035	0.051	0.079			
Blocks, Boards, and Pipe Insulations																
Asbestos paper, laminated and corrugated	920	120											0.051	0.065	0.087	

4-ply	420	190										0.078	0.082	0.098
6-ply	420	255										0.071	0.074	0.085
8-ply	420	300										0.068	0.071	0.082
Magnesia, 85%	590	185									0.051	0.055	0.055	0.061
Calcium silicate	920	190									0.055	0.059	0.063	
Cellular glass	700	145	0.046	0.048	0.051	0.052	0.055	0.058	0.062	0.069	0.079	0.075	0.089	0.104
Diatomaceous silica	1145	345									0.092	0.098		0.104
	1310	385									0.101	0.100		0.115
Polystyrene, rigid														
Extruded (R-12)	350	56	0.023	0.022	0.023	0.023	0.025	0.025	0.026	0.027	0.027	0.029		
Extruded (R-12)	350	35	0.023	0.023	0.025	0.025	0.026	0.026	0.027	0.027	0.029	0.029		
Molded beads	350	16	0.026	0.030	0.033	0.035	0.036	0.036	0.038	0.038	0.040			
Rubber, rigid foamed	340	70			0.029	0.030	0.029	0.030	0.032	0.032	0.033	0.033		
Insulating Cement														
Mineral fiber (rock, slag or glass)														
With clay binder	1255	430							0.071	0.079	0.088	0.105	0.123	
With hydraulic setting binder	922	560								0.108	0.115	0.123	0.123	0.137
Loose Fill														
Cellulose, wood or paper pulp	—	45	0.036	0.039	0.039	0.042	0.038	0.042	0.042	0.043	0.046	0.049		
Perlite, expanded	—	105		0.036	0.042	0.043	0.046	0.049	0.051	0.053	0.056			
Vermiculite, expanded	—	122	0.049	0.056	0.058	0.061	0.063	0.065	0.068	0.071	0.066			
	—	80	0.049	0.055	0.058	0.061	0.063	0.066						

(continued)

543

Table A.6 (*continued*)

Other materials

Description/ composition	Temperature (K)	Density, ρ (kg/m^3)	Thermal conductivity, k (W/m $\cdot$ K)	Specific heat, c_p (J/kg $\cdot$ K)
Asphalt	300	2115	0.062	920
Bakelite	300	1300	1.4	1465
Brick, refractory				
Carborundum	872	—	18.5	—
	1672	—	11.0	—
Chrome brick	473	3010	2.3	835
	823		2.5	
	1173		2.0	
Diatomaceous	478	—	0.25	—
silica, fired	1145	—	0.30	
Fire clay, burnt 1600 K	773	2050	1.0	960
	1073	—	1.1	
	1373	—	1.1	
Fire clay, burnt 1725 K	773	2325	1.3	960
	1073		1.4	
	1373		1.4	
Fire clay brick	478	2645	1.0	960
	922		1.5	
	1478		1.8	
Magnesite	478	—	3.8	1130
	922	—	2.8	
	1478		1.9	
Clay	300	1460	1.3	880
Coal, anthracite	300	1350	0.26	1260
Concrete (stone mix)	300	2300	1.4	880
Cotton	300	80	0.06	1300
Foodstuffs				
Banana (75.7% water content)	300	980	0.481	3350
Apple, red (75% water content)	300	840	0.513	3600
Cake, batter	300	720	0.223	—
Cake, fully baked	300	280	0.121	—
Chicken meat, white	198	—	1.60	—
(74.4% water content)	233	—	1.49	
	253		1.35	
	263		1.20	
	273		0.476	
	283		0.480	
	293		0.489	
Glass				
Plate (soda lime)	300	2500	1.4	750
Pyrex	300	2225	1.4	835
Ice	273	920	1.88	2040
	253	—	2.03	1945

Table A.6 *(continued)*

Other materials (continued)

Description/ composition	Temperature (K)	Density, ρ (kg/m^3)	Thermal conductivity, k (W/m · K)	Specific heat, c_p (J/kg · K)
Leather (sole)	300	998	0.159	—
Paper	300	930	0.180	1340
Paraffin	300	900	0.240	2890
Rock				
Granite, Barre	300	2630	2.79	775
Limestone, Salem	300	2320	2.15	810
Marble, Halston	300	2680	2.80	830
Quartzite, Sioux	300	2640	5.38	1105
Sandstone, Berea	300	2150	2.90	745
Rubber, vulcanized				
Soft	300	1100	0.13	2010
Hard	300	1190	0.16	—
Sand	300	1515	0.27	800
Soil	300	2050	0.52	1840
Snow	273	110	0.049	—
		500	0.190	—
Teflon	300	2200	0.35	—
	400		0.45	—
Tissue, human				
Skin	300	—	0.37	—
Fat layer (adipose)	300	—	0.2	—
Muscle	300	—	0.5	—
Wood, cross grain				
Balsa	300	140	0.055	—
Cypress	300	465	0.097	—
Fir	300	415	0.11	2720
Oak	300	545	0.17	2385
Yellow pine	300	640	0.15	2805
White pine	300	435	0.11	—
Wood, radial				
Oak	300	545	0.19	2385
Fir	300	420	0.14	2720

Adapted from Reference [2]

Table A.7 Thermophysical Properties of Saturated Fluids

Saturated liquids

T (K)	ρ (kg/m³)	c_p (kJ/kg·K)	$\mu \cdot 10^2$ (N·s/m²)	$\nu \cdot 10^6$ (m²/s)	$k \cdot 10^3$ (W/m·K)	$\alpha \cdot 10^7$ (m²/s)	Pr	$\beta \cdot 10^3$ (K⁻¹)
Engine Oil (Unused)								
273	899.1	1.796	385	4280	147	0.910	47,000	0.70
280	895.3	1.827	217	2430	144	0.880	27,500	0.70
290	890.0	1.868	99.9	1120	145	0.872	12,900	0.70
300	884.1	1.909	48.6	550	145	0.859	6400	0.70
310	877.9	1.951	25.3	288	145	0.847	3400	0.70
320	871.8	1.993	14.1	161	143	0.823	1965	0.70
330	865.8	2.035	8.36	96.6	141	0.800	1205	0.70
340	859.9	2.076	5.31	61.7	139	0.779	793	0.70
350	853.9	2.118	3.56	41.7	138	0.763	546	0.70
360	847.8	2.161	2.52	29.7	138	0.753	395	0.70
370	841.8	2.206	1.86	22.0	137	0.738	300	0.70
380	836.0	2.250	1.41	16.9	136	0.723	233	0.70
390	830.6	2.294	1.10	13.3	135	0.709	187	0.70
400	825.1	2.337	0.874	10.6	134	0.695	152	0.70
410	818.9	2.381	0.698	8.52	133	0.682	125	0.70
420	812.1	2.427	0.564	6.94	133	0.675	103	0.70
430	806.5	2.471	0.470	5.83	132	0.662	88	0.70
Ethylene Glycol [C₂H₄(OH)₂]								
273	1130.8	2.294	6.51	57.6	242	0.933	617	0.65
280	1125.8	2.323	4.20	37.3	244	0.933	400	0.65
290	1118.8	2.368	2.47	22.1	248	0.936	236	0.65
300	1114.4	2.415	1.57	14.1	252	0.939	151	0.65
310	1103.7	2.460	1.07	9.65	255	0.939	103	0.65
320	1096.2	2.505	0.757	6.91	258	0.940	73.5	0.65
330	1089.5	2.549	0.561	5.15	260	0.936	55.0	0.65
340	1083.8	2.592	0.431	3.98	261	0.929	42.8	0.65
350	1079.0	2.637	0.342	3.17	261	0.917	34.6	0.65
360	1074.0	2.682	0.278	2.59	261	0.906	28.6	0.65
370	1066.7	2.728	0.228	2.14	262	0.900	23.7	0.65
373	1058.5	2.742	0.215	2.03	263	0.906	22.4	0.65
Glycerin [C₃H₅(OH)₃]								
273	1276.0	2.261	1060	8310	282	0.977	85,000	0.47
280	1271.9	2.298	534	4200	284	0.972	43,200	0.47
290	1265.8	2.367	185	1460	286	0.955	15,300	0.48
300	1259.9	2.427	79.9	634	286	0.935	6780	0.48
310	1253.9	2.490	35.2	281	286	0.916	3060	0.49
320	1247.2	2.564	21.0	168	287	0.897	1870	0.50

Table A.7 (*continued*)

Saturated liquids (continued)

T (K)	ρ (kg/m^3)	c_p (kJ/kg$\cdot$K)	$\mu \cdot 10^2$ (N$\cdot$s/m^2)	$\nu \cdot 10^6$ (m^2/s)	$k \cdot 10^3$ (W/m$\cdot$K)	$\alpha \cdot 10^7$ (m^2/s)	Pr	$\beta \cdot 10^3$ (K^{-1})
Refrigerant-134a (C$_2$H$_2$F$_4$)								
230	1426.8	1.249	0.04912	0.3443	112.1	0.629	5.5	2.02
240	1397.7	1.267	0.04202	0.3006	107.3	0.606	5.0	2.11
250	1367.9	1.287	0.03633	0.2656	102.5	0.583	4.6	2.23
260	1337.1	1.308	0.03166	0.2368	97.9	0.560	4.2	2.36
270	1305.1	1.333	0.02775	0.2127	93.4	0.537	4.0	2.53
280	1271.8	1.361	0.02443	0.1921	89.0	0.514	3.7	2.73
290	1236.8	1.393	0.02156	0.1744	84.6	0.491	3.5	2.98
300	1199.7	1.432	0.01905	0.1588	80.3	0.468	3.4	3.30
310	1159.9	1.481	0.01680	0.1449	76.1	0.443	3.3	3.73
320	1116.8	1.543	0.01478	0.1323	71.8	0.417	3.2	4.33
330	1069.1	1.627	0.01292	0.1209	67.5	0.388	3.1	5.19
340	1015.0	1.751	0.01118	0.1102	63.1	0.355	3.1	6.57
350	951.3	1.961	0.00951	0.1000	58.6	0.314	3.2	9.10
360	870.1	2.437	0.00781	0.0898	54.1	0.255	3.5	15.39
370	740.3	5.105	0.00580	0.0783	51.8	0.137	5.7	55.24
Refrigerant-22 (CHClF$_2$)								
230	1416.0	1.087	0.03558	0.2513	114.5	0.744	3.4	2.05
240	1386.6	1.100	0.03145	0.2268	109.8	0.720	3.2	2.16
250	1356.3	1.117	0.02796	0.2062	105.2	0.695	3.0	2.29
260	1324.9	1.137	0.02497	0.1884	100.7	0.668	2.8	2.45
270	1292.1	1.161	0.02235	0.1730	96.2	0.641	2.7	2.63
280	1257.9	1.189	0.02005	0.1594	91.7	0.613	2.6	2.86
290	1221.7	1.223	0.01798	0.1472	87.2	0.583	2.5	3.15
300	1183.4	1.265	0.01610	0.1361	82.6	0.552	2.5	3.51
310	1142.2	1.319	0.01438	0.1259	78.1	0.518	2.4	4.00
320	1097.4	1.391	0.01278	0.1165	73.4	0.481	2.4	4.69
330	1047.5	1.495	0.01127	0.1075	68.6	0.438	2.5	5.75
340	990.1	1.665	0.00980	0.0989	63.6	0.386	2.6	7.56
350	920.1	1.997	0.00831	0.0904	58.3	0.317	2.8	11.35
360	823.4	3.001	0.00668	0.0811	53.1	0.215	3.8	23.88
Mercury (Hg)								
273	13,595	0.1404	0.1688	0.1240	8180	42.85	0.0290	0.181
300	13,529	0.1393	0.1523	0.1125	8540	45.30	0.0248	0.181
350	13,407	0.1377	0.1309	0.0976	9180	49.75	0.0196	0.181
400	13,287	0.1365	0.1171	0.0882	9800	54.05	0.0163	0.181
450	13,167	0.1357	0.1075	0.0816	10,400	58.10	0.0140	0.181
500	13,048	0.1353	0.1007	0.0771	10,950	61.90	0.0125	0.182
550	12,929	0.1352	0.0953	0.0737	11,450	65.55	0.0112	0.184
600	12,809	0.1355	0.0911	0.0711	11,950	68.80	0.0103	0.187

(*continued*)

Table A.7 (*continued*)

Saturated liquid—vapor, 1 atm					
Fluid	T_{sat} (K)	h_{fg} (kJ/kg)	ρ_f (kg/m^3)	ρ_g (kg/m^3)	$\sigma \cdot 10^3$ (N/m)
Ethanol	351	846	757	1.44	17.7
Ethylene glycol	470	812	1111^c	—	32.7
Glycerin	563	974	1260^c	—	63.0^c
Mercury	630	301	12,740	3.90	417
Refrigerant R-134a	247	217	1377	5.26	15.4
Refrigerant R-22	232	234	1409	4.70	18.1

Adapted from Reference [2]

Table A.8 Thermophysical Properties of Saturated Water

Temperature, T (K)	Pressure, p (bars)^b	Specific volume (m³/kg)		Heat of vaporization, h_{fg} (kJ/kg)	Specific heat (kJ/kg·K)		Viscosity (N·s/m²)		Thermal conductivity (W/m·K)		Prandtl number		Surface tension, $\sigma_f \cdot 10^3$ (N/m)	Expansion coefficient, $\beta_f \cdot 10^6$ (K⁻¹)	Temperature, T (K)
		$v_f \cdot 10^3$	v_g		$c_{p,f}$	$c_{p,g}$	$\mu_f \cdot 10^6$	$\mu_g \cdot 10^6$	$k_f \cdot 10^3$	$k_g \cdot 10^3$	Pr_f	Pr_g			
273.15	0.00611	1.000	206.3	2502	4.217	1.854	1750	8.02	569	18.2	12.99	0.815	75.5	−68.05	273.15
275	0.00697	1.000	181.7	2497	4.211	1.855	1652	8.09	574	18.3	12.22	0.817	75.3	−32.74	275
280	0.00990	1.000	130.4	2485	4.198	1.858	1422	8.29	582	18.6	10.26	0.825	74.8	46.04	280
285	0.01387	1.000	99.4	2473	4.189	1.861	1225	8.49	590	18.9	8.81	0.833	74.3	114.1	285
290	0.01917	1.001	69.7	2461	4.184	1.864	1080	8.69	598	19.3	7.56	0.841	73.7	174.0	290
295	0.02617	1.002	51.94	2449	4.181	1.868	959	8.89	606	19.5	6.62	0.849	72.7	227.5	295
300	0.03531	1.003	39.13	2438	4.179	1.872	855	9.09	613	19.6	5.83	0.857	71.7	276.1	300
305	0.04712	1.005	29.74	2426	4.178	1.877	769	9.29	620	20.1	5.20	0.865	70.9	320.6	305
310	0.06221	1.007	22.93	2414	4.178	1.882	695	9.49	628	20.4	4.62	0.873	70.0	361.9	310
315	0.08132	1.009	17.82	2402	4.179	1.888	631	9.69	634	20.7	4.16	0.883	69.2	400.4	315
320	0.1053	1.011	13.98	2390	4.180	1.895	577	9.89	640	21.0	3.77	0.894	68.3	436.7	320
325	0.1351	1.013	11.06	2378	4.182	1.903	528	10.09	645	21.3	3.42	0.901	67.5	471.2	325
330	0.1719	1.016	8.82	2366	4.184	1.911	489	10.29	650	21.7	3.15	0.908	66.6	504.0	330
335	0.2167	1.018	7.09	2354	4.186	1.920	453	10.49	656	22.0	2.88	0.916	65.8	535.5	335
340	0.2713	1.021	5.74	2342	4.188	1.930	420	10.69	660	22.3	2.66	0.925	64.9	566.0	340
345	0.3372	1.024	4.683	2329	4.191	1.941	389	10.89	668	22.6	2.45	0.933	64.1	595.4	345
350	0.4163	1.027	3.846	2317	4.195	1.954	365	11.09	668	23.0	2.29	0.942	63.2	624.2	350
355	0.5100	1.030	3.180	2304	4.199	1.968	343	11.29	671	23.3	2.14	0.951	62.3	652.3	355
360	0.6209	1.034	2.645	2291	4.203	1.983	324	11.49	674	23.7	2.02	0.960	61.4	697.9	360
365	0.7514	1.038	2.212	2278	4.209	1.999	306	11.69	677	24.1	1.91	0.969	60.5	707.1	365
370	0.9040	1.041	1.861	2265	4.214	2.017	289	11.89	679	24.5	1.80	0.978	59.5	728.7	370
373.15	1.0133	1.044	1.679	2257	4.217	2.029	279	12.02	680	24.8	1.76	0.984	58.9	750.1	373.15
375	1.0815	1.045	1.574	2252	4.220	2.036	274	12.09	681	24.9	1.70	0.987	58.6	761	375
380	1.2869	1.049	1.337	2239	4.226	2.057	260	12.29	683	25.4	1.61	0.999	57.6	788	380
385	1.5233	1.053	1.142	2225	4.232	2.080	248	12.49	685	25.8	1.53	1.004	56.6	814	385
390	1.794	1.058	0.980	2212	4.239	2.104	237	12.69	686	26.3	1.47	1.013	55.6	841	390
400	2.455	1.067	0.731	2183	4.256	2.158	217	13.05	688	27.2	1.34	1.033	53.6	896	400

(continued)

Table A.8 (*continued*)

Temperature, T (K)	Pressure, p (bars)[b]	Specific volume (m³/kg)		Heat of vaporization, h_fg (kJ/kg)	Specific heat (kJ/kg·K)		Viscosity (N·s/m²)		Thermal conductivity (W/m·K)		Prandtl number		Surface tension, $\sigma_f \cdot 10^3$ (N/m)	Expansion coefficient, $\beta_f \cdot 10^6$ (K⁻¹)	Temperature, T (K)
		$v_f \cdot 10^3$	v_g		$c_{p,f}$	$c_{p,g}$	$\mu_f \cdot 10^6$	$\mu_g \cdot 10^6$	$k_f \cdot 10^3$	$k_g \cdot 10^3$	Pr_f	Pr_g			
410	3.302	1.077	0.553	2153	4.278	2.221	200	13.42	688	28.2	1.24	1.054	51.5	952	410
420	4.370	1.088	0.425	2123	4.302	2.291	185	13.79	688	29.8	1.16	1.075	49.4	1010	420
430	5.699	1.099	0.331	2091	4.331	2.369	173	14.14	685	30.4	1.09	1.10	47.2		430
440	7.333	1.110	0.261	2059	4.36	2.46	162	14.50	682	31.7	1.04	1.12	45.1		440
450	9.319	1.123	0.208	2024	4.40	2.56	152	14.85	678	33.1	0.99	1.14	42.9		450
460	11.71	1.137	0.167	1989	4.44	2.68	143	15.19	673	34.6	0.95	1.17	40.7		460
470	14.55	1.152	0.136	1951	4.48	2.79	136	15.54	667	36.3	0.92	1.20	38.5		470
480	17.90	1.167	0.111	1912	4.53	2.94	129	15.88	660	38.1	0.89	1.23	36.2		480
490	21.83	1.184	0.0922	1870	4.59	3.10	124	16.23	651	40.1	0.87	1.25	33.9	—	490
500	26.40	1.203	0.0766	1825	4.66	3.27	118	16.59	642	42.3	0.86	1.28	31.6	—	500
510	31.66	1.222	0.0631	1779	4.74	3.47	113	16.95	631	44.7	0.85	1.31	29.3	—	510
520	37.70	1.244	0.0525	1730	4.84	3.70	108	17.33	621	47.5	0.84	1.35	26.9	—	520
530	44.58	1.268	0.0445	1679	4.95	3.96	104	17.72	608	50.6	0.85	1.39	24.5	—	530
540	52.38	1.294	0.0375	1622	5.08	4.27	101	18.1	594	54.0	0.86	1.43	22.1	—	540
550	61.19	1.323	0.0317	1564	5.24	4.64	97	18.6	580	58.3	0.87	1.47	19.7	—	550
560	71.08	1.355	0.0269	1499	5.43	5.09	94	19.1	563	63.7	0.90	1.52	17.3	—	560
570	82.16	1.392	0.0228	1429	5.68	5.67	91	19.7	548	76.7	0.94	1.59	15.0	—	570
580	94.51	1.433	0.0193	1353	6.00	6.40	88	20.4	528	76.7	0.99	1.68	12.8	—	580
590	108.3	1.482	0.0163	1274	6.41	7.35	84	21.5	513	84.1	1.05	1.84	10.5	—	590
600	123.5	1.541	0.0137	1176	7.00	8.75	81	22.7	497	92.9	1.14	2.15	8.4	—	600
610	137.3	1.612	0.0115	1068	7.85	11.1	77	24.1	467	103	1.30	2.60	6.3	—	610
620	159.1	1.705	0.0094	941	9.35	15.4	72	25.9	444	114	1.52	3.46	4.5	—	620
625	169.1	1.778	0.0085	858	10.6	18.3	70	27.0	430	121	1.65	4.20	3.5	—	625
630	179.7	1.856	0.0075	781	12.6	22.1	67	28.0	412	130	2.0	4.8	2.6	—	630
635	190.9	1.935	0.0066	683	16.4	27.6	64	30.0	392	141	2.7	6.0	1.5	—	635
640	202.7	2.075	0.0057	560	26	42	59	32.0	367	155	4.2	9.6	0.8	—	640
645	215.2	2.351	0.0045	361	90	—	54	37.0	331	178	12	26	0.1	—	645
647.3[c]	221.2	3.170	0.0032	0	∞	∞	45	45.0	238	238	∞	∞	0.0	—	647.3[c]

Adapted from Reference [2]

550

Table A.9 Thermophysical Properties of Liquid Metals

Composition	Melting point (K)	T (K)	ρ (kg/m^3)	c_p (kJ/kg·K)	$\nu \cdot 10^7$ (m^2/s)	k (W/m·K)	$\alpha \cdot 10^5$ (m^2/s)	Pr
Bismuth	544	589	10,011	0.1444	1.617	16.4	0.138	0.0142
		811	9739	0.1545	1.133	15.6	1.035	0.0110
		1033	9467	0.1645	0.8343	15.6	1.001	0.0083
Lead	600	644	10,540	0.159	2.276	16.1	1.084	0.024
		755	10,412	0.155	1.849	15.6	1.223	0.017
		977	10,140	—	1.347	14.9	—	—
Potassium	337	422	807.3	0.80	4.608	45.0	6.99	0.0066
		700	741.7	0.75	2.397	39.5	7.07	0.0034
		977	674.4	0.75	1.905	33.1	6.55	0.0029
Sodium	371	366	929.1	1.38	7.516	86.2	6.71	0.011
		644	860.2	1.30	3.270	72.3	6.48	0.0051
		977	778.5	1.26	2.285	59.7	6.12	0.0037
NaK, (45%/55%)	292	366	887.4	1.130	6.522	25.6	2.552	0.026
		644	821.7	1.055	2.871	27.5	3.17	0.0091
		977	740.1	1.043	2.174	28.9	3.74	0.0058
NaK, (22%/78%)	262	366	849.0	0.946	5.797	24.4	3.05	0.019
		672	775.3	0.879	2.666	26.7	3.92	0.0068
		1033	690.4	0.883	2.118	—	—	—
PbBi, (44.5%/55.5%)	398	422	10,524	0.147	—	9.05	0.586	—
		644	10,236	0.147	1.496	11.86	0.790	0.189
		922	9835	—	1.171	—	—	—
Mercury	234			See Table A.5				

Adapted from Reference [2]

nal (n) or Hemispherical (h) Emissivity of Selected Surfaces

xides

		\multicolumn{11}{c}{Emissivity, ε_n or ε_h, at various temperatures (K)}
		100	200	300	400	600	800	1000	1200	1500	2000	2500
	(h)	0.02	0.03	0.04	0.05	0.06						
	(h)	0.06	0.06	0.07								
	(h)			0.82	0.76							
...ium												
Polished or plated	(n)	0.05	0.07	0.10	0.12	0.14						
Copper												
Highly polished	(h)			0.03	0.03	0.04	0.04	0.04				
Stably oxidized	(h)					0.50	0.58	0.80				
Gold												
Highly polished or film	(h)	0.01	0.02	0.03	0.03	0.04	0.05	0.06				
Foil, bright	(h)	0.06	0.07	0.07								
Molybdenum												
Polished	(h)					0.06	0.08	0.10	0.12	0.15	0.21	0.26
Shot-blasted, rough	(h)					0.25	0.28	0.31	0.35	0.42		
Stably oxidized	(h)					0.80	0.82					
Nickel												
Polished	(h)					0.09	0.11	0.14	0.17			
Stably oxidized	(h)					0.40	0.49	0.57				
Platinum												
Polished	(h)						0.10	0.13	0.15	0.18		
Silver												
Polished	(h)			0.02	0.02	0.03	0.05	0.08				
Stainless steels												
Typical, polished	(n)			0.17	0.17	0.19	0.23	0.30				
Typical, cleaned	(n)			0.22	0.22	0.24	0.28	0.35				
Typical, lightly oxidized	(n)						0.33	0.40				
Typical, highly oxidized	(n)						0.67	0.70	0.76			
AISI 347, stably oxidized	(n)					0.87	0.88	0.89	0.90			
Tantalum												
Polished	(h)								0.11	0.17	0.23	0.28
Tungsten												
Polished	(h)							0.10	0.13	0.18	0.25	0.29

Table A.10 (*continued*)

Nonmetallic substances

Description/composition		Temperature (K)	Emissivity ε
Aluminum oxide	(*n*)	600	0.69
		1000	0.55
		1500	0.41
Asphalt pavement	(*h*)	300	0.85–0.93
Building materials			
Asbestos sheet	(*h*)	300	0.93–0.96
Brick, red	(*h*)	300	0.93–0.96
Gypsum or plaster board	(*h*)	300	0.90–0.92
Wood	(*h*)	300	0.82–0.92
Cloth	(*h*)	300	0.75–0.90
Concrete	(*h*)	300	0.88–0.93
Glass, window	(*h*)	300	0.90–0.95
Ice	(*h*)	273	0.95–0.98
Paints			
Black (Parsons)	(*h*)	300	0.98
White, acrylic	(*h*)	300	0.90
White, zinc oxide	(*h*)	300	0.92
Paper, white	(*h*)	300	0.92–0.97
Pyrex	(*n*)	300	0.82
		600	0.80
		1000	0.71
		1200	0.62
Pyroceram	(*n*)	300	0.85
		600	0.78
		1000	0.69
		1500	0.57
Refractories (furnace liners)			
Alumina brick	(*n*)	800	0.40
		1000	0.33
		1400	0.28
		1600	0.33
Magnesia brick	(*n*)	800	0.45
		1000	0.36
		1400	0.31
		1600	0.40
Kaolin insulating brick	(*n*)	800	0.70
		1200	0.57
		1400	0.47
		1600	0.53
Sand	(*h*)	300	0.90
Silicon carbide	(*n*)	600	0.87
		1000	0.87
		1500	0.85
Skin	(*h*)	300	0.95
Snow	(*h*)	273	0.82–0.90
Soil	(*h*)	300	0.93–0.96

(*continued*)

Table A.10 (*continued*)

Nonmetallic substances			
Description/composition		Temperature (K)	Emissivity ε
Rocks	(*h*)	300	0.88–0.95
Teflon	(*h*)	300	0.85
		400	0.87
		500	0.92
Vegetation	(*h*)	300	0.92–0.96
Water	(*h*)	300	0.96

Adapted from Reference [2]

Table A.11   Solar Radiative Properties for Selected Materials

Description/composition	α_s	ε	α_s/ε	τ_s
Aluminum				
Polished	0.09	0.03	3.0	
Anodized	0.14	0.84	0.17	
Quartz overcoated	0.11	0.37	0.30	
Foil	0.15	0.05	3.0	
Brick, red (Purdue)	0.63	0.93	0.68	
Concrete	0.60	0.88	0.68	
Galvanized sheet metal				
Clean, new	0.65	0.13	5.0	
Oxidized, weathered	0.80	0.28	2.9	
Glass, 3.2-mm thickness				
Float or tempered				0.79
Low iron oxide type				0.88
Metal, plated				
Black sulfide	0.92	0.10	9.2	
Black cobalt oxide	0.93	0.30	3.1	
Black nickel oxide	0.92	0.08	11	
Black chrome	0.87	0.09	9.7	
Mylar, 0.13-mm thickness				0.87
Paints				
Black (Parsons)	0.98	0.98	1.0	
White, acrylic	0.26	0.90	0.29	
White, zinc oxide	0.16	0.93	0.17	
Plexiglas, 3.2-mm thickness				0.90
Snow				
Fine particles, fresh	0.13	0.82	0.16	
Ice granules	0.33	0.89	0.37	
Tedlar, 0.10-mm thickness				0.92
Teflon, 0.13-mm thickness				0.92

Adapted from Reference [2].

Table A.12 Thermophysical Properties of Fluids (Heat Exchangers)

Engine oil

T (°C)	ρ (kg/m^3)	c_p (J/KgK)	k (W/mK)	$\mu \times 10^2$ (N.s/m^2)	Pr
0	899	1796	0.147	384.8	47100
20	888	1880	0.145	79.92	10400
40	876	1964	0.144	21.02	2870
60	864	2047	0.14	7.249	1050
80	852	2131	0.138	3.195	490
100	840	2219	0.137	1.705	276
120	828	2307	0.135	1.027	175
140	816	2395	0.133	0.653	116
160	805	2483	0.132	0.451	84

50% Ethylene glycol

T (°C)	ρ (kg/m^3)	c_p (J/KgK)	k (W/mK)	$\mu \times 10^2$ (N.s/m^2)	Pr
0	1083	3180	0.379	1.029	86.3
20	1072	3310	0.319	0.459	47.6
40	1061	3420	0.404	0.238	20.1
60	1048	3520	0.417	0.139	11.8
80	1034	3590	0.429	0.099	8.3
100	1020	3650	0.442	0.080	6.6
120	1003	3680	0.454	0.066	5.4

Ethylene glycol

T (°C)	ρ (kg/m^3)	c_p (J/KgK)	k (W/mK)	$\mu \times 10^2$ (N.s/m^2)	Pr
0	1130	2294	0.242	6.501	615
20	1116	2382	0.249	2.140	204
40	1101	2474	0.256	0.957	93
60	1087	2562	0.26	0.516	51
80	1077	2650	0.261	0.321	32.4
100	1058	2742	0.263	0.215	22.4

Glycerin

T (°C)	ρ (kg/m^3)	c_p (J/KgK)	k (W/mK)	$\mu \times 10^2$ (N.s/m^2)	Pr
0	1276	2261	0.282	1060.4	84700
10	1270	2319	0.284	381.0	31000
20	1264	2386	0.286	149.2	12500
30	1258	2445	0.286	62.9	5380
40	1252	2512	0.286	27.5	2450
50	1244	2583	0.287	18.7	1630

Table A.12 (*continued*)

Water

T ($^\circ$C)	ρ (kg/m^3)	c_p (J/KgK)	k (W/mK)	$\mu \times 10^6$ (N.s/m^2)	Pr
0	1002	4217	0.552	1792	13.6
20	1000	4181	0.597	1006	7.02
40	994	4178	0.628	654	4.34
60	985	4184	0.651	471	3.02
80	974	4196	0.668	355	2.22
100	960	4216	0.68	282	1.74
120	945	4250	0.685	233	1.45
140	928	4283	0.684	199	1.24
160	909	4342	0.67	173	1.10
180	889	4417	0.675	154	1.00
200	866	4505	0.665	139	0.94
220	842	4610	0.572	126	0.89
240	815	4756	0.635	117	0.87
260	785	4949	0.611	108	0.87
280	752	5208	0.58	102	0.91
300	714	5728	0.54	96	1.11

Table A.13   Standard Spectral Irradiations

λ μm	AM1.5G W/m²μm	AM1.5D	λ μm	AM1.5G W/m²μm	AM1.5D	λ μm	AM1.5G W/m²μm	AM1.5D
0.305	9.5	3.4	0.740	1211.2	971.0	1.520	262.6	239.3
0.310	42.3	15.6	0.753	1193.9	956.3	1.539	274.2	248.8
0.315	107.8	41.1	0.758	1175.5	942.2	1.558	275.0	249.3
0.320	181.0	71.2	0.763	643.1	524.8	1.578	244.6	222.3
0.325	246.8	100.2	0.768	1030.7	830.7	1.592	247.4	227.3
0.330	395.3	152.4	0.780	1131.1	908.9	1.610	228.7	210.5
0.335	390.1	155.6	0.800	1081.6	873.4	1.630	244.5	224.7
0.340	435.3	179.4	0.816	849.2	712.0	1.646	234.8	215.9
0.345	438.9	186.7	0.824	785.0	660.2	1.678	220.5	202.8
0.350	483.7	212.0	0.832	916.4	765.5	1.740	171.5	158.2
0.360	520.3	240.5	0.840	959.9	799.8	1.800	30.7	28.6
0.370	666.2	324.0	0.860	978.9	815.2	1.860	2.0	1.8
0.380	712.5	362.4	0.880	933.2	778.3	1.920	1.2	1.1
0.390	720.7	381.7	0.905	748.5	630.4	1.960	21.2	19.7
0.400	1013.1	556.0	0.915	667.5	565.2	1.985	91.1	84.9
0.410	1158.2	656.3	0.925	690.3	586.4	2.005	26.8	25.0
0.420	1184.0	690.8	0.930	403.6	348.1	2.035	99.5	92.5
0.430	1071.9	641.9	0.937	258.3	224.2	2.065	60.4	56.3
0.440	1302.0	798.5	0.948	313.6	271.4	2.100	89.1	82.7
0.450	1526.0	956.6	0.965	526.8	451.2	2.148	82.2	76.2
0.460	1599.6	990.8	0.980	646.4	549.7	2.198	71.5	66.4
0.470	1581.0	998.0	0.993	746.8	630.1	2.270	70.2	65.0
0.480	1628.3	1046.1	1.040	690.5	582.9	2.360	62.0	57.6
0.490	1539.2	1005.1	1.070	637.5	539.7	2.450	21.2	19.8
0.500	1548.7	1026.7	1.100	412.6	366.2	2.494	18.5	17.0
0.510	1586.5	1066.7	1.120	108.9	98.1	2.537	3.2	3.0
0.520	1484.9	1011.5	1.130	189.1	169.5	2.941	4.4	4.0
0.530	1572.4	1084.9	1.137	132.2	118.7	2.973	7.6	7.0
0.540	1550.7	1082.4	1.161	339.0	301.9	3.005	6.5	6.0
0.550	1561.5	1102.2	1.180	460.0	406.8	3.056	3.2	3.0
0.630	1434.1	1062.1	1.320	250.2	223.4	3.245	3.2	3.0
0.650	1419.9	1061.7	1.350	32.5	30.1	3.317	13.1	12.0
0.670	1392.3	1046.2	1.395	1.6	1.4	3.344	3.2	3.0
0.690	1130.0	859.2	1.443	55.7	51.6	3.450	13.3	12.2
0.710	1316.7	1002.4	1.463	105.1	97.0	3.573	11.9	11.0
0.718	1010.3	816.9	1.477	105.5	97.3	3.765	9.8	9.0
0.724	1043.2	842.8	1.497	182.1	167.1	4.045	7.5	6.9

Adapted from Reference [4] and [5]

Table A.14   Absorption Coefficients for Si

λ (μm)	α (cm^{-1})	λ (μm)	α (cm^{-1})	λ (μm)	α (cm^{-1})	λ (μm)	α (cm^{-1})
0.218	1.80E+06	0.458	3.58E+04	0.698	2318	0.938	139.2
0.226	1.84E+06	0.466	3.62E+04	0.706	2091	0.946	121.4
0.234	1.80E+06	0.474	3.04E+04	0.714	1931	0.954	105.9
0.242	1.77E+06	0.482	2.39E+04	0.722	1896	0.962	92.53
0.25	1.83E+06	0.49	2.01E+04	0.73	1715	0.97	81.14
0.258	1.93E+06	0.498	1.86E+04	0.738	1702	0.978	71.41
0.266	2.09E+06	0.506	1.72E+04	0.746	1526	0.986	63
0.274	2.29E+06	0.514	1.44E+04	0.754	1505	0.994	55.43
0.282	2.38E+06	0.522	1.33E+04	0.762	1373	1.002	48.04
0.29	2.28E+06	0.53	1.02E+04	0.77	1302	1.01	40.5
0.298	1.83E+06	0.538	1.12E+04	0.778	1249	1.018	33.24
0.306	1.55E+06	0.546	1.03E+04	0.786	1110	1.026	26.72
0.314	1.38E+06	0.554	7227	0.794	1110	1.034	21.42
0.322	1.27E+06	0.562	7622	0.802	991.2	1.042	17.5
0.33	1.19E+06	0.57	7449	0.81	885.7	1.05	14.37
0.338	1.13E+06	0.578	5889	0.818	838.2	1.058	11.43
0.346	1.08E+06	0.586	6523	0.826	765.8	1.066	8.812
0.354	1.07E+06	0.594	5958	0.834	691	1.074	6.908
0.362	1.01E+06	0.602	5214	0.842	621.6	1.082	5.681
0.37	6.91E+05	0.61	5119	0.85	557.6	1.09	4.791
0.378	3.69E+05	0.618	4557	0.858	498.7	1.098	3.972
0.386	2.21E+05	0.626	4423	0.866	444.6	1.106	3.115
0.394	1.51E+05	0.634	3676	0.874	395.1	1.114	2.167
0.402	1.13E+05	0.642	3455	0.882	350.1	1.122	1.333
0.41	9.05E+04	0.65	3142	0.89	309.2	1.13	0.9011
0.418	7.45E+04	0.658	3060	0.898	272.3	1.138	0.9011
0.426	6.09E+04	0.666	2847	0.906	239.1	1.146	1.001
0.434	5.35E+04	0.674	2604	0.914	209.5	1.154	0.8774
0.442	4.69E+04	0.682	2385	0.922	183.1	1.162	0.5019
0.45	4.08E+04	0.69	2373	0.93	159.7	1.17	0.16

Adapted from reference [6]

REFERENCES

1. R.E. Sonntag, C. Borgnakke, and G.J.V. Wylen, *Fundamentals of Thermodynamics* (5th ed.). New York: John Wiley and Sons, 1998, p. 650.

2. F.P. Incropera, D.P. Dewitt, T.L. Bergman, and A.S. Lavine, *Fundamentals of Heat and Mass Transfer*. New York: John Wiley and Sons, 2007.

3. D.A. Reay and P.A. Kew, Appendix I, *Heat Pipes* (5th ed.). Oxford, UK: Butterworth-Heinemann (Elsevier), 2006.

4. Standard ASTM E891-92, Standard for Terrestrial Solar Direct Normal Solar Spectral Irradiance Tables for Air Mass 1.5. American Society for Testing and Materials, West Conshocken, PA.

5. Standard ASTM E892-92, Standard for Terrestrial Solar Spectral Irradiance Tables at Air Mass 1.5 for a 37° Tilted Surface. West Conshocken, PA, American Society for Testing and Materials.

6. Edward D. Palik, *Handbook of Optical Constants of Solids*. New York: Academic Press, 1998.

Appendix B

Thermoelectrics

B.1 THERMOELECTRIC EFFECTS

The thermoelectric effect consists of three effects: the Seebeck effect, the Peltier effect, and the Thomson effect.

Seebeck Effect

The *Seebeck effect* is the conversion of a temperature difference into an electric current. As shown in Figure B.1, wire A is joined at both ends to wire B and a voltmeter is inserted in wire B. Suppose that a temperature difference is imposed between two junctions, it will generally found that a potential difference $\Delta\phi$ or voltage V will appear on the voltmeter. The potential difference is proportional to the temperature difference. The potential difference V is

$$V = \alpha_{AB}\Delta T \tag{B.1}$$

where $\Delta T = T_H - T_L$ and α is called the *Seebeck coefficient* (also called *thermopower*) which is usually measured in $\mu V/K$. The sign of α is positive if the electromotive force, *emf*, tends to drive an electric current through wire A from the hot junction to the cold junction as shown in Figure B.1. The relative Seebeck coefficient is also expressed in terms of the absolute Seebeck coefficients of wires A and B as

$$\alpha_{AB} = \alpha_A - \alpha_B \tag{B.2}$$

In practice one rarely measures the absolute Seebeck coefficient because the voltage meter always reads the relative Seebeck coefficient between wires A and B. The absolute Seebeck coefficient can be calculated from the Thomson coefficient.

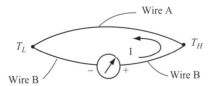

Figure B.1   Schematic basic thermocouple

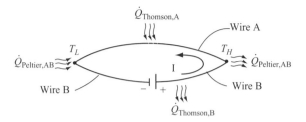

Figure B.2 Schematic for the Peltier effect and the Thomson effect

Peltier Effect

When current flows across a junction between two different wires, it is found that heat must be continuously added or subtracted at the junction in order to keep its temperature constant, which is shown in Figure B.2. The heat is proportional to the current flow and changes sign when the current is reversed. Thus, the Peltier heat absorbed or liberated is

$$\dot{Q}_{Peltier} = \pi_{AB} I \tag{B.3}$$

where π_{AB} is the *Peltier coefficient* and the sign of π_{AB} is positive if the junction at which the current enters wire A is heated and the junction at which the current leaves wire A is cooled. The Peltier heating or cooling is *reversible* between heat and electricity. This means that heating (or cooling) will produce electricity and electricity will produce heating (or cooling) without a loss of energy.

Thomson Effect

When current flows in a wire with a temperature gradient, heat is absorbed or librated across the wire depending on the material and the direction of the current. The Thomson heat is proportional to both the electric current and the temperature gradient, which is illustrated in Figure B.2. Thus, the Thomson heat absorbed or librated across a wire is

$$\dot{Q}_{Thomson} = \tau I \Delta T \tag{B.4}$$

where τ is the *Thomson coefficient* and the sign of τ is positive if heat is absorbed as shown in wire A. The sign of τ is negative if heat is librated as shown in wire B. The Thomson coefficient is unique among the three thermoelectric coefficients because it is the only thermoelectric coefficient directly measurable for individual materials. There is other form of heat, called *Joule heating*, which is irreversible and is always generated as current flows in a wire. The Thomson heat is *reversible* between heat and electricity. This heat is not the same as Joule heating, or $I^2 R$.

B.2 THOMSON (OR KELVIN) RELATIONSHIPS

The interrelationships between the three thermoelectric effects are important in order to understand the basic phenomena. In 1854, Thomson [3] studied the relationships thermodynamically and provided two relationships by applying the *first and second laws of thermodynamics* with an assumption that the reversible and irreversible processes in thermoelectricity are separable. The necessity for the assumption remained an objection to the theory until the advent of the new thermodynamics. The relationships were later completely confirmed by experiment being essentially a consequence of Onsager's Principle [4] in 1931. It is not surprising that, since the thermal energy and electrostatic energy in quantum mechanics are often reversible, those in thermoelectrics are reversible.

For a small temperature difference, the Seebeck coefficient is expressed in terms of the potential difference per unit temperature, as shown in Equation (B.1).

$$\frac{\Delta\phi}{\Delta T} = \frac{d\phi}{dT} = \alpha_{AB} \tag{B.5}$$

where ϕ is the electrostatic potential and α is the Seebeck coefficient that is a driving force and is referred to as thermopower.

Consider two dissimilar wires A and B constituting a closed circuit (Figure B.3), in which the colder junction is at a temperature T and the hotter junction is at T+ΔT and both are maintained by heat reservoirs. Two additional reservoirs are positioned at the midpoints of wires A and B. Each of these reservoirs is maintained at a temperature that is the average of those at the hotter and colder junctions [5].

It is assumed that the thermoelectric properties such as the Seebeck, Peltier and Thomson coefficients remain constant with the passage of a small current. In order to study the relationships between the thermoelectric effects, a sufficiently small temperature difference is to be applied between the two junctions as shown in Figure B.3. It is then expected that a small counterclockwise current flows (this is the way that the dissimilar wires A and B are arranged). We learned earlier that thermal energy is converted to electrical energy and that the thermal energy constitutes the Peltier heat and the Thomson heat, both of which are reversible. The difficulty is that the circuit inevitably involves irreversibilities. The current flow and the temperature difference are always accompanied by *Joule heating* and *thermal conduction*, respectively, both of which are irreversible and make no contribution to the thermoelectric effects. Thomson assumed that the reversible and irreversible processes are separable. So the reversible

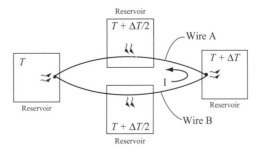

Figure B.3 Closed circuit for the analysis of thermoelectric phenomena [5]

electrical and thermal energies can be equated using the first law of thermodynamics.

$$\underbrace{I\Delta\phi}_{\text{Electrical Energy}} = \underbrace{I\pi_{AB}(T+\Delta T)}_{\text{Peltier heat at hot junction}} - \underbrace{I\pi_{AB}(T)}_{\text{Peltier heat at cold junction}}$$

$$+ \underbrace{\tau_B I \Delta T}_{\text{Thomson heat in wire B}} - \underbrace{\tau_A I \Delta T}_{\text{Thomson heat in wire A}} \qquad \text{(B.6)}$$

For a small potential difference due to the small temperature difference, the potential difference is expressed mathematically as

$$\Delta\phi = \frac{d\phi}{dT}\Delta T \qquad \text{(B.7)}$$

Using this and dividing Equation (B.6) by the current I provides

$$\frac{d\phi}{dT}\Delta T = \pi_{AB}(T+\Delta T) - \pi_{AB}(T) + (\tau_B - \tau_A)\Delta T \qquad \text{(B.8)}$$

Dividing this by ΔT gives

$$\frac{d\phi}{dT} = \frac{\pi_{AB}(T+\Delta T) - \pi_{AB}(T)}{\Delta T} + (\tau_B - \tau_A) \qquad \text{(B.9)}$$

Since ΔT is small, we conclude

$$\frac{d\phi}{dT} = \frac{d\pi_{AB}}{dT} + (\tau_B - \tau_A) \qquad \text{(B.10)}$$

which is nothing more than the Seebeck coefficient as defined in Equation (B.5). This is the fundamental thermodynamic theorem for a closed thermoelectric circuit; it shows the energy relationship between the electrical Seebeck effect and the thermal Peltier and Thomson effects. The potential difference, or in other words the electromotive force (*emf*), results from the Peltier heat and Thomson heat or vice versa. More importantly the Seebeck effect is directly caused by both to the Peltier and Thomson effects.

The assumption of separable reversibility permits that the net change of entropy of the surroundings (reservoirs) of the closed circuit is equal to zero. The results are in excellent agreement with experimental findings [6]. The net change of entropy of all reservoirs is

$$\Delta S = -\frac{Q_{\text{at junction } T+\Delta T}}{T+\Delta T} + \frac{Q_{\text{at junction } T}}{T} - \frac{Q_{\text{along wire B}}}{T+\Delta T/2} + \frac{Q_{\text{along wire A}}}{T+\Delta T/2} = 0 \quad \text{(B.11)}$$

Using Equations (B.3) and (B.4)

$$-\frac{I\pi_{AB}(T+\Delta T)}{T+\Delta T} + \frac{I\pi_{AB}(T)}{T} - \frac{\tau_B I \Delta T}{T+\Delta T/2} + \frac{\tau_A I \Delta T}{T+\Delta T/2} = 0 \qquad \text{(B.12)}$$

Dividing this by $-I\Delta T$ yields

$$\frac{\frac{\pi_{AB}(T+\Delta T)}{T+\Delta T} - \frac{\pi_{AB}(T)}{T}}{\Delta T} + \frac{\pi_B}{T+\Delta T/2} - \frac{\tau_A}{T+\Delta T/2} = 0 \qquad \text{(B.13)}$$

Since ΔT is small and $\Delta T/2 << T$, we have

$$\frac{d\left(\frac{\pi_{AB}}{T}\right)}{dT} + \frac{\tau_B - \tau_A}{T} = 0 \tag{B.14}$$

By taking the derivative of the first term, we have

$$\frac{T\frac{d\pi_{AB}}{dT} - \pi_{AB}}{T^2} + \frac{\tau_B - \tau_A}{T} = 0 \tag{B.15}$$

which reduces to

$$\frac{\pi_{AB}}{T} = \frac{d\pi_{AB}}{dT} + \tau_B - \tau_A \tag{B.16}$$

Combining Equations (B.10) and (B.16) yields

$$\frac{d\phi}{dT} = \frac{\pi_{AB}}{T} \tag{B.17}$$

Using Equation (B.5), we have a very important relationship as

$$\pi_{AB} = \alpha_{AB} T \tag{B.18}$$

Combining Equations (B.14) and (B.17) gives

$$\tau_A - \tau_B = T\frac{d\alpha_{AB}}{dT} \tag{B.19}$$

Equations (B.18) and (B.19) are well known to be called the *Thomson (Kelvin) relationships*. By combining Equations (B.3) and (B.18), the thermal energy caused solely by the Seebeck effect is derived,

$$\dot{Q}_{Peltier} = \alpha_{AB} T I \tag{B.20}$$

which is a very important thermoelectric equation and reversible thermal energy.

By taking the integral of Equation (B.19) after dividing it by T, the Seebeck coefficient has an expression in terms of the Thomson coefficients. Since the Thomson coefficients are measurable, the Seebeck coefficient can be calculated from them.

$$\alpha_{AB} = \int_0^T \frac{\tau_A}{T}dT - \int_0^T \frac{\tau_B}{T}dT \tag{B.21}$$

B.3 HEAT BALANCE EQUATION

Suppose that a thermocouple consists of a p-type material (positive α) and an n-type material (negative α) as shown in Figure B.4 (a).

The steady-state heat balance at the junction at temperature T_1 gives

$$\dot{Q}_1 = q_p + q_n \tag{B.22}$$

The heat flow at the p-type material constitutes the Peltier heat and Fourier's law of conduction. Thus, we have

$$q_p = \alpha_p T_1 I + \left(-k_p A_p \left. \frac{dT}{dx} \right|_{x=0} \right) \tag{B.23}$$

Likewise, for the n-type material, we have

$$q_n = \alpha_n T_1 I + \left(-k_n A_n \left. \frac{dT}{dx} \right|_{x=0} \right) \tag{B.24}$$

Now we consider a differential element to obtain the temperature gradient along x, which is shown in Figure B.4 (b). The heat balance with a heat generation (Joule heating) in the differential element becomes

$$\underbrace{q_x - \left(q_x + \frac{dq_x}{dx}dx \right)}_{\text{heat transfer across the surface of the element}} + \underbrace{\frac{I^2 \rho_p}{A_p}dx}_{\text{Joule heating}} = 0 \tag{B.25}$$

$$-\frac{d}{dx}\left(-k_p A_p \frac{dT}{dx} \right) + \frac{I^2 \rho_p}{A_p} = 0 \tag{B.26}$$

$$\frac{d}{dx}\left(\frac{dT}{dx} \right) = -\frac{I^2 \rho_p}{A_p^2 k_p} \tag{B.27}$$

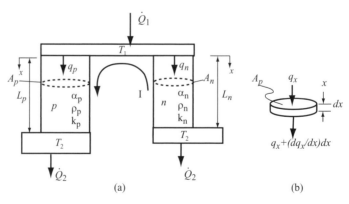

(a)   (b)

Figure B.4 (a) Thermoelectric couple with a pair of dissimilar materials, (b) Differential element

By integrating this, we have

$$\frac{dT}{dx} = -\frac{I^2 \rho_p}{A_p^2 k_p} x + C_1 \tag{B.28}$$

Taking the integral again from 0 to L_p as

$$\int_{T_1}^{T_2} dT = \int_0^{L_p} \left(-\frac{I^2 \rho_p}{A_p^2 k_p} x + C_1 \right) dx \tag{B.29}$$

$$T_2 - T_1 = -\frac{I^2 \rho_p}{2 A_p^2 k_p} L_p^2 + C_1 L_p \tag{B.30}$$

$$C_1 = \frac{I^2 \rho_p L_p}{2 A_p^2 k_p} - \frac{(T_1 - T_2)}{L_p} \tag{B.30a}$$

From Equation (B.28), we obtain the temperature gradient at $x = 0$ as

$$\left. \frac{dT}{dx} \right|_{x=0} = \frac{I^2 \rho_p L_p}{2 A_p^2 k_p} - \frac{(T_1 - T_2)}{L_p} \tag{B.31}$$

Inserting this into Equation (B.23) gives

$$q_p = \alpha_p T_1 I - \frac{1}{2} I^2 \frac{\rho_p L_p}{A_p} + \frac{k_p A_p}{L_p} (T_1 - T_2) \tag{B.32}$$

In a similar way, we have the heat transport at the n-type material as

$$q_p = \alpha_n T_1 I - \frac{1}{2} I^2 \frac{\rho_n L_n}{A_n} + \frac{k_n A_n}{L_n} (T_1 - T_2) \tag{B.33}$$

Equation (B.22) for the thermocouple now becomes

$$\dot{Q}_1 = (\alpha_p - \alpha_n) T_1 I - \frac{1}{2} I^2 \left(\frac{\rho_p L_p}{A_p} + \frac{\rho_n L_n}{A_n} \right) + \left(\frac{k_p A_p}{L_p} + \frac{k_n A_n}{L_n} \right) (T_1 - T_2) \tag{B.34}$$

Finally, the heat balance equation at the junction at temperature T_1 is expressed as

$$\dot{Q}_1 = \alpha T_1 I - \frac{1}{2} I^2 R + K (T_1 - T_2) \tag{B.35}$$

where

$$\alpha = \alpha_p - \alpha_n \tag{B.36}$$

$$R = \frac{\rho_p L_p}{A_p} + \frac{\rho_n L_n}{A_n} \tag{B.37}$$

$$K = \frac{k_p A_p}{L_p} + \frac{k_n A_n}{L_n} \tag{B.38}$$

In a similar way, we can obtain the heat balance equation at the junction at temperature T_2 as

$$\dot{Q}_2 = \alpha T_2 I + \frac{1}{2} I^2 R + K (T_1 - T_2) \tag{B.39}$$

B.4 FIGURE OF MERIT AND OPTIMUM GEOMETRY

An important characteristic parameter was found from the analyses of thermoelectrics, which is called the figure of merit Z.

$$Z = \frac{\alpha^2}{KR} \tag{B.40}$$

A good thermoelectric material must have a large Seebeck coefficient to produce the required voltage, high electrical conductivity to minimize the thermal noise, and a low thermal conductivity to decrease thermal losses from the thermocouple junctions. As the expression shows, the higher figure of merit exhibits the high performance. Either increasing the Seebeck coefficient or decreasing KR increases the figure of merit. However, it is normally very challenging to decrease KR because increasing thermal conductance K generally decreases electric resistance R. For a given pair of materials, R rises and K falls as the ratio of length to cross-sectional area increases and, indeed, a thermocouple can be designed for a given heat load and electric current by altering the ratio in both arms.

Using Equations (B.37) and (B.38), KR becomes

$$KR = \left(\frac{k_p A_p}{L_p} + \frac{k_n A_n}{L_n} \right) \left(\frac{\rho_p L_p}{A_p} + \frac{\rho_n L_n}{A_n} \right)$$

$$= k_p \rho_p + k_p \rho_n \frac{L_n / A_n}{L_p / A_p} + k_n \rho_p \frac{L_p / A_p}{L_n / A_n} + k_n \rho_n \tag{B.41}$$

Maximizing Z is equivalent to minimizing KR since α is not a function of geometry. We set the differentiation of KR with respect to $\frac{L_n / A_n}{L_p / A_p}$ to be zero.

$$\frac{d(KR)}{d\left(\frac{L_n / A_n}{L_p / A_p} \right)} = 0 \tag{B.42}$$

Using Equation (B.41), this becomes

$$k_p \rho_n + k_n \rho_p \frac{-1}{\left(\frac{L_n / A_n}{L_p / A_p} \right)^2} = 0 \tag{B.43}$$

which leads to an important geometric relationship

$$\frac{L_n / A_n}{L_p / A_p} = \left(\frac{\rho_p k_n}{\rho_n k_p} \right)^{\frac{1}{2}} \tag{B.44}$$

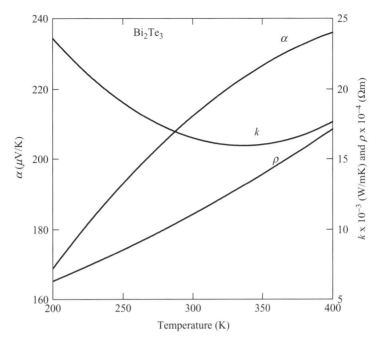

Figure B.5   Thermoelectric properties versus temperature

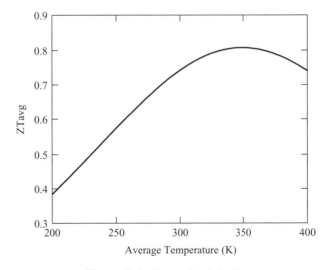

Figure B.6   Generalized charts

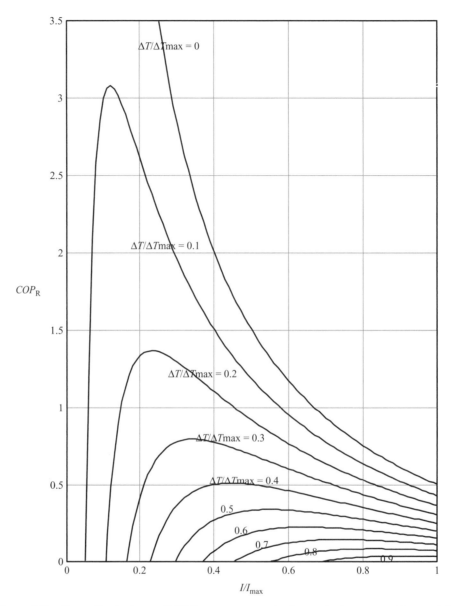

Figure B.7 Generalized chatrs of COP vs current ratio for a single-stage peltier module, $ZT_c = 1$

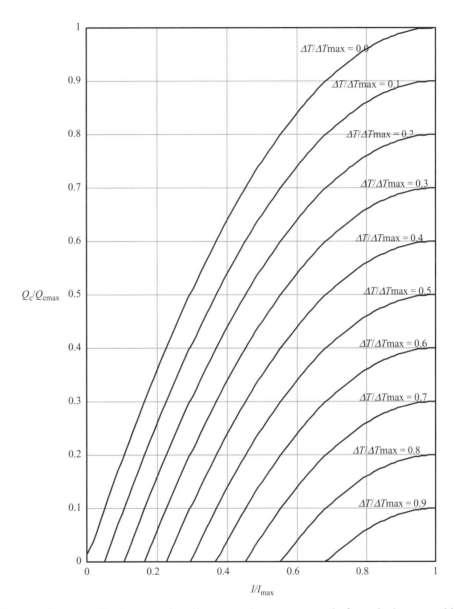

Figure B.8   Generalized chates of cooling rate ratio vs. current ratio for a single-stage peltier module (independent of ZT_c)

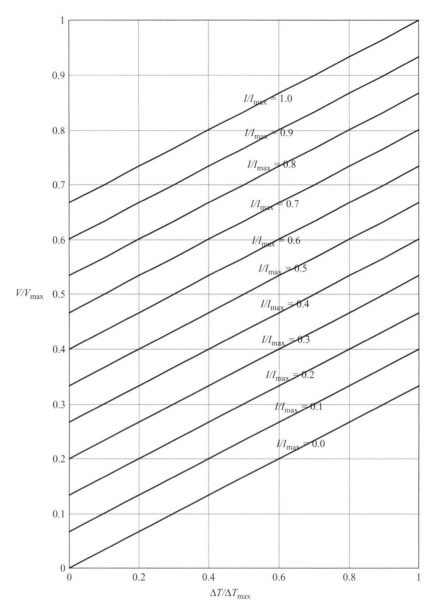

Figure B.9 Universal charts of voltage rate ratio vs. temperature difference ratio for a single-stage peltier module, $ZT_c = 1$

Inserting Equation (B.44) into Equation (B.41) yields

$$KR = k_p\rho_p + k_p\rho_n \left(\frac{\rho_p k_n}{\rho_n k_p}\right)^{\frac{1}{2}} + k_n\rho_p \left(\frac{\rho_n k_p}{\rho_p k_n}\right)^{\frac{1}{2}} + k_n\rho_n \qquad (B.45)$$

Eventually,

$$KR = \left[\left(k_p\rho_p\right)^{\frac{1}{2}} + \left(k_n\rho_n\right)^{\frac{1}{2}}\right]^2 \qquad (B.46)$$

which holds in the optimum geometry.

REFERENCES

1. T. J. Seebeck, Ueber den magnetismus der galvenische kette, *Abh. K. Akad. Wiss Berlin*, 289, 1821.

2. J. C. A. Peltier, Nouvelles experiences sur la caloricite des courants electrique, *Ann. Chem. Phys.*, Vol. 56, pp. 371, 1834.

3. W. Thomson, Account of Researches in Thermo-Electricity, *Philos. Mag.* [5], 8, 62, 1854.

4. L. Onsager, *Phys. Rev.*, 37, 405–526, 1931.

5. D. M. Rowe, *CRC Handbook of Thermoelectrics*. Boca Raton, FL: CRC Press, 1995.

6. R.P. Benedict, *Fundamentals of Temperature, Pressure, and Flow Measurements*. (3rd ed.). Chapter 7, Thermoelectric Thermometry, pp. 74–126. New York: Wiley-Interscience, 1984.

7. D. M. Rowe, *Thermoelectrics Handbook, Micro to Nano*. Boca Raton, FL: CRC Press, 2006.

8. G. Min and D. M. Rowe, Optimization of Thermoelectric Module Geometry for "Waste Heat" Electric Power Generation, *Journal of Power Sources*, Vol. 38. pp. 253–259, 1992.

Appendix C

Pipe Dimensions

Table C.1 Pipe Dimensions

			Nominal pipe size				
NPS (in.)	DN (mm)	O.D. (in.)	O.D. (mm)	Schedule	I.D. (in.)	I.D. (mm)	O.D/I.D
1/8	6	0.405	10.29	10	0.307	7.80	1.32
				40	0.269	6.83	
				80	0.215	5.46	
1/4	8	0.540	13.72	10	0.410	10.41	1.32
				40	0.364	9.24	
				80	0.302	7.67	
3/8	10	0.675	17.15	40	0.493	12.52	1.37
				80	0.423	10.74	
1/2	15	0.840	21.34	40	0.622	15.80	1.35
				80	0.546	13.87	
				160	0.464	11.79	
3/4	20	1.050	26.67	40	0.824	20.93	1.27
				80	0.742	18.85	
1	25	1.315	33.40	40	1.049	26.64	1.25
				80	0.957	24.31	
1 1/4	32	1.660	42.16	40	1.380	35.05	1.20
				80	1.278	32.46	
1 1/2	40	1.900	48.26	40	1.610	40.89	1.18
				80	1.500	38.10	
2	50	2.375	60.33	40	2.067	52.50	1.15
				80	1.939	49.25	
2 1/2	65	2.875	73.03	40	2.469	62.71	1.16
				80	2.323	59.00	
3	80	3.500	88.90	40	3.068	77.93	1.14
				80	2.900	73.66	
3 1/2	90	4.000	101.60	40	3.548	90.12	1.13
				80	3.364	85.45	
4	100	4.500	114.30	40	4.026	102.26	1.12
				80	3.826	97.18	
5	125	5.563	141.30	10 S	5.295	134.49	1.05
				40	5.047	128.19	
				80	4.813	122.25	

Table C.1   (*Continued*)

NPS (in.)	DN (mm)	O.D. (in.)	O.D. (mm)	Schedule	I.D. (in.)	I.D. (mm)	O.D/I.D
				Nominal pipe size			
6	150	6.625	168.28	10 S	6.357	161.47	1.04
				40	6.065	154.05	
				80	5.761	146.33	
8	200	8.625	219.08	10 S	8.329	211.56	1.04
				30	8.071	205.00	
				80	7.625	193.68	
10	250	10.750	273.05	10 S	10.420	264.67	1.03
				30	10.192	258.88	
				Extra heavy	9.750	247.65	
12	300	12.750	323.85	10 S	12.390	314.71	1.03
				30	12.090	307.09	
				Extra heavy	11.750	298.45	
14	350	14.000	355.60	10	13.500	342.90	1.04
				Standard	13.250	336.55	
				Extra heavy	13.000	330.20	
16	400	16.000	406.40	10	15.500	393.70	1.03
				Standard	15.250	387.35	
				Extra heavy	15.000	381.00	
18	450	18.000	457.20	10 S	17.624	447.65	1.02
				Standard	17.250	438.15	
				Extra heavy	17.000	431.80	

Appendix D

Curve Fitting of Working Fluids

Curve Fit for Working Fluids Chosen

Table D.1 Working Fluid Properties

	Acetone			Ammonia			Flutec PP9		
	0°C	40°C	80°C	0°C	40°C	80°C	0°C	30°C	90°C
Latent heat, kJ/kg	564	536	495	1263	1101	891	98.4	94.5	86.1
Liquid density, kg/m^3	812	768	719	638.6	579.5	505.7	2029	1960	1822
Vapor density, kg/m^3	0.26	1.05	4.30	3.48	12.0	34.13	0.01	0.12	1.93
Liquid thermal conductivity, W/m°C	0.813	0.175	0.160	0.298	0.272	0.235	0.059	0.057	0.054
Liquid viscosity, Ns/m^2 × 10^3	0.395	0.269	0.192	0.25	0.20	0.15	3.31	1.48	0.65
Vapor viscosity, Ns/m^2 × 10^5	0.78	0.86	0.95	0.92	1.16	1.40	0.90	1.06	1.21
Vapor pressure, bar	0.10	0.60	2.15	4.24	15.34	40.90	0.00	0.01	0.12
Vapor specific heat, kJ/kg °C	2.11	2.22	2.34	2.125	2.160	2.210	0.87	0.94	1.09
Liquid surface tension, N/m × 10^2	2.62	2.12	1.62	2.48	1.833	0.767	2.08	1.80	1.24

D.1 CURVE FITTING FOR WORKING FLUID PROPERTIES CHOSEN

D.1.1 MathCad Format:

$$\text{ORIGIN} := 1$$

1 : Acetone
2 : Ammonia
3 : Flutec PP9

$$tx := \begin{pmatrix} 0 \\ 40 \\ 80 \end{pmatrix}$$

Latent Heat, kJ/kg $hy := \begin{pmatrix} 564 & 2263 & 98.4 \\ 536 & 2101 & 94.5 \\ 495 & 892 & 86.2 \end{pmatrix}$

$hy1 := hy^{\langle 1 \rangle}$ $hy2 := hy^{\langle 2 \rangle}$ $hy3 := hy^{\langle 3 \rangle}$

$hs1 := 1spline(tx, hy1)$ $h_1(t) := interp(hs1, tx, hy1, t)$

$hs2 := 1spline(tx, hy2)$ $h_2(t) := interp(hs2, tx, hy2, t)$

$hs3 := 1spline(tx, hy3)$ $h_3(t) := interp(hs3, tx, hy3, t)$

Liquid density, kg/m^3 $dly := \begin{pmatrix} 812 & 638.6 & 2029 \\ 768 & 579.5 & 1960 \\ 719 & 505.7 & 1822 \end{pmatrix}$

$dly1 := dly^{\langle 1 \rangle}$ $dly2 := dly^{\langle 2 \rangle}$ $dly3 := dly^{\langle 3 \rangle}$

$dls1 := 1spline(tx, dly1)$ $\rho_{l1}(t) := interp(dls1, tx, dly1, t)$

$dls2 := 1spline(tx, dly2)$ $\rho_{l2}(t) := interp(dls2, tx, dly2, t)$

$dls3 := 1spline(tx, dly3)$ $\rho_{l3}(t) := interp(dls3, tx, dly3, t)$

Vapor density, kg/m^3 $dvy := \begin{pmatrix} 0.26 & 3.48 & 0.01 \\ 1.05 & 12.0 & 0.12 \\ 4.30 & 34.13 & 1.93 \end{pmatrix}$

$dvy1 := dvy^{\langle 1 \rangle}$ $dvy2 := dvy^{\langle 2 \rangle}$ $dvy3 := dvy^{\langle 3 \rangle}$

$dvs1 := 1spline(tx, dvy1)$ $\rho_{v1}(t) := interp(dvs1, tx, dvy1, t)$

$dvs2 := 1spline(tx, dvy2)$ $\rho_{v2}(t) := interp(dvs2, tx, dvy2, t)$

$dvs3 := 1spline(tx, dvy3)$ $\rho_{v3}(t) := interp(dvs3, tx, dvy3, t)$

Liquid thermal conductivity, W/m*C $ky := \begin{pmatrix} 0.813 & 0.298 & 0.059 \\ 0.175 & 0.272 & 0.057 \\ 0.160 & 0.235 & 0.054 \end{pmatrix}$

$$\mathrm{ky1} := \mathrm{ky}^{\langle 1 \rangle} \qquad \mathrm{ky2} := \mathrm{ky}^{\langle 2 \rangle} \qquad \mathrm{ky3} := \mathrm{ky}^{\langle 3 \rangle}$$

$$\mathrm{ks1} := \mathrm{lspline(tx, ky1)} \qquad \mathrm{k_1(t)} := \mathrm{interp(ks1, tx, ky1, t)}$$

$$\mathrm{ks2} := \mathrm{lspline(tx, ky2)} \qquad \mathrm{k_2(t)} := \mathrm{interp(ks2, tx, ky2, t)}$$

$$\mathrm{ks3} := \mathrm{lspline(tx, ky3)} \qquad \mathrm{k_3(t)} := \mathrm{interp(ks3, tx, ky3, t)}$$

Liquid viscosity, $\mathrm{N^* s/m^2}$ $\qquad \mathrm{vly} := \begin{pmatrix} 0.395 & 0.25 & 3.31 \\ 0.269 & 0.20 & 1.48 \\ 0.192 & 0.15 & 0.65 \end{pmatrix}$

$$\mathrm{vly1} := \mathrm{vly}^{\langle 1 \rangle} \qquad \mathrm{vly2} := \mathrm{vly}^{\langle 2 \rangle} \qquad \mathrm{vly3} := \mathrm{vly}^{\langle 3 \rangle}$$

$$\mathrm{vls1} := \mathrm{lspline(tx, vly1)} \qquad \mu_{11}(t) := \mathrm{interp(vls1, tx, vly1, t)} \cdot 10^{-3}$$

$$\mathrm{vls2} := \mathrm{lspline(tx, vly2)} \qquad \mu_{12}(t) := \mathrm{interp(vls2, tx, vly2, t)} \cdot 10^{-3}$$

$$\mathrm{vls3} := \mathrm{lspline(tx, vly3)} \qquad \mu_{13}(t) := \mathrm{interp(vls3, tx, vly3, t)} \cdot 10^{-3}$$

Vapor viscosity, $\mathrm{N^* s/m^2}$ $\qquad \mathrm{vvy} := \begin{pmatrix} 0.78 & 0.92 & 0.90 \\ 0.86 & 1.16 & 1.06 \\ 0.95 & 1.40 & 1.21 \end{pmatrix}$

$$\mathrm{vvy1} := \mathrm{vvy}^{\langle 1 \rangle} \qquad \mathrm{vvy2} := \mathrm{vvy}^{\langle 2 \rangle} \qquad \mathrm{vvy3} := \mathrm{vvy}^{\langle 3 \rangle}$$

$$\mathrm{vvs1} := \mathrm{lspline(tx, vvy1)} \qquad \mu_{v1}(t) := \mathrm{interp(vvs1, tx, vvy1, t)} \cdot 10^{-5}$$

$$\mathrm{vvs2} := \mathrm{lspline(tx, vvy2)} \qquad \mu_{v2}(t) := \mathrm{interp(vvs2, tx, vvy2, t)} \cdot 10^{-5}$$

$$\mathrm{vvs3} := \mathrm{lspline(tx, vvy3)} \qquad \mu_{v3}(t) := \mathrm{interp(vvs3, tx, vvy3, t)} \cdot 10^{-5}$$

Vapor pressure, bar $\qquad \mathrm{vpy} := \begin{pmatrix} 0.10 & 4.24 & 0.00 \\ 0.60 & 15.34 & 0.01 \\ 2.15 & 40.90 & 0.12 \end{pmatrix}$

$$\mathrm{vpy1} := \mathrm{vpy}^{\langle 1 \rangle} \qquad \mathrm{vpy2} := \mathrm{vpy}^{\langle 2 \rangle} \qquad \mathrm{vpy3} := \mathrm{vpy}^{\langle 3 \rangle}$$

$$\mathrm{vps1} := \mathrm{lspline(tx, vpy1)} \qquad \mathrm{p_{v1}(t)} := \mathrm{interp(vps1, tx, vpy1, t)}$$

$$\mathrm{vps2} := \mathrm{lspline(tx, vpy2)} \qquad \mathrm{p_{v2}(t)} := \mathrm{interp(vps2, tx, vpy2, t)}$$

$$\mathrm{vps3} := \mathrm{lspline(tx, vpy3)} \qquad \mathrm{p_{v3}(t)} := \mathrm{interp(vps3, tx, vpy3, t)}$$

Vapor specific heat, kJ/kg.C $cvy := \begin{pmatrix} 2.11 & 2.125 & 0.87 \\ 2.22 & 2.160 & 0.94 \\ 2.34 & 2.210 & 1.09 \end{pmatrix}$

$cvy1 := cvy^{\langle 1 \rangle}$ $cvy2 := cvy^{\langle 2 \rangle}$ $cvy3 := cvy^{\langle 3 \rangle}$

$cvs1 := \text{lspline}(tx, cvy1)$ $c_{v1}(t) := \text{interp}(cvs1, tx, cvy1, t)$

$cvs2 := \text{lspline}(tx, cvy2)$ $c_{v2}(t) := \text{interp}(cvs2, tx, cvy2, t)$

$cvs3 := \text{lspline}(tx, cvy3)$ $c_{v3}(t) := \text{interp}(cvs3, tx, cvy3, t)$

Surface tension, N/m $sty := \begin{pmatrix} 2.62 & 2.48 & 2.08 \\ 2.12 & 1.833 & 1.80 \\ 1.62 & 0.767 & 1.24 \end{pmatrix}$

$sty1 := sty^{\langle 1 \rangle}$ $sty2 := sty^{\langle 2 \rangle}$ $sty3 := sty^{\langle 3 \rangle}$

$sts1 := \text{lspline}(tx, sty1)$ $\sigma_1(t) := \text{interp}(sts1, tx, sty1, t) \cdot 10^{-2}$

$sts2 := \text{lspline}(tx, sty2)$ $\sigma_2(t) := \text{interp}(sts2, tx, sty2, t) \cdot 10^{-2}$

$sts3 := \text{lspline}(tx, sty3)$ $\sigma_3(t) := \text{interp}(sts3, tx, sty3, t) \cdot 10^{-2}$

Appendix E

Tutorial I for 2-D

Problem Description for Tutorial I

When an electrically insulated stranded wire is subjected to a DC current across the wire, heat is generated in the wire, simultaneously dissipated into the ambient air due to natural convection and radiation. The wire is anticipated to reach a steady-state temperature. The purpose of the laboratory project No. 1 is an experimental and analytical study that includes the computer simulations on the thermal behavior of the electrical wire. An 18-gauge (wire)-16 mil (insulation thickness) electrical wire is laid horizontally in the quiescent ambient air and subjected to three different assigned DC currents, separately. The project is divided into three parts: Part I, Part II, and Part II.

Part I: Develop an analytical model for the horizontal wire in open air with a heat generation (W/m^3) of the wire conductor using an empirical correlation for the heat transfer coefficient at the wire surface. You are required to use mathematical software MathCad for the calculations. The effects of radiation heat transfer, as well as conduction and convection, should be included in the analysis. Compute the temperature rises of the conductor for three different wire sizes (14, 18, and 20 gauges), along with a range of current 1–60 Amps. The model can provide the heat generations in the conductors for Part III.

Part II: Take measurements for each assigned current using a data acquisition system, which consists mainly of a LabView virtual instrument software, a NI data acquisition board, a current controller (DC power supply), and a shunt resistor. The acquisition system provides a spreadsheet containing transient times, ambient temperatures, voltages across the test wire, and voltages across the shunt resistor in turn with an interval of 0.5 seconds. Convert the voltage of shunt resistor (RS1) into the applied current by dividing by the shunt resistance of 0.007951 that is very durable with temperature variation. And also convert the voltage of the wire conductor (wire 1) into the temperature rise of the wire conductor by using a calibration curve ($T = (R-0.0566595405)/0.0002554547$) that was pretested in a temperature bath and by subtracting the measured ambient temperature from the obtained temperature of the wire conductor. Provide the analytical solution (MathCAD format), including a professional graph for the current versus the temperature rise, along with time for the three wire sizes superimposed with the three sets of the measurements.

Part III: Perform numerical computations according to the experimental data using design tools, Gambit and Fluent, to simulate the thermal behavior of the wire. A tutorial for the computation is provided to help students.

E.1   TUTORIAL I: USING GAMBIT AND FLUENT FOR THERMAL BEHAVIOR OF AN ELECTRICAL WIRE

Radius of copper wire = 1.17 mm/2 = 0.585 mm

PVC insulation thickness = 0.016 in. × 25.4 mm/in. = 0.406 mm

Outer radius of wire = 0.585 mm + 0.406 mm = 0.991 mm

(see Figure E.1)

E.1.1   Creating Geometry in Gambit

1. Starting Gambit
   a. To open Gambit program, you may go to **Start/Programs/Simulation/Fluent Inc Products/Gambit**
   b. Alternatively, to open a Gambit file, you may open the **MS-DOS Command Prompt**. This is usually located under **Start/Programs/Accessories/Command Prompt**. Once you have the **MS-DOS Command Prompt** open

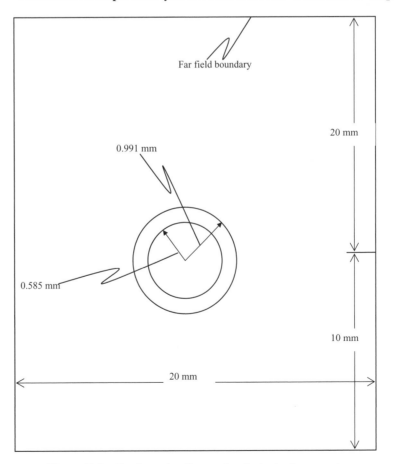

Figure E.1   Configuration for an electrical wire in open air

type **d:** and press Enter. Then type **cd fluent.inc\ntbin\ntx86** and press Enter. This is the directory that you need to be in to run gambit. Now type **gambit –id singlewire** (you can use any name that you like in place of singlewire) and press **Enter.**

2. Gambit Use

Most buttons are activated by left-clicking on them. The buttons that have a small arrow underneath them have menus that you can activate by right-clicking and then selecting the option of your choice. One thing to notice is that when your mouse is over a button, the name of that button appears in the Description window at the bottom of your screen.

Geometry

Vertex

Creat
Vertex

3. Vertex Creation

As shown at the beginning of this tutorial, you will create concentric circles that represent a wire with insulation as well as a large rectangular boundary to represent ambient conditions. To begin with the circles, you need to specify the center and two points on its radius. To create these points, click on **Geometry**, **Vertex**, **Create Vertex**. The Create Real Vertex window opens. Make sure that **Coordinate System** is set at **c_sys.1** and **Type** is set at **Cartesian**. Now in the **Global** column enter your points to create the first vertex as **x:0, y:0, z:0** and then click **Apply**. Always confirm the execution you just did in the **Transcript** window at the lower left. Continue for two points on the radius of each circle and the four corners of the boundary as follows, clicking **Apply** after each.

Vertex 2	x:0.585	y: 0	z: 0	**Apply**
Vertex 3	x:0	y: 0.585	z: 0	**Apply**
Vertex 4	x:0.991	y: 0	z: 0	**Apply**
Vertex 5	x:0	y: 0.991	z: 0	**Apply**
Vertex 6	x:10	y: 20	z: 0	**Apply**
Vertex 7	x:−10	y: 20	z: 0	**Apply**
Vertex 8	x:−10	y: −10	z: 0	**Apply**
Vertex 9	x:10	y: −10	z: 0	**Apply**

Now all the needed vertices are created.

NOTE: Be sure to create the points in this order, or you will have to do a lot of extra thinking later in the tutorial.

4. Controlling the Graphics Area

Some basic controls for the graphics area are as follows.

Mouse button (click and hold)	Horizontal movement of mouse	Vertical movement of mouse
Left	Rotates image about a vertical axis through the center of the graphics area.	Rotates image about a horizontal axis through the center of the graphics area.
Center	Translates image horizontally.	Translates image vertically.
Right	Rotates image about an axis perpendicular to the screen and through the center of the graphics area.	Zooms image in and out.

Orient
Model

Fit to
Window

Geometry

Edge

Create
Edge

Circle

Geometry

Edge

Creating
Edge

Creating
Straight
Edge

You can also return to a standard view by simply clicking on **Orient Model** and **Fit to Window**.

5. Creating Circular Edges from the Vertices

Make sure that you are still in **Geometry** and click on **Edge**. Right-click on **Create Edge,** and select **Circle**. The Create Real Full Circle dialog box will appear.

- To draw the first circle for the wire, click on the pull-down bar next to the **Center** list box. Select vertex.1 in the left column and use the arrows to move it into the right column. Select **Close**. Now select the pull-down bar next to the **End-Point** list box. Select vertex.2 and vertex.3 and click on **Close**. Then click **Apply**.

- To draw the other circle for the insulation, select vertex.1 for the **Center** and for the **End-point** select vertex.4 and vertex.5. Click **Apply**.

You have now created two yellow circles corresponding to the wire and insulation.

6. Creating Straight Lines from the Vertices

To draw the lines that create the rectangular boundary you will again need to create edges. While still in **Geometry** and **Edge**, right-click on **Create Edge** and select **Create Straight Edge**. The Create Straight Edge dialog box will appear. Make sure that **Type** is set to **Real**. In the **Vertices** pull-down menu select vertex.6, vertex.7, vertex.8, and vertex.9. Click **Apply**. Then select vertex.6 and vertex.9 and click **Apply**. You should now have a yellow rectangle corresponding to the far-field boundary conditions.

7. Creating Faces from the Edges

To create faces in Gambit from the edges that you have already drawn, you need to click on **Geometry, Face, Form Face**. To create the face on the circle representing the wire, be sure that **Type** is set to **Real**. Then click on the **Edges** pull-down menu and select edge.1 like you did when creating the edges. Then click **Apply**. Notice that the line color changed from yellow to blue. Do the same to create the face that represents the insulation, but this time select edge.2. Finally, create the face that represents the far-field boundary condition. Select edge.3, edge.4, edge.5, and edge.6 for this face. Now you should have created three faces, but they are superimposed on each other. Therefore, you need to subtract one from the other to represent a specific area.

8. Subtracting Faces

While still in **Geometry**, select **Face, Boolean Operations, Subtract Real Faces**. In the Subtract Real Faces dialog box, click on the **Face** pull-down menu and select face.3 and click **Close**. Then in the **Subtract Faces** pull-down menu select face.2 and click **Close**. Click on **Retain** by the Subtract Faces menu and click **Apply**. Then to subtract the wire from the insulation, select face.2 in the **Face** pull-down menu and select face.1 in the **Subtract Faces** menu. Make sure that **Retain** is still selected and click on **Apply**.

9. Connecting Vertices and Edges

Finally you have to get rid of all redundant geometry, so while still in **Geometry**, click on **Vertex**, and **Connect Vertices**. In the Connect Vertices dialog box

Geometry

Face

Form Face

Geometry

Face

Boolean
Operations

Substract
Real Faces

Geometry

Vertex

Connect
Vertices

change **Pick** to **All** and make sure that **Real** is selected. Click on **Apply**. Then choose **Edge**, and **Connect Edges**. In the Connect Edges dialog box change **Pick** to **All** and make sure that **Real** is selected. Click on **Apply**.

You have now created the geometry of the single wire and should move on to creating the mesh.

Mesh Creation in Gambit

1. Define Zones

   From the menu bar, select **Solver** from the main menu and choose **fluent 5/6**.

2. Define Edge Conditions

   You will now give each of the edges that you created in Gambit a name and a type. Click on **Zones** and then **Specify Boundary Types**.

   - For the inner diameter, make sure that **Action** is set to **Add**. Type in **wire-id** in the **Name** box. Set **Type** to **Wall** and change **Entity** to **Edges**. In the **Edges** pull-down menu, select edge.1, the edge that is associated with the innermost circle, and click **Close**. Then click **Apply**.
   - For the outer diameter, in the **Name** box put **wire-od**. Make sure that **Type** is still set to **Wall** and that **Entity** is on **Edges**. Select edge.2, the circle for the insulation diameter and click **Close** and **Apply**.
   - For the far-field boundary, change the **Name** to **far-field** and **Type** to **Pressure-Outlet**. Keep everything else the same and select edge.3. edge.4, edge.5, and edge.6. Click **Apply**.

3. Define Area Conditions

   Here you will specify the type of material that each face is made out of. Click on **Zones** and **Specify Continuum Types.**

   - To add the air inside of the far-field boundary, make sure that **Action** is set to **Add**. Set **Name** equal to air and **Type** to **Fluid**. Also change **Entity** to **Faces**. In the **Faces** pull-down menu, select face.3, the face that corresponds to the area inside of the rectangle, and click **Close** and **Apply**.
   - To set the insulation material to PVC, change the **Type** to **Solid**, **Name** to pvc, and leave everything else as it is. Select face.2, the area that represents the insulation, and click **Apply**.
   - Finally, to set the wire material to copper, change **Name** to copper and leave all other settings as they are. Select face.1, the area of the wire itself, and click **Apply**.

4. Create Mesh Generation

   To mesh the edges that you have created, click on **Mesh**, **Edge**, **Mesh Edges**. Select edge.1 and edge.2, the two circles that you created. Leave everything alone except **Spacing**. Change **Spacing** from 1 to 0.2 and click **Apply**. The red lines and white points will change to blue. Now select edge.3, edge.4, edge.5, and edge.6, the rectangular edges, and change **Spacing** to 1. Click **Apply**.

*The reason that the type for the far-field boundary is set to Pressure-Outlet is to simulate the ambient conditions within the boundaries of the box. If you were to set the type to Wall, the air that is inside the box would be constrained and would not be able to flow freely.

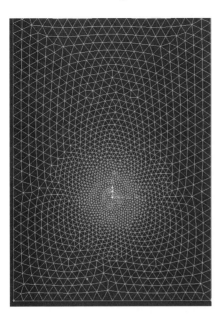

Figure E.2   An example of the mesh generation

Edge

Connect
Edges

Zones

Specify
Boundary
Types

Zones

To mesh the faces, click on **Mesh**, **Face**, **Mesh Faces**. Select all of the faces in the **Mesh Faces** pull-down box. Change **Element** from **Quad** to **Tri** and in **Spacing** make sure that **Apply** is not selected. Then click **Apply** at the bottom of the window.

The mesh should now be created as shown in Figure E.2.

5. Save the Mesh

Click on **File** and **Save**.

Then to export the mesh, click on **File**, **Export**, **Mesh,** and click on **Export 2-D (x-y) Mesh**, and save the file as filename.msh and click **Accept**.

Creation of the geometry and mesh are now complete. You can exit Gambit and begin in Fluent.

Computing in Fluent

1. Opening the File

Open Fluent 6.2 through the Start Menu and select 2d. Then click on **File**, **Read**, **Case** and select the filename.msh that you created in Gambit.

2. Scaling the Model

Click on **Grid** and **Scale**. Change the **Grid Was Created In** from m to mm and click on **Change Length Units**. Then click on **Scale** and **Close**.

Click on **Grid** and **Check** and scroll up and check the following

Domain Extents: x-coordinate: min $= -10$ mm, max $= 10$ mm

y-coordinate: min $= -10$ m, max $= 20$ mm

3. Defining the Model

- Click **Define**, **Models**, **Solver**. Keep everything set to default for a steady solution by clicking **OK**.

Tutorial I for 2-D

Specify
Continuum
Types

Mesh

Edge

Mesh
Edges

Mesh

Face

Mesh
Faces

- Click **Define**, **Models**, **Energy**. Make sure that **Enable Energy** is checked and click **OK**.
- Click **Define**, **Models**, **Radiation**. Select **Discrete Ordinates** and leave everything set to default and click **OK**. You will get an information window reading: "Available material properties or methods have changes, please confirm the property values before continuing." Click **OK**.

4. Defining Materials

   Click on **Define** and then on **Materials**. The Materials window will open as air.
   - For air change the following:
     Density[kg/m3]: **constant** to **boussinesq**
     Density[kg/m3]: 1.16
     Thermal Expansion Coeffecient [1/k]: 0.00333
   Click **Change/Create**.
   - For copper of the wire, click **Database**.
     Change **Material Type** from **Fluid** to **Solid**.
     Choose **copper[cu]** from the **Solid Materials** box.
     Click on **Copy** and then **Close**.
   - For the pvc of the insulation, you will have to create a material, because pvc does not exist in the system.
     In the **Materials** window, go to the **Solid Materials** pull-down menu and select **aluminum**.
     Change **Name** from **aluminum** to **pvc**
     Change **Chemical Formula** from **al** to **pvc**.
     Density[kg/m^3]: from 2719 to 1480
     Cp[j/kg-k]: from 871 to 1050
     Thermal conductivity[w/m-k]: from 202.4 to 0.16
     Click **Change/Create**.
     You will be asked to "**Overwrite Aluminum?**," click Yes
     Click **Close**.

5. Defining Operating Conditions

   Click **Define, Operating Conditions**.

   Click to select **Gravity** and change **Y[m/s2]** from 0 to −9.81.

   Also set **Operating Temperature[k]** to 300. This is a good approximate value; if you would like, enter the ambient temperature from the experiment.

   Click **OK**.

6. Defining Boundary Conditions

   - Click **Define, Boundary Conditions.**

---

*This value of 3244772 is the heat generation from a current of 13 A applied to a wire of 18 gage. Calculations and a table will be shown in the appendix.

Under **Zone**, select **copper** and click **Set**. Under **Material Name**, select **copper**. Click to select **Source Terms**. Set **Energy[w/m3]** to 3244772 and click **OK**.

- Select **wire-od-shadow** from under **Zone** and click **Set**.

  Change **Internal Emissivity** to 1 and click **OK**.

- Select **far-field** and click **Set**.

  Change **Internal Emissivity** to 1 and click **OK** and then **Close**.

7. Getting a Solution
   - Click Solve, Controls, Solution.

     Change **Momentum and Energy** in the **Discretization** box from **First Order Upwind** to **Second Order Upwind.** Click **OK**.

   - Click **Solve, Initialize, Initialize.**

     Select **All-zones** in the **Compute From** pull-down menu.

     Make sure that the temperature matches what you put in for operating temperature earlier (300 K?).

     Click Init, Apply, Close.

   - Click **Solve, Monitors, Residual.**

     Click to select **Plot** in the **Options** and then click **OK**.

   - Click **Solve, Iterate.**

     Change **Number of Iterations** to 1,500, then click **Iterate**.

     When the iterations are completed, click **Close**.

     You may get a message saying that the solution is converged after just one iteration is completed. This is an error. Click on Iterate again until the iterations run continuously and let Fluent run through the iterations. The about 300 iterations should take approximately 60 seconds, depending on the speed of the computer used.

8. Displaying Results

   To display the results, click **Display**, **Contours.**

   In the **Contours Of** box, select **Temperature** and also select **Filled** in **Options**.

   Click **Display** and a picture of the temperature distribution will appear.

   To control the picture location, left-click and you can re-center the picture wherever you see fit.

   To zoom in and out on the picture, use the center mouse button to click and drag a box around the specified area. To zoom in, click and drag the box from the upper-left corner to the lower right. To zoom out, click and drag the box from the lower right to the upper left.

   To determine the temperature value at a node, right-click at the location that you would like the temperature. The fluent window will then report a value for the temperatures at the edges of that node.

   To display the velocity vectors, click on **Display**, **Vectors**. Click on **Display** and a picture of Velocity Vectors will appear.

E.2   CALCULATIONS FOR HEAT GENERATION

Description	Symbol	Units
Current	I	[Amperes]
Stranded Wire Radius	a	[m]
Solid Wire Radius	r	[m]
Unit Length	L = 1	[m]
Pure Copper Resistivity	$\zeta = 1.7 \times 10^{-8}$	[$\Omega \cdot$m]
Insulation Thickness	$\delta = 0.0004064$	[m]
Stranded Wire Outer Radius	$b = a + \delta$	[m]
Pure Copper Radius	$A = \pi \cdot r^2$	[m^2]
Resistance	$R_c = \zeta \cdot \dfrac{L}{A}$	$\left[\dfrac{\Omega}{m}\right]$
Volume of Stranded Wire	$V_c = \pi \cdot a^2 \cdot L$	[m^3]
Heat Generation	$U = \dfrac{I^2 \cdot R_c}{V_c}$	$\left[\dfrac{W}{m^3}\right]$

Heat Generation Table (W/m^3)

Current (amps)	Gage number			
	14	16	18	20
0	0	0	0	0
5	80233.4	184258.1	480000	1111137
10	320933.7	737032.2	1920000	4444548
13	542378	1245584	**3244772**	7511286
15	722100.9	1658323	4320000	10000000

The results for Tutorial I are displayed in Figure E.3 and E.4.

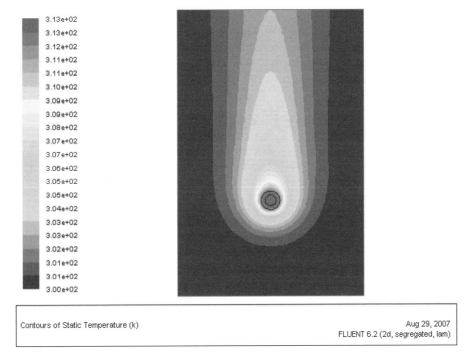

Contours of Static Temperature (k)	Aug 29, 2007 FLUENT 6.2 (2d, segregated, lam)

Figure E.3   Temperature contour of isotherms for natural convection on a heated wire

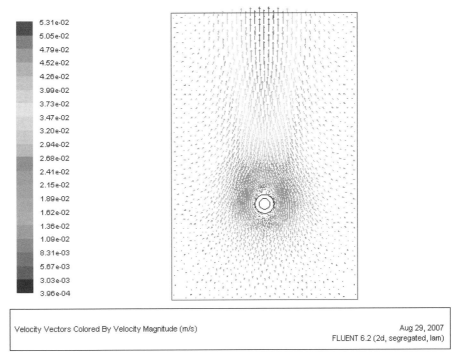

Velocity Vectors Colored By Velocity Magnitude (m/s)	Aug 29, 2007 FLUENT 6.2 (2d, segregated, lam)

Figure E.4   Velocity vector for natural convection on a heated wire

Appendix F

Tutorial II for 3-D

Problem Description for Tutorial II

The device most frequently used for transfer of energy (heat) is the heat exchanger. A heat exchanger affects the heat transfer from one fluid to another. There are many types of heat exchangers, including double-pipe, shell-and-tube, cross-flow, and plate-and-frame. Specific applications may be found in space heating and air-conditioning, power production, waste heat recovery, and chemical processing. In this laboratory project, we consider two types of heat exchangers: *double-pipe heat exchanger* and *shell-and-tube heat exchanger*. The project is divided into three parts: Part I, II, and III.

Part I: Take measurements for both parallel-flow and counterflow of the *double-pipe heat exchanger,* as indicated/assigned in the attached Experiment numbers 1 and 2. Also, take measurements for counter flow of the *shell-and-tube heat exchanger,* as indicated/assigned in the attached Experiment No. 3. The three sets of original experimental data sheets should be attached in the Appendix of the final report.

Part II: Obtain the analytical solutions corresponding to the experiments performed. You are required to use mathematical software, MathCAD, for prediction of outlet temperatures of both warm and cold fluids of the heat exchangers. The analysis for the *double-pipe heat exchanger* should include both the LMTD and effective-NTU methods that must be identical, while the analysis for the *shell-and-tube heat exchanger* may include only the effective-NTU Method. Provide two graphs of the computed temperature distributions along the pipe for both parallel flows and counterflows of the *double-pipe heat exchanger*. Discuss the results for comparison of all the heat exchangers between the measurements and analytical predictions, including the effectiveness and the rate of heat transfer. The copies of the MathCAD's files of the original work should be attached as an appendix of the final report.

Part III: Perform numerical simulations using design tools (SolidWorks, Gambit and Fluent) for only the *double-pipe heat exchanger* according to the experimental results (inlet temperatures of both warm and cold fluids). Tutorials for these simulations are available to help students. Show properly the isothermal contours and velocity vectors of a symmetry plane of the quarter pipe for both the parallel flows and counterflows, and also show the computed surface temperature profiles along the pipe with the measurements. Discuss the results with a summary table and an error analysis for comparison among the measurements, analytical predictions, and numerical computations in the final report.

F.1   TUTORIAL II DOUBLE-PIPE HEAT EXCHANGER: USING SOLIDWORKS, GAMBIT, AND FLUENT

F.1.1   Double-Pipe Heat Exchanger

Copper:

(a) Material (copper tube)

UNS# = c12200

$k = 339.2$ W/mK

$\rho = 8954$ kg/m^3

(b) Dimensions

Overall Length $L = 48.0$ in.

Inner Tube $D_o = 0.625$ in.

$D_i = 0.545$ in.

Outer Tube $D_o = 0.875$ in.

$D_i = 0.785$ in.

Aluminum:

(c) Material (aluminum tube)

Alloy # = 6061-T6

$K = 154.0$ W/mK

$\rho = 2707$ kg/m^3

(d) Dimensions

Overall Length $L = 48.0$ in.

Inner Tube $D_o = 0.625$ in.

$D_i = 0.496$ in.

Outer Tube $D_o = 0.875$ in.

$D_i = 0.785$ in.

F.1.2   Construct Model in SolidWorks

Open SolidWorks 2000 on your computer.

Create Outer Pipe

1. To create the Outer Pipe Document, click **New** on the Standard toolbar, or click **File, New.**

The New **SolidWorks Document** dialog box appears. Select the **Part** icon and click **OK**.

New

Set Dimensions to English

1. Click **Tools, Options**... select the **Document Properties** tab, select **Units**. Set **Linear units** to Inches, and **Decimal Places** to 4, and click **OK**.

Sketch

Open a Sketch

1. To open a sketch, click the **Sketch** button on the Sketch toolbar, or click **Insert, Sketch** on the menu bar.

Create Circles Representing Tube Wall

Circle

1. Click **Circle** on the Sketch Tools toolbar, or click **Tools, Sketch Entity, Circle**.
2. Move the pointer to the sketch origin, and hold down left mouse button. Drag the pointer to create a circle. Release the mouse button to complete the circle.
3. Repeat this to create a second circle with a different radius. Actual radii are not important at this time.

Adding Dimensions

Dimension

Dimension

1. Click **Dimension** on the Sketch Relations toolbar, or right-click anywhere in the graphics area and select **Dimension** from the shortcut menu.
2. Click one of the circles and then click the location of the dimension. Repeat for second circle. Click **Dimension** on the Sketch Relations toolbar again, or right-click anywhere in the graphics area and select **Dimension** from the shortcut menu again.
3. Right-click on either dimension, select **Properties**... on the shortcut menu, set **Value:** to 0.875 in. (This is the outside diameter of the outer pipe), and press Enter or click **OK**. Repeat this for other dimension, setting it equal to the inner radius. (0.785 in.)

Extruding the Outer Pipe Volume

Extrude
Boss/Base

1. Click **Extrude Boss/Base** on the Features toolbar, or click **Insert, Base, Extrude**. The **Extrude Feature** dialog box appears, and the view of the sketch changes to isometric.
2. Specify the type and depth of the extrusion: Make sure that **Type** is set to **Blind** and set **Depth** to 48 in. Either use the arrows to increment the value or enter the value.
3. Make sure that **Extrude as** is set to **Solid Feature**.
4. Click **OK** to create the extrusion
5. Click **Rebuild** on the Standard toolbar, and save part. (suggested name **Outer Pipe.SLDPRT**).

Rebuild

Zoom to
Fit

Changing View and Part Color

1. Click **Zoom to Fit** or click **View, Modify, Zoom to Fit**.

2. Click the **Outer Pipe** icon at the top of the FeatureManager design tree, (the FeatureManager design tree is located to the left of the graphics area).

3. Click **Edit Color** on the Standard toolbar. The **Edit Color** dialog box appears.

Edit Color

4. Click the desired color on the palette, then click **OK.** In **Shaded mode**, the part is displayed in the new color.

5. Save the part.

Shaded

To save time, only one quarter of the pipe will be used in fluent.

Remove Three Quarters of the Pipe.

1. Click **Plane1** in the FeatureManager design tree.

2. Click **Sketch**, or click **Insert, Sketch**.

Sketch

3. Click **Normal To** on the Standard Views toolbar. The part will be oriented such that Plane1 faces you.

Normal to

4. Click **Line** or click **Tools, Sketch Entity, Line**. Move the pointer to the sketch origin, and hold down left mouse button. Drag to the right to create a horizontal line that extends beyond the outside of the pipe. Repeat to create a vertical line through the top of the pipe. (It may be necessary to click **Zoom to Fit** at this time.)

Line

5. Beginning at the top of the vertical line, create connected lines in a counter-clockwise direction enclosing the pipe and ending at the right side of the horizontal line. Suggested pattern is shown in Figure F.1.

Zoom to Fit

6. Click **Extrude Cut**on the Features toolbar, or click **Insert, Cut, Extrude**.

7. In the **Extrude Cut Feature** dialog box, select **Through All** in the **Type** list, and click **OK.** (If an error is displayed, Toggle the **Reverse Direction** checkbox.)

Extrude Cut

8. Save the part.

Repeat Construct Model in SolidWorks to create Inner Pipe, Outer Water, and Inner Water. (Make sure to use dimensions associated with your model.)

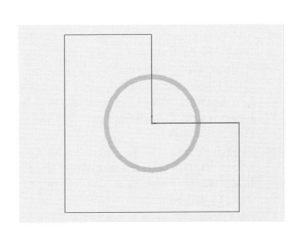

Figure F.1   Remove three quaters of a pipe

Assemble Parts to Create Heat Exchanger

1. To create the Outer Pipe Document, click **New** on the Standard toolbar, or click **File, New.**

   The New **SolidWorks Document** dialog box appears. Select the **Assembly** icon and click **Ok**.

New

Set Dimensions to English

1. Same as parts.

Place the Four Parts into the Assembly

1. Open **Outer Pipe. SLDPRT**, **Inner Pipe.SLDPRT**, **Outer Water. SLDPRT**, and **Inner Water. SLDPRT**. (Names of files will not match if you named them differently.)

2. To display all the open files click **Window, Tile Horizontally** on the menu bar.

3. Move the pointer to the **Outer Pipe** icon in the corresponding Feature Manager design tree. Click and hold to drag into the Feature Manager design tree of the assembly window. The pointer changes to ⟵⊕. Adding a part to an assembly this way results in the part automatically inferencing the assembly origin. This means that the origins are coincident and the three default planes are aligned.

4. For the other parts, click and drag their image in the graphics area into the assembly graphics area (suggested order: Inner Tube, Outer Water, Inner Water).

Isometric

5. Maximize the assembly window and click **Isometric**.

6. To align all the parts, mates must be added. Click **Mate** on the Assembly toolbar, or click **Insert, Mate**. The **Assembly Mating** dialog box appears.

Mate

7. Click the outside face of **Outer Pipe** and the outside face of **Outer Water**. (You may need to zoom in to see better, click **Zoom to Area** and drag pointer to create a rectangle around what you want to zoom in on.) Then select **Concentric**. Clicking **Preview** allows you to see what the mate does. End a mate by clicking **Apply**. To keep from having to click **Mate** for every mate, click the pushpin in the upper-left corner of the **Assembly Mating** dialog box.

Zoom to Area

8. Add the following Mates: outside face of **Outer Pipe** and the outside face of **Inner Water, Concentric,** and outside face of **Outer Pipe** and the outside face of **Inner Pipe, Concentric.**

Mate

9. Click **Front** then add the following relations: end face of **Outer Pipe** and end face of **Outer Water, Coincident**, end face of **Outer Pipe** and end face of **Inner Water, Coincident**, and end face of **Outer Pipe** and end face of **Inner Pipe, Coincident.**

Front

10. Click **Left** then add the following relations: left face of **Outer Pipe** and left face of **Outer Water, Coincident**, left face of **Outer Pipe** and left face of **Inner Water, Coincident**, and left face of **Outer Pipe** and left face of **Inner Pipe, Coincident.**

Left

11. Click **Rebuild** on the Standard toolbar and Save assembly (suggested name, **Double Pipe.SLDASM**).

Rebuild

F.1.3   Meshing the Double Pipe Heat Exchanger in Gambit

Import Double Pipe Assembly into Gambit

Orient
Model

Isometric

Fit To
Window

Geometry

Face

Connect/
Disconnect
Faces

1. Open **Double Pipe.SLDASM** in SolidWorks.

2. To import a file containing volumes to Gambit, it must be in ACIS format (*.sat). It must also be Version 5.0 or older. Click **File, Save As. . .** to open the **Save As** dialog box. Set **Save as type** to **ACIS Files (*.sat),** set **Version:** to **5.0**, and set **Units:** to **inches**. Change the name to **DoublePipe** because spaces in file names cause errors in Gambit.

3. Click **Save.**

4. To open a Gambit program (suggested name **DoublePipe**): You may open **MS-DOS Command Prompt** switch to the D drive and open appropriate files. Once you have the **MS-DOS Command Prompt** open type **d:** and press Enter. Then type **cd fluent.inc\ntbin\ntx86** and press enter. Finally type **gambit –id DoublePipe** and press Enter.

5. In Gambit, click **File, Import, Acis** this will open the **Import ACIS File** dialog box. Click **Browse.** This opens the **Select File** dialog box. Under **Filter,** enter the path to where you saved the file **DoublePipe.sat**. Select **DoublePipe.sat** under **Files** and click **Accept**. Click **Accept** in the **Import ACIS File** dialog box to open file. (It may be easier to save the ACIS file **DoublePipe.sat** on a base drive to prevent errors caused by folder names that confuse Gambit.)

6. Right-click on **Orient Model** and select **Isometric**. Then click **Fit To Window**. This will give you an isometric view of the complete geometry.

7. Because the geometry was imported as volumes, redundant faces exist. (This is when two or more faces lie exactly on one another.) To remove these faces, click **Geometry, Face, Connect/Disconnect Faces**. This will open the **Connect Faces** dialog box. Set **Faces** to **All**, make sure **Real** is selected, and click **Apply**. The geometry is now ready to mesh.

Mesh Curved Edges

Mesh

Edge

Mesh
Edges

1. To construct a mesh, it is best to start with the more critical areas. Also a mesh of an edge will shape a mesh of a face, which will shape a mesh of a volume. This means that meshing edges first, faces second, and volumes last will frequently give you the most control. To set how many radial divisions you want, you can mesh the curved edges on the ends of the Heat Exchanger. Click **Mesh, Edge, Mesh Edges**. This opens the **Mesh Edges** dialog box. Click the **Edges** list box to make it the active field. The list box will be yellow and any edges selected will be entered into this list box. To select an edge, you must position the pointer on the line, hold the shift key, and click the left mouse button. (This operation will be referred to as *Shift-left-click*.) It may be easier to select these edges if you zoom in on the end shown in the picture below. Controlling the graphics area is as follows:

Mouse button (click and hold)	Horizontal movement of mouse	Vertical movement of mouse
Left	Rotates image about a vertical axis through the center of the graphics area.	Rotates image about a horizontal axis through the center of the graphics area.
Center	Translates image horizontally.	Translates image vertically.
Right	Rotates image about an axis perpendicular to the screen and through the center of the graphics area.	Zooms image in and out.

Mesh

2. To mesh the curved lines, select all eight curved lines. (Four are shown in Figure F.2, and other four are on other end of heat exchanger.) Once the lines are selected change **Interval Size** to **Interval Count**, leave the value in the box to the left of **Interval Count** as 10, and click **Apply**.

Edge

Mesh Edges

Mesh Linear Edges to Further Define Mesh Requirements

1. Meshing the nine edges that run the length of the heat exchanger will set the length of the elements, as shown in Figure F.3. Click **Mesh, Edge, Mesh Edges**. Setting **Interval Count** to 384 will make the elements 1/8 of an inch long. Select the nine edges shown below, and click **Apply**. The mesh is shown directly in Figure F.4 showing which lines to select.

Mesh Faces to Define Final Mesh Specifications

Specify Model Display Attributes

1. To set the thickness of the elements in the pipes, a mesh of the face representing the end of the pipe must be created. Hiding the meshes created up to this point will make this easier. Click **Specify Model Display Attributes**, select **Off** from the selection menu to the right of **Mesh,** and click **Apply**. Click **Mesh, Face, Mesh Faces**. This opens the **Mesh Faces** dialog box. Select the face that

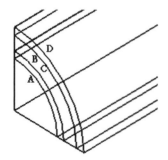

Figure F.2   Four curved lines on one end of the pipe

Mesh

Face

Mesh
Faces

represents the end of the outer water on both ends of the heat exchanger. If another face is highlighted, continue holding shift and click the middle mouse button until the correct face is outlined. This will toggle through the faces that contain this edge. Under **Elements:** the option **Quad** will default. Also, under **Type:** the option **Map** will default. Change **Interval Size** to **Interval Count**. Change the value in the box to the left of **Interval Count** to 5, and click **Apply**. The face with correct mesh is shown in Figure F.5.

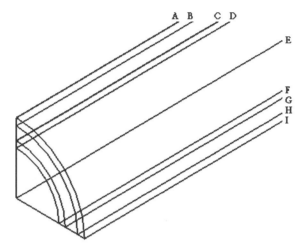

Figure F.3 Linear edges of the pipe

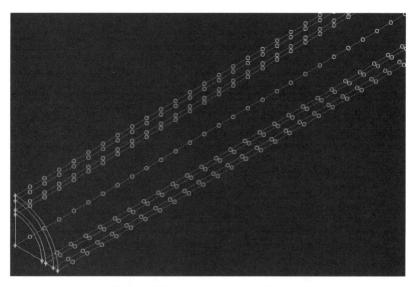

Figure F.4 Meshed edges of the pipe

Figure F.5 A meshed face of the pipe

2. To mesh the faces that represent the ends of the tubes in the heat exchanger, you must first select these two faces on both ends of the heat exchanger. Next, select **Apply** under **Spacing:**, set the value in the box to the left of **Interval Count** to 3, and click **Apply**.

3. For the face that represents the end of the inner water, you must first select the face on **ONE** end of the heat exchanger. Then set the value in the box to the left of **Interval Count** to 10 and click **Apply**. The end should now look like Figure F.6.

Mesh

Volume

Mesh
Volumes

Meshing the Volumes

1. Creating the mesh of the four volumes is easy at this point, because all the previous meshes will control its shape. To create the meshes in all four volumes, click **Mesh, Volume, Mesh Volumes**. This opens the **Mesh Volumes** dialog box. *Shift-Left-Click* to select all four volumes. Verify that **Elements:** defaulted to **Hex/Wedge** and **Type:** defaulted to **Cooper**. Deselect **Apply** under **Spacing** and click **Apply**. The construction of the mesh is now complete and you should save the file at this time.

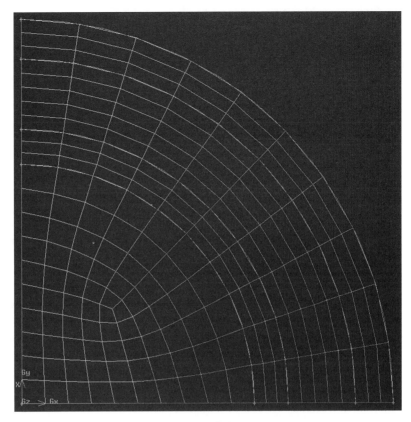

Figure F.6

Setting Zones and Boundary Conditions

Zones

Specify
Continuum
Types

1. In this step you define how the faces interact. For example, the face that represents the inside of the inner pipe will separate a solid from a liquid with zero thickness and zero thermal resistance. The first step will be to tell Gambit which solver will be used. This will change the list of possible face and volume options. To do this click **Solver, Fluent5**.

2. We will begin by defining the four volumes. Click **Zones, Specify Continuum Types**. This will open the **Specify Continuum Types** dialog box. *Shift-Left-Click* the volume that represents the inner water. (It may be easier if the mesh is turned off.) Click the text box to the right of **Name:** and enter "innerwater." Make sure that **Type:** is set to **Fluid**, and click **Apply**. Next, select the volume that represents the outer water. Set **Name:** to "outerwater" and leave **Type:** set to **Fluid**. Click **Apply**. Select the volume that represents the inner pipe. Set **Name:** to "innerpipe", set **Type:** to **Solid** and click **Apply**. Finally, select the volume that represents the outer pipe. Leave **Type:** set to **Solid**, set **Name:** to "outerpipe", and click **Apply**.

Zones

Specify
Boundary
Types

3. There are several boundary conditions that need to be set for this model. The first group will be the faces that were created when the heat exchanger was cut axially. To set a boundary layer, click **Zones, Specify Boundary Types**. This will open the **Specify Boundary Types** dialog box. There are two symmetry faces. Select a face that was created in the axial cut, set **Type:** to **Symmetry**, set **Name:** to "symmetry 1" and click **Apply**. Repeat the above procedure and give a name as "symmetry 2." This will make the analysis software solve the calculations as if the heat exchanger was complete. To control flow rate in both the inner and outer water it is easiest to set one of the faces that represent the ends of these volumes to velocity inlets. For parallel flow, place both velocity inlets on the same end of the heat exchanger. For counter flow, place the velocity inlets on opposite ends of the heat exchanger. (This tutorial will explain the modeling of parallel flow.) First, place the velocity inlet for the inner water. Select one of the faces that represent the ends of the inner water. Set **Type:** to **Velocity_Inlet**, set **Name:** to "innerinlet", and click **Apply**. Second select the face that represents the end of the outer water on the same end of the heat exchanger (see velocity inlets in Figure).

Leave **Type:** set to **Velocity_Inlet**, set **Name:** to "outerinlet", and click **Apply**. Now that we can control the flow rate in the heat exchanger, we need to allow the water entering to escape. Setting the other end of the heat exchanger to pressure outlets will keep the pressure in the system constant. To do this, you must select the face representing the end of the inner water that was not set as the velocity inlet. Set **Type:** to **Pressure_Outlet**, set **Name:** to "inneroutlet", and click **Apply**. Next, select the face that represents the end of the outer water not set as a velocity inlet, leave **Type:** set to **Pressure_Outlet**, set **Name:** to "outeroutlet", and click **Apply**.

4. The work in Gambit is now complete. Save the file and export the mesh. To export a mesh click **File, Export, Mesh** set **File Name:** to **DopblePipe.msh**, and click **Accept**.

F.1.4 Analysis of Heat Exchanger in Fluent

Opening the Heat Exchanger Mesh in Fluent

1. First open the fluent program located on the desktop, select **3d** as the version, and click **Run**.

2. To import the mesh, click **File, Read, Case**, select the file **DoublePipe.msh**, and click **OK**.

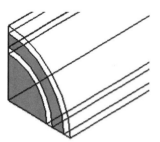

Figure F.7 Faces in an inner pipe and an outer pipe

Scaling and Cleaning the Mesh

1. Fluent has a default scale in meters, but the geometry was created in inches. To correct for this click **Grid, Scale**. This will open the **Scale Grid** dialog box. Set **Grid Was Created In** to **in**, click **Scale, Close**. Make sure that the dimension is correct: The length from the minimum to maximum is 48 inch.

2. To simplify the mesh click **Grid, Smooth/Swap**. This will open the **Smooth/Swap Grid** dialog box. Click **Smooth, Swap** to clean the mesh. Repeat this until number of faces swapped = 0, then click **Close**.

3. Finally, click **Grid, Check** to have Fluent look for errors in the mesh.

Defining the Mesh

1. The First thing you define is the solver. This sets what Fluent will calculate. To do this click **Define, Models, Solver**. This will open the **Solver** dialog box. Make sure **Solver** is set to **Segregated**, **Space** is set to **3D**, **Velocity Formulation** is set to **Absolute**, **Formulation** is set to **Implicit**, **Time** is set to **Steady**, and then click **OK**.

2. To have Fluent solve an energy balance, energy must be enabled. To enable energy, click **Define, Models, Energy**. This will open the **Energy** dialog box. Make sure **Enable Energy** is selected and then click **OK**.

3. Fluent has several sets of equations that model viscosity. The set you will use is called **k-epsilon**. To turn this viscosity model on, click **Define, Models, Viscous**. This will open the **Viscous Model** dialog box. Select **k-epsilon [2 eqn]**, make sure **Viscous Heating** is not selected, leave all the default values alone, and click **OK**.

4. The properties of all the materials used in the calculations must be set. To do this, click **Define, Materials**. This will open the **Materials** dialog box. Fluent has a database of materials that you can select. For the heat exchanger you will need water and copper. (Aluminum may also be required if that is the heat exchanger you are modeling.) To add water to the list of available materials for the calculations, click **Database**. This will open the **Database Materials** dialog box. Leave **Material Type** set to **fluid**, select **water-liquid [h2o<l>]** from the choices bellow **Fluid Materials**, and click **Copy**. This will add water to the list of **Fluid Materials** in the **Materials** dialog box. While still in the **Database Materials** dialog box, change **Material Type** to **solid**, select **copper [cu]** from the choices below **Solid Materials**, and click **Copy, Close**. In the **Materials** dialog box, set **Material Type** to **solid**, **Solid Materials** to **copper [cu]**, **Density [kg/m3]** to **8954**, **Thermal Conductivity [w/m-k]** to **339.2**, and click **Change/Create**. (For heat exchangers containing aluminum, set the density and thermal conductivity of aluminum to match the values on the first page of this tutorial.) To close the dialog box, click **Close**.

5. The next thing that needs to be set is the operating condition. This sets the ambient conditions. Click **Define, Operating Conditions** to open the **Operating Conditions** dialog box. Select **Gravity** (the dialog box will expand). Under **Gravitational Acceleration** set **Y [m/s2]** to **−9.81**. Set **Operating Temperature [k]** to **300** and click **OK**.

6. The last area that needs to be defined is the boundary conditions. This is where you set the temperature of the water entering the heat exchanger and the flow rate of the water. To do this click **Define, Boundary Conditions**. This will open the **Boundary Conditions** dialog box. The first step is to specify the material for each of the four volumes. Click **innerwater** (this is located under **Zone**). **Fluid** should be highlighted under **Type**, and click **Set**. This will open the **Fluid** dialog box. Under **Material Name** change the default **air** to **water-liquid**, leave all other values as default, and click **OK**. Repeat this process for **outerwater**. Now select **innerpipe** in the **Boundary Conditions** dialog box (**solid** should now be highlighted under **Type**) and click **Set**. This will open the **Solid** dialog box. Under **Material Name** change the default **aluminum** to **copper**, leave all other settings as default, and click **OK**. Repeat this process for **outerpipe** (for heat exchanger using aluminum inner pipe, you must specify the materials accordingly).

7. The flow rate is set at the inlet boundary condition. To set these, click on **inner-inlet** in the **Boundary Conditions** dialog box and click **Set**. The value of **Velocity Magnitude [m/s]** must be calculated from the flow rate modeled and the cross-sectional area modeled. An example for 2 gal/min in the inner pipe of a copper heat exchanger is

$$\text{Velocity} = \frac{2 \text{ gal}}{1 \text{ min}} * \frac{231 \text{in.}^3}{1 \text{gal}} * \frac{1 \text{ min}}{60 \text{ sec}} \div \left(\text{area} = .2333 \text{ in.}^2\right) * \frac{1\text{m}}{39.37 \text{ in.}}$$

$$= 0.8384 \, \frac{\text{m}}{\text{s}}$$

Change **Temperature [K]** to the value entering the inner pipe, set **Turb. Kinetic Energy [m²/s²]** to **0.01**, **Turb. Dissipation Rate [m²/s³]** to **0.1**, leave all other values as default, and click **OK**. For **outer-inlet**, set **Velocity Magnitude [m/s]** to the value calculated for flow rate and cross-sectional area modeled, change **Temperature [K]** to the value entering the outer pipe, set **Turb. Kinetic Energy [m²/s²]** to **0.01**, **Turb. Dissipation Rate [m²/s³]** to **0.1**, leave all other values as default, and click **OK**. The values for all the other boundaries are correct on their default settings, so close the **Boundary Conditions** dialog box. Save the case at this time.

Running Calculations on Heat Exchanger in Fluent

1. The heat exchanger is fully defined at this point and ready to have calculations made. To run the calculations, click **Solve, Initialize, Initialize**. This will open the **Solution Initialization** dialog box. Select **all-zones** under **Compute From**, then click **Init, Apply,** and **Close**. Next click **Solve, Monitors, Residual**. This will open the **Residual Monitors** dialog box. Select both **Print** and **Plot** under **Options**, and click **OK**. Finally, click **Solve, Iterate**. This will open the **Iterate** dialog box. Set **Number of Iterations** to **500** and click **Iterate**. This step will take some time. After about 80 iterations, there should be a message that states "solution is converged." If all 500 iterations are made without a solution, then an error was made somewhere in the tutorial. Save case and data at this time.

Viewing Results

1. Now that Fluent has completed the calculations, several things that can be viewed. One is the temperature gradient at the exit. To view this, click **Display, Contours**. This will open the **Contours** dialog box. Make sure **Filled** is selected under **Options**, change **Pressure** to **Temperature** under **Contours Of**, under **Surfaces** select both **inner-outlet** and **outer-outlet**, set **Levels** to **100**, and click **Display**. (For counter-flow cases, both exits cannot be viewed at the same time. Only one end of the heat exchanger can be viewed at a time this way. Thus, you must view **inner-outlet** with **outer-inlet** and **inner-inlet** with **outer-outlet**.)

2. The view will be small and only of the quarter of the heat exchanger that was modeled. To obtain a better view of this display, click **Display, Views**. This will open the **Views** dialog box. Click **Define Plane** to open the **Mirror Planes** dialog box. Set **X** to **1** and click **Add**. This will create a plane of symmetry called **x = 0**. A plane of symmetry at **y=0** must also be created so set **X** back to **0**, set **Y** to **1**, then click **Add**. These will be all the planes of symmetry you will need, so click **OK**. The two planes that were just created should be highlighted in the box under **Mirror Planes**. Clicking **Apply** will mirror the image about the planes of symmetry you just defined, but the image will still be small. To correct for this, click **Auto Scale**. Now you should have a good view of the exit of the heat exchanger.

3. Another good view is one of just the planes of symmetry. This gives a view created by a slice lengthwise through the heat exchanger. To see this view, go back to the **Contours** dialog box, deselect **outer-inlet** and **inner-inlet**, select the four surfaces that have symmetry in their name, and click **Display**. The display will not show correctly at this point. To correct the display, go back to the **Views** dialog box. Select either **right** or **left** under **Views**, click **Restore,** and then click **Auto Scale**. This will show you the temperature gradient as you move along the heat exchanger.

4. There are other characteristics that can be seen in this view. To see the velocity profile, change **Temperature** to **Velocity** in the **Contours** dialog box and click **Display**. You may also view the velocity as vectors. Click **Display, Velocity Vectors** in the original fluent window. This will open the **Velocity Vectors** dialog box. Select the four surfaces with symmetry in their name and click **Display**. This display will be hard to see, and will require zooming in to see clearly. To zoom in, create a box around the point of interest by clicking and holding the center mouse button while dragging the mouse to create a box. A box created from left to right will zoom in and a box created from right to left will zoom out. To see the fully developed velocity profile, zoom in on the exit of the heat exchanger.

5. Fluent will also give information that is quite useful if one knows where to get it. To calculate the amount of energy transferred one can simply look at the heat flux through the inner pipe. Because the inner pipe was modeled with no energy flux through the ends, all energy flux through the inner wall equals energy flux through the outer wall, which equals energy flux through the pipe itself. To obtain these values, click (in the original fluent window) **Report, Fluxes**. This will open the **Flux Reports** dialog box. Select **Total Heat Transfer Rate** under

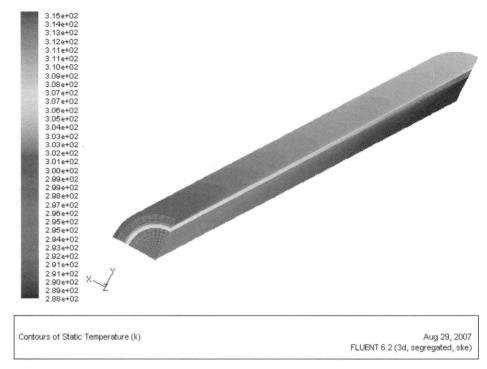

3.15e+02
3.14e+02
3.13e+02
3.12e+02
3.11e+02
3.11e+02
3.10e+02
3.09e+02
3.08e+02
3.07e+02
3.07e+02
3.06e+02
3.05e+02
3.04e+02
3.03e+02
3.03e+02
3.02e+02
3.01e+02
3.00e+02
2.99e+02
2.99e+02
2.98e+02
2.97e+02
2.96e+02
2.95e+02
2.95e+02
2.94e+02
2.93e+02
2.92e+02
2.91e+02
2.91e+02
2.90e+02
2.89e+02
2.88e+02

Contours of Static Temperature (k)   Aug 29, 2007
                                                                                           FLUENT 6.2 (3d, segregated, ske)

Figure F.8   Temperature contours for internal flow in a double-pipe heat exchanger

Options. There will be three boundaries with wall and shadow in their name. The flux across two should be almost identical, and the flux across the other will be almost zero. To obtain the fluxes, simply select one of these boundaries and click **Compute**. The two that are almost identical represent the inner and outer walls of the inner pipe, thus show the energy transfer rate of the heat exchanger. (Note, this is only one quarter of the total energy transfer rate.) The difference in these numbers is caused by the transient conditions in the pipe as the calculations approach equilibrium. From this value, the flow rates, and the specific heat, you can easily obtain outlet temperatures. These values should be comparable to the values obtained through experimentation and analytical calculations.

6. To calculate the average outlet temperatures of the cold or hot water, click **Report, Surface Integrals**. This will open the **Surface Integrals** dialog box. Select **Average** under **Options**, **Temperature** under **Field Variable** and **inner-outlet** or **outer-outlet** under **Surfaces**. Then, click **Compute**. You will see the average temperature under **Area-Weighted Average (K)**.

Appendix G

Computational Work of Heat Pipe

G.1 A HEAT PIPE AND HEAT SINK

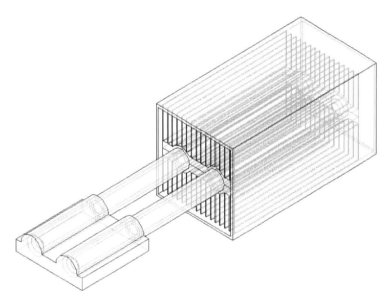

Figure G.1 A sketch of heat pipes and a heat sink

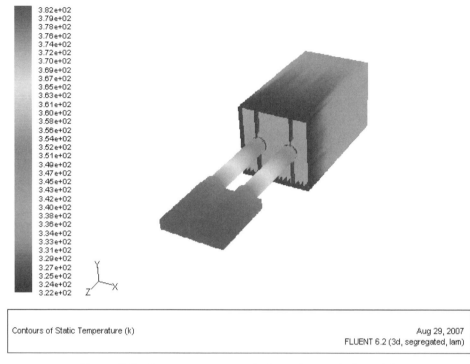

3.82e+02
3.79e+02
3.78e+02
3.76e+02
3.74e+02
3.72e+02
3.70e+02
3.69e+02
3.67e+02
3.65e+02
3.63e+02
3.61e+02
3.60e+02
3.58e+02
3.56e+02
3.54e+02
3.52e+02
3.51e+02
3.49e+02
3.47e+02
3.45e+02
3.43e+02
3.42e+02
3.40e+02
3.38e+02
3.36e+02
3.34e+02
3.33e+02
3.31e+02
3.29e+02
3.27e+02
3.25e+02
3.24e+02
3.22e+02

Contours of Static Temperature (k)

Aug 29, 2007
FLUENT 6.2 (3d, segregated, lam)

Figure G.2 Temperature contour for heat pipes and a heat sink

Appendix H

Computational Work of a Heat Sink

H.1 ELECTRONIC PACKAGE COOLING

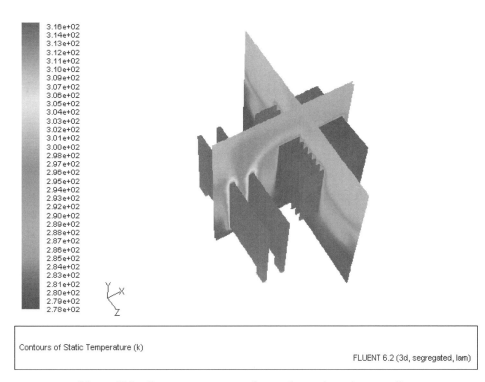

Contours of Static Temperature (k)

FLUENT 6.2 (3d, segregated, lam)

Figure H.1 Temperature contour for an electronic package cooling

Appendix I

Tutorial for MathCAD

I.1 TUTORIAL PROBLEM FOR MathCAD

A huge water reservoir is drained with a 4-nominal schedule 80 copper pipe, 60 m long. The piping system is shown in Figure I.1. The fittings are regular and threaded. Determine the volume flow rate through the system.

$$\text{Water Property}: \quad \rho := 1000\frac{\text{kg}}{\text{m}^3} \quad \mu := 0.89 \cdot 10^{-3}\frac{\text{N} \cdot \text{s}}{\text{m}^2}$$

Tips: To have the symbol ρ, type **r** first and put the cursor right behind the **r**, then press **[Ctrl]** and **g** simultaneously. Likewise, we can have μ with **m** and ε with **e**.

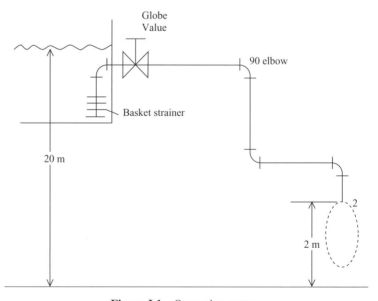

Figure I.1   Open pipe system

Tips: Try typing **D:9.718*cm** and you will see on screeen **D:=9.718·cm,** where the definition sign := is different with the Boolean sign = in functioning.

$$4 \text{ nom. sch.80}: \quad D: = 9.718 \text{ cm} \quad A: = 74.17 \text{cm}^2$$

$$\text{Copper Pipe:} \quad \varepsilon: = 0.00015 \text{ cm}$$

$$\text{Length of pipe:} \quad L = :60 \text{ m}$$

Reynolds Number

$$Re(v,D): = \frac{\rho \cdot v \cdot D}{\mu} \quad \text{where only the velocity } \mathbf{v} \text{ (lower case) is unknown.}$$

Tips: Use [**Spacebar**] repeatedly to control the **blue editing line** under the expression. For example, r[Ctrl]g*v*D[Spacebar][Spacebar]/m[Ctrl]g will give you the above expression.

Tips: For example, type r[Ctrl]g*v^2[Spacebar][Spacebar]/2*g and you will see $\frac{\rho \cdot v^2}{2 \cdot g}$

Modified Bernoulli Equation

$$\left(\frac{P}{\rho \cdot g} + \frac{v^2}{2 \cdot g} + z \right)_1 = \left(\frac{P}{\rho \cdot g} + \frac{v^2}{2 \cdot g} + z \right)_2 + \left(\frac{f \cdot L}{D_h} + K \right) \cdot \frac{v^2}{2 \cdot g}$$

Property evaluation:

$$P_1 = P_2 = P_{atm} \ v_1 = 0 \ v_2 = 0 \ z_1 := 20 \cdot m \ z_2 := 2 \cdot m$$

Tips: Try typing a period (.) to have the subscript. For example, type P.1 and you will see on screen P_1.

Tips : To have the **Boolean equal sign '='**, type [**Ctrl**] and = simultaneously.

Chen's Equation for Friction Factor (Moody Diagram) for turbulent flow

$$Chen(v,D) := \left[-2.0 \log \left[\frac{\frac{\varepsilon}{D}}{3.7065} - \frac{5.0452}{Re(v,D)} \cdot \log \left[\frac{1}{2.8257} \cdot \left(\frac{\varepsilon}{D} \right)^{1.1098} \right. \right. \right.$$

$$\left. \left. \left. + \frac{5.8506}{Re(v,D)^{0.8981}} \right] \right] \right]^{-2}$$

In general, we may have an expression for the friction factor including laminar flow as:

$$f(v,D) := \left| \begin{array}{ll} Chen(v,D) & \text{if } Rev(v,D) > 2100 \\ \frac{64}{Re(v,D)} & \text{otherwise} \end{array} \right.$$

Tips: In order to have the vertical line after the definition sign, select **View, Toolbars, Programming, Add Line** from the **Menu.** Select **if** in the **Programming** and also select > in the **Boolean.**

From Table 3.2

$$K_{basket} := 1.3 \quad K_{90elbow} := 1.4 \quad K_{globevalve} := 10 \quad K_{exit} := 1.0$$

$$K: = K_{basket} + 4 \cdot K_{90elbow} + K_{globevalve} + K_{exit}$$

Tips : Type K . basket : 1.3 and you will see $K_{basket}:=1.3$.

$$z_1 = z_2 + \frac{f(v,D) \cdot L}{D} \cdot \frac{v^2}{2 \cdot g} + K \cdot \frac{v^2}{2 \cdot g} \quad \text{where the velocity v is only unknown}$$

variable to be sought.

This problem is attempted to solve using the "Find" function, or, equivalently, the "root" function can be used.

Steps

1. Make an initial guess, try to give a value as close as possible.
2. Type "Given" following by the expression, in which the Boolean equal sign "=" must be used to make equal both sides of the expression.
3. Enter the "Find " function.

$$v: = 0.1 \cdot \frac{m}{s} \quad \text{Initial Guess}$$

Given

$$z_1 = z_2 + \frac{f(v,D) \cdot L}{D} \cdot \frac{v^2}{2 \cdot g} + K \cdot \frac{v^2}{2 \cdot g} \quad \text{where the Boolean sign = is used.}$$

$$v: = Find(v) \quad \text{Find function which was built in MathCad}$$

$$v = 3.65 \frac{m}{s} \quad \text{Highlight : Choose \textbf{Properties} from the \textbf{Format} menu}$$

The volume flow rate is then,

$$Q: = v \cdot A$$

$$Q = 0.027 \frac{m^3}{s} \quad \text{Answer}$$

Check the assumption of turbulent flow by Reynolds number:

$$Re(v,D) = 3.986 \times 10^5$$

OK because Re > 2,100

Conversion Factors

Dimension	Metric	English
Acceleration	$1 \text{ m/s}^2 = 100 \text{ cm/s}^2$	$1 \text{ m/s}^2 = 3.2808 \text{ ft/s}^2$
		$1 \text{ ft/s}^2 = 0.3048 \text{ m/s}^2$
Area	$1 \text{ m}^2 = 104 \text{ cm}^2 = 10^6 \text{ mm}^2$	$1 \text{ m}^2 = 1550 \text{ in}^2 = 10.764 \text{ ft}^2$
	$= 10^{-6} \text{ km}^2$	$1 \text{ ft}^2 = 144 \text{ in}^2 = 0.0929034 \text{ m}^2$

Dimension	Metric	English
Density	$1 \text{ g/cm}^3 = 1 \text{ kg/L} = 1000 \text{ kg/m}^3$	$1 \text{ g/cm}^3 = 62.428 \text{ lbm/ft}^3$ $= 0.036127 \text{ lbm/in}^3$ $1 \text{ lbm/in}^3 = 1728 \text{ lbm/ft}^3$ $1 \text{ kg/m}^3 = 0.062428 \text{ lbm/ft}^3$
Energy, Heat, Work, Internal Energy, Enthalpy	$1 \text{ kJ} = 1000 \text{ J} = 1000$ $\text{N} \cdot \text{m} = 1 \text{ kPa} \cdot \text{m}^3$ $1 \text{ kJ/kg} = 1000 \text{ m}^2/\text{s}^2$ $1 \text{ kWh} = 3600 \text{ kJ}$ $1 \text{ Wh} = 3600 \text{ J}$ $1 \text{ cal} = 4.1868 \text{ J}$ $1 \text{ Cal} = 4.1868 \text{ kJ}$	$1 \text{ kJ} = 0.94782 \text{ Btu}$ $1 \text{ Btu} = 1.055056 \text{ kJ}$ $= 5.40395 \text{ psia·ft}^3$ $= 778.169 \text{ lbf·ft}$ $1 \text{ Btu/lbm} = 25.037 \text{ ft}^2/\text{s}^2$ $= 2.326 \text{ kJ/kg}$ 1 kJ/kg $= 0.430 \text{ Btu/lbm}$ $1 \text{ kWh} = 3412.14 \text{ Btu}$ $1 \text{ therm} = 10^5 \text{ Btu} = 1.055$ $\times 10^5 \text{kJ(natural gas)}$
Force	$1 \text{ N} = 1 \text{ kg} \cdot \text{m/s}^2 = 10^5 \text{ dyne}$ $1 \text{ kgf} = 9.80665 \text{ N}$	$1 \text{ N} = 0.22481 \text{ lbf}$ $1 \text{ lbf} = 32.174 \text{ lbm·ft/s}^2 = 4.44822 \text{ N}$
Heat flux	$1 \text{ W/cm}^2 = 10^4 \text{ W/m}^2$	$1 \text{ W/m}^2 = 0.3171 \text{ Btu/h} \cdot \text{ ft}^2$
Heat Transfer Coefficient	$1 \text{ W/m}^2 \cdot^\circ \text{C} = 1 \text{ W/m}^2 \cdot \text{K}$	$1 \text{ W/m}^2 \cdot^\circ \text{C} = 0.17612$ $\text{Btu/h} \cdot \text{ft}^2 \cdot^\circ \text{F}$
Length	$1 \text{ m} = 100 \text{ cm} = 1000 \text{ mm}$ $= 10 \text{ } \mu\text{m}$ $1 \text{ km} = 1000 \text{ m}$	$1 \text{ m} = 39.370 \text{ in.} = 3.2808 \text{ ft}$ $= 1.0926 \text{ yd}$ $1 \text{ ft} = 12 \text{ in.} = 0.3048 \text{ m}$ $1 \text{ mile} = 5280 \text{ ft} = 1.6093 \text{ km}$ $1 \text{ in} = 2.54 \text{ cm} = 25.4 \text{ mm}$ $1 \text{ yard} = 0.9144 \text{ m}$
Mass	$1 \text{ kg} = 1000 \text{ g}$ $1 \text{ metric ton} = 1000 \text{ kg}$	$1 \text{ kg} = 2.2046226 \text{ lbm}$ $1 \text{ lbm} = 0.45359237 \text{ kg}$ $1 \text{ ounce} = 28.3495 \text{ g}$ $1 \text{ slug} = 32.174 \text{ lbm} = 14.5939 \text{ kg}$ $1 \text{ short ton} = 2000 \text{ lbm} = 907.1847 \text{ kg}$
Power, Heat Transfer Rate	$1 \text{ W} = 1 \text{ J/s}$ $1 \text{ kW} = 1000 \text{ W} = 1.341 \text{ hp}$ $1 \text{ hp} = 745.7 \text{ W}$	$1 \text{ kW} = 3412.14 \text{ Btu/h}$ $= 737.56 \text{ lbf·ft/s}$ $1 \text{ hp} = 550 \text{ lbf·ft/s}$ $= 0.7068 \text{ Btu/s} = 42.41 \text{ Btu/min}$ $= 2544.5 \text{ Btu/h} = 0.74570 \text{ kW}$ $1 \text{ boiler hp} = 33,475 \text{ Btu/h}$ $1 \text{ Btu/h} = 1.055056 \text{ kJ/h}$ $1 \text{ ton of refrigeration} = 200 \text{ Btu/min}$
Pressure	$1 \text{ Pa} = 1 \text{N/m}^2$ $1 \text{ kPa} = 10^3 \text{ Pa} = 10^{-3} \text{ Mpa}$ $1 \text{ bar} = 10^5 \text{Pa}$ $1 \text{ atm} = 101.325 \text{ kPa}$ $= 1.01325 \text{ bars}$ $= 760 \text{ mm Hg at } 0 \text{ }^\circ\text{C}$ $= 1.03323 \text{ kfg/cm}^2$ $1 \text{ mm Hg} = 0.1333 \text{ kPa}$ $1 \text{ torr} = 133.322 \text{ Pa}$ $1 \text{ dyne cm}^2 = 10 \text{ Pa}$	$1 \text{ Pa} = 1.4504 \times 10^{-4}\text{psia}$ $= 0.020886 \text{ lbf/ft}^2$ $1 \text{ psi} = 144 \text{ lbf/ft}^2$ $= 6.894757 \text{ kPa}$ $1 \text{ in Hg} = 3.387 \text{ kPa}$ $1 \text{ atm} = 14.696178 \text{ psia}$ $1 \text{ in H}_2\text{O } (60^\circ\text{F}) = 248.84 \text{ Pa}$

Dimension	Metric	English
Specific Heat	$1 \text{ kJ/kg·°C} = 1 \text{ kJ/kg·K}$ $= 1 \text{ J/g·°C}$	$1 \text{ Btu/lbm·°F} = 4.1868 \text{ kJ/kg·°C}$ $1 \text{ Btu/lbmol·R} = 4.1868 \text{ kJ/kmol·K}$ $1 \text{ kJ/kg·°C} = 0.23885 \text{ Btu/lbm·°F}$ $= 0.23885 \text{ Btu/lbm·R}$
Specific Volume	$1 \text{ m}^3/\text{kg} = 1000 \text{ L/kg}$ $= 1000 \text{ cm}^3/\text{g}$	$1 \text{ m}^3/\text{kg} = 16.02 \text{ ft}^3/\text{lbm}$ $1 \text{ ft}^3/\text{lbm} = 0.062428 \text{ m}^3/\text{kg}$
Temperature	$T(K) = T(°C) + 273.15$ $\Delta T(K) = \Delta T(°C)$	$T(R) = T(°F) + 459.67 = 1.8 \, T(K)$ $T(°F) = 1.8 T(°C) + 32$ $\Delta T(°F) = \Delta T(R) = 1.8 \, \Delta T(K)$
Thermal Conductivity	$1 \text{ W/m·°C} = 1 \text{ W/m·K}$	$1 \text{ W/m·°C} = 0.57782 \text{ Btu/h·ft·°F}$
Velocity	$1 \text{ m/s} = 3.60 \text{ km/h}$	$1 \text{ m/s} = 3.2808 \text{ ft/s} = 2.237 \text{ mi/h}$ $1 \text{ mi/h} = 1.46667 \text{ ft/s}$ $1 \text{ mi/h} = 1.6093 \text{ km/h}$
Absolute Viscosity	$1 \text{ centipoise (cp)} 10^{-3} \text{N·s/m}^2$ $1 \text{ Poise (P)} = 100 \text{ centipoise}$	$1 \text{ N·s/m}^2 = 2.088543 \text{ lbf·s/ft}^2$ $= 0.671969 \text{ lbm/ft·s}$
Kinematic Viscosity	$1 \text{ centistoke} = 10^6 \text{m}^2/\text{s}$ $1 \text{ Stoke} = 100 \text{ centistoke}$	$1 \text{ m}^2/\text{s} = 10.7639104 \text{ ft}^2/\text{s}$
Volume	$1 \text{ m}^3 = 1000 \text{L} = 10^6 \text{cm}^3 \text{(cc)}$	$1 \text{ m}^3 = 6.1024 \times 10^4 \text{in}^3 = 35.315 \text{ ft}^3$ $= 264.17 \text{ gal (U.S)}$ $1 \text{ U.S. gallon} = 231 \text{ in}^3 = 3.7854 \text{ L}$ $= 0.1336805 \text{ ft}^3$ $1 \text{ fl ounce} = 29.5735 \text{ cm}^3$ $= 0.0295735 \text{ L}$ $1 \text{ U.S. gallon} = 128 \text{ fl ounces}$
Volume Flow Rate	$1 \text{ m}^3/\text{s} = 60,000 \text{ L/min} =$ $10^6 \text{ cm}^3/\text{s}$	$1 \text{ m}^3/\text{s} = 15,850 \text{ gal/min (gpm)}$ $= 35.315 \text{ ft}^3/\text{s}$ $= 2118. \text{ft}^3/\text{min(cfm)}$

Index